D0483965

VERTEBRATE PHYSIOLOGY

William J. McCauley—*Professor of Biological Sciences, University of Arizona*

1971 W. B. SAUNDERS COMPANY • PHILADELPHIA • LONDON • TORONTO

W. B. Saunders Company: West Washington Square
Philadelphia, Pa. 19105

12 Dyott Street
London, WC1A 1DB

1835 Yonge Street
Toronto 7, Ontario

Vertebrate Physiology SBN 0-7216-5878-4

 Made in the United States of America. Press of W. B. Saunders Company. Library of Congress catalog card number 75-132177.

Print number: 9 8 7 6 5 4 3 2 1

4/73

This tube, an enigmatic pipe,
Whose end was laid before begun,
That lengthens, broadens, shrinks and breaks
—Puzzle, machine, automaton.

—SIR RICHARD BURTON (From *The Kasidah* by Sir Richard Burton, New York, David McKay Company, Inc., 1931.)

PREFACE

Teachers of college physiology have long recognized the need for a textbook that treats the comparative physiology of the vertebrates. Existing textbooks are either directed specifically toward the human, medical point of view or, when they are comparative in nature, cover the entire animal kingdom. Such broad coverage inevitably results in a concentration on the invertebrate phyla. I have felt so strongly about the need for a textbook dealing exclusively with comparative vertebrate physiology that I have collected materials of this nature for a number of years and have evolved a course about them. Finally, at the urging of students and friends, I decided to attempt the writing of such a book. Despite the fact that two or three other books with broad vertebrate coverage have appeared during the preparation of the manuscript of this book, I feel that the approach here is sufficiently different to make it useful in many college courses.

The course that I teach follows the topic sequence of this book, but the book has been arranged so that, hopefully, it will be easy for individual instructors to select portions which implement a course set up in almost any sequence. Subject matter has been extensively cross-referenced throughout the book so that students can easily find explanatory material that has been placed in another chapter. The diversity of background in the average college classroom is such that some students will need to cross-check these references more frequently than others. The author has assumed some background in general biology or zoology and some in general chemistry.

Classic mammalian physiology is, of course, included, but the medical or clinical aspects have been omitted except when they add materially to the understanding of the physiological mechanisms involved. Unavoidably, since the bulk of established physiological data is mammalian, these portions occupy a major part of this book. There is, however, a rapidly increasing literature that deals with the physiology of non-mammalian vertebrates—extensive enough to provide an appreciation of evolutionary patterns. This, I believe, is the most appropriate point of view for the student majoring in biology or zoology to adopt.

The author wishes to express his sincere appreciation to Mrs. Joanna McComb for rendering the illustrations and to Colin Campbell, John Eads, G. Clay Mitchell, Dawn Mitchell, David Salter, and Carl Tomoff for critical reading of the manuscript.

WM. J. MCCAULEY

University of Arizona
Tucson, Arizona

CONTENTS

CHAPTER 1

INTRODUCTION

1.1 THE LIFE PROCESS

Certain minimal processes are fundamental to life of any kind. Growth, development, and reproduction require that each organism exploit some source of energy and use it to power the synthesis of living matter within itself. The green plants convert the photic energy of the sun to chemical energy by means of the light-absorbing properties of chlorophyll and complex, coupled chemical reactions. They utilize only simple materials (water and carbon dioxide) to build complex organic molecules, which are then used in the construction of the plant's own tissues. Animals, lacking this photosynthetic ability, must ingest plants or other animals to obtain ready-made organic substances. These substances are then used both as building materials and as sources of chemical energy.

In general, the synthetic processes of plants are successive chemical reductions leading to the storage of energy in the bonding of organic molecules. Animals release this energy for their own use by essentially reversing the process. A series of chemical oxidations breaks the organic molecules into simple components and releases the energy stored in their bonds. The overall chemical equations are:

1. Green plants:

$$\text{Solar energy} + H_2O + CO_2 \rightarrow \text{Organic compounds} + O_2$$

2. Animals:

$$\text{Organic compounds} + O_2 \rightarrow \text{Chemical energy} + H_2O + CO_2$$

Actually, both series of reactions take place in plants since they also break down some of the compounds which they have previously synthesized in order to obtain chemical energy at times when solar energy is not available.

All living organisms, then, are constantly taking substances from the environment, utilizing them in some way, and returning other materials to the environment. This is a dynamic process which must be continuous even when the organisms are, apparently, undergoing no change. In other words, it is necessary merely to maintain life itself. The life process is often spoken of, therefore, as a dynamic steady state. It has been compared with a whirlpool that is constantly dragging materials into its vortex and throwing them out again, more or less changed, while the whirlpool itself remains, apparently, unchanged. If the dynamic process stopped, the whirlpool, like life, would cease to exist. The analogy would be improved if whirlpools grew, developed, and produced little whirlpools almost, but not quite, like themselves.

To carry out the minimal processes of life, then, an animal must take in both oxygen and organic food and return to the environment carbon dioxide, water, and certain chemical wastes that it cannot utilize. As long as the animal is small—perhaps the size of a single cell—these exchanges with the environment can be carried out by diffusion of substances through the surface of the organism.

1.2 DIFFUSION

Diffusion is the random distribution of substances throughout all of the space available to them. It takes place because molecules of all substances are in constant motion in random directions. The amount of this motion is greatest in gases, somewhat less in liquids, and still less in solids, but it occurs to some degree in all three physical states. The amount of motion is also temperature dependent; it occurs more rapidly at high temperatures and disappears completely at absolute zero (−273° C). The random motion of a given molecule tends to be in a straight line as long as nothing blocks the way. Consequently, when molecules are closely packed (high concentration), collisions take place frequently, and the net movement of a molecule in a given direction is small (Fig. 1-1). Molecules on the periphery of such a region of concentration, however, can more easily move (diffuse) away from the region since movements in that direction are likely to be longer. In moving away, these molecules expose other molecules which in turn follow them. In this way concentrations of molecules tend to disperse until they are equally distributed throughout the available space. This process of dispersion is called diffusion. At equilibrium, when the molecules in question are equally distributed, the likelihood of collision is the same for all molecules so that the net movements are everywhere random and equal.

The rate at which diffusion takes place depends, as we have seen, upon the temperature. Two other factors are also of major importance: (1) the measured distance along which the diffusion occurs and (2) the difference in concentration of the diffusing molecules along this measured distance. The shape and size of the diffusing molecules, the nature of the medium through which the diffusion takes place, patterns of molecular charge, and other factors also affect diffusion rates. The diffusion distance and the concentration difference together establish

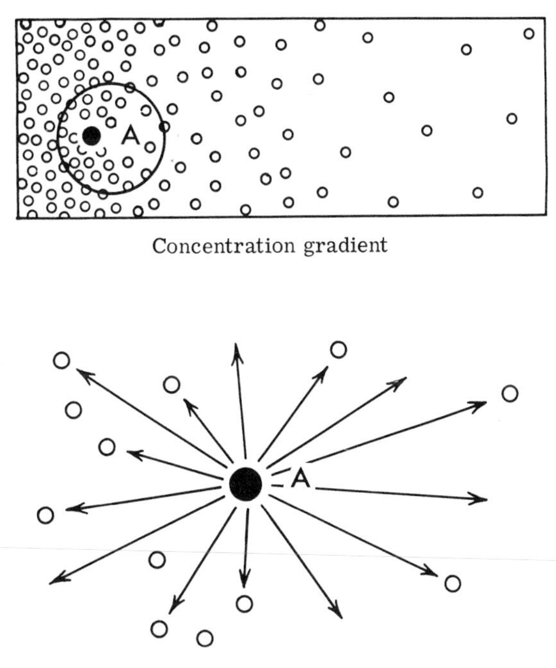

Figure 1–1 A given molecule or particle is less likely to have its motion interrupted by collision with another molecule or particle when its random motion is in a direction away from a region of concentration of that molecule or particle. (From Cockrum and McCauley: *Zoology.* Philadelphia, W. B. Saunders Company, 1965.)

the diffusion gradient (Fig. 1-2), and the diffusion rate is proportional to the steepness of this gradient.

It has been calculated that oxygen would diffuse from the head to the toes of an average-size human in something over 100 years but, with the same concentration difference, oxygen will diffuse through a membrane 1.5 microns thick in only 0.003 second. Similarly, a cylinder of animal tissue one centimeter in diameter would, if exposed to a 100 per cent oxygen atmosphere, become 90 per cent saturated with oxygen in about three hours. Under identical conditions, however, a single cell with a diameter of seven microns would become equally saturated with oxygen in only 0.006 second. These examples illustrate the importance of a short diffusion distance in the establishment of a steep diffusion gradient and, therefore, rapid diffusion.

As cells group together to form larger organisms, the distance from the environmental medium to the center of the organism becomes greater, and diffusion times are correspondingly prolonged. Unless the metabolism (oxygen demand) of the organism is accordingly decreased, some other mechanism must be provided to transport the oxygen. Of course, such a decrease in metabolism is nearly always disadvantageous to an organism.

In certain lower multicellular organisms, such as sponges, the body structure is such that virtually all cells remain on the surface. Diffusion distances are, therefore, short, and no additional transporting mechanism is necessary. The environmental water is propelled over and through the organism, however, so that high concentration differences can be maintained. If it were not for this activity of special flagellated

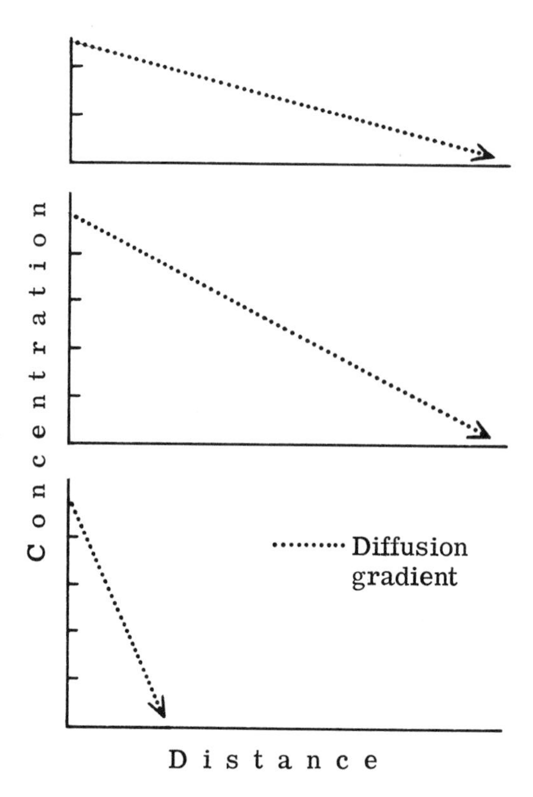

Figure 1–2 The rate of diffusion depends upon the steepness of the diffusion gradient, and this in turn depends upon the distance through which diffusion occurs and the difference in concentration of the diffusing substance.

cells, the diffusion gradient would become less steep as local environmental supplies of oxygen were depleted.

What has been said for oxygen applies, of course, equally well to the inward diffusion or transport of nutrients and the outward movement of carbon dioxide and other cell wastes.

1.3 SPECIALIZATION

In larger, more complex animals, it is impractical to have all cells situated at or near the body surface. Consequently, distances over which oxygen and other substances must move are necessarily long, and diffusion alone is inadequate to meet the metabolic needs of the organisms. Some sort of additional transporting mechanism is, therefore, essential, and this role has been assumed by the circulatory system. The circulatory system consists of a fluid that flows from vessels near the animal surface to vessels located near all of the internal cells of the organism. This fluid takes up, by diffusion, oxygen and nutrients from the environment and then physically transports them to the immediate vicinity of the internal cells. There these substances pass, again by diffusion, to the cells. At the same time, carbon dioxide and other wastes from the cells diffuse into the circulating fluid. These are physically transported back to the animal surface and lost, by diffusion, into the environment. Exchanges at both the animal's surface and at the site of internal cells takes place over short distances and, in both places, high concentration differences are maintained by the circulation of the fluid. Thus, diffusion remains an important process even in these larger multicellular animals.

In the higher animals specialized "surface" areas for gas exchange (gills, lungs), for nutrient absorption (intestinal lining), and for waste excretion (kidneys) exist. From these surfaces and from the internal cells of the animal the circulating fluid is separated only by thin, permeable membranes.

More active organisms have higher demands for oxygen and nutrients and consequently they produce greater quantities of carbon dioxide and other wastes. In such organisms, then, the rate at which the transporting fluid is circulated must be greater than in more sedentary organisms. An increased rate of circulation requires that the fluid be pumped rapidly through an essentially closed system of tubes and at relatively high pressures. Greater demand for oxygen and nutrients also requires that the circulating fluid have an increased capacity for these substances. As we shall see, a number of animal adaptations have evolved to meet these physical requirements.

When nutrients are first absorbed, they are not always in a form that can be utilized by the animal. Moreover, all of the required nutrients may not be simultaneously available in the environment. Consequently, certain substances require further processing or storage within the animal body, or both processing and storage. Special physiological systems are needed to carry out these functions and to maintain appropriate concentrations of various substances in the circulating fluids where they are available to the metabolizing cells.

Most animals must engage in physical activity of some sort, and this requires movement. Movement is produced by the activity of special contractile tissues (muscles). These must be arranged in the body in such a way that their attachments to body parts result in movements of those parts. Moreover, these movements must be under the control of the animal so that the movements are directed and purposeful. This means that each muscle must respond to some particular controlling factor or factors, which are either nervous or chemical in nature.

Further specialized structures are required to carry out the reproductive functions of the animal, *i.e.*, the production of sex cells from which a new generation can be derived. In higher animals this reproductive function also requires the production of special containers, either outside or inside the animal body, in which the early development of the offspring may take place.

A host of other specialized functions requiring specialized tissues and organs also appear as animals become, through evolution, more and more complex. Finally, such complex arrays of specialized organs and functions must be integrated in meaningful ways if the animal as a whole is to operate efficiently. This means that an internal communication system must be present to coordinate all of the subprocesses of life and to alter them as conditions may demand. Changing conditions both outside the animal and within it constantly require modification of the various physiological processes we have mentioned. These changing conditions must be sensed by means of specially adapted sensory organs, and information regarding these changes must be sent to a central integrator or computer (central nervous system). There "decisions" can be made based upon the incoming sensory information and, often, upon experience or learning. Once a decision has been made, signals can be sent along nervous or chemical pathways to modify the action of various organs in appropriate ways.

A useful analogy may be drawn between the evolutionary development of a higher animal and the growth of a human society. A lone primitive man presumably did his own hunting, shelter building, weapon making, and all other tasks necessary to his well-being, much as a single cell performs all of the physiological processes of life for itself. With the advent of group living it became possible for those most skilled at a particular task to perform it for the entire group and, with practice, to become even more skilled. Such specialization has increased to such an extent that today an individual may carry out only a small part of a given task. Perhaps he may only install headlights in cars in an automobile factory. To make such specialization possible it has been necessary for society to develop transportation systems (circulatory system), communication systems (nervous system), and a system of management and government (central integrating systems). Currently the increasing size of society is posing new problems of all kinds, and unless better mechanisms evolve, it will soon be necessary to drastically limit the size of the "organism."

1.4 SEMIPERMEABILITY AND OSMOSIS

Many kinds of membranes, both living and non-living, are said to be semipermeable or selectively permeable. This means that certain kinds of molecules pass through them rapidly and with apparent ease, others pass more slowly, and still others not at all. This effect, which is due to the nature of the membrane, is superimposed upon the factors governing diffusion (Sect. 1.2). All semipermeable membranes are more or less permeable to water, the biological solvent, but they exhibit differential permeability to various solutes (dissolved substances). Moreover, the degree of permeability to water and to various solutes may be altered from time to time for a given living membrane. This greatly influences the diffusion of substances into and out of cells.

Let us consider now an illustration of a semipermeable membrane and a solute to which it is not permeable. If this substance is present in different concentrations on the two sides of the membrane, it influences the flow of the solvent (water) through that membrane. The following simple illustration is not a completely accurate description of the true physical situation, but it is a useful way of thinking of it since it does accurately predict the direction of water flow through the membrane. It also provides a comparative estimate of the rate and quantity of water flow through the membrane (Fig. 1-3).

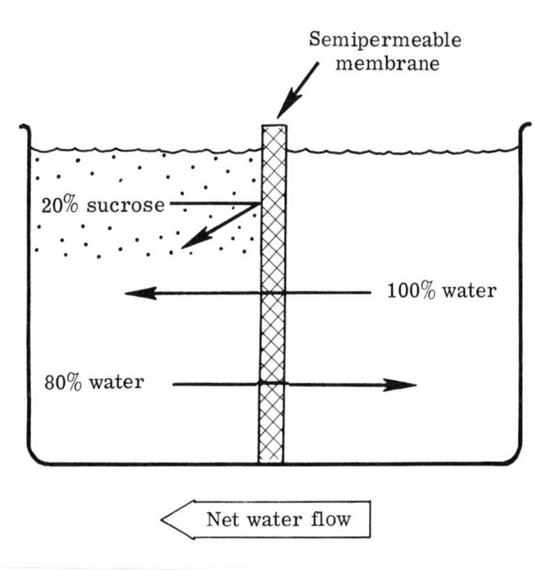

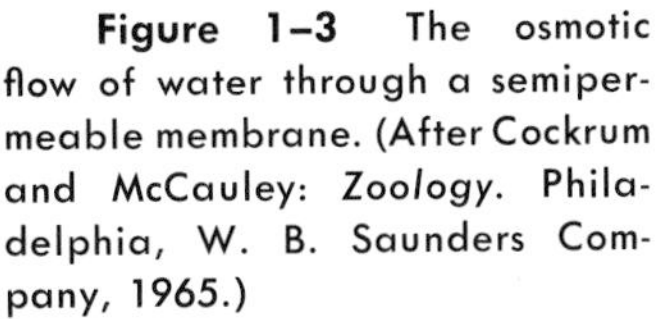
Figure 1–3 The osmotic flow of water through a semipermeable membrane. (After Cockrum and McCauley: *Zoology*. Philadelphia, W. B. Saunders Company, 1965.)

In the example shown, the membrane is not permeable to sucrose molecules, but it is completely permeable to water molecules. On the left side of the membrane 20 per cent of the molecules present are sucrose, whereas the remaining 80 per cent are water molecules. All are, of course, in constant motion so that some molecules will be striking the membrane at all times. If the membrane is permeable to a given molecule striking it, then that molecule is likely to pass on through the membrane; if the membrane is not permeable to it, then the molecule will be deflected back into the medium from which it came. Thus, 80 per cent of the molecules striking the left side of the membrane pass through it and the remaining molecules bounce back.

On the right side of the membrane there are only water molecules; therefore, 100 per cent of the molecules striking the membrane from that side are likely to pass on through it. If we assume that the same total number of molecules strike the membrane from each side, then more will pass through it from right to left than in the opposite direction. Thus, there will be a net flow of water from right to left through the membrane. This process of differential flow of solvent is called osmosis.

Osmosis would occur if sucrose was present on both sides of the membrane as long as the concentration was not the same on each side. The greater the difference in concentration of such a non-penetrating solute, the greater will be the consequent flow of water through the membrane. A large difference in solute concentration on the two sides of the membrane constitutes a steep osmotic gradient.

To the extent that it causes a differential flow of water through a semipermeable membrane, a solute of given concentration is said to exert osmotic pressure. The amount of osmotic pressure exerted is directly proportional to the *number* of molecules or particles of solute, not to the *kind* of solute. That is, if a membrane is impermeable to either sucrose or urea, a thousand molecules of sucrose per milliliter of solution will have the same osmotic pressure as a thousand molecules of urea per milliliter of solution. Either solution placed on one side of such a membrane with distilled water on the other side will cause an equal flow of water through that membrane.

In a simple example such as that shown in Figure 1-3, an equilibrium will be eventually reached. Since more water flows from right to left, the total volume of solution on the left side of the membrane will increase, whereas the volume of water on the right side will decrease (Fig. 1-4). The increased weight of solution on the left side will oppose the continued influx of water from the right. As this situation gradually develops, the rate of water influx slowly decreases. When the two forces are in balance, the flow of water through the membrane will be the same in both directions, and the system will be in equilibrium. The weight of water necessary to accomplish this is a measure of the osmotic pressure.

The concentration of solutes in a solution affects not only the osmotic pressure but also the freezing point, boiling point, and vapor pressure of the solution. All of these are called colligative properties of solutions and they depend upon the numbers of molecules or particles of solute present in a given volume of solution. Thus, the colligative properties of a solution are mathematically related to each other. If any one such property is known, the others can be calculated. Generally the freezing point or the vapor pressure is experimentally determined, since the osmotic pressure is technically more difficult to measure directly.

If a given solution has a higher osmotic pressure than the contents

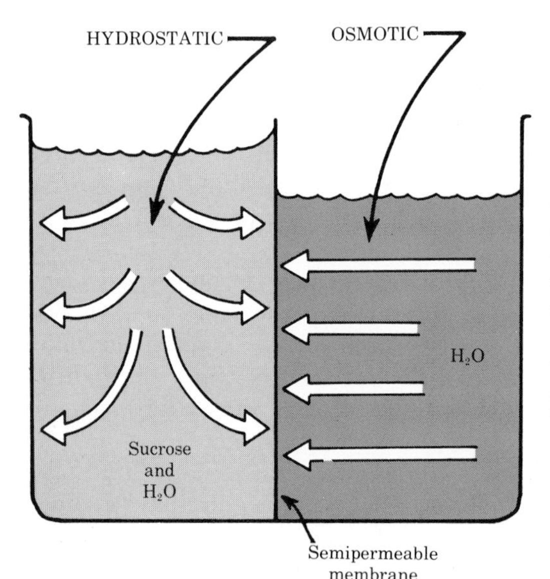

Figure 1–4 The forces of hydrostatic pressure oppose the forces of osmotic pressure. An equilibrium is reached when the two forces become equal, and thereafter the movement of water through the semipermeable membrane is the same in both directions. (After Cockrum and McCauley: *Zoology*. Philadelphia, W. B. Saunders Company, 1965.)

of a cell, it is said to be hyperosmotic with respect to that cell. If it has a lower osmotic pressure, it is said to be a hypoosmotic solution.

It must be remembered that although the osmotic pressure of a solution depends only upon its solute concentration, the osmotic effect varies from membrane to membrane. The semipermeability characteristics of living membranes vary, not only from membrane to membrane, but often from time to time. Moreover, the opposing osmotic pressure of fluids contained within cells is inconstant. Cellular metabolism constantly varies the concentrations of various solutes within the cell. The actual osmotic effect of a given solution upon a cell is spoken of as its tonicity. Immersing a cell in a hypertonic solution causes the cell to lose water and, therefore, to shrink. Immersing it in a hypotonic solution causes it to gain water and to swell.

As we shall see in the next chapter, membranes not only separate the cell contents from the surrounding medium but they also segregate portions within living cells. Thus, the semipermeability characteristics of various membranes are of extreme importance to many of the physiological functions of cells and cell parts.

1.5 ACTIVE TRANSPORT

The simple diffusion of substances and the osmotic flow of water through membranes occur in response to physical laws and require no expenditure of energy on the part of the living cell. This kind of natural movement of substances through membranes is called passive transport. Certain membranes, however, are capable of actively "pumping" materials through themselves. This active transport, as it is called, may even be in a direction opposite to the diffusion gradient for the substance in question. The cell actually expends measurable amounts of energy to operate such "pumps." Changes in the oxygen consumption of a cell accompany periods of active transport, and by determining these changes the energetics of the process can be measured. The mechanisms responsible for active transport are not completely understood but they probably involve biochemical transformations within the membrane as well as intra-membrane "carrier" molecules. Presumably these carrier molecules combine with a given solute on one side of the membrane, and then the combined molecules diffuse to the other membrane surface. There the molecules separate, the solute passes out, and the carrier diffuses back to pick up another molecule of solute. More detailed discussions of the mechanisms of active transport may be found in any textbook of cellular physiology.

Substances are sometimes transported across cell membranes by another process, called pinocytosis. Pinocytosis, or "cell drinking," consists of the formation of small vesicles at the cell surface (Fig. 2–1). Small quantities of the environmental medium are engulfed in a manner reminiscent of the engulfment of prey by an ameba. Since this also requires that the cell expend energy, it can be thought of as another form of active transport.

1.6 CONTROL

Each specialized part of the life process must be controlled so that it can be speeded, slowed, turned off, or modified as the well-being of the total organism may require. In part, the purpose of such controlling mechanisms is the maintenance within the animal of fairly constant internal conditions. Changing environmental conditions, such as temperature, acid-base balance, salt concentrations, and so forth, tend to produce similar changes within the animal. In many cases the continuation of life depends upon the successful resistance to such internal changes. Constant monitoring and adjustment are needed to prevent them from occurring. The maintenance of relatively constant internal conditions is known as homeostasis, and the science of physiology is largely a study of homeostatic mechanisms. In addition, physiological systems must control purposeful internal changes associated with growth, development, repair, and reproduction.

Most homeostatic mechanisms operate on a "feedback" principle. In an air-conditioned room, the thermostat monitors the room temperature and acts to turn the heating or refrigeration units on and off to maintain room temperature within a preset range. Thus, the results of the operation of the heating or cooling unit are "fed back" to these units to control their further activity. Under certain conditions it is desirable to reset the thermostat to maintain a warmer or a cooler room than usual. Similar situations exist in physiological control systems. Blood pressure, for example, is generally maintained within given limits, but during exercise it is desirable to reset blood pressure at higher levels for a time. The physiological mechanisms controlling blood pressure have a normal resting range but can be reset to permit these changes to occur. Like the thermostat in the foregoing analogy, these mechanisms operate through "feedback" principles (Chapter 4).

The life process is, then, a continually changing thing in each individual animal—altering to meet the needs of the moment and influenced by the current status of the environment, both within and without the animal body. Changes also occur to permit and direct growth, development, and the periodic involvement of the animal in reproductive function. Such controls, both of homeostasis and of orderly change, involve all of the subprocesses of life including internal chemical activity, physical movement, and even the behavior of the individual toward others of its own kind.

1.7 THE ORIGINS OF LIFE

Before the time of Lazzaro Spallanzani, about 200 years ago, the concept of continual spontaneous generation of living things was almost universally accepted. Van Helmont (1577-1644) even published a recipe for the "generation" of mice. Required ingredients included a few grains of rice and a dirty shirt! He expressed amazement at the similarity between his artificially produced mice and those which he knew to be the result of natural breeding! The work of Spallanzani and others, however, proved conclusively that life cannot arise spontaneously. Life only comes about through the natural reproduction of pre-existing life. This new point of view was an important step toward the development of modern ideas of organic evolution and it also demonstrated another important fact. Since life is present on Earth and since spontaneous generation is not possible under present conditions, then conditions must have been different at the time when life first appeared on this planet.

The development of astronomy and geology yielded evidence that conditions were, indeed, very different more than a billion years ago. During recent years experiments have demonstrated, moreover, that the spontaneous generation of simple living things was almost inevitable, given the environmental conditions of that period and sufficient time for random chemical combinations to occur.

Primordial Earth was composed of intensely hot vaporized material. As it cooled, the heavier materials naturally settled toward the center of the mass and established a molten core of iron, nickel, and similar materials. Most of the carbon was carried into this core where it existed largely as C_2 rather than in the form of compounds. Oxygen and nitrogen were also carried into the core as oxides and nitrides of heavy metals. Helium and hydrogen, on the other hand, remained in the gaseous atmosphere and probably were partly lost into space.

Over the molten core a layer of cooling and solidifying lava formed and separated the deeper molten materials from the early atmosphere of the planet. This atmosphere probably consisted mostly of superheated water vapor. The moon was undoubtedly already present and exerting its tidal effects upon the earth. The tides produced in the molten materials under the thin solidifying crust must have frequently resulted in ruptures of the crust and consequent volcanic outflows. In such places the metallic carbides and nitrides were exposed to the superheated atmosphere, and chemical reactions occurred producing hydrocarbons (from carbides) and ammonia (from nitrides). The hydrocarbons were probably all of the simple aliphatic types such as methane and ethane.

The atmospheres of certain other planets in our solar system contain simple hydrocarbons and ammonia which most likely came about in the same way. It is probable that the earth would have remained in this same state had it not been for the unique subsequent events leading to and including the appearance of life. Life has continuously acted to modify the earth and its atmosphere and, indeed, it continues to do so today.

At about the same time in the formation of our planet, photic (light) and ionizing (x-ray) radiations from the sun were numerous and intense, and the atmosphere of the earth provided little shelter from them. Violent electrical storms were also common. When such conditions are experimentally provided in the presence of ammonia, simple hydrocarbons, and superheated water, many kinds of familiar organic compounds are formed. These include all of the component parts of the most complex organic substances associated with life. Given further time for random recombinations of these components, it was inevitable that complex substances such as chlorophyll, hemoglobin, proteins, and so forth would appear. Since the period leading up to this stage may have lasted for more than two billion years, adequate time for such random recombinations was certainly available.

With further cooling, water condensed from the atmosphere and formed the seas. Run-off from the land carried materials of all kinds to the seas, and very likely ocean water became a concentrated solution of simple hydrocarbons and ammonia. Authorities estimate that the concentrations of these substances in sea water may have amounted to as much as 10 per cent. A solution of this kind could not accumulate today because living things would absorb and utilize the organic molecules as rapidly as they appeared. In the absence of life, however, such accumulation undoubtedly occurred.

The gradual escape of light gases such as hydrogen and helium left, as we have noted, water and some oxygen in the atmosphere. The effects of radiation probably converted much of the oxygen to ozone and this, together with water, shielded the earth with increasing effectiveness from solar radiation. Although such radiation was probably essential to the formation of simple organic molecules, it must also have caused the breakdown of more complex combinations. Thus, while radiation played an important early role, the decrease in radiation was equally essential to the formation of more complex organic molecules.

Amino acids (Sect. 9.4) in solution tend to link together to form long chains, which, by definition, are called proteins. Furthermore, proteins tend to align themselves in groups to form larger particles because of cohesive attractions between them. These larger particles are capable of binding to themselves comparatively large amounts of water. The water is not chemically combined with the protein but is held physically as water of hydration or bound water (Sect. 8.1). The hydrated groups of protein molecules are, because of their size, classed as colloids, i.e., they are larger than dissolved materials but still small enough to remain suspended in the medium. When hydrated colloids come in contact with each other, they sometimes remain together and share the surrounding bound water. Under such conditions the surface of the bound water is compacted and may become membranelike. Accumulations of colloids of this kind are called coacervates and they exhibit some of the characteristics of living cells. The "membrane," because of its high surface tension, accumulates other substances that may then pass into the "cell." Because internal conditions there are different, chemical reactions may occur, and, therefore, diffusion gradients may be established across the membrane. Thus, such a nonliving "cell" may exhibit growth. Due to purely physical forces, such a "cell" may even divide to form essentially equal daughter "cells."

The proteins of coacervates may well have contained molecules that acted as catalysts or enzymes (Sect. 9.8). Such substances speed and direct chemical reactions in definite ways. Under their influence a great deal of internal modification of existing molecules undoubtedly took place, and this led to even greater organic complexity.

Those cell-like associations which "functioned" most rapidly and efficiently must have grown and divided at such a rate that they came to dominate in the seas where life began. Thus, selection and survival—the important factors of animal evolution—were operative before life, as such, appeared on Earth. Compounds of great complexity made their appearance and functioned as non-living systems capable of reproductive duplication and of energy utilization before organisms, as such, ever existed.

It is impossible and essentially unimportant to draw any sharp line between the self-duplicating but non-living molecular associations and the first primitive but true organisms. Whatever we wish to call such "organisms" at this level, they must have resembled bacteria, although they were certainly much less complex than modern bacteria. Certainly they were anaerobic—that is, they required no oxygen—as are many lower organisms today. Oxygen was present in the atmosphere in only low concentrations and the solubility of oxygen in sea water was, and still is, low. Probably these organisms obtained energy from the partial breakdown of the more complex organic molecules that were already present in the sea water. It may have been that some of them also made use of solar energy to a degree. If so, the process was not like the photosynthesis carried on by modern green plants. Rather, it was only an acceleration of organic chemical reactions within the organism. Even such primitive processes as those mentioned, however, require the presence of an array of complex substances like the cytochrome enzymes (Sect. 10.7), but, as we have shown, there is no reason to believe that such enzymes were not available at that time.

With the passage of time and an increase in the numbers of living things in the seas, the concentration of preformed organic materials there must have become depleted. Ordinarily modern waters contain only 2 to 20 milligrams of dissolved organic matter per liter of water. This quantity is probably insufficient for any organism to live upon. In certain special instances, of course, natural waters may contain higher concentrations of dissolved organic matter and are, therefore, capable of supporting populations of organisms. Parasitic animals living in the intestines or blood of vertebrate hosts are also provided with high concentrations of organic matter. These, however, are unusual circumstances.

At the time when pre-existing dissolved organic compounds were being depleted there may have developed some organisms which, by chance, contained chlorophyll. We have already mentioned that this, as well as the chemically related cytochrome enzymes, probably already existed. The coupling of such compounds in organized systems permits the trapping and utilization of solar energy through the process of photosynthesis. This would be a big but not impossible step forward, and it undoubtedly occurred at this time, if not earlier. These photosynthesizing organisms, then, became independent of pre-existing organic materials and, instead, produced their own from simple, inorganic sources. Most importantly, as a by-product of this synthetic activity they produced molecular oxygen and released it into the atmosphere. Organisms that can accomplish such syntheses are called autotrophic organisms.

The activity of autotrophic organisms (plants and their predecessors) set the stage for the appearance of heterotrophic organisms, those that cannot utilize solar energy. The heterotrophs were then able to maintain themselves by exploiting the autotrophs. Actually, there is no sharp distinction today between primitive autotrophs and heterotrophs. Many lower organisms utilize both means of support.

Thus, each stage of the physical development of Earth and of the appearance and diversification of life has been a necessary sequential phase. The proper conditions for the generation of life once existed, but today they are no longer extant. Conditions that have been altered since the appearance of life have influenced and, in a sense, directed the further evolution of living things leading to the modern diversity of life. Life continues to influence the physical conditions of the atmosphere today. Indeed, the burning of fossil fuels (coal, oil) by man is increasing the carbon dioxide and decreasing the oxygen of the atmosphere. Man's industry and automobile traffic is adding unnatural chemical substances to it and causing the formation of smog. These changes result in a trapping of solar heat, and this will, if continued, perhaps raise the average temperature of our planet. This, in turn, will alter environmental conditions in other ways so that tropical plants may grow further from the equator than before and polar ice caps may melt. The melting of the ice caps would drastically change the existing coastlines by raising the water level of the seas. By his use of long-lasting insecticides man is also destroying basic forms of life of many kinds. Among these are the minute green plants of the seas which produce a major portion of the earth's oxygen. It may well be, in fact, that man will ultimately succeed in destroying life altogether through pollution, insecticides, or nuclear warfare!

1.8 EVOLUTION

Life began as a process with the first living things on Earth or, depending upon the point of view, even before that. In any case, it has been continuous to the present time. Life does not end with the death of an individual but passes on in an unbroken sequence from generation to generation. Viewed in this light, it is apparent that the moment by moment changes in the physiology of the individual are but a small part of the whole story. Long-term modifications of each species also occur and these are correlated with long-term changes in the environment. Such long-term changes in morphology are, of course, correlated with long-term changes in function. Indeed, changes in form and function are often inseparable.

There is not space in a book such as this to discuss the mechanisms of evolutionary change; that is a voluminous science in its own right. As a result of such changes, however, various modern vertebrates have become very different from each other. Vertebrates are found over very wide ranges of environmental temperature (−2° C to about +40° C), pressure (0.5 atmospheres to more than 1000 atmospheres), salt concentration (near zero to considerably more than sea water), and other factors.

In this book we shall be particularly interested in the evolution of physiological function among the vertebrates but we shall often need to consider the parallel changes in structure as well. We shall deal with function primarily at the organ and system level of observation but, to place it in proper perspective, we shall occasionally study physiology at the cellular level as well. We shall be almost entirely concerned with vertebrate animals but sometimes we shall also compare them with invertebrates.

By far the greatest amount of information is available for the mammals, especially for man and the laboratory mouse and rat. Such studies have had more immediate application to problems of human health and disease. Consequently, research of this kind has been well financed. To an increasing degree, however, curiosity had led investigators to examine the physiology of lower vertebrates. There is today a growing awareness that an understanding of homeostasis in lower forms contributes richly to the understanding of human physiological systems.

CHAPTER 2

CELLS

2.1 INTRODUCTION

All living things are composed, at least at some stage of their life history, of one or more units called cells (Fig. 2-1). Each cell is a living unit both structurally and functionally. Each has its own internal complement of functional parts (organelles) and its own independent physiology, at least to a degree. To a greater or lesser extent in multicellular organisms the physiology of a single cell is also influenced by the functional activities of other cells. It is perhaps not altogether incorrect to think of multicellular animals as colonies of tiny, one-celled animals. Each cell of such a "colony" is specialized in form and function to play a particular role in the economy of the colony. Whether cells are completely independent, as a one-celled animal, or whether they are parts of a multicellular organism, they all have many things in common. Generally similar internal structures play like roles in the living processes of all cells. In all but the lowest of multicellular animals, cells of similar specialized types combine to form similar tissues, and these tissues combine to form similar organs. The stomach, for example, consists of glandular, epithelial, muscular, and connective tissues which are generally associated in similar ways to form a hollow organ in which particular stages of food digestion can be carried out.

Organs, in turn, become associated with each other to form organ systems such as the digestive system. The stomach, the intestine, esophagus, oral-pharyngeal apparatus, and certain associated glandular organs (pancreas, liver) form such an organ system devoted to the total task of food processing. Finally, organ systems are associated, together with interlocking controls, to form the total organism. No organ system, organ, tissue, or cell operates alone or independently of the rest of the animal in all ways, but many cellular functions are carried out independently by each cell.

In this book we shall be primarily concerned with the functions of organs and organ systems of vertebrate animals, but since functions at

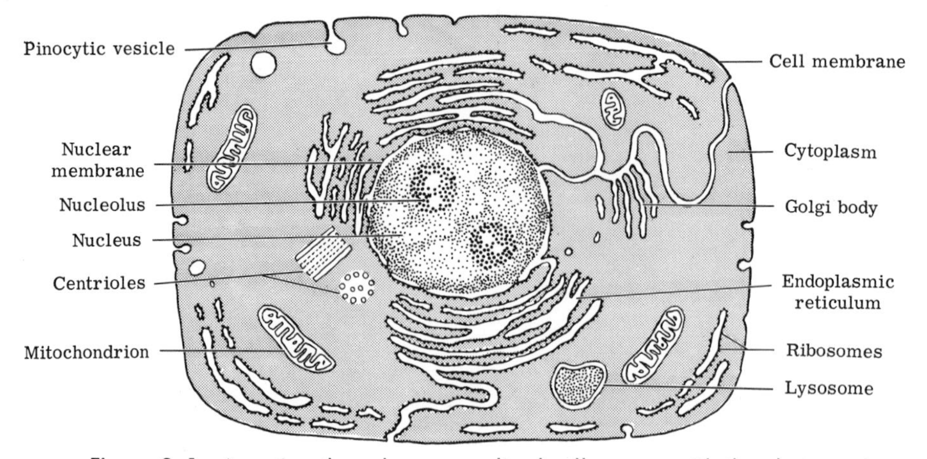

Figure 2–1 A section through a generalized cell as seen with the electron microscope (schematic). (From Cockrum and McCauley: *Zoology*. Philadelphia, W. B. Saunders Company, 1965.)

that level are but the summation of cellular functions, it is important to acquire some basic concepts of cell structure and physiology.

Cell structures vary in so many ways that there is really no such thing as a "typical" cell. A generalized animal cell, however, contains a number of recognizable parts or organelles. Typically there is a prominent nucleus that contains nucleoplasm and is enclosed by a nuclear membrane. The remainder of the cell consists of cytoplasm, and the whole cell is enclosed by a plasma membrane. The nucleoplasm and cytoplasm, together with their membranes and inclusions, are collectively spoken of as protoplasm. In other words, protoplasm is the general name for all living matter. Protoplasm is largely composed of water, about 70 to 80 per cent, but contains in addition electrolytes, proteins, carbohydrates, lipids, and various other organic materials. Some of these substances form structural parts of the cell and some are merely present in solution. Some function in the controlling of various biochemical processes and some are the reactants or products of these processes.

2.2 CELL ORGANELLES

Besides the plasma membrane and the nuclear membrane already mentioned, many other membranes subdivide the cell's interior. These segregate portions of the cell so that various internal biochemical processes are also somewhat segregated. The membranes are, of course, semipermeable, and selected and controlled movements of various substances occur through them (Sect. 1.4).

The structure of all cell membranes is, apparently, quite similar but of sufficient complexity to elude, thus far, positive understanding. The structure for the plasma membrane originally suggested by Danielli was later applied by others to all cell membranes and referred to as the

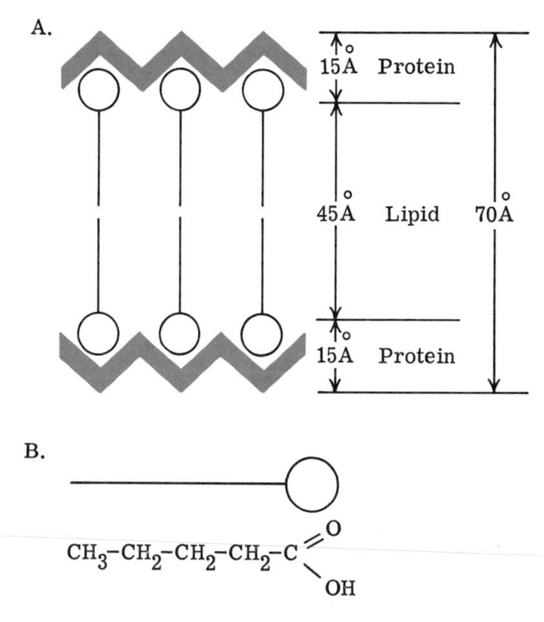

Figure 2-2 The structure of the plasma membrane of a cell according to Danielli. *A*, structure and dimensions of the membrane; *B*, chemical structure of a typical lipid molecule (fatty acid).

unit membrane (Fig. 2-2). This concept pictured all cell membranes as having, fundamentally, a double layer of lipid molecules overlaid on both sides by a layer of protein molecules. The outermost layer of the plasma membrane, at least, is now known to contain a considerable amount of carbohydrate. Except for this, however, the Danielli membrane composition is still generally accepted.

There is abundant evidence to suggest that typical cell membranes have tiny pores penetrating through them. Moreover, these pores can apparently vary in size from time to time. Such variation in pore size results in changes in the permeability of the membrane to various substances. Attempts to explain the changing porosity of membranes have led to a somewhat altered version of the Danielli membrane structure (Fig. 2-3). According to this modern concept, the lipid portion of the membrane consists of a number of "pillars" with pores between them. Variations in the shape of the pillars results in changes in the pore size.

The plasma membrane enclosing the entire cell is almost certainly of the type just described. The nuclear membrane, however, is actually composed of two such unit membranes separated by a relatively wide, fluid-filled space.

Much of the cytoplasm of the typical cell is filled with a series of interconnected membranes collectively referred to as the endoplasmic reticulum (ER). These membranes form tubules, spaces, and vesicles that are filled with an endoplasmic matrix (Fig. 2-4). The endoplasmic matrix is a fluid different in composition from the general cytoplasmic fluids that surround these spaces. Some portions of the endoplasmic reticulum are marked by the presence of small granular particles called ribosomes. These particles are apparently attached to the endoplasmic

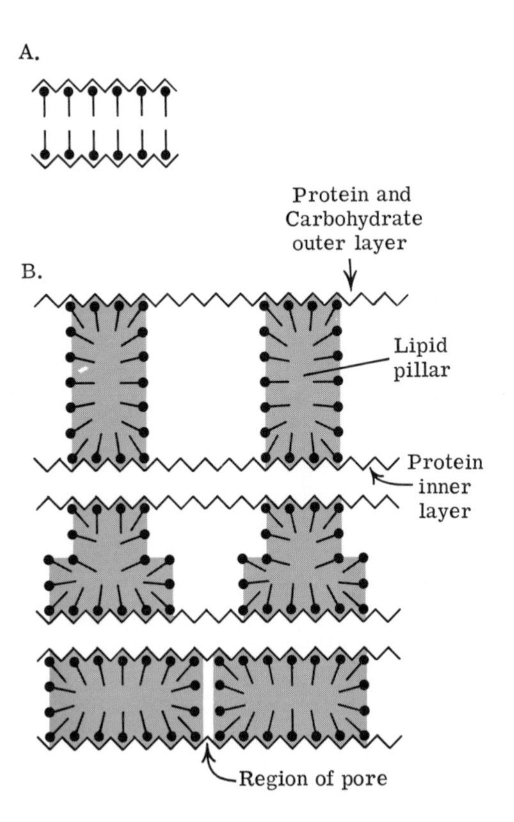

Figure 2–3 A recent concept of membrane structure. *A*, the Danielli membrane (See Figure 2–2); *B*, the lipid pillar membrane alternating from open pores (*above*) to closed pores (*below*).

reticulum, giving it a roughened appearance. This portion of the endoplasmic reticulum is accordingly called the granular, or rough, ER. Other portions lack ribosomes and are differentiated by the name smooth ER.

Portions of the smooth ER are expanded to form larger vesicles referred to as the Golgi complex. The Golgi complex is most prominent in secretory cells of various kinds. It apparently functions in the temporary storage of materials that are to be secreted.

The membranes forming the endoplasmic reticulum and Golgi complex are continuous with the nuclear membrane (Fig. 2-9) and, perhaps, with the plasma membrane as well.

It now seems certain that the rough portions of the endoplasmic reticulum function in the synthesis of proteins from amino acids (Sect. 9.4). The specificity (amino acid sequence) of the proteins is dictated, at

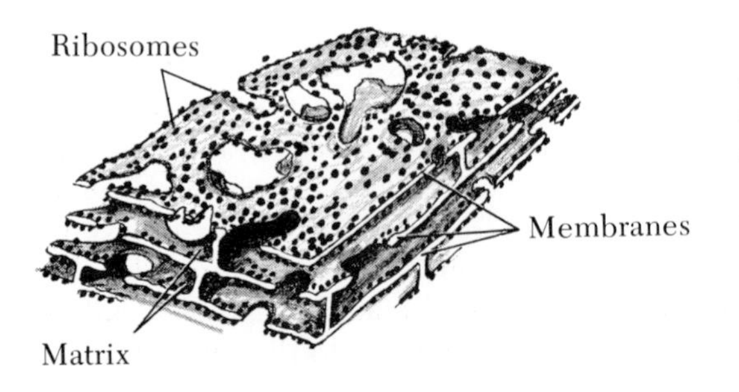

Figure 2–4 The structure of the endoplasmic reticulum of a cell. (From Guyton: *Textbook of Medical Physiology*, 4th Edition. Philadelphia, W. B. Saunders Company, 1971. Redrawn from De Robertis, Nowinski, and Saez: *Cell Biology*, 4th Edition. Philadelphia, W. B. Saunders Company, 1965.)

least in part, by the ribosomes. Ribosomes, in turn, are constructed in the nucleus of the cell under the control of the genetic material of the chromosomes. The process is more complex than this statement would imply, and the interested student is referred to any recent textbook of cytology or cellular physiology for a more complete discussion.

The smooth portions of the endoplasmic reticulum probably function in the synthesis of non-protein materials, especially lipids and various secretion products of cells. The secretion products may also include proteins formed on the rough ER. In either case, they are then transferred to the Golgi complex for storage, concentrated to form secretion granules, and finally passed from the cell.

Variable numbers of mitochondria occur in the cytoplasm of all animal cells (Fig. 2-5). They are present in greater numbers in cells with larger energy requirements and, from this as well as from other evidence, the mitochondria have been identified as the "powerhouses" of the cell. Sugar can be broken down in the cytoplasm to yield some energy and smaller hydrocarbon molecules. This process is known as anaerobic glycolysis and involves a series of biochemical reactions known collectively as the Embden-Meyerhof pathway (Sect. 10.5). In the mitochondria these smaller hydrocarbons are oxidized to form carbon dioxide and water and to yield much larger amounts of energy. These reactions within the mitochondria involve a cyclic biochemical process known as the Krebs cycle (Sect. 10.6) and a series of cytochrome enzymes (Sect. 10.7). All the many steps of these reactions require specific organic catalysts (enzymes) in order to proceed at appropriate rates (Sect. 9.8). It is thought that these enzymes are arranged sequentially on the innermost of the two mitochondrial membranes (Fig. 2-5).

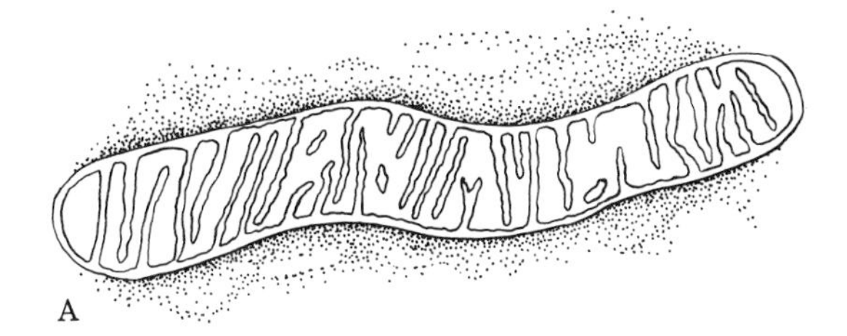

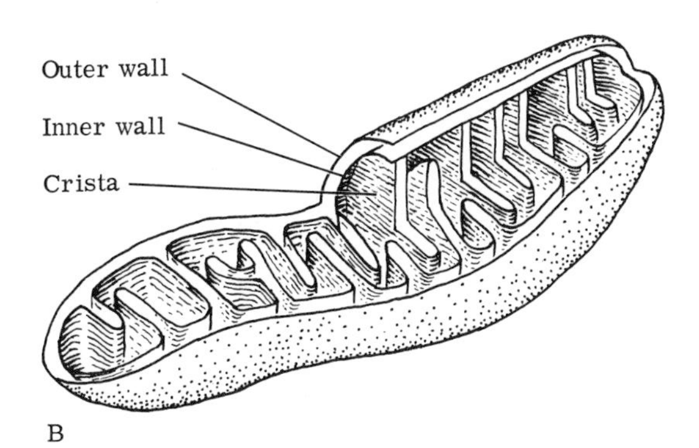

Figure 2–5 The structure of a mitochondrion. (From Cockrum and McCauley: *Zoology.* Philadelphia, W. B. Saunders Company, 1965.)

This inner membrane is thrown into folds, or cristae, which greatly increase its surface area.

Many cells contain accumulations of enzymes that are enclosed within membranes to form structures called lysosomes. In one-celled animals the lysosomes function by discharging their contained enzymes into food vacuoles. There the enzymes work by catalyzing the molecular breakdown of food materials so that the products can be absorbed into the cytoplasm of the cell. Thus, they act as important parts of the digestive "system" of these one-celled animals. In vertebrate animals some of the lysosomes function in the same general way since their contained enzymes are discharged into the digestive tract (Sect. 9.8). Lysosomes probably also act in extensive tissue remodeling processes, such as the metamorphosis of amphibians, the resorption of bone by osteoclasts (Sect. 2.6), and the atrophy of unused muscles. The lysosomes are undoubtedly derived from the Golgi complex, in which the newly synthesized enzymes are first stored.

Most cells contain two small structures called centrioles that play important roles in cell division. Each is a cylinder composed of nine parallel fibrils (Fig. 2-6). The two centrioles generally lie at right angles to each other in resting cells. At the beginning of prophase of cell division one centriole migrates to each pole of the dividing cell and there acts as a center for the organization of the spindle fibers. Following the completion of cell division, one centriole remains in each of the two daughter cells. They undergo self-replication at once so that each daughter cell contains a pair of centrioles.

Some cells have extensions called cilia or flagella, and these contain nine parallel fibrils, as do centrioles. They are called the outer fibers of the cilium or flagellum and are enclosed by an extension of the plasma membrane of the cell (Fig. 2-7). Each outer fiber, like those of the centrioles, appears to be doubled when seen in cross section. In addition, cilia and flagella (but not centrioles) have a doubled central fiber and sometimes additional fibers as well.

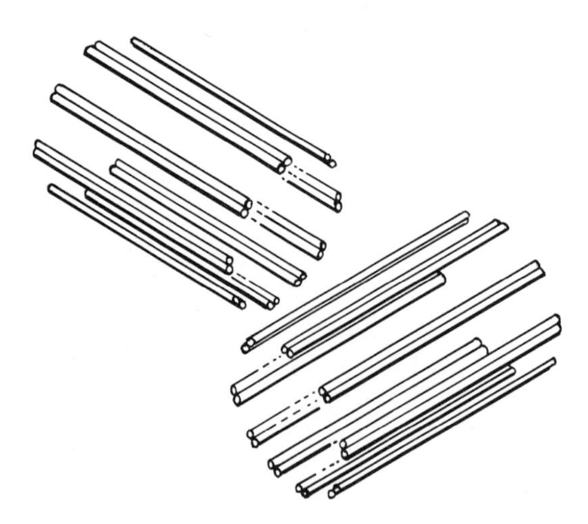

Figure 2–6 A pair of centrioles. Characteristically they lie at right angles to each other, as shown, except during cell division. (From Cockrum and McCauley: *Zoology*. Philadelphia, W. B. Saunders Company, 1965.)

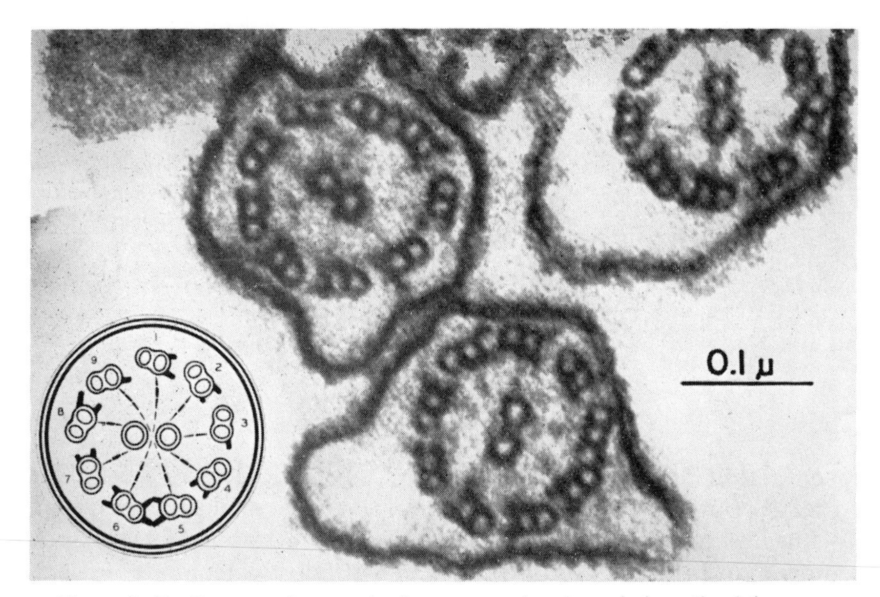

Figure 2–7 Electron micrograph of a cross section through the tails of three spermatozoa showing the typical arrangement of fibrils in a flagellum. (From De Robertis, Nowinski, and Saez: *Cell Biology,* 4th Edition. Philadelphia, W. B. Saunders Company, 1965. Courtesy of B. Afzelius.)

At the base of each cilium or flagellum is a structure called the basal granule. It lies just beneath the cell surface and has a structure much like that of a centriole, *i.e.*, the nine fibrils are present but there is no central fiber. Moreover, the nine outer fibrils appear to be tripled rather than doubled. It is assumed that the basal granule functions in some way to control the activity of cilia and flagella and that the actual contractile elements lie within the cilia and flagella themselves.

Cilia that function in cell motility or in propelling environmental media past fixed cells are found in respiratory passages, sinus cavities, oviducts, and elsewhere in vertebrate animals. In addition, certain cells have highly modified, non-motile cilia which function as sensory organelles. Such cells occur in the vertebrate retina (Sect. 15.9) and in the organ of Corti of the inner ear (Sect. 14.5).

In addition to the organelles already mentioned, cells often contain other identifiable structures such as vacuoles and crystals. Vacuoles are usually enclosed by membranes derived from the plasma membrane and contain materials engulfed from the surrounding medium or produced by the Golgi complex of the cell. Specialized cell types contain still other organelles; for example, the contractile fibers and microtubules present in muscle cells (Sect. 2.3).

The nucleoplasm contains, of course, the genetic materials of the chromosomes as well as proteins and small amounts of other substances. Morphologically nuclear inclusions are more variable than those of the cytoplasm—not only from cell to cell, but within the same cell at different times. Chromosomes as such are seen clearly only during active cell division. The most frequently identifiable structure

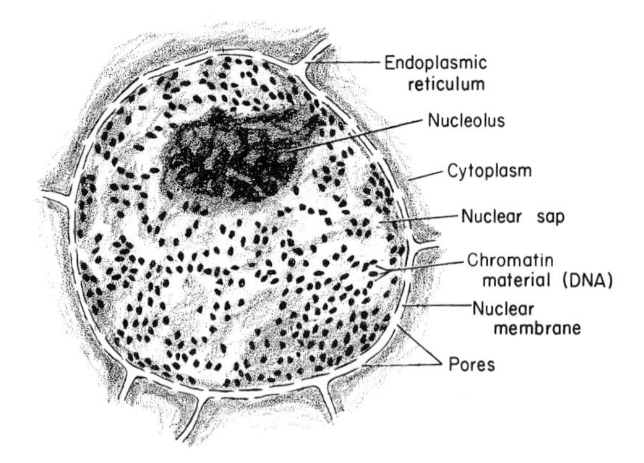

Figure 2–8 The structure of a cell nucleus and its membrane. (From Guyton: *Textbook of Medical Physiology*, 4th Edition. Philadelphia, W. B. Saunders Company, 1971.)

present, other than the chromosomes, is the nucleolus (Fig. 2-8). This is a small clump of ribonuclear protein lacking an enclosing membrane and attached to two or more of the chromosomes. Its specific function is unknown, and in some cells more than a single nucleolus is present.

2.3 SKELETAL MUSCLE CELLS

Muscle tissue, most of which is of the skeletal type, makes up the greatest amount of the soft tissues of most vertebrates. Moreover, muscle contraction is the greatest energy-consuming activity of most animals. Consequently, the physiology of contraction has received a great deal of attention from investigators (Chapter 16). With the advent of the electron microscope the fine structure of skeletal muscle cells became known, and it was possible to draw tentative conclusions regarding the mechanics of contraction. Less is known of the physiology of smooth muscle, but there is evidence to suggest that both types are fundamentally similar. Indeed, it may be that all protoplasmic movement is based upon similar biochemical processes but is variously modified by the different mechanical orientations of cell parts.

Many of the organelles discussed in the preceding sections are present in muscle cells, *e.g.*, membranes, mitochondria, and nuclei. Some of these are modified in structure to play specialized roles. The cells of skeletal muscle are comparatively large; they may range up to several centimeters in length and from 10 to 100 microns in diameter. Each cell is called a muscle fiber, and groups of cells are bound together by connective tissues to form fasciculi (Sect. 16.2).

Inside each skeletal muscle cell are several irregularly spaced nuclei that are otherwise not very different in appearance from typical cell nuclei. Skeletal muscle cells are strikingly different, however, in that they contain large numbers of longitudinal myofibrils which, when seen through the optical microscope, have prominent cross-striations (Fig. 2-9). The electron microscope reveals that the myofibrils are, in reality, composed of filaments of two general diameters, thick and thin. These interdigitate with each other in a repeating pattern, and it is this

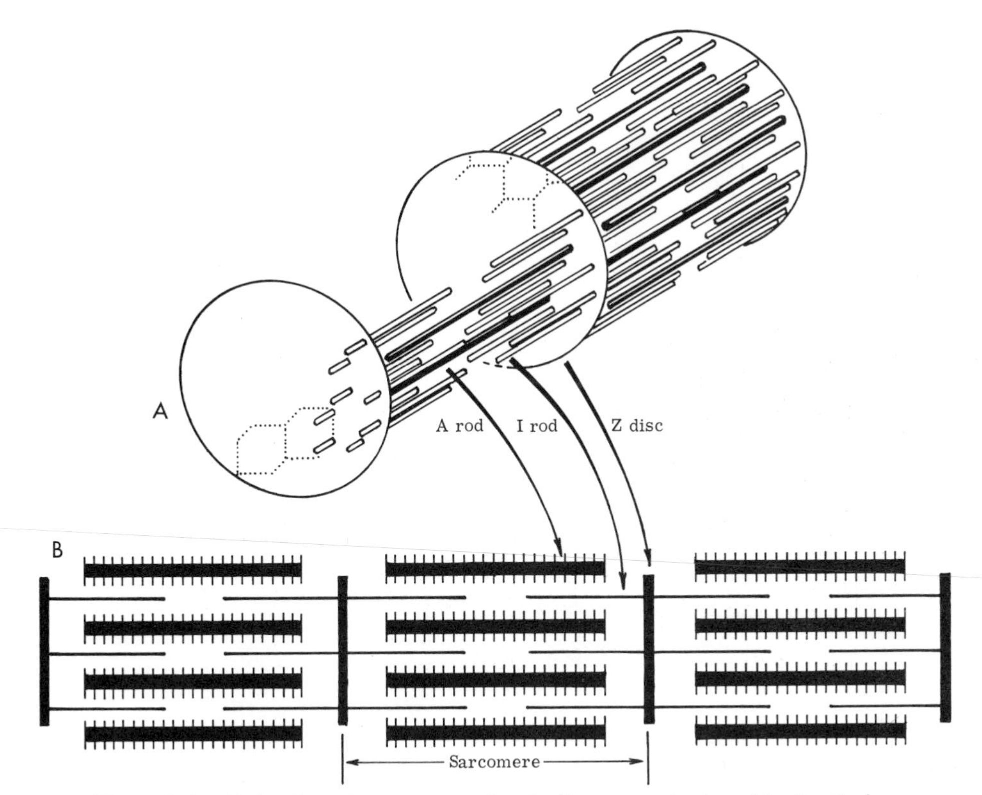

Figure 2–9 Skeletal muscle. *A*, perspective; *B*, diagrammatic. *A* and *B* after Cockrum and McCauley: *Zoology*. Philadelphia, W. B. Saunders Company, 1965.

RELAXED

CONTRACTED

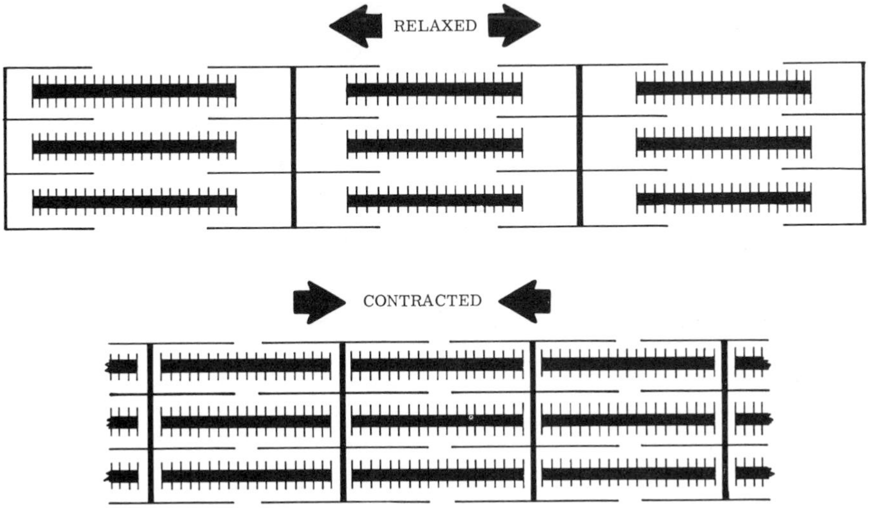

Figure 2–10 The shortening of striated muscle sarcomeres during contraction. See labels in Figure 2–9. (After Cockrum and McCauley: *Zoology*. Philadelphia, W. B. Saunders Company, 1965.)

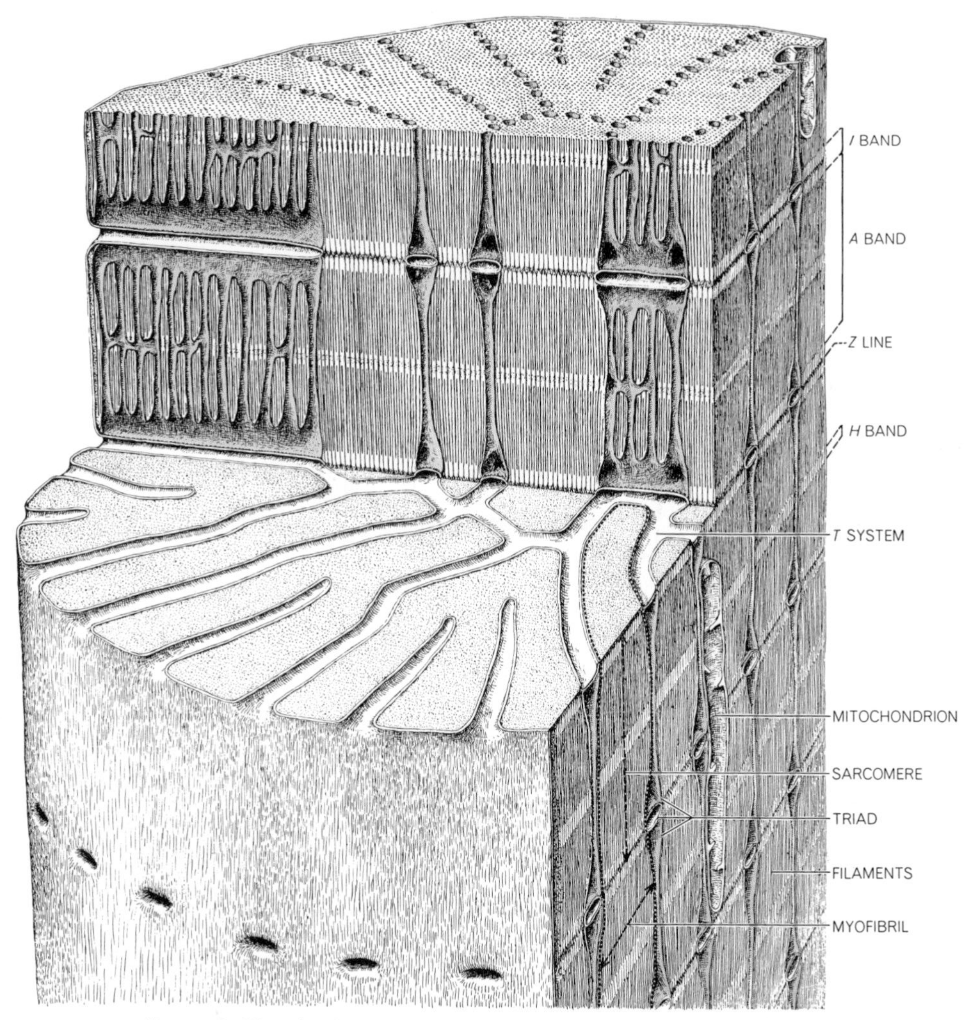

Figure 2–11 The fine structure of striated muscle. (From Florey: *An Introduction to General and Comparative Animal Physiology*. Philadelphia, W. B. Saunders Company, 1966. Reprinted from Porter and K. R. C. Franzini-Armstrong: The sarcoplasmic reticulum. Scient. Amer., *212*:73, 1965.)

orderly arrangement of filaments that gives the striated appearance to the myofibrils. Each unit of the pattern is called a sarcomere and terminates at each end at a Z disc. The thinnest of the filaments are attached to the Z discs, whereas the thicker filaments are suspended, apparently unattached, between them.

Chemical extraction has shown that the thick filaments are composed of a protein called myosin and that the thin filaments contain a different protein called actin (Sect. 16.3). Chemical reactions between these two proteins during contraction cause the two kinds of filaments to interdigitate further with each other thus inducing a shortening of the sarcomere (Fig. 2-10). Since all of the sarcomeres of the cell shorten simultaneously, the entire cell shortens.

The energy-storing molecule adenosine triphosphate (Sect. 10.3), present in large amounts in skeletal muscle, is associated with the actin filaments, and there are numerous mitochondria within the cell.

The endoplasmic reticulum of skeletal muscle cells is highly specialized and has been given the name sarcoplasmic reticulum. It consists of longitudinally oriented microtubules that lie between and parallel to the myofibrils of the cell. In the region of the interdigitation of filaments these tubules are associated with enlarged vesicles called triads (Fig. 2-11).

As we shall see in Chapter 11, nervous impulses are electrical in nature and are transferred to the membrane of skeletal muscles at their points of innervation. It is believed that the sarcoplasmic reticulum carries these electrical disturbances throughout the muscle mass and that this causes a release of calcium ions at the triads. The calcium ions, in turn, trigger the internal chemical events associated with actual contraction of the cells. Cardiac muscle cells (Sect. 16.6) and smooth muscle cells (Sect. 16.7) are somewhat different in general appearance, but these cells contain a similar contractile apparatus and, apparently, function in much the same way as do skeletal muscles.

2.4 NERVE CELLS

Neurons, or nerve cells, differ widely in appearance in different parts of the vertebrate nervous system. Most are designed to transmit nervous impulses from place to place, but others apparently serve as structural elements to hold portions of the system together, or they may function in still different roles. Indeed, we are at present a long way from understanding the functions of a number of cell types within the central nervous system. There is really no "typical" neuron, but we may base our discussion on a generalized conducting neuron of the type located in the longer central nervous tracts or outside the central nervous system. Such a neuron is composed of a cell body, a few short dendrites, and a single, long axon (Fig. 2-12). Ordinarily a dendrite picks up an impulse from a sensory receptor or from another neuron; the impulse is then transmitted through the cell body and along the axon to an effector organ or to another neuron. The point of contact

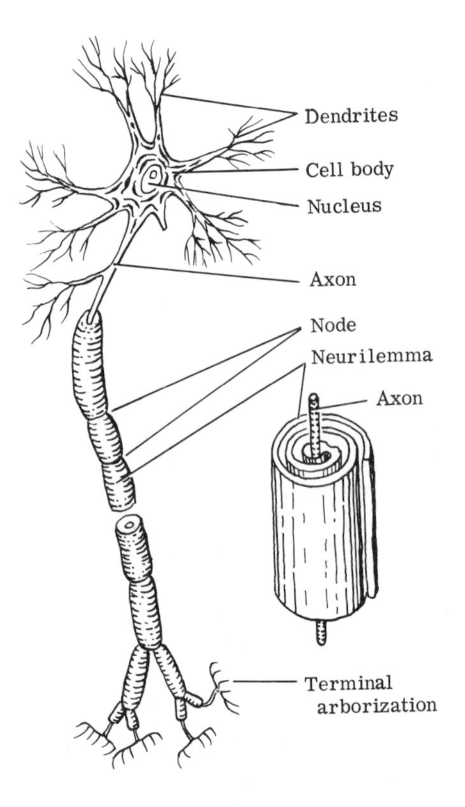

Figure 2–12 A generalized neuron. The inset shows the way in which the neurilemma is wrapped about the axon. (From Cockrum and McCauley: *Zoology*. Philadelphia, W. B. Saunders Company, 1965.)

between one neuron and another is called a synapse, and much of the specificity and versatility of the nervous system is due to the properties of its many synapses (Sect. 11.4).

Compared with other cell types the length of a single neuron is sometimes very great. For example, a motor neuron passing from the spinal cord to muscles in the foot of a man may be more than three feet in length. If the cell body were enlarged to the size of a tennis ball, the axon would then be a little less than a half-inch in diameter and about a mile in length!

The cell body contains a large nucleus and variable amounts of chromatin material scattered in the form of granules (Fig. 2-13). Other typical organelles, such as the Golgi complex, mitochondria, and centrioles, are also present. The cell body and the dendrites and axon also contain a network of fine neurofibrils (Fig. 2-14).

Neuron processes lying outside the central nervous system are often covered with complex layers of a fatty substance called myelin and a membrane called the neurilemma or sheath of Schwann. These are wrapped about the axon in a spiral fashion (Fig. 2-12) because of the way in which the Schwann cells wrap extensions of their plasma membranes about the neuron process. At intervals the neurilemma is interrupted and a short segment of bare axon exposed. These points are known as nodes of Ranvier, and they play an important role in the rapid transmission of nerve impulses (Sect. 11.3).

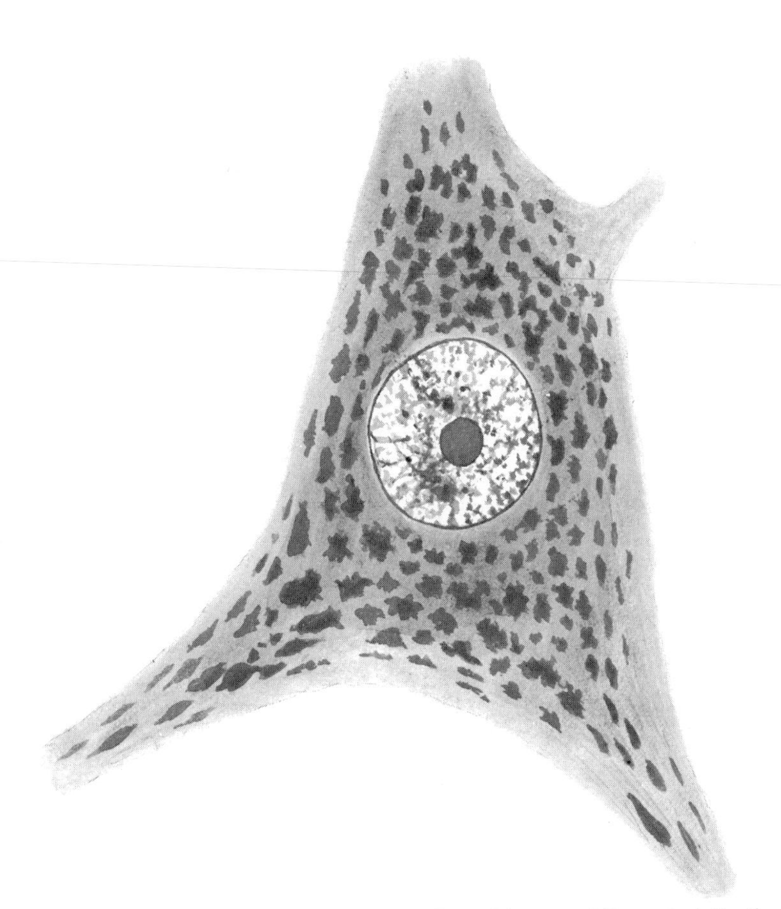

Figure 2–13 The cell body of a neuron. (From Bloom and Fawcett: *A Textbook of Histology*, 9th Edition. Philadelphia, W. B. Saunders Company, 1968.)

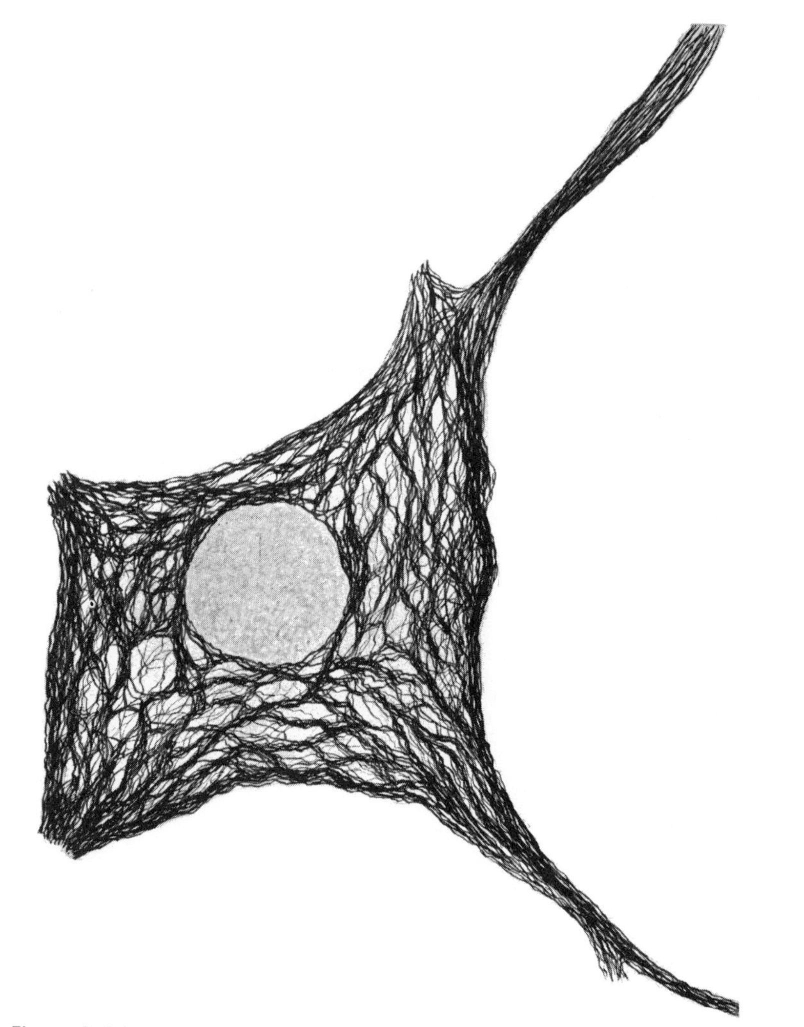

Figure 2–14 Neurofibrils within the cell body of a neuron. (From Bloom and Fawcett: *A Textbook of Histology*, 9th Edition. Philadelphia, W. B. Saunders Company, 1968.)

2.5 EPITHELIAL CELLS

Epithelium is the name applied to tissues that cover the body (skin) or line the cavities of hollow organs. These tissues contain several different kinds of cells that function in protection, absorption, and secretion. A few of them also serve as special sensory receptors. The cells of epithelial tissues are arranged in single layers (simple epithelium) or in multiple layers (stratified epithelium). Consequently, they occur in various shapes ranging from flattened plates (squamous cells) to cuboid and columnar forms (Figs. 2–15 and 2–16).

In simple epithelium the cells tend to be polarized; that is, the superficial end of the cell is different in appearance from the deep end. The deep end rests on a basement membrane or lamina (Fig. 2-15) which is composed of extracellular materials and separates the epithelium from deeper tissues. The superficial, exposed end of epithelial cells may bear cilia (Fig. 2-15) or they may have a so-called brush border. The striated appearance of a brush border results from the presence of numerous cell processes known as microvilli (Fig. 2-17). These differ from cilia in that they do not contain the characteristic fibrillar structures (Sect. 2.2). In still other cases the superficial ends of epithelial cells are covered with a tough extracellular material known as keratin.

Some epithelial cells are glandular in nature; that is, they produce secreted materials that are discharged onto the surface of the tissue. Such secreted materials include protective and lubricating mucus as well as digestive enzymes. Glandular cells of the lining of the digestive tract produce both kinds of secretions. Sometimes glandular cells occur singly and sometimes in groups, forming organized glands. An example of the former is the goblet cell (Fig. 2-15), which is a mucus-producing cell.

Typically, glandular epithelial cells contain many mitochondria and a well-developed Golgi complex; this would be expected from

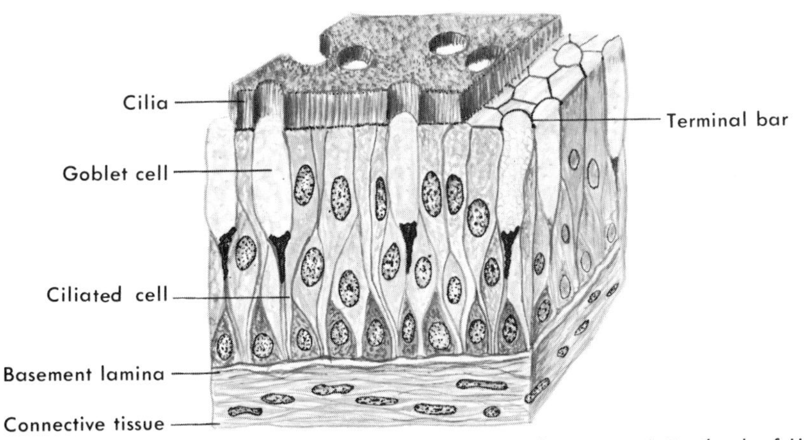

Figure 2–15 Columnar epithelium. (From Bloom and Fawcett: *A Textbook of Histology*, 9th Edition. Philadelphia, W. B. Saunders Company, 1968.)

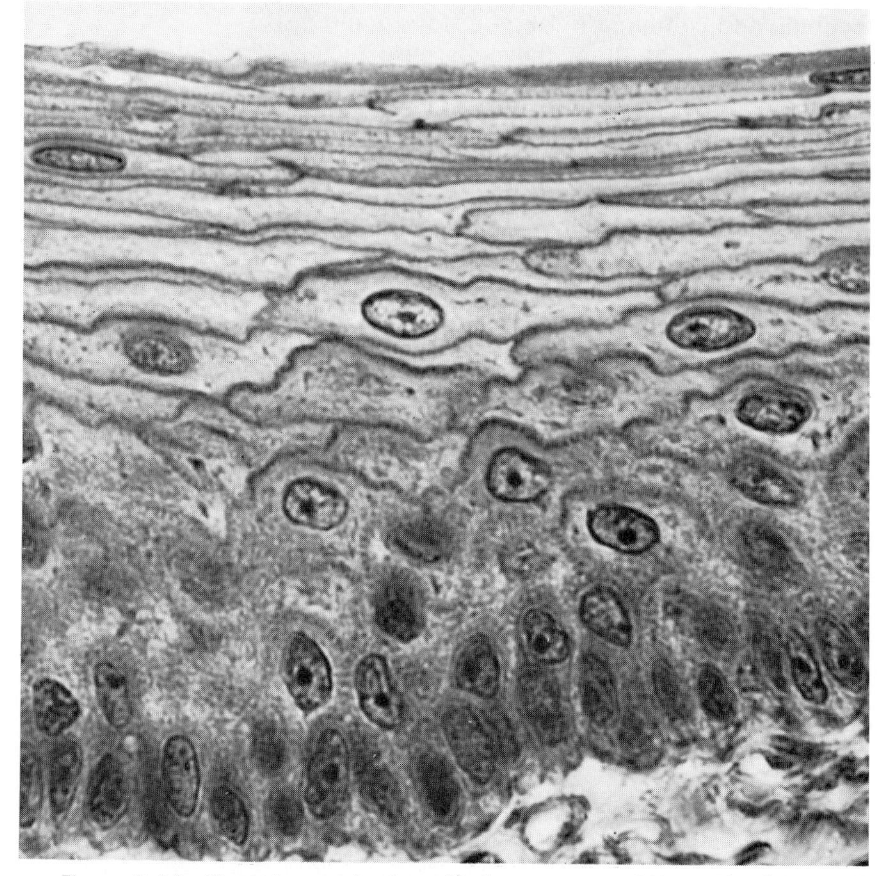

Figure 2–16 Photomicrograph of stratified squamous epithelium. The deeper cells are characteristically cuboid in shape, and there is progressive flattening of cells in the more superficial layers. (From Bloom and Fawcett: *A Textbook of Histology*, 9th Edition. Philadelphia, W. B. Saunders Company, 1968.)

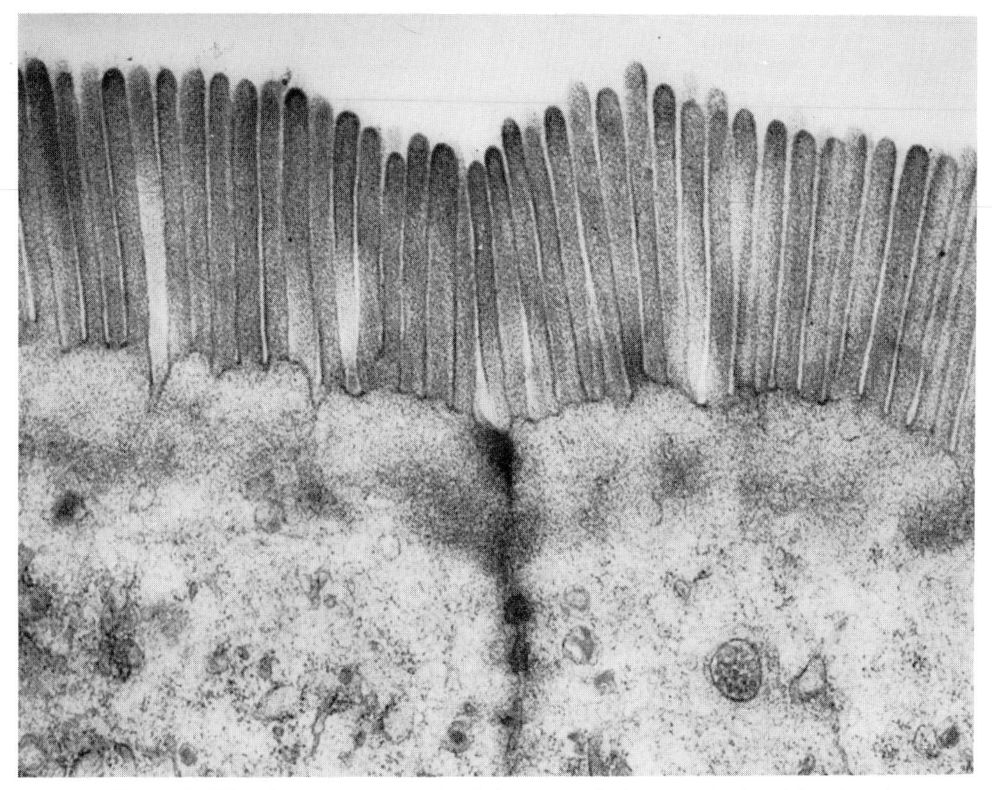

Figure 2–17 Electron micrograph of the microvilli forming the brush border of the intestinal epithelium of a hamster. (From Bloom and Fawcett: *A Textbook of Histology,* 9th Edition. Philadelphia, W. B. Saunders Company, 1968.)

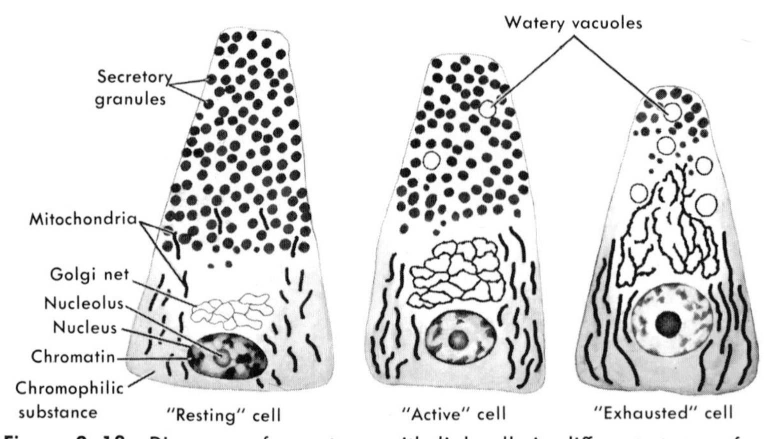

Figure 2–18 Diagrams of secretory epithelial cells in different stages of activity (see text). (From Bloom and Fawcett: *A Textbook of Histology*, 9th Edition. Philadelphia, W. B. Saunders Company, 1968.)

their high level of activity (Sect. 2.2). Secretory droplets or granules derived from the Golgi complex are commonly present near the superficial end of the cell. The appearance of such cells varies according to whether they are relatively inactive, active, or "exhausted" (Fig. 2-18).

2.6 CONNECTIVE TISSUE CELLS

Several types of tissue fall into the general category of connective tissues. These include bone, cartilage, tendons, ligaments, and the finer tissues (fascia) that bind together other kinds of tissues. Blood is commonly classed as a connective tissue, even though it does not play the same kinds of roles, because of similarities between the embryonic origins of blood cells and other types of connective tissue. Blood cells are discussed in more detail elsewhere in this book (Chapter 7).

More typical connective tissues consist of cells plus greater or lesser amounts of extracellular materials secreted by those cells. Extracellular materials consist of fibrous or elastic strands, deposits of a homogenous matrix, or both. The matrix may be soft or semifluid as in hyaline cartilage (Fig. 2-19), or it may be hard and mineralized as in bone. The cells in cartilage and bone lie comparatively far apart in spaces or lacunae with a matrix filling between the cells. In other kinds of connective tissue such as tendons, ligaments, and fascia (Fig. 2-20) there are more fibrous or elastic strands, less homogenous matrix, and the cells lie scattered throughout the tissue.

Bone cells are of three general types: (1) osteoblasts, which lay down the mineralized matrix of hydroxyapatite ($Ca_{10}(PO_4)_6(OH)_2$); (2) osteocytes, which lie "trapped" in the lacunae of the completed bone; and (3) osteoclasts, which lie on the surfaces of bone and cause resorption of the mineralized matrix. Osteoblasts and osteoclasts working together make possible the continual remodeling of bone structure that

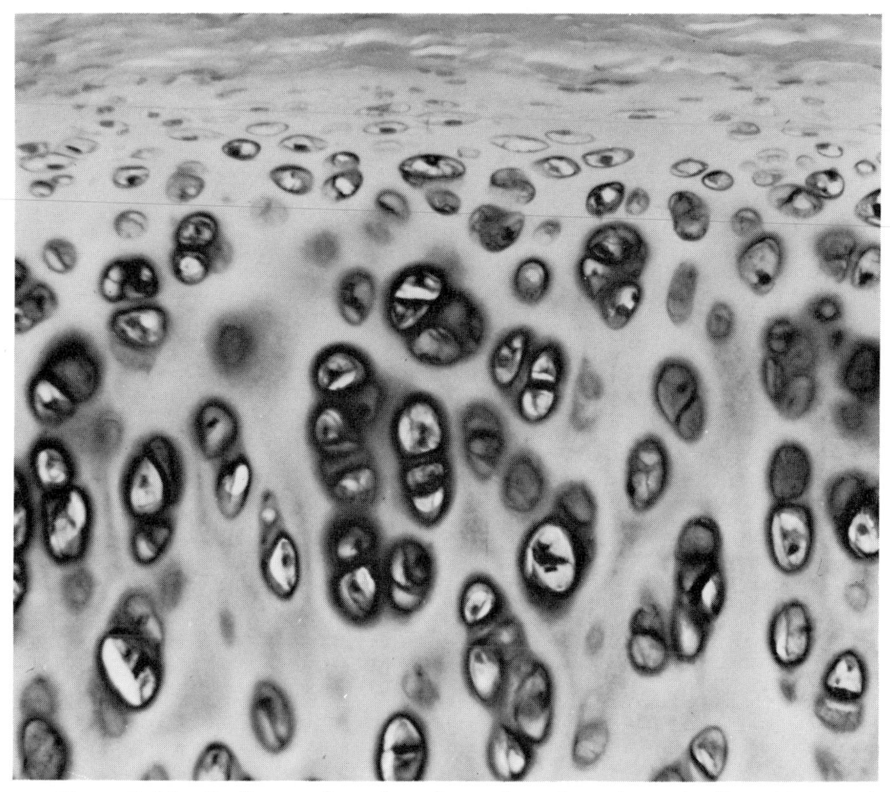

Figure 2–19 Hyaline cartilage from the trachea of a guinea pig. (From Bloom and Fawcett: *A Textbook of Histology,* 9th Edition. Philadelphia, W. B. Saunders Company, 1968.)

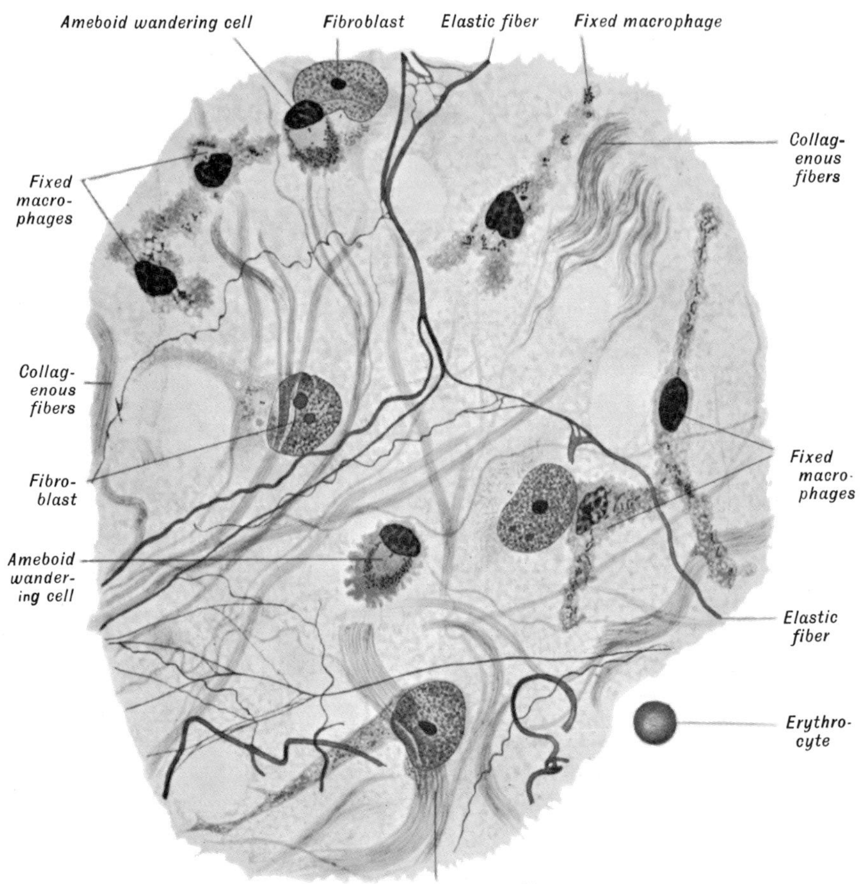

Figure 2–20 Section through loose connective tissue from the human thigh. The red blood corpuscle (erythrocyte) in the lower right is for size comparison. It is 7 microns in diameter. (From Bloom and Fawcett: *A Textbook of Histology*, 9th Edition. Philadelphia, W. B. Saunders Company, 1968.)

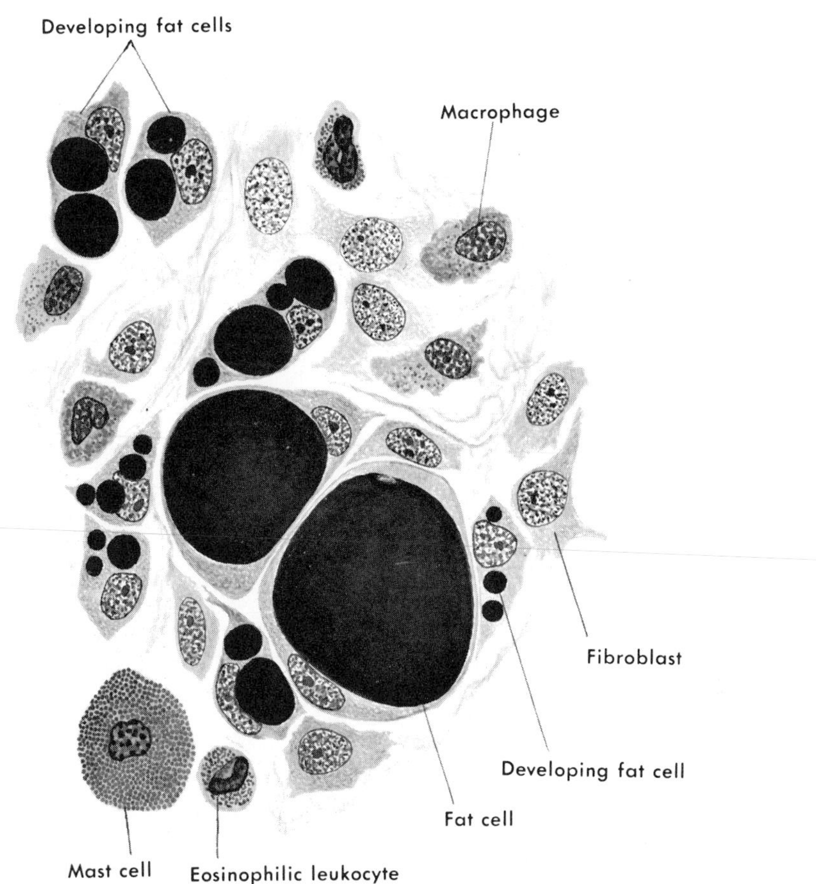

Figure 2–21 Fat cells in the loose connective tissue of a rat. The fat has been stained black by osmic acid. (From Bloom and Fawcett: *A Textbook of Histology*, 9th Edition. Philadelphia, W. B. Saunders Company, 1968.)

accompanies growth and development and that adapts completed bones to new stresses.

Tendons, ligaments, and fascia contain cells known as fibroblasts (Fig. 2-20). These are spindle-shaped (fusiform), flattened, or stellate cells which form the fibrous or elastic strands of such tissues. Also present are macrophages, which are phagocytic cells; that is, they are capable of engulfing in ameboid fashion various kinds of foreign materials or cell detritus. Some macrophages are fixed in position and some are wandering cells capable of moving about through the tissues like amebae.

Adipose cells, or fat cells (Fig. 2–21), are also classed as connective tissues. These function in the storage of fats and consist of little more than large vacuoles enclosed by a plasma membrane. A flattened nucleus lies at the edge of the cell. This kind of tissue occurs mostly in the subcutaneous portions of the body or in association with the supporting mesenteries of abdominal organs.

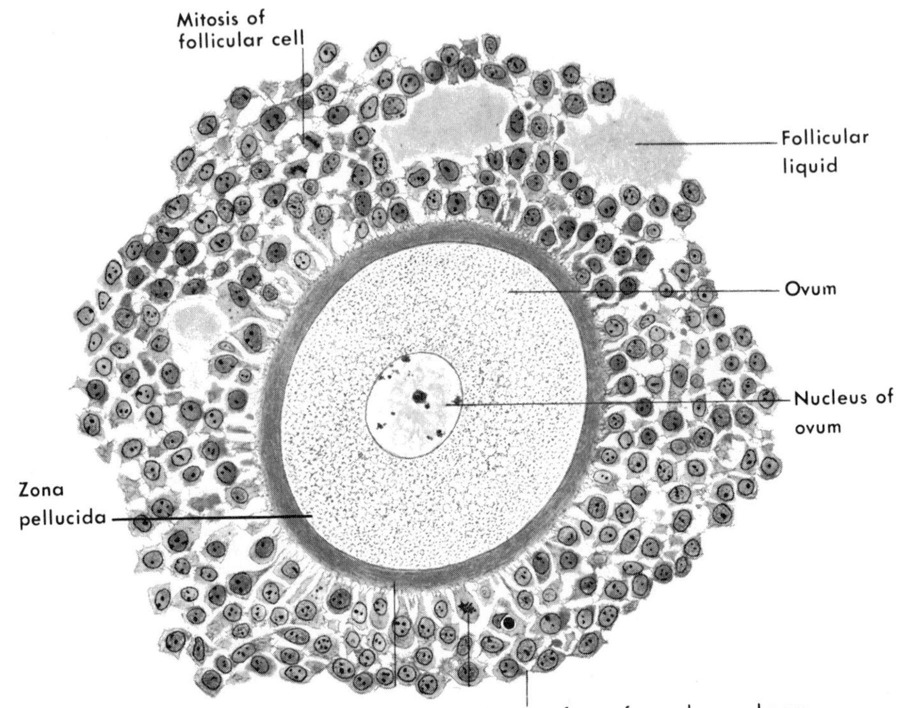

Figure 2–22 A mature mammalian ovum surrounded by follicular cells. (From Bloom and Fawcett: *A Textbook of Histology*, 9th Edition. Philadelphia, W. B. Saunders Company, 1968.)

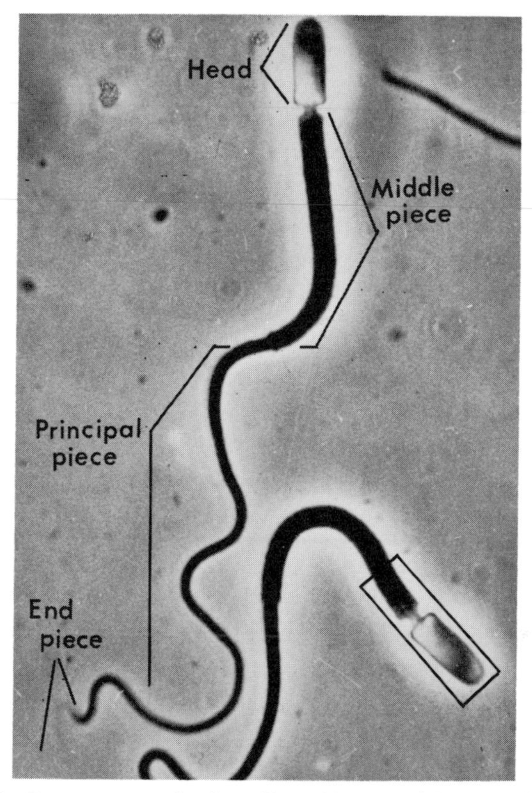

Figure 2–23 Spermatozoa of a bat. (From Bloom and Fawcett: *A Textbook of Histology*, 9th Edition. Philadelphia, W. B. Saunders Company, 1968.)

2.7 MISCELLANEOUS SPECIAL CELLS

The gametes, or sex cells, are highly specialized cells containing only half as much nuclear genetic material as other cells. The female sex cells, or ova, are relatively large, spherical cells containing variable amounts of yolk material and often surrounded by numbers of protective follicular cells (Fig. 2–22). The male sex cells, or spermatozoa, are more complex (Fig. 2–23). In these, the genetic materials (nucleus) are encased in a head, which often has a protective cap lying over it. Attached to the head is a tail consisting of a neck, middle piece, principal piece, and end piece. The middle piece has a sheath wrapped about it that also encloses numerous mitochondria. The tail contains the typical set of fibrils found in cilia or flagella (Sect. 2.2), and indeed this is a highly specialized flagellum. The shape of the head varies greatly in different vertebrate species, but the tail is quite similar in all.

Other highly specialized cells of the vertebrate body include the receptor cells of the retina of the eye (Sect. 15.9), the taste buds (Sect. 13.7), and the receptor cells of the inner ear (Chapter 14). Many other specialized receptor cells are discussed in various places throughout this book.

CHAPTER 3

RESPIRATION

3.1 INTRODUCTION

All animals must obtain oxygen from the environment and return carbon dioxide to it. The standard atmospheric concentration of oxygen is 21 per cent, and a very low animal consumption rate of oxygen would be about 0.1 milliliter per gram of tissue per hour. With this availability and at this consumption rate, oxygen would diffuse into tissues to a depth of about 0.5 millimeters before it was all used up. Remember that diffusion over such a distance is quite slow (Sect. 1.2). Consequently, an animal with even a very low oxygen consumption rate, depending entirely upon surface diffusion to meet its requirements could be only about one millimeter in thickness. If it were any thicker or if its oxygen demands were any greater, then some additional transporting mechanism would have to be provided to carry oxygen to the deeper cells. Diffusion alone would be inadequate.

The oxygen transporting mechanism common to all larger animals is, of course, the circulatory system. The circulating fluid passes close to the animal surface where it exchanges, by diffusion, carbon dioxide for oxygen. The fluid then flows to deeper regions where it gives up its oxygen to the metabolizing tissue cells and receives from them carbon dioxide in exchange. Again, the exchange is carried out by diffusion. This means that the diffusion distances both at the animal surface and deep within the animal must be extremely small.

In vertebrates special "surface" areas are provided in the gills or lungs where the circulating blood can pass within three microns of the environmental medium (water or air); comparable distances exist between the blood and each internal tissue cell. Ventilation of the gills with water or of the lungs with air provides a constantly changing environmental medium adjacent to these surfaces. Circulation of the blood continually brings blood that is poor in oxygen and rich in carbon dioxide adjacent to the same respiratory surfaces. Thus, in the gills or lungs, steep diffusion gradients are constantly maintained. This favors the rapid diffusion of carbon dioxide outward and oxygen inward so that the blood leaves the respiratory surface rich in oxygen and poor in carbon dioxide. Similar physical relationships favor efficient gas exchanges between the blood and the metabolizing tissue cells deeper within the animal body.

The process of ventilating the gills with water or the lungs with air and the diffusion exchanges of gases between these surfaces and the blood constitute what is known as external respiration. The gas exchanges between the blood and the metabolizing tissue cells and the utilization of oxygen and production of carbon dioxide by those cells constitute the process of internal respiration.

All other factors being equal, the volume of metabolizing tissue determines the oxygen demand of the animal. This entire demand must be met by gas exchange through the available surface area of the gills or lungs. Thus, where the demand is greater, the gill or lung area must also be greater. As an organism grows larger, its volume increases approximately as a cube, whereas its total surface area increases only as a square (Fig. 3-1). Even if the entire surface area of the animal could function in respiratory exchanges with the environment, this would be insufficient to meet the oxygen demands of larger animals. Moreover, the general surface (skin) of a vertebrate animal is not satisfactory for respiratory exchanges. It cannot be thin enough to permit efficient gas diffusion through it and still be tough enough to provide a protective covering for the animal. This is why special respiratory surfaces with

$A = 4\pi r^2$ $V = \frac{4}{3}\pi r^3$

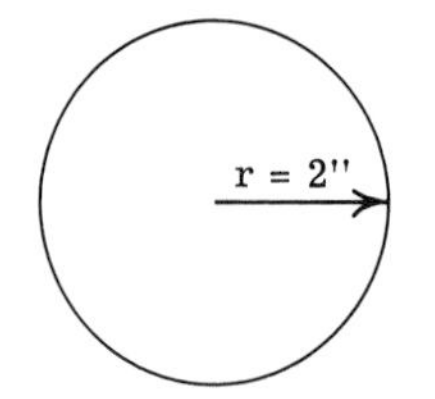

A = 4 x 3.14 x 4 = 50.24 square inches.

V = 1.33 x 3.14 x 8 = 33.44 cubic inches.

r = 4"

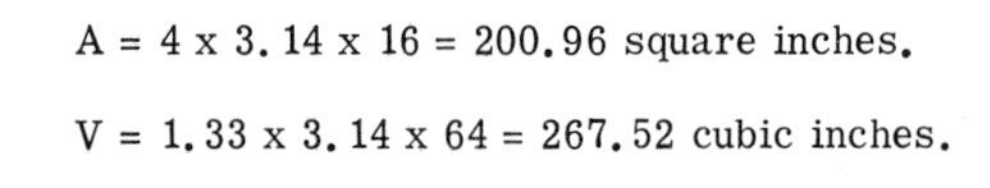

A = 4 x 3.14 x 16 = 200.96 square inches.
V = 1.33 x 3.14 x 64 = 267.52 cubic inches.

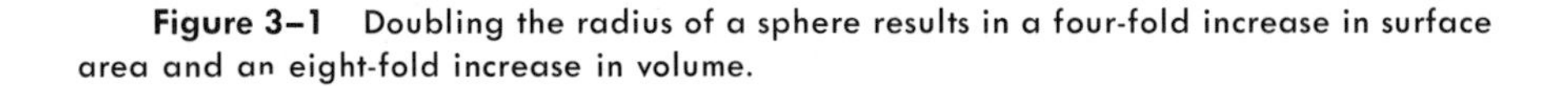

Figure 3–1 Doubling the radius of a sphere results in a four-fold increase in surface area and an eight-fold increase in volume.

greatly increased areas for exchange must be provided in the form of gills or lungs. The total respiratory area of human lungs, for example, is about 40 times as great as the total skin area!

If the adjacent environmental medium (water or air) remained stationary, it would soon be depleted of its oxygen and saturated with carbon dioxide. Consequently, it must be constantly changed. Vertebrates either pump water past the gills or they repeatedly fill and empty the lungs with new air. Both processes are known as ventilation. Similarly, if the circulating blood remained stationary, that portion adjacent to the respiratory surfaces would soon become saturated with oxygen and unable to accept more. It would also become thoroughly depleted of carbon dioxide so that no further diffusion exchange of gases could occur. Consequently, the blood must continuously circulate past the respiratory surfaces in order to maintain continuous gas exchanges by diffusion. Within limits, increases in either the ventilation rate or the circulation rate will result in increased rates of gas exchange with the environment. Both these mechanisms are used by vertebrates to increase the availability of oxygen to their tissues during exercise or other periods of heightened oxygen demand. It is a matter of common experience that when exercising we breathe more deeply and more rapidly and our heart rates increase.

3.2 THE ENVIRONMENT

Oxygen diffuses through air more than 3 million times faster than it diffuses through water. Moreover, the solubility of oxygen in water is so low that water contains only one twentieth as much oxygen as does air. This means that oxygen is far less available to aquatic organisms than it is to air-breathers. This is a limiting factor even to air-breathers, however, since oxygen must be dissolved in the fluids of the respiratory membrane and in the blood itself during the process of diffusion.

Given the facts mentioned in the preceding paragraph, we can calculate that an aquatic animal must ventilate the respiratory surface with 20 liters of water in order to obtain the same amount of oxygen available to an air-breathing animal in a single liter of air. Since the density and viscosity of water are much higher than the density and viscosity of air, ventilation with water requires that the animal do more work. In a man at rest, about 1 or 2 per cent of the oxygen uptake is required to produce the energy used in ventilating the lungs. For a fish the figure is probably about 20 per cent under comparable conditions.

Increased salinity and temperature each reduces the oxygen concentration of water. Thus, marine fishes and fishes living in warm tropical waters have less oxygen available to them than do freshwater fishes. Moreover, increased temperature results in an increase in the metabolic rate of cold-blooded vertebrates. In warmer waters, then, the oxygen demand of a fish is increased at the same time and by the same factors that decrease the availability of oxygen.

Some bodies of water are extremely poor in oxygen because of poor circulation of the water (stagnation) or because oxygen-consuming processes, like the rotting of submerged vegetable matter, are present. The decay of plant materials also produces carbon dioxide, and waters in which such decay occurs are generally rich in that gas. In such bodies of water only the immediate surface may contain sufficient oxygen to meet the needs of aquatic vertebrates.

All of the factors mentioned thus far in this section make it clear that there are many advantages to breathing air instead of water. Thus, the development of air-breathing apparatus was favored long before terrestrial vertebrates evolved. Although air is a better source of oxygen than water, the terrestrial environment is more demanding in other ways. Air is desiccating and far more susceptible to variations in temperature and other physical conditions. The transition from aquatic to terrestrial life would seem, therefore, to have been an attractive move but a difficult one to achieve.

The transition from aquatic to terrestrial life may well have been preceded by the adaptation of certain fishes to oxygen-poor waters. In such situations many modern fishes rise regularly to the surface to gulp in air. This behavior is frequently seen, for example, in ordinary goldfish kept in an unaerated aquarium. Vascular membranes of the pharynx, esophagus, stomach, and even of the intestine serve as accessory respiratory surfaces for such air breathing. Certainly any increase in the area of respiratory surface would greatly benefit a fish, and thus the development of an esophageal diverticulum, such as a swim bladder or a lung, would be favored. This is apparently the adaptation that actually appeared in more than a single line of descent, including the ancestors of the modern lungfishes (Dipnoi).

The Dipnoi arose in the tropical swamps characteristic of the Devonian period, a little more than 400 million years ago. The warm climate and abundant vegetation in these swamps must have greatly reduced the oxygen content of the water. Moreover, periodic droughts occurred, and portions of the swamps must have dried up, leaving only scattered pools. The animal that could breathe air could survive either by migrating overland to remaining ponds or by estivating until the next rainy season. Because the Dipnoi possessed lungs in addition to gills and because they were able to reduce their rate of metabolism in order to survive long periods of estivation, the lungfishes were able to employ successfully both of these survival measures.

Of the lungfishes living today, *Neoceratodus* (Australia) has well-developed gills and only a single, poorly developed lung; it cannot remain long out of water. *Protopterus* (Africa) has paired lungs and less well-developed gills; it can get along about equally well in either water or air. *Lepidosiren* (South America) has highly developed, paired lungs, but the gills are so poorly developed that the animal will drown if held under water! The three living groups, then, illustrate three stages in the probable evolutionary conversion from aquatic to aerial respiration.

As we shall see (Chapter 6), it is not sufficient for an animal merely to have lungs in order to breathe air. Appropriate changes in the circulatory system must accompany the development of the lungs, and these changes also appeared in the ancestral lungfishes. To be able to migrate overland from pool to pool also presupposed the development of some method of locomotion. The fins of the lungfishes could be used as levers (primitive legs) for this purpose (Sect. 16.8). Likewise, the ability to estivate required many changes in physiological control (Chapter 18).

The elimination of carbon dioxide by animals is not ordinarily a limiting factor even in water. Carbon dioxide is 28 times more soluble in water than is oxygen. Moreover, when dissolved in water carbon dioxide forms a hydrate, carbonic acid (H_2CO_3). This dissociates to form hydrogen and bicarbonate ions, which often react with other dissolved substances to form various carbonates and bicarbonates. Such reactions reduce the concentration of carbon dioxide, as such, and this helps to maintain steep diffusion gradients between the animal surface and the immediately surrounding environment.

3.3 AQUATIC VERTEBRATES

Quite likely the earliest aquatic vertebrates had external gills much like those of modern tadpoles. These are filamentous structures that hang outside the pharyngeal region and are ventilated by the fish's movements through the water. Such external gills, however, are subject to predation and other injuries. Their gas exchange efficiency is much reduced when the animal is not swimming, and they interfere with the fish's streamlining when it is swimming. The development of internal gills, in gill pouches or in opercular cavities, removed many of these disadvantages. The overlying operculum (gill cover) of typical fishes provides protection for the gills and communication between the gills and the oral-pharyngeal cavity provides a means of pumping water over them even when the animal is at rest.

The generalized structure of fish gills is shown in Figure 3-2. Branchial arches bridge across the opercular cavity, which opens between the oral-pharyngeal cavity and the outside gill opening. These branchial arches bear successive gill filaments, or primary lamellae that appear as flat plates lying parallel with each other. Each primary lamella, in turn, bears smaller secondary lamellae at right angles to itself. Together these lamellae form a sievelike structure through which water passes. Blood capillaries in the lamellae pass within 0.5 to 0.3 micron of the water-bathed surface, so diffusion distances between water and blood are very short.

Blood flow through the secondary lamellae is in a direction opposite to that of the passing water, and this greatly increases the efficiency of gas exchange. If water and blood passed adjacent to each other in the same direction, a given "drop" of water and "drop" of blood would be contiguous longer than necessary to complete all of the possible diffusion of gases between them. During this countercurrent exchange, however, each "drop" of water is constantly brought adjacent to differ-

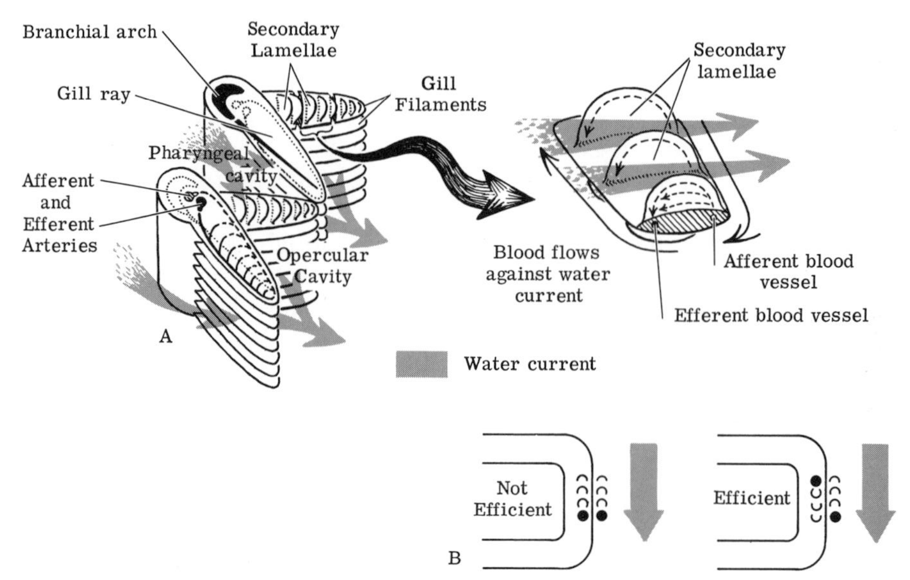

Figure 3–2 The structure of typical fish gills and the principle of counter-current exchange. *A*, the structure of fish gills showing the direction of water and blood flow; *B*, the principle of counter-current exchange. (After Hoar: *General and Comparative Physiology*. Englewood Cliffs, N.J., Prentice-Hall, 1966 and Hughes: *Comparative Physiology of Vertebrate Respiration*. Cambridge, Harvard University Press, 1963.)

ent "drops" of blood. Experimentally it has been found that reversing the direction of water flow through the gills of a fish reduces the oxygen uptake to about one eighth of its normal value simply because, under these conditions, blood and water are traveling in the same direction.

The countercurrent exchange principle applies to a number of physiological mechanisms other than gill respiration. It also occurs in the secretion of gases into swim bladders (Sect. 3.4), in the control of renal water loss (Sect. 8.6), and in the extremities of certain vertebrates that require special thermoregulatory controls (Sect. 18.4).

The primary and secondary lamellae of typical fish gills form a meshwork or sieve having up to a quarter of a million pores. This provides a total surface area for gas exchange that is about ten times the fish's skin area. Because the pores are very small, all of the water passing through them passes very close to the respiratory surfaces of the gill lamellae. As a result, about 80 per cent of the oxygen available in the water is taken up by the blood passing through the gills.

An interesting adaptation of the gills is present in the South American lungfish *Lepidosiren*. This animal, as we have noted (Sect. 3.2), respires mostly by means of its lungs. It lives in waters that are very poor in oxygen and lays its eggs on the bottom of the pool. Thus, sufficient oxygen is not available for the developing young. During the breeding season, the adult male develops special enlarged external gills on its underside. The male broods over the nest in which the eggs are developing and thus protects them from predators and provides them with extra oxygen. Periodically the male rises to the surface of the water and gulps air into the lungs and then returns to the nest. The blood, having taken oxygen from the lungs, contains more of this gas than does the surrounding water. Consequently, oxygen diffuses out through the special external gills into the water surrounding the eggs.

The ventilation of the gills of a teleost fish (bony fish) is accomplished by means of a double pump (Fig. 3-3). In the first phase of pumping, the mouth opens and water is drawn into the oral-pharyngeal cavity when the floor of this cavity is depressed. The operculum remains closed during this phase, but the opercular (gill) cavity beneath it is actively expanded. This draws water from the oral-pharyngeal cavity through the gills and into the opercular cavity. In the second phase of pumping, the mouth closes and the opercular cover opens. The oral-pharyngeal floor is raised again, driving more water through the gills and back to the outside through the opercular opening. This sequence of events, involving as it does both an oral-pharyngeal pump and an opercular cavity pump, results in an almost continuous flow of water through the gills. Intermittent flow, such as would result from the action of a single pump, of course would mean reduced efficiency of gas exchange.

In the elasmobranch fishes (sharks, rays, and their relatives) the process of gill ventilation is similar except that water enters the oral-pharyngeal cavity through a special spiracle, located on each side of the head, rather than through the mouth. Also, instead of being located in a

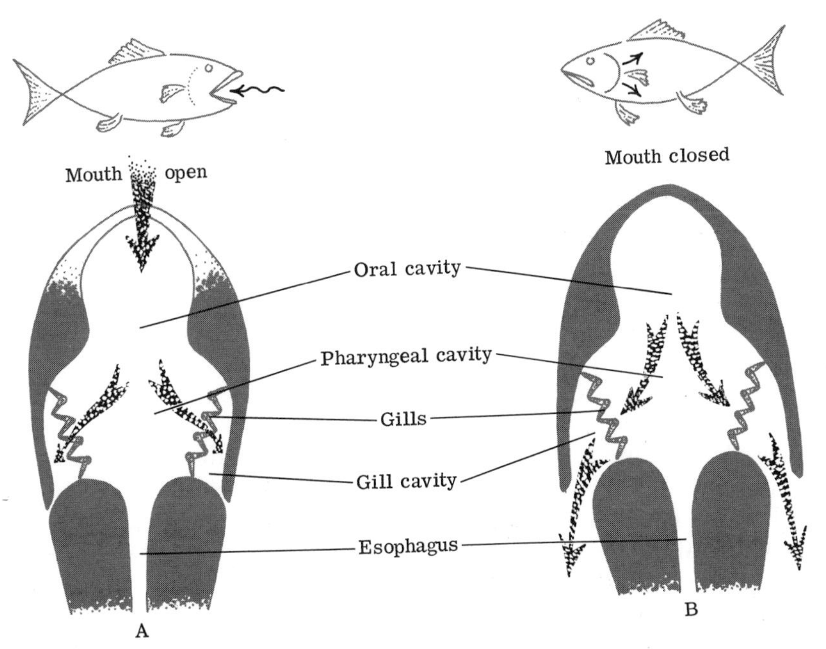

Figure 3–3 Gill ventilation in a teleost fish. *A*, with the mouth open water is drawn into the oral and pharyngeal cavities. At the same time expansion of the opercular cavity draws water through the gills. *B*, with the mouth closed water is forced through the gills and out through the opened operculum.

single opercular cavity, the gills are located in each of several gill pouches that open independently onto the surface. The spiracle is actually a modified first gill pouch which contains no gills. The others each contain a single branchial arch and attached lamellae. The substitution of a spiracle for the mouth in the process of ventilation is probably related to the fact that many sharks and rays are bottom feeders—indeed, all the ancestral elasmobranchs were—and water entering the mouth would be contaminated with bottom detritus.

Both teleosts and elasmobranchs sometimes suspend active water pumping when they are swimming, since the forward motion forces a sufficient amount of water over the gills through the opened mouth. It has been reported that some fishes can never stop swimming forward because they cannot achieve sufficient ventilation by oral-pharyngeal and opercular pumping alone.

During quiet swimming the tips of the primary lamellae of adjacent gills meet so that all of the water must pass through the gill pores. When the fish swims more actively, however, adductor muscles associated with the primary lamellae separate the tips slightly to allow some of the increased water flow to bypass the gill pores. Swift forward swimming might otherwise result in too strong a current through the gills, which would damage them.

3.4 THE SWIM BLADDER

Many, but not all, teleost fishes possess a gas-filled bladder that arises during embryonic development as a diverticulum of the digestive tract. These bladders bear a superficial resemblance to a simple lung. In some species the bladder retains a tubular connection with the gut and it can be used as an accessory organ of respiration, as we have already mentioned (Sect. 3.2). In other species the bladder loses its connection with the digestive tract but remains filled with gas. In both cases it is possible for the fish to inflate the bladder by secreting into it oxygen from the blood.

The structure of the swim bladder is fairly uniform in all species (Fig. 3-4). It is an oblong structure that has a smaller side chamber, known as the oval, communicating with it. Often muscles in the wall are present that can close off the opening between the swim bladder proper and the chamber of the oval. In the walls are two vascular areas; the gas gland, which actively inflates the swim bladder, and an area in the wall of the oval through which oxygen can be reabsorbed to deflate the bladder. The relative amounts of blood passing through these two areas determine the degree of inflation at any given time. Moreover, closing off the opening into the oval completely prevents gas reabsorption and, thus, favors inflation of the bladder.

To transport oxygen from the blood into the swim bladder means moving it against a steep concentration gradient, since the content of the swim bladder is usually pure oxygen. The mechanisms involved in this process are not yet completely understood but enough is known to facilitate some conjecture.

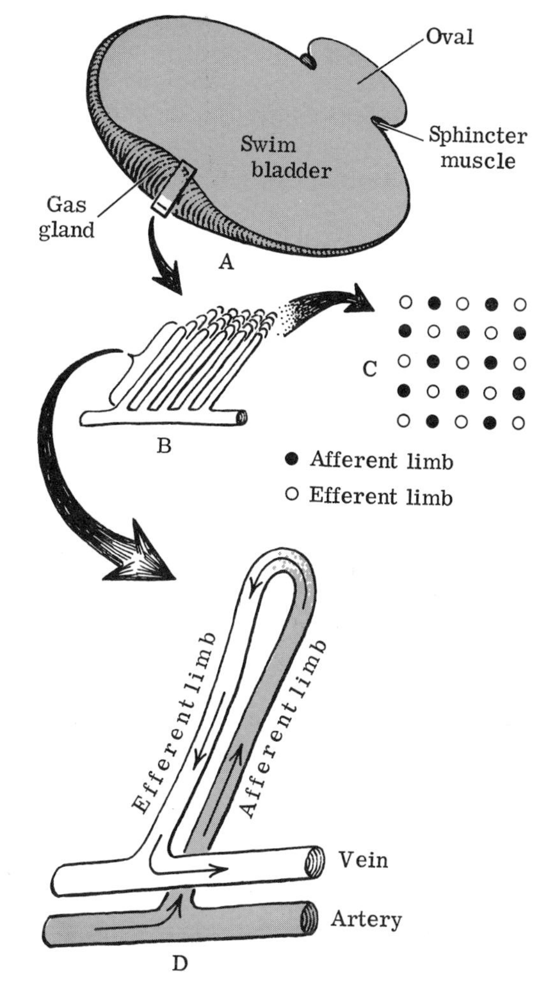

Figure 3–4 The swim bladder. *A*, structure of a typical swim bladder; *B*, the vascular hairpin loops of the gas gland; *C*, the relationship of afferent and efferent limbs of the vascular loops; *D*, the structure of a single vascular hairpin loop from the gas gland. (After Hughes: *Comparative Physiology of Vertebrate Respiration*. Cambridge, Harvard University Press, 1963.)

The gas gland consists of a network of blood vessels arranged so that each makes a hairpin turn near the lining of the swim bladder. Blood flows into each hairpin loop and gives up some of its oxygen before flowing back out through the efferent limb of the loop. This oxygen passes into the swim bladder. During active inflation of the swim bladder the metabolism of glucose increases markedly in the region of the hairpin turns, and this results in an accumulation there of carbon dioxide and lactic acid (Chapter 10). These substances both lower the ability of the blood to retain oxygen (Sect. 7.4), and consequently it is released from the blood in that region.

The swim bladder primarily serves to adjust the buoyancy of the fish so that it may remain at a particular depth without constantly needing to overcome any natural tendency to either rise or sink. The density of a fish averages about 1.076, so without such a gas-filled bladder it would tend to sink slowly. It has been calculated that in fresh water a fish weighing 100 grams should require a swim bladder with a capacity of about 7 milliliters to provide the necessary buoyancy to offset the fish's density. Since sea water has a greater specific gravity, a swim bladder of only a 5-milliliter capacity would be adequate to accomplish the same purpose. These are approximately the sizes of the swim bladders actually found in such fishes. Teleosts with heavy skeletal structures and those inhabiting fresh water are most benefitted by swim bladders, and it is in these fish that the organ most often is found. Marine fishes with lighter skeletons are less often so equipped. Elasmobranch fishes do not have swim bladders; instead, these animals store low-specific gravity fats, especially in the liver, to achieve the necessary buoyancy.

Swim bladders have a disadvantage that is not shared by the fat-storage buoyancy system of elasmobranchs. Just as the bubbles in a glass of soda increase in size and in rate of ascent as they rise, so the swim bladder expands and becomes increasingly buoyant as the fish approaches the surface of the water. This can be offset only by reabsorption of oxygen from the swim bladder as the fish rises toward the surface. It has been calculated that fish at a depth of 20 meters could rise rapidly to a depth of 15 meters without difficulty. Any further rise, however, would increase the buoyancy of the fish so much that it would be incapable of stopping its ascent; the fish would then be forced to the surface. If the fish stopped for a time at 15 meters, reabsorption of oxygen from the swim bladder could take place, and then further controlled ascent would be possible. This disadvantage would be offset if the swim bladder was a rigid, non-expansible structure. It is interesting to note that the swim bladder of *Latimeria,* a deep-sea coelocanth fish, is ossified. *Latimeria* apparently ranges over great differences in depth, since it sometimes appears relatively close to the surface. Other deep-sea fishes sometimes explode when brought rapidly to the surface in fishing nets. Presumably this is primarily due to overexpansion of the swim bladder.

Swim bladders also sometimes function in connection with hearing (Sect. 14.7) and with sound production or amplification in various fishes.

Some authorities believe that the evolution of the swim bladder has been homologous with the development of lungs; others disagree with this contention. The argument against homology is based upon the fact that the swim bladder is typically a dorsal outpocketing from the gut, is a single, unpaired structure, and is supplied by blood circulation in series with the general body circulation. The arteries supplying it, moreover, are derived from the arteries supplying the gut in other vertebrates. Lungs, on the other hand, are typically paired outpocketings from the ventral side of the gut and are supplied by arteries that are in parallel with the general body circulation. These arteries are derived from those supplying the branchial (gill) circulation of other fishes. There are exceptions, however, to each of the generalizations mentioned. It may well be that swim bladders are homologous with lungs in some lines of evolutionary descent but not in others. In any case, it is clear that there is a "physiological homology" in the development of swim bladders and lungs. Both are gut diverticula and both serve to exchange oxygen between the blood and a gaseous medium. Moreover, in those fishes which retain a tubular connection between the gut and the swim bladder, the bladder can be inflated by gulping air through the mouth. The mechanism for such air gulping is very similar to the mechanisms for ventilating the lungs of lower vertebrates such as lungfishes, amphibians, and reptiles.

3.5 AMPHIBIAN RESPIRATION

Except for the lungfishes, the lowest vertebrate species with true lungs are the amphibians. The urodele amphibians (salamanders) include both totally aquatic species with external gills and mainly terrestrial species with lungs. The anuran amphibians (frogs and toads) are mainly terrestrial as adults, although a few, such as the African clawed toad *Xenopus laevis,* remain totally aquatic. All anurans are air-breathers, however, and even *Xenopus* rises periodically to the surface to fill its lungs. The vascular skins of those amphibians that are partially aquatic also serve as respiratory membranes, especially when the animal is submerged. Consequently, these amphibians can remain under water for extended periods of time so long as they are not very active. Activity increases the oxygen needs of the tissues and necessitates more frequent filling of the lungs with air. *Rana pipiens,* the common grass frog of North America, can remain submerged for several hours, and *Xenopus* can stay under water for two weeks or more.

All amphibian larvae are aquatic and equipped with external gills. Skin respiration, especially in the long, thin tails of tadpoles, is probably an important adjunct to gill respiration in these larvae.

The lungs of adult amphibians are, like those of lungfishes, relatively simple, uncomplicated sacs without internal subdivisions. In this feature they resemble swim bladders more closely than they resemble typical vertebrate lungs. Evolution of lungs has led to greater and greater degrees of internal subdivision (Fig. 3–5) and thus to greater total surface areas for the exchange of gases with the blood.

The ventilation of amphibian lungs depends upon an oral-pharyngeal pump reminiscent of that used by teleost fishes to ventilate the gills (Sect. 3.3). Air is drawn in through the nostrils rather than through the mouth, but this is accomplished by depression of the oral-pharyngeal floor (Fig. 3-6). The nostrils then close and the oral-pharyngeal floor is again raised driving the air through the open glottis into the lungs. The glottis at the entrance to the trachea then closes, holding the air in the lungs for a time. This allows the air in the lungs to be more thoroughly depleted of its oxygen, since turbulence there brings more of the air into direct contact with the lung surfaces. While the air is being held in the lungs, there is a repeated oral-pharyngeal pumping of air in and out through the nostrils. Authorities differ in their opinions of the purposes served by this. Probably it is purely olfactory (smelling) behavior, but it may also serve a significant respiratory role. The lining

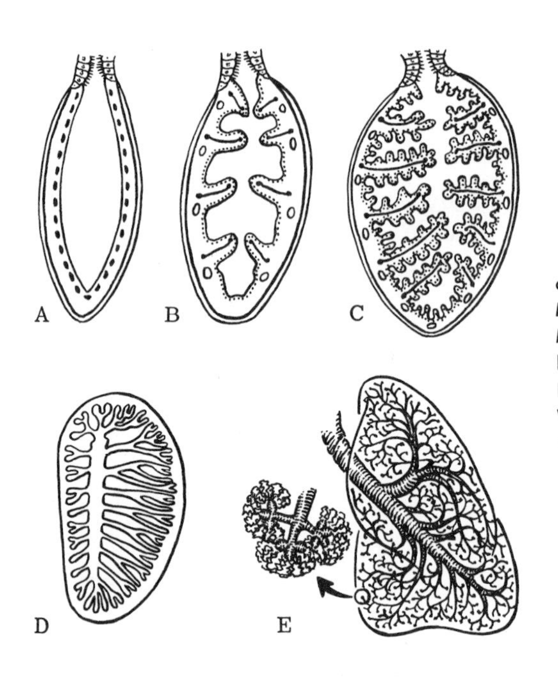

Figure 3–5 The increasing complexity of vertebrate lungs. *A, B,* and *C,* amphibians; *D,* a turtle; *E,* a mammal. (For a diagram of the bird lung see Figure 3–8.) (After Hughes: *Comparative Physiology of Vertebrate Respiration.* Cambridge, Harvard University Press, 1963.)

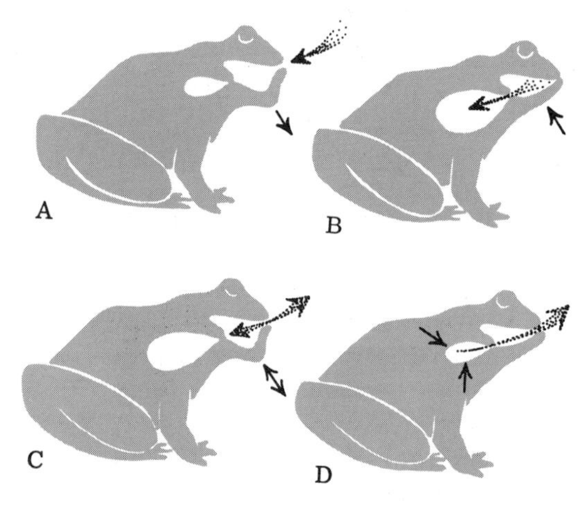

Figure 3–6 Amphibian ventilation. *A,* air is drawn in through the open nostrils; *B,* air is forced into the lungs by compression of the oral-pharyngeal cavity; *C,* air is passed back and forth through the nostrils in olfactory movements while air in the lungs remains stationary; *D,* the glottis is opened and air leaves the lungs. (After Gordon, *et al.*: *Animal Function: Principles and Adaptations.* New York, The MacMillan Company, 1968.)

membranes of the mouth and pharynx are highly vascular, so some gas exchange with the blood undoubtedly occurs. Finally, the glottis re-opens and the lungs are emptied by contraction of smooth muscles in their walls. Skeletal muscles in the thoracic walls may also aid in emptying the lungs.

3.6 REPTILIAN RESPIRATION

The lungs of most reptiles are inflated by active expansion of the thoracic walls while both the nostrils and the glottis are held open. After lung inflation, the glottis is closed, and the muscles of the thoracic walls relax again. This places the air in the lungs under some positive pressure, which may increase the rate of oxygen extraction from it. In any case, holding the air for a time presumably favors more complete oxygen extraction, as it does in amphibian lung ventilation. Also, as is true in the amphibians, repeated oral-pharyngeal pumping of air occurs while the glottis is closed. Again, this may be simply olfactory activity or it may contribute to respiratory gas exchange through the oral and pharyngeal walls. Finally, the glottis is reopened, and the lungs are emptied by the movements of smooth muscles in the walls of the lungs and, possibly, by the contraction of skeletal muscles of the thoracic walls.

This represents an improvement over the amphibian mechanism for lung ventilation since active thoracic expansion, rather than oral-pharyngeal pumping, is used to inflate the lungs. The oral-pharyngeal pump still operates while air is being held in the lungs, however. Also, the oral-pharyngeal pump can be employed by some reptiles to super-inflate the lungs. Following normal inhalation, oral-pharyngeal activity can be used to drive larger quantities of air into the lungs. In these animals the lung cavities frequently communicate with extensive, non-respiratory air sacs that can be filled by this means. Superinflation causes the entire animal to puff up to a larger size, and this behavior is used to frighten predators, in mating displays, and to wedge the animal into holes or cracks so that it cannot easily be captured. The possession

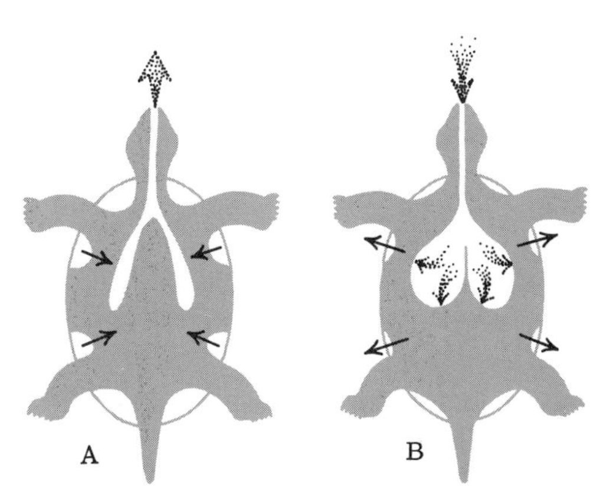

Figure 3–7 Turtle ventilation. *A*, air is expelled from the lungs by contraction of body wall muscles and muscle within the lung itself; *B*, air is taken in by movements of the body walls. (After Gordon, *et al.*: *Animal Function: Principles and Adaptations.* New York, The Macmillan Company, 1968.)

of the accessory air sacs in certain reptiles preceded in evolution the very important air sacs of birds (Sect. 3.7).

Because the major portion of the body wall of turtles is rigid, general expansion and contraction of the thoracic walls cannot occur. In these animals there are specially adapted skeletal muscles, lying adjacent to the limbs, that produce movements of the soft parts of the thoracic wall to draw air into the lungs (Fig. 3-7). Probably movements of the limbs themselves also aid in ventilation. There is evidence to suggest that the muscles involved are derived from the intercostal muscles and diaphragm; these are the same muscles used for respiration in other reptiles and higher vertebrates.

3.7 BIRD RESPIRATION

The lungs and the extensive non-respiratory air sacs of birds are both ventilated by active expansions and contractions of the thoracic and abdominal walls. Oral-pharyngeal pumping, like that seen in fishes, amphibians, and reptiles, is not present. During flight the respiratory movements may be correlated with the wingbeats so that the same muscular movements can accomplish both tasks. Inspired air is not held for a time as in amphibians and reptiles but is, instead, moved rhythmically in and out of the lungs.

Except for the birds and, to a lesser extent, some reptiles, the lungs of vertebrates are cul-de-sacs, *i.e.*, blind pockets connected with the outside atmosphere through the trachea and other respiratory passages (Fig. 3-5). In such lungs even the most forceful expiration does not completely empty the system of air. The respiratory passages remain filled, and with the next inspiration this "dead air" is drawn back into the lungs. Fresh air from the outside, thus, is mixed with air that has not been exhaled at all. At best, the air in the lungs is a mixture of fresh air and dead air from which most of the oxygen had already been removed.

The obvious inefficiency of such a system is corrected in the birds. During inspiration the dead air is drawn *through* the lungs and into the air sacs (Fig. 3-8). The lungs are then filled with the incoming fresh air. Without this increase in respiratory efficiency it would be difficult, if not impossible, for birds to meet the great oxygen demands imposed by the activity of flying. To do so would require that the lungs be very much larger. Actually, the benefits acquired with the addition of air sacs are so great that bird lungs are, in general, comparatively smaller than those of other vertebrates. The common crow has a respiratory surface in its lungs of about 0.6 square centimeter per gram of body weight, whereas man has 7.0 square centimeters per gram of body weight. The little brown bat *Myotis lucifugus* is a mammal and therefore has no air sacs, yet it must meet the oxygen demands of flight like a bird. The respiratory surface of this animal amounts to about 100 square centimeters per gram of body weight!

Besides improving respiratory efficiency, the air sac system prob-

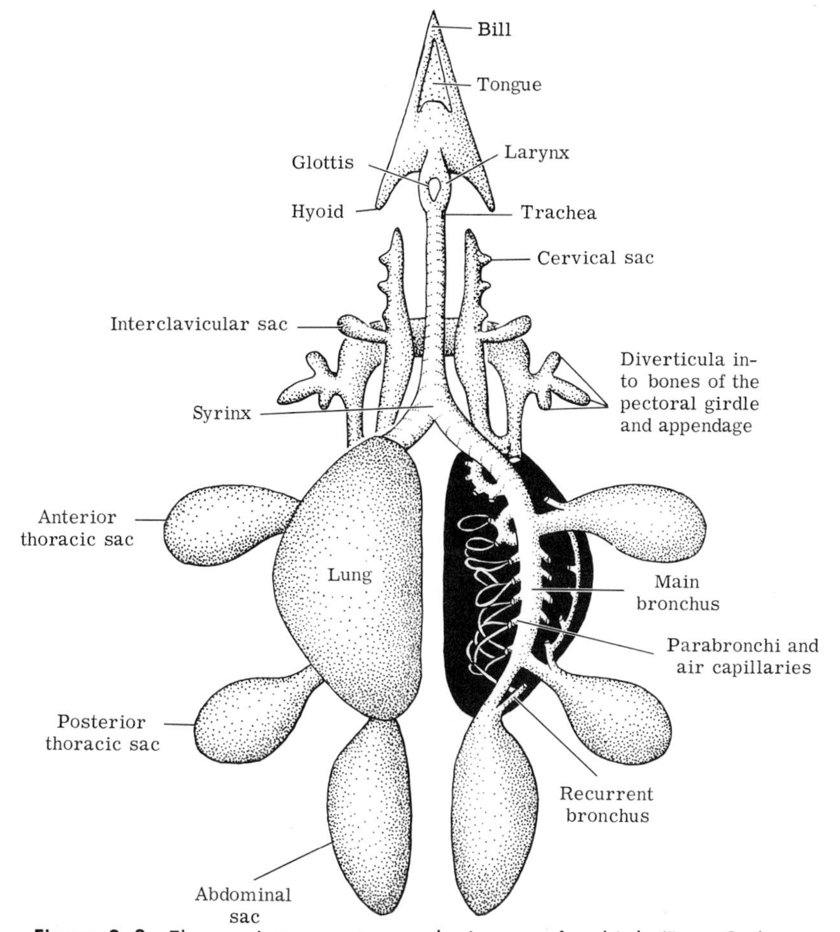

Figure 3–8 The respiratory system and air sacs of a bird. (From Cockrum and McCauley: *Zoology*. Philadelphia, W. B. Saunders Company, 1965.)

ably also functions in the control of body temperature. It has been calculated that for cooling purposes a pigeon in active flight would require a volume of air three times as great as that required for respiratory gas exchange. The air sac system ramifies throughout the body, and undoubtedly ventilation of this system effectively carries away much of the heat generated by active flight muscles. The air sacs, occupying as they do a substantial percentage of the total volume of the bird, reduce its weight, and, of course, this is another important contribution to flight.

Because the air sacs remove the problem of dead air space in the respiratory passages, the volume of the dead air space is no longer a critical factor; it can be much larger without interfering with efficient gas exchange in the lungs. Frequently the tracheas of birds are doubled upon themselves lengthwise and they often contain large, resonating vocal chambers. By comparison, even the addition of a small amount of dead air space in man makes lung ventilation much more difficult. This is a familiar fact to anyone who has made use of a snorkel in skin diving. Since the snorkel is effectively an extension of the respiratory passage, it, therefore, adds its own volume to the dead air space. It is also familiar to any boy who has attempted to emulate the heros of western books and movies by trying to breath through a hollow reed while hiding submerged in the river!

The internal structure of bird lungs is quite different from that of most other vertebrates. Air is drawn from the trachea through a main bronchus and into the air sacs. Smaller tubules called parabronchi branch off from the main bronchus, and still smaller air capillaries branch off from the parabronchi. Presumably air currents and, perhaps, physical movements of the lungs ventilate all of these passages while air is passing through the main bronchus. Ventilation of the air capillaries occurs, then, both during inspiration and expiration. This undoubtedly results in a very complete extraction of the oxygen contained in the air.

3.8 MAMMALIAN RESPIRATION

Air is conducted between the pharynx and the lungs of all terrestrial vertebrates by the trachea and its major branches, the right and the left bronchi (Fig. 3-9). Within the human lung each bronchus divides

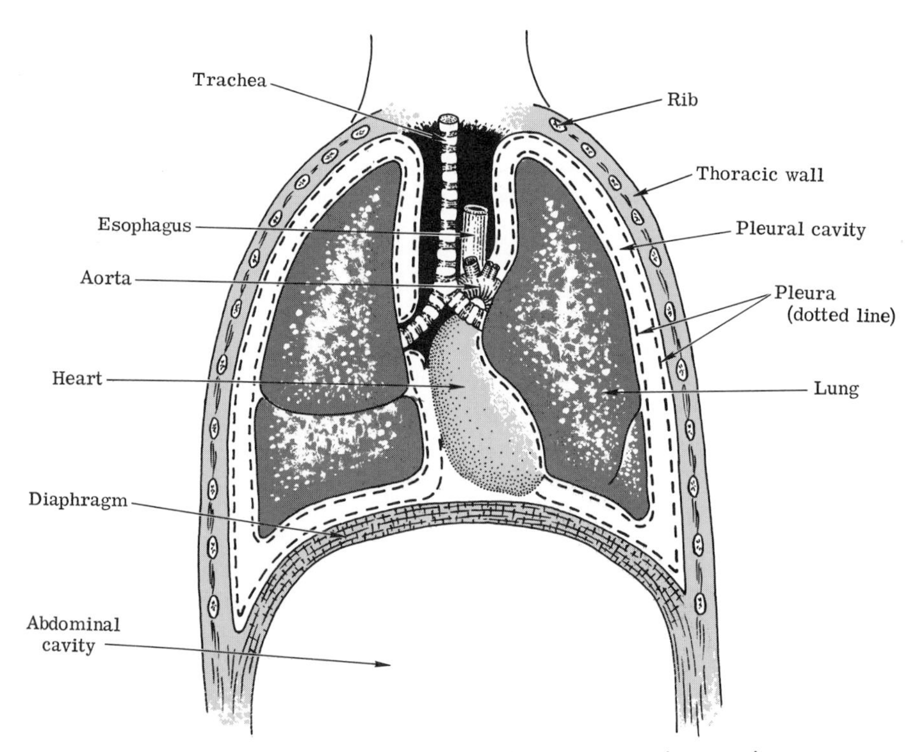

Figure 3-9 Structure and relationships of the human lungs and surrounding organs.

again and again through a total of more than 20 successive bifurcations. This gives rise, ultimately, to about a million terminal passages called alveolar ducts. The walls of the trachea, bronchi, and major branches contain cartilaginous rings that hold them open and smooth muscles that control their diameters. As the branches become smaller and smaller, however, the cartilage disappears and the walls become progressively thinner.

Each of the million or more alveolar ducts of the human lung ends in a cluster of bubble-like spaces, or alveoli. An individual alveolus has a diameter of 75 to 300 microns. Collectively the total of some 300 alveoli in the two lungs has a surface area of about 70 square meters, or more than 40 times the average area of the human skin. The alveolar walls are richly vascular and extremely thin. In places less than a single micron separates the air in the alveolus from the blood in the associated blood capillaries. Thus, exchange of oxygen and carbon dioxide by diffusion takes place rapidly and with ease.

Each lung lies in the lateral portion of the thoracic cavity and is attached only at its root medially where the bronchi and blood vessels enter its substance. Between the two lungs in the medial portion of the thoracic cavity is a mass of tissue known as the mediastinum, which is composed of the heart and its major vessels, the esophagus, the trachea and its bronchi, and various nerves and other structures. Each lung is almost completely surrounded, except at its root, by a potential space known as the pleural "cavity" (Fig. 3-9). This space is completely lined by a pleural membrane. In other words, the pleural membrane covers the surface of the lungs and also lines the inner surface of the thoracic walls, the upper surface of the diaphragm, and the lateral surface of the mediastinum. The pleural "space" enclosed by pleural membrane is a *potential* space rather than an actual space; that is, the pleura covering the lung and that which lines the thoracic cavities are in actual contact with each other so that no real space exists between them. These layers of pleura are not fastened together, however, so a potential space does exist. If, for example, air was injected between the pleural membrane layers, they would be forced apart and an actual space would be present. No such thing occurs under normal conditions, however.

The thoracic walls are elastic but tend to resist either expansion or contraction of the thoracic volume. Both expansion and contraction actually occur, of course, as results of muscle action during respiration. When these muscles relax, however, the thoracic walls return to the natural resting position as a result of their own elasticity.

The lungs themselves are also elastic structures. Like toy balloons, the lungs would nearly empty themselves of air if allowed to do so. The elastic pull of the lungs is opposed by that of the thoracic walls; this creates a negative pressure (*i.e.*, less than atmospheric air pressure) in the potential space between the layers of pleural membrane. If, for example, a milliliter of air at atmospheric pressure was injected into the potential pleural space, the elastic pull of the lungs would expand it sufficiently to reduce the air pressure by a factor of four or five millimeters of mercury. If a sufficient amount of air is injected in this way so that the lung completely collapses, the pleural space will be filled with air at atmospheric pressure. Under abnormal conditions this can actually occur. If, for example, a stab wound admits air through the thoracic wall, the elastic pull of the lungs will draw air into the pleural cavity and fill it, and as a result, the lung will collapse. Such an abnormal situation is known as a pneumothorax. When the stab wound is closed, the air in the pleural cavity is absorbed by the pleura; after a period of several days, the lungs become reinflated, and the normal negative pressure is reestablished in the pleural space.

The lungs of mammals are inflated during inspiration by active expansion of the thoracic volume; this is accomplished by the contraction of various muscles. The external layer of the intercostal muscles raises the ribs to increase the transverse diameter of the thoracic cavity (Fig. 3-10), and the muscles of the diaphragm contract to increase the cranial-to-caudal dimension of the thorax (Fig. 4-7). Both actions, by increasing thoracic volume, also decrease the pressure in the pleural space still further. This causes the lungs to expand, and, as a result, air enters them through the respiratory passages. The respiratory function of the diaphragm is unique to mammals since no complete diaphragm exists in birds or other lower vertebrates.

Deflation of mammalian lungs is usually accomplished by merely relaxing the muscles of inspiration and allowing elastic recoil of the thoracic walls and lungs to return them to their normal resting positions. During labored respiration, however, the process can be hastened and the amount of expelled air increased by the utilization of other muscles. The action of the muscles of expiration, which are used during labored conditions, results in compression of the thoracic volume and, consequently, greater and faster emptying of the lungs. The internal layer of intercostal muscles pulls the ribs downward, thus decreasing the transverse diameter of the thorax. Also, muscles of the

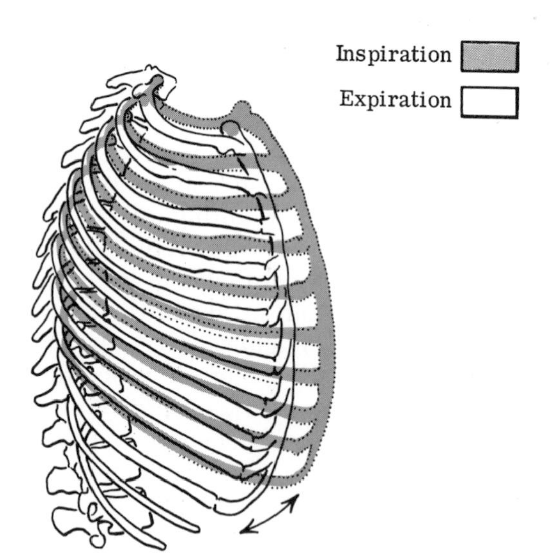

Figure 3–10 Action of the ribs in human ventilation.

abdominal wall squeeze inward on the abdominal contents which, in turn, crowd the diaphragm in a cranial direction. This reduces the cranial-to-caudal dimension of the thorax. Both actions reduce thoracic volume and supplement the elastic emptying of the lungs.

Ordinary respiratory movements, then, consist of an active thoracic expansion (inspiration) alternating with a passive return to the resting thoracic volume (expiration). During exercise, times of excitement, or other situations calling for increased ventilation of the lungs, additional, active compression of the thorax occurs. At the same time, of course, inspiration is also greater and the rate of alternating inspirations and expirations is increased. All these processes result in increased total ventilation of the lungs.

Under quiet, resting conditions, the average human moves about 500 milliliters of air in and out of the lungs with each respiratory cycle. This quantity of air moves in and out as the ocean tides move in and out, and for this reason the cycle is spoken of as tidal respiration (Fig. 3–11). By increasing muscular effort it is possible to inhale an additional 3000 milliliters of air over and above the normal tidal intake. This additional air is called the inspiratory reserve volume. Also, by active expiration it is possible for the average human to exhale about 1100 milliliters of air beyond the normal tidal expiration. This quantity is called the expiratory reserve volume. The total—tidal volume plus inspiratory reserve volume plus expiratory reserve volume—represents the maximum amount of air that can be inspired and expired with the greatest physical effort. It amounts to 500 plus 3000 plus 1100, or about 4600 milliliters. This total is called the vital capacity.

Even following the most forceful of expirations, the lungs still contain about 1200 milliliters of air. This is called the residual volume, and when added to the vital capacity, it gives a total lung capacity of approximately 5800 milliliters for the average human.

In addition to the volumes of air thus far discussed, the respiratory passages contain about 150 milliliters of dead air space (Sect. 3.7). During the usual respiratory cycle, then, the first 150 milliliters of the tidal intake is composed of air from the dead space—air which has not

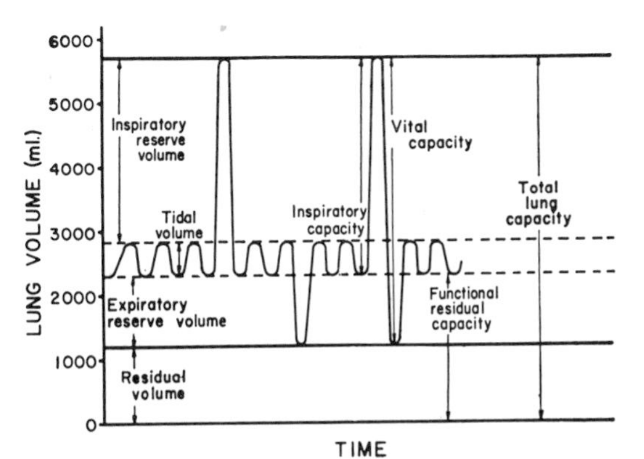

Figure 3–11 Diagram showing respiratory excursions during tidal respiration and during maximal inspiration or expiration. (From Guyton: *Textbook of Medical Physiology*, 4th Edition. Philadelphia, W. B. Saunders Company, 1971.)

been exchanged with pure atmospheric air. Only about 350 milliliters of the tidal inspiration, therefore, is actually effective in lung ventilation. This imposes an obvious inefficiency in mammalian ventilation; it will be recalled that this inefficiency is not present in the bird lung because of the extensive air sacs.

The average human ventilation rate is about 12 complete cycles per minute. Assuming normal tidal volume, this amounts to six liters of ventilation per minute. Deducting 150 milliliters of dead air from each cycle leaves a total of 4200 milliliters of effective air exchange with the atmosphere per minute. During periods of exercise the total ventilation rate can be increased to more than 100 cycles per minute. As we have seen, the total depth of respiration can be increased to as much as 4600 milliliters per cycle. Total lung ventilation, then, can increase to more than 400 liters per minute in the average human. It is doubtful that these limits are ever actually achieved, but they are theoretically possible.

3.9 CONTROL OF MAMMALIAN VENTILATION

Few studies have been made of non-mammalian respiratory controls, and much remains to be learned about mammalian mechanisms. Nervous pathways in the central nervous system are involved and these have not been completely mapped to date. It is known, however, that the nervous control centers reside in the medulla of the brain. It appears that certain neurons located there stimulate the muscles of inspiration to contract and thus inflate the lungs. These neurons are collectively spoken of as the inspiratory center of the medulla (Fig. 3-12). Another group of neurons in the same general area of the medulla constitutes an expiratory center. During normal tidal respiration, the expiratory center acts by merely inhibiting the activity of the inspiratory center. This results in passive relaxation of the muscles of inspiration and, as we have seen, consequent expiration. During periods of exercise or emotional excitement, however, the expiratory center may also stimulate muscles of expiration. This will increase both the rate and degree of lung expiration (Sect. 3.8). Thus, together these two nervous centers control the alternate filling and emptying of the lungs.

It is also known that special receptors responsive to the stretching of surrounding tissues are located in the passages of the lungs. As the lungs are filled during inspiration, the tissues are stretched and these receptors respond by sending trains of nervous impulses to the brain. The more the lungs are filled, the greater is the degree of stretching, and the greater is the frequency of these nervous impulses. Thus, the medulla is constantly "informed" of the amount with which the lungs are filled at any moment. This frequency change controls the alternate activity of the inspiratory and expiratory centers. Inspiration is stopped when the lungs have been filled to the tidal level; then expiration takes place until the tidal air has been expired.

As we have seen, it is possible to increase the ventilation volume

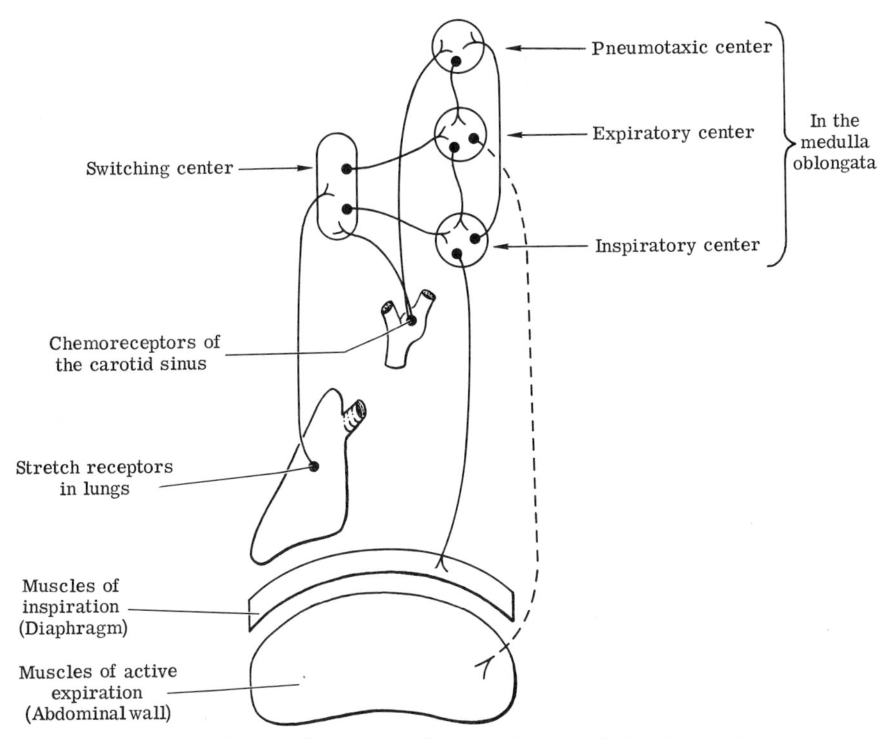

Figure 3–12 The control of mammalian ventilation (see text).

beyond the tidal levels, and we shall see presently how this is controlled. Under normal resting conditions, however, only tidal exchange occurs. This is due to the fact that the muscular effort involved in greater depth of respiration is disproportionate to the extra volume of oxygen made available by it. Consequently, it is energetically uneconomical to breathe more deeply than is necessary to meet the oxygen needs of the moment.

It is interesting to note in passing that receptors similar to the stretch receptors of mammalian lungs also occur in the pharyngeal walls of the dogfish shark and perhaps in other fishes as well. These control the degree of ventilation of the gills with water—a similar control system acting on an analogous mechanism.

Although the mechanisms just described are adequate for normal quiet respiration in mammals, it is sometimes necessary to increase the depth of respiration to meet the increased oxygen demands of exercise or for other reasons. Consequently, mechanisms must be provided to "reset" the nervous centers of the medulla of the brain. When these centers are reset, they allow greater lung inflation with each cycle before inspiration is stopped and expiration is allowed to occur. Resetting of the centers occurs in the following manner:

1. Receptor endings located in the arch of the aorta and in the carotid body near the bifurcation of the external and internal carotid arteries in the neck are sensitive to the oxygen concentration of the

blood. During exercise the active muscle tissues extract larger quantities of oxygen from the blood, and so its concentration in these areas falls. The oxygen-sensing receptors respond to changing oxygen concentration by altering the frequency of nervous impulses sent to the medulla of the brain. Thus the medulla is also constantly "informed" of the blood oxygen concentration existing at the moment. When that concentration falls below normal levels, the inspiratory and expiratory centers of the medulla are "reset" to permit filling and emptying of the lungs beyond the usual tidal limits. Exercise also results in an increase in the concentration of carbon dioxide in the blood. Increased carbon dioxide concentration makes the oxygen-sensitive receptors even more sensitive to falling oxygen concentrations. Thus, increasing carbon dioxide augments the effect of falling oxygen levels.

2. Receptors sensitive to chemical changes in either the blood or the cerebrospinal fluid (Sect. 12.2) are present in the medulla of the brain itself. The increased carbon dioxide concentration in blood that accompanies exercise does not result in any appreciable change in blood pH because the blood contains adequate buffer systems to prevent the buildup of acid. Cerebrospinal fluid, however, lacks such buffers, and there the increase in carbon dioxide concentration results in decreased pH. This apparently has an effect similar to that of falling oxygen concentration in the blood, *i.e.*, it initiates a resetting of the medullary control centers for inspiration and expiration to permit deeper ventilation of the lungs with each respiratory cycle.

In any case, of course, the deeper ventilation of the lungs continues until normal oxygen and carbon dioxide concentrations have been reestablished in the blood and cerebrospinal fluid. Then normal, resting tidal respiration is again resumed.

The medulla also contains a third group of neurons concerned with control of lung ventilation in mammals. This is called the pneumotaxic center (Fig. 3-12). Whereas the mechanisms discussed in preceding paragraphs control the *depth* of each respiratory cycle, the pneumotaxic center is concerned with the *rate* of respiratory cycling, that is, the number of such cycles per minute. The pneumotaxic center is a kind of timing device. When the inspiratory center begins its activity to initiate the act of inspiration, it also sends nervous impulses to the pneumotaxic center "informing" it that the respiratory cycle is beginning. The pneumotaxic center then begins to measure an appropriate length of time, and at the end of this time it acts by stimulating the expiratory center. It will be recalled that the expiratory center acts by inhibiting the inspiratory center. Thus, at some appropriate moment after inspiration begins, the activity of the pneumotaxic center causes inspiration to cease again. Quite clearly, the mechanisms that control depth of respiration and those that control the rate of respiration must be coordinated by way of other interconnecting nervous pathways. Even though we know these coordinating pathways must be present, they are complex and incompletely understood.

The pneumotaxic center is able to vary the length of time during which inspiration occurs so that cycling can be speeded during exercise and slowed during quiet periods of tidal respiration. Such changes are mediated by information input from the oxygen receptors of the carotid sinus and aortic arch as well as by those in the medulla itself. These are the same receptors that are involved in the controls of respiratory depth already discussed in this section. The mechanism involved in the actual timing depends upon the phenomenon of synaptic delay (Sect. 11.4). The incoming impulses from the inspiratory center are routed through varying numbers of synapses within the pneumotaxic center, and since each synapse results in a delay in transmission, the number of such synapses determines the total time required before these impulses reach the expiratory center.

The mechanisms described in preceding paragraphs explaining the controls of respiratory depth and rate are theoretical. They are models based upon what is known about the connections within and principles governing the central nervous system. We have insufficient knowledge of the precise nervous connections to be able to determine the actual mechanisms that may be involved. Whatever they are, they cannot be vastly different from those described, so such theoretical models give us a reasonable basis for discussion. It is quite certain that the actual mechanisms are more complex than the model but they certainly involve similar principles.

Certain emotions such as fear, anger, and excitement produce changes in the depth and rate of respiration even though no simultaneous changes in blood oxygen or carbon dioxide occur. Such respiratory changes are apparently in anticipation of exercise that might result from the situation that caused the emotion. For example, if a bear enters the room, the respiratory rate and depth of humans in the room increase in anticipation of having to escape from the bear. Such changes demonstrate the fact that higher brain centers are also capable of "resetting" the medullary control centers. In addition, of course, conscious levels of the brain can exert limited controls over respiration. If this were not true, it would be impossible to speak, sing, or play any wind musical instrument. We are able either to hold our breaths or to increase our respiration for limited periods of time simply by consciously willing it to happen. Such conscious control is limited in its extent, however. For example, even with the greatest will power we can hold our breaths only until we lose consciousness, after which mormal respiration will again commence.

A number of subconscious respiratory reflex controls are also well known. The special acts of coughing, sneezing, and hiccoughing occur without our volition. Similarly, the momentary suspension of breathing while food is being swallowed is beyond conscious control. The sudden, involuntary intake of breath when we are startled or when we experience sudden pain or a dash of cold water are similar reflex actions. These are similar in every way to the many muscular reflexes that are involved in posture and locomotion (Sect. 11.2) even though they involve the spinal cord as well as the medulla of the brain.

In diving birds and reptiles there are carbon-dioxide sensitive receptors in the lungs, and increased concentrations of carbon dioxide there result in an inhibition of respiratory activity. This probably helps these animals to suspend breathing when diving. Mammals have no such receptors, and prolonged diving is difficult or impossible for most mammalian species. Humans cannot remain submerged long before the central nervous system centers force inspiration to occur. Mammals that can remain submerged for long periods possess a lowered sensitivity to the decrease in blood oxygen, and there are other special physiological mechanisms involved as well (Sect. 6.5). Increasing the carbon dioxide from the normal atmospheric level of 0.04 per cent to a level of 0.2 per cent results in a doubling of the rate of human ventilation. In the harbor seal, however, the concentration of atmospheric carbon dioxide must be increased to 10.0 per cent to produce the same increased rate of ventilation.

Both the high pressures encountered in diving and the low pressures that exist at high elevations impose further restrictions on animals. The adaptive physiological mechanisms that permit vertebrates to enter these regions of extreme pressures involve both the respiratory and the cardiovascular systems, which are discussed in Chapter 6.

CHAPTER 4

CIRCULATION: FUNDAMENTALS

4.1 INTRODUCTION

We have seen that blood must circulate; it must flow from the surfaces specialized for gas exchange and for the exchange of wastes for nutrients between the blood and the environment to the immediate vicinity of all the cells of the animal body and back again. At the specialized surfaces and at the internal tissue sites, short diffusion distances and high concentration differences must be maintained so that diffusion gradients may be steep and the exchange of substances correspondingly fast and efficient (Sect. 1.2). To accomplish these ends requires that blood circulation be rapid and that thin, permeable membranes separate the blood from the specialized surface areas and from the internal metabolizing cells. Rapid blood circulation becomes increasingly important in animals that are active and have, therefore, greater demands for oxygen and nutrients than have less active animals. All vertebrate animals, therefore, require strong blood pumps (hearts) and closed vascular systems in which considerable pressure can be developed.

The typical vertebrate cardiovascular system consists of a heart that pumps blood through two separate vascular circuits. One of these is the pulmonary (lung), or branchial (gill), circuit and the other is the systemic circuit. Blood, then, flows from the heart to the respiratory organs (lungs or gills), back to the heart, to the other portions of the body, and finally back to the heart again. The vessels carrying it away from the heart are arteries, which typically have thick, muscular walls. Those transporting blood back toward the heart are veins, which have thinner, less muscular walls. Between the two in both circuits blood flows from arteries to veins through tiny, thin-walled blood capillaries.

4.2 THE LAW OF LAPLACE

On first consideration it might seem that the necessity of maintaining a high pressure is inconsistent with the presence of thin capillary walls at the points of diffusion exchange. One might well suppose that such pressures could be contained only by thick, strong walls, and that thin, permeable membranes would be ruptured by the internal distending pressure. It turns out, however, that no great wall strength is necessary as long as the radius of the vessel is small. As the radius becomes larger, the wall thickness must also increase because under these conditions the distending pressure produces a greater strain or tension on the wall of the vessel. This is why the aorta must be very thick and muscular and the walls of capillaries can be thin. An average capillary has a radius of only about four microns, whereas the radius of the aorta is about 3000 times greater.

The relationship between vessel diameter and wall tension is expressed by the law of Laplace:

$$T = PR$$

where T = tension on the vascular wall in dynes per square centimeter, P = distending pressure in millimeters of mercury, and R = radius of the vessel in centimeters.

Almost everyone has observed that the tip of an expanded toy balloon is not inflated (stretched) to the same extent as the larger main portion of the balloon (Fig. 4-1). The entire balloon is made of the same material and has nearly the same initial wall thickness. Quite obviously, both portions of the balloon must contain air at the same pressure since they are continuous with each other. The differences in

THE RELATION BETWEEN PRESSURE, WALL TENSION AND RADIUS IN HOLLOW ORGANS

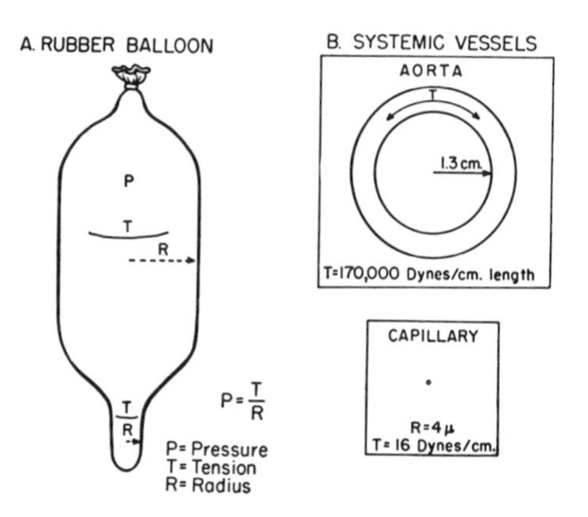

Figure 4–1 The law of Laplace: relationships between pressure, radius, and wall tension in hollow vessels. (From Rushmer: *Cardiovascular Dynamics,* 3rd Edition. Philadelphia, W. B. Saunders Company, 1970.)

radius, however, result in a very great difference in wall tension, or stretching, of the rubber in the two portions of the balloon. This effect is even more clearly marked in the type of toy balloon that consists of a series of inflated portions alternating with constricted portions. If one squeezes with his hand one of the inflated segments so as to reduce its radius, the air is forced into other segments. If the radius of the squeezed segment is reduced sufficiently, it will remain small even when the hand releases it. Under such conditions the tension, or stretching, force on the wall of this segment is so reduced that the segment does not reinflate to its former size.

In the human vascular system an average blood capillary with a radius of four microns supports a blood pressure of about 30 millimeters of mercury and is exposed to only 16 dynes per square centimeter of wall tension. The aorta, by comparison, has a radius about 3000 times greater, a distending pressure only about three to four times greater, and is exposed to a wall tension of 170,000 dynes per square centimeter. Common facial tissue will withstand, without tearing, tensions about 3000 times greater than those to which blood capillaries are exposed.

4.3 VOLUME FLOW OF BLOOD

The circulating blood of vertebrate animals is contained within a system of closed tubes which directs it from place to place by means of a pumping force applied by the beating heart. Since the system is a closed one, equal volumes of blood must pass each segment of the system during equal time periods (Fig. 4-2). That is, the same total volume of blood which passes through the aorta each minute must also pass through the capillaries and through the veins in the same length of time. This is also the quantity of blood which passes through the heart each minute.

The total cross-sectional area of the vascular bed is quite different in the arteries, capillaries, and veins. It is greatest in the capillaries and least in the arteries. The velocity of blood flow, then, must vary from place to place as does the velocity of water flow in a river. River water flows faster in a region of smaller cross-sectional area (rapids) than it does in a region of large cross-sectional area (ponds) even though the total volume of water passing through each is the same per unit of time. The velocity of blood flow through the aorta is about 40 to 50 centimeters per second, whereas in the capillaries it is only 0.07 centimeters per second under usual resting conditions. This is due to the fact that the total cross-sectional area of the capillaries is some 600 times greater than that of the aorta. Since individual capillaries are quite short, this reduction in the rate of flow is necessary in order to provide sufficient time for the exchange of materials by diffusion. Even then, a given "drop" of blood remains in a capillary for only a moment before passing on into the associated vein.

Even though equal volumes of blood must flow through the aorta

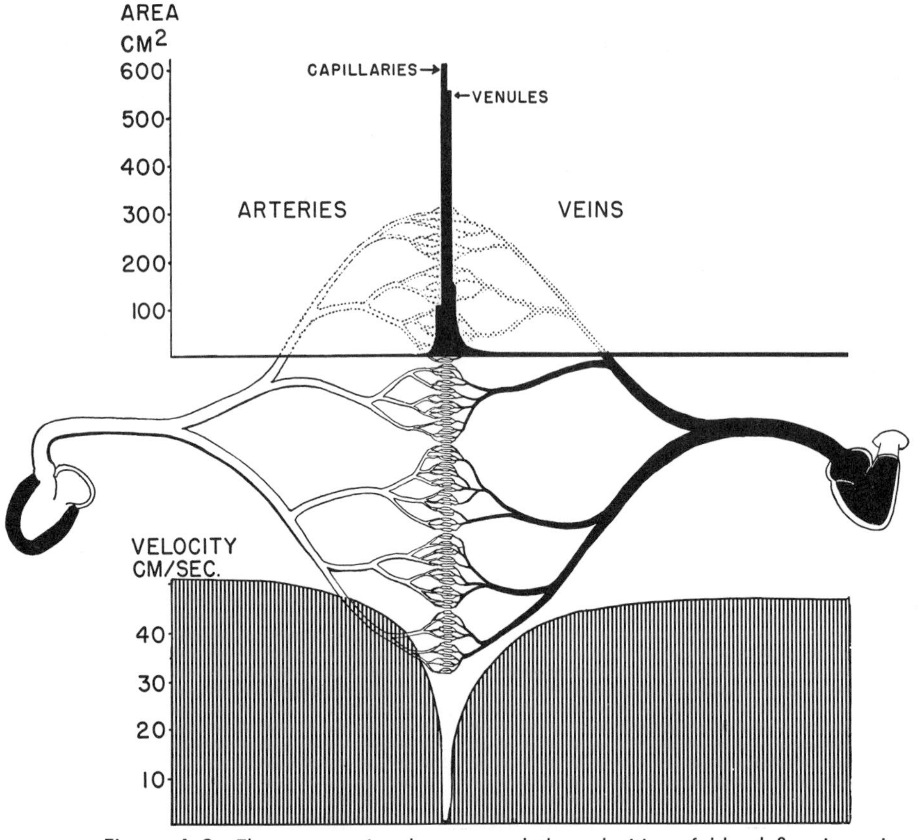

Figure 4–2 The cross-sectional areas and the velocities of blood flow in various parts of the human circulatory system. (From Rushmer: *Cardiovascular Dynamics,* 3rd Edition. Philadelphia, W. B. Saunders Company, 1971.)

and through the total capillary bed in equal time periods, it is not necessary for each capillary to carry the same volume as every other capillary. Certain capillaries may even be closed off completely (Sect. 5.7) and thus force other capillaries to carry greater volumes of blood. In fact, at any given moment only a fraction of the capillaries of the human body are open and functional. Individual capillaries open for variable lengths of time and then close again as the demands of the associated tissues may require. In a similar way, dilation and constriction of smaller arteries (arterioles) can direct greater or lesser volumes of blood to the capillary beds of various organs or regions, and in this manner blood can be shunted to those tissues where the demand is greatest. For example, the skeletal muscles may receive greater amounts of blood during physical exercise, or the digestive tract may receive greater amounts during the active digestion of a meal. The total blood volume of the body is not sufficient to give maximum supply to all parts of the body at once, however. For this reason we are warned as children not to go swimming for a time after eating simply because the bulk of the blood has been diverted to the digestive tract, and the skeletal muscles may not receive enough to support the exercise of active swimming. As a result, the skeletal muscles may fail to function normally, cramps may occur, and drowning may result.

4.4 RESISTANCE AND PRESSURE DROP

The volume of fluid passing through a tube depends, in part, upon the pressure gradient, *i.e.*, the pressure difference at the two ends of the tube. If this pressure difference is great, then more fluid is forced through the tube in a given length of time. If the pressure difference is less, then smaller amounts of fluid pass through the tube. Quite obviously, if the pressure difference is zero, no fluid passes through the tube at all.

The walls of the tube exert a frictional resistance to the passage of fluid through it, so the pressure remaining as the fluid leaves the tube is lower than the pressure applied to it at the other end. This change in pressure, resulting from the amount of force "used up" in overcoming frictional resistance, is known as the pressure drop of the system. The longer the tube, of course, the greater is the resistance drop.

Frictional resistance is greatest where molecules of liquid are actually in contact with the walls of the tube and least where the molecules are most distant from the walls of the tube (Fig. 4–3). Thus the liquid flows most easily and, therefore, most rapidly in the center of the tube lumen. It follows that a tube with a large radius exerts less frictional resistance than one with a smaller radius. This is true because in a large tube a greater proportion of the liquid molecules are passing well away from the walls of the tube. It also follows that the pressure drop characteristic of the system is inversely proportional to the tube radius. In a tube the size of a blood capillary virtually all the fluid is near the vessel wall, so the frictional resistance is very high and the consequent pressure drop is great.

Although the foregoing considerations hold for homogenous fluids, the presence of blood cells in vertebrate blood adds other factors of importance. In the first place, the blood cells tend to be swept along near the center of the vessel lumen so that only the plasma is in contact with the vascular walls. In the capillaries, however, the lumen diameter is inadequate to permit the passage of a red blood corpuscle in its normal shape. Red blood corpuscles are actually deformed during passage through capillaries and as a result they are in contact with all of the capillary wall. This adds greatly to the frictional resistance.

Frictional resistance in any blood vessel increases as the velocity of flow increases; it requires more energy to force fluid rapidly through a tube than to pass it more slowly. Consequently, the pressure drop is directly proportional to the velocity of flow.

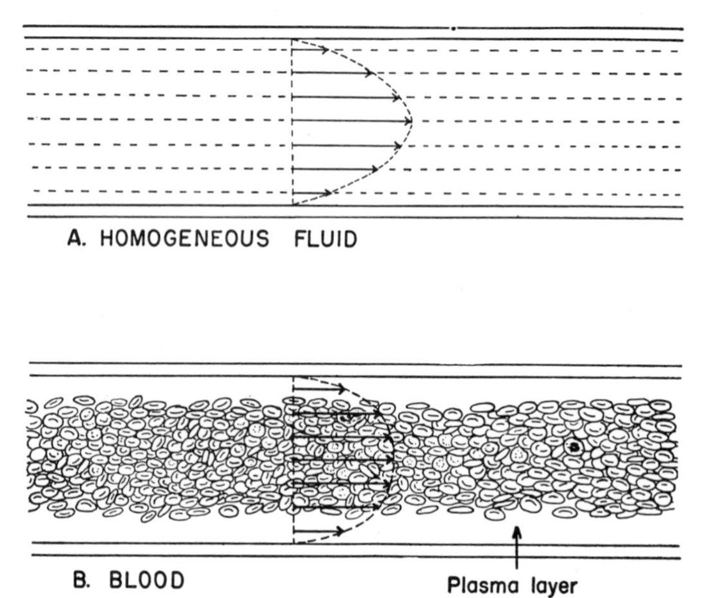

Figure 4–3 Laminar flow of homogeneous liquids (*A*) and of blood (*B*) through a tube. (From Rushmer: *Cardiovascular Dynamics,* 2nd Edition. Philadelphia, W. B. Saunders Company, 1961.)

Finally, the viscosity of a liquid is directly proportional to the resistance (and consequent pressure drop), since it takes more energy to force a "thick" fluid like syrup through a tube than to pass a "thin" fluid like water. Blood is more viscous than water, since it contains dissolved proteins and other substances, but its viscosity is normally unchanging. This is a constant factor in vertebrate circulation.

All the resistance factors discussed in the preceding paragraphs are expressed by the equation of Poiseuille:

$$\Delta P = FV \frac{8L}{\pi R^4}$$

where ΔP = the pressure drop, F = the velocity of flow, V = the viscosity of the fluid, L = the length of the tube, and R = the radius of the tube.

Note that doubling the tube length (L), the viscosity (V), or the velocity of flow (F) will each result in a doubling of the pressure drop (ΔP). A decrease of tube radius (R) by one half, however, will produce a 16-fold increase in pressure drop (ΔP). Tube radius, then, is by far the most important of the factors involved.

The pressure drop which occurs in the capillaries is much greater than that which occurs in any other portion of the vascular system (Fig. 4-4). This is true because of the small radius, because of the very large

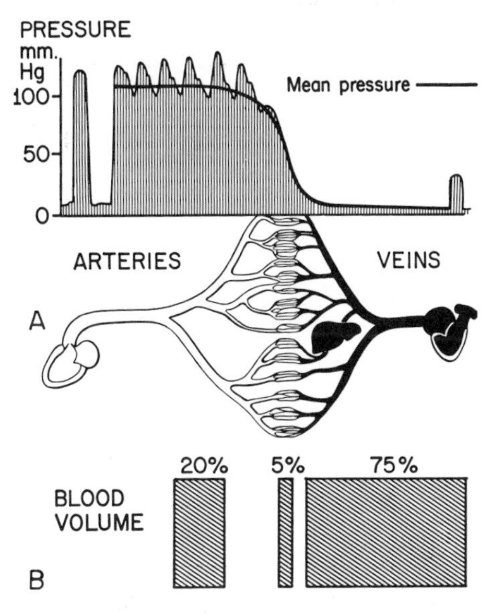

Figure 4–4 The blood pressures and volumes normally associated with the various portions of the human circulatory system. (From Rushmer: *Cardiovascular Dynamics,* 3rd Edition. Philadelphia, W. B. Saunders Company, 1971.)

total frictional area, and because of the necessary deformation of red blood corpuscles. The large surface area of all the capillaries taken together, as compared with the surface area of the aorta, is necessary in order to provide a large surface for the diffusion exchange of gases, nutrients, and wastes with the tissue cells.

4.5 CAPILLARY-TISSUE FLUID EXCHANGE

A typical capillary is composed of an endothelial cell layer about one micron in thickness. Surrounding this endothelium is a narrow, fluid-filled perivascular space, which is encircled by a second thin cellular membrane (Fig. 4-5). Surrounding the whole structure is the interstitial matrix in which metabolizing tissue cells are embedded. Exchanges of substances between the blood in the capillary lumen and the tissue cells, thus, must take place through the two cellular layers of the capillary wall, the perivascular space, and varying amounts of interstitial matrix. If such exchanges were entirely dependent upon diffusion, the distances involved would often be limiting factors. Therefore, in addition to diffusion there is also an actual flow of fluid from each capillary lumen through the capillary wall to the interstitial matrix and then back to the capillary lumen. Fluids leave the capillary near its arterial end and return to it near its venous end; en route, the fluids pass close to the metabolizing tissue cells in the interstitial matrix.

Both the cellular layers surrounding the capillary lumen are extremely permeable to water, dissolved gases, nutrients, and wastes. They are much less permeable, however, to dissolved proteins and totally

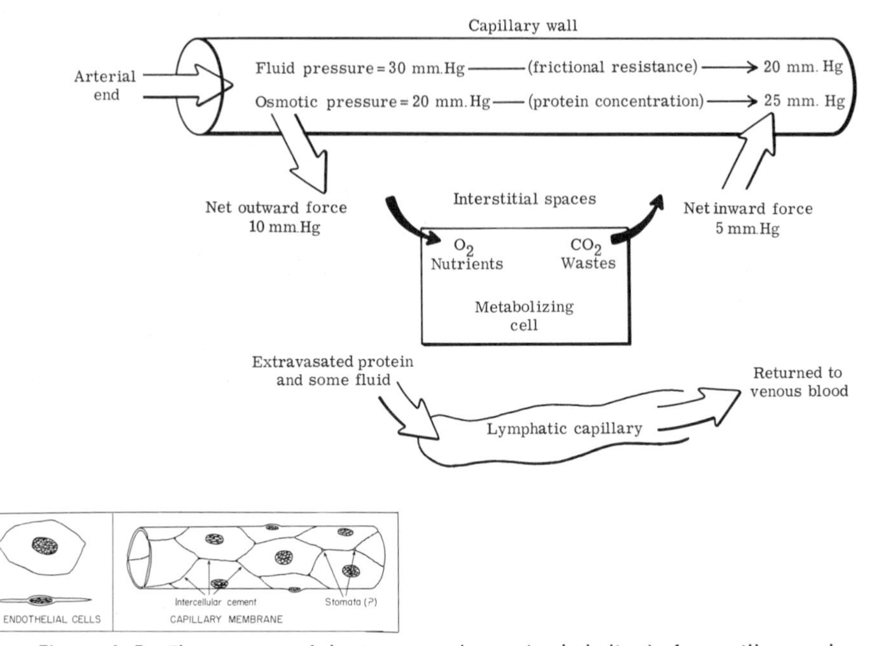

Figure 4–5 The structure of the inner membrane (endothelium) of a capillary and a diagrammatic representation of the forces involved in fluid exchanges between capillaries and tissue spaces. (From Rushmer: *Cardiovascular Dynamics,* 3rd Edition. Philadelphia, W. B. Saunders Company, 1970, and from Cockrum and McCauley: *Zoology.* Philadelphia, W. B. Saunders Company, 1965.)

impermeable to red blood corpuscles. Thus, the red blood corpuscles and most of the proteins are retained within the capillary while a portion of the blood plasma, together with various dissolved substances, passes freely through the capillary walls.

The movement of fluid in and out of the capillary depends upon delicate balances between the hydrostatic pressures (the pumping pressure of the heart, gravity, and so forth) and the osmotic pressures (primarily due to the presence of dissolved proteins). We can assume that both kinds of pressure are constant outside the capillary in the interstitial spaces. Both kinds of pressure within the capillary, however, undergo changes as the blood flows through the vessel (Fig. 4-5). As blood enters the arterial end of an average human capillary, it has a hydrostatic pressure of about 30 millimeters of mercury. As a result of the high frictional resistance of the capillary, however, there is a pressure drop of about 10 millimeters of mercury, so a pressure of only 20 millimeters of mercury remains by the time the blood reaches the venous end of the capillary. Hydrostatic pressure is an outwardly directed force; that is, it tends to force fluids out from the capillary lumen and into the interstitial matrix. In the arterial end of an average capillary the osmotic pressure amounts to about 20 millimeters of mercury (Sect. 1.4). Since the hydrostatic pressure drives fluid, but not dissolved protein, out from the capillary, the concentration of dissolved protein increases in the blood. This causes osmotic pressure to increase to about 25 millimeters of mercury at the venous end of the capillary. It will be recalled that osmotic pressure is proportional to the concentration of such dissolved molecules in a fluid. The osmotic pressure is an inwardly directed force; that is, it tends to draw fluids from the interstitial spaces into the capillary.

Now, at the arterial end of the capillary there is an outward (hydrostatic) force of 30 millimeters of mercury and an inward (osmotic) force of 20 millimeters of mercury. This leaves a net outward force of 10 millimeters of mercury; thus fluid leaves the capillary. At the venous end of the capillary, however, the outward (hydrostatic) force has decreased to 20 millimeters of mercury and the inward (osmotic) force has increased to 25 millimeters of mercury. Thus, at this end of the capillary there is a net inward force of 5 millimeters of mercury; thus fluid reenters the capillary from the interstitial matrix.

We have mentioned that most of the dissolved protein remains in the capillary and this accounts for the increase in osmotic pressure there. The capillary wall is not entirely impermeable to protein, however, so a small amount does pass out with the fluids and enters the interstitial matrix. If this escaped protein were allowed to accumulate in the interstitial matrix, it would exert osmotic pressure there and soon upset the delicate balance of forces described in the preceding paragraph. The escaped protein does not reenter the capillary at its venous end; some other means is required to remove it from the interstitial matrix. This role is played by the lymphatic system. Lymphatic capillaries, lying scattered throughout the tissues, have blind beginnings and extremely permeable walls. The proteins that have escaped from blood capillaries, together with small amounts of fluid, easily enter the lymphatic capillaries through their walls. The content of lymphatic capillaries flows into larger lymphatic vessels and ultimately, through the thoracic duct or the right lymphatic duct, it is passed back into the venous blood at points near the heart. Thus the escaped protein is ultimately returned to the venous blood by way of the lymphatic route. In a sense the lymphatic system can be thought of as an accessory venous system. The return of lost blood protein was undoubtedly its primitive role, although it has also acquired other functions in vertebrate evolution. During vertebrate embryology it arises in conjunction with the venous system.

Any circumstance which upsets the balance of hydrostatic and osmotic forces will also upset the proper exchange of fluids between the capillaries and the interstitial matrix. Proteins may collect in the interstitial matrix because of injury which increases the permeability of the capillary walls to protein. This may also occur if venous or lymphatic drainage is interfered with. In such cases, the osmotic pressures outside the capillary soon overbalance those inside and as a result fluids do not reenter the capillary in a normal fashion. This leads to an accumulation of fluids in the interstitial matrix, a condition known as edema. Edema is characterized by swelling of inflamed areas or regions in which circulation is sluggish. Protein starvation can also effect capillary-tissue exchanges of fluid by reducing the normal quantities of protein dissolved in the blood plasma. This also leads to edema because the osmotic pressure of the blood is too low to effectively withdraw fluids from the interstitial matrix.

4.6 VENOUS RETURN TO THE HEART

By the time the circulating blood has reached the veins the total pressure drop has been so great that very little hydrostatic pressure remains (Fig. 4-4). Virtually all of the pumping force of the heart has been "used up" in overcoming the frictional resistance to flow through the arteries and capillaries. Further force is required to raise the blood back to the level of the heart (venous return), and this additional force must be supplied from some other source. This fact is of little importance to the lower vertebrates since most parts of the body are nearly on the same level with the heart and the veins are large, low-resistance vessels. In the higher quadruped vertebrates, however, blood in the extremities may be well below the level of the heart. In a human in a standing position, moreover, a large part of the trunk is also below the heart.

When a skeletal muscle contracts, it shortens but does not decrease in volume, and consequently the belly of the muscle swells. Within the restricted space in a limb such swelling of muscles results in compression of veins (Fig. 4-6). The veins of the extremities are equipped with valves which permit the movement of blood in the upward direction only. The result is that each muscular contraction in the extremities

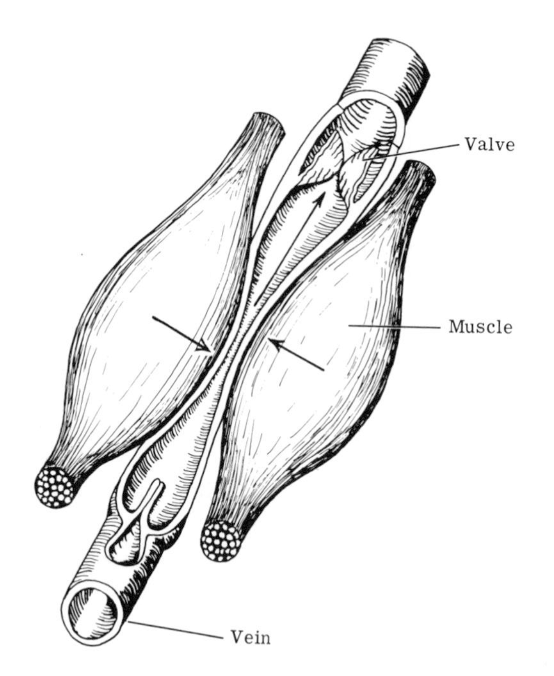

Figure 4-6 The return of venous blood from the extremities produced by the milking action of muscles. (From Cockrum and McCauley: *Zoology*. Philadelphia, W. B. Saunders Company, 1965.)

boosts some blood toward the trunk. Once it has reached the trunk of a quadruped animal, it is approximately on the same level as the heart. For these animals, then, the milking action of muscles in the extremities is generally sufficient to account for the return of venous blood to the heart.

In quadrupeds with long, thin legs like those of the horse, for example, the greater part of each extremity contains tendons but no actual muscular tissue. The muscle bellies are located only in the proximal (upper) portion of the extremity. In such cases the milking action of muscles is less effective and is supplemented by a different mechanism. The hoof of the horse contains a compressible vascular bed, and when the animal places its weight on a hoof, the blood is squeezed from this vascular tissue upward into the veins. Valves in the veins prevent its return as in other quadrupeds. For this reason a sweating, overheated horse should be walked about to maintain circulation and thus permit efficient cooling of the animal.

Respiratory activity also aids in the return of venous blood to the heart. This mechanism is significant only in mammals, since these are the only vertebrates with a well-developed diaphragm separating the thoracic and 'abdominal cavities. It is of particular importance in man, since standing maintains the trunk in a vertical position. In this position, blood returned from the lower extremities by the milking action of muscles reaches a part of the trunk that is still far below the level of the heart.

It will be recalled (Sect. 3.8) that inhalation of air occurs when the thoracic volume is actively increased by the contraction of muscles which raise the ribs and by the contraction of the diaphragm (Fig. 4-7). This action lowers the pressure within the thorax and, consequently,

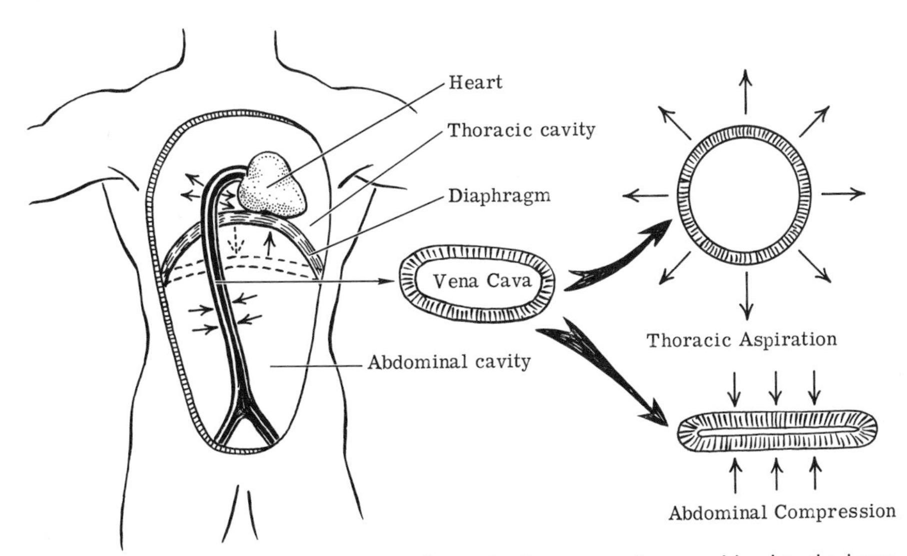

Figure 4–7 The role of lung ventilation in the return of venous blood to the heart (see text).

expands the lungs. The lowering of thoracic pressure also expands the major veins leading to the heart and, therefore, draws blood into them from outside the thoracic cavity. Also, the lowering of the diaphragm compresses the contents of the abdominal cavity (intestines) so that the veins in this region are compressed as well. Thus, at the same time that blood is drawn into the expanding thoracic veins it is driven out of the compressed abdominal veins. Both actions serve to boost it upward through the trunk to the level of the heart.

The lymphatic vessels (Sect. 4.5) constitute another low-pressure, low-resistance system much like the venous system. The same forces that aid the return of venous blood to the heart also propel the lymph in the same direction. Lower vertebrates may possess small, pulsating accessory lymph hearts that are located in either the pelvic or the pectoral regions or in both places.

CHAPTER 5

CIRCULATION: VERTEBRATE CARDIOVASCULAR SYSTEMS

THE MAMMALIAN CARDIOVASCULAR SYSTEM

5.1 INTRODUCTION

We have seen that the mammalian cardiovascular system, like that of other vertebrates, consists of a heart that pumps blood through two closed circuits, the pulmonary circuit and the systemic circuit, each of which consists of arteries carrying blood away from the heart, veins returning it to the heart, and capillaries connecting the arteries to the veins.

The portion of the vertebrate circulatory system that has always attracted major attention is, of course, the heart. It, even more than the brain, has classically been regarded as the center of life and personality. Throughout much of history it was supposed that bravery, morality, compassion, and perhaps even the soul itself resided in the heart. Modern investigation has dispelled such ideas, but the heart remains in the spotlight of both popular and scientific interest. The study of the hearts of lower vertebrates, moreover, has demonstrated an evolutionary history of great importance in the development of modern vertebrate forms. Most of what is known of cardiac function, however, has been learned from the study of mammals, so we shall begin with a discussion of the mammalian heart.

As a brief comparison will demonstrate, the human heart is a most impressive pump. The fuel pump of an automobile is a pump of comparable size but far less efficiency. If an automobile consumes gasoline at the rate of one gallon every 20 miles, at 60 miles per hour the fuel pump will handle about 72 gallons in 24 hours of continuous driving. Probably the average fuel pump will require replacement after 30 or 40 thousand miles, but even if it lasts for 100 thousand miles it will only have pumped about 5000 gallons of gasoline. By comparison, the human heart pumps about 70 milliliters of blood per beat, and it beats about 70 times per minute. At this rate it pumps more than 140,000 gallons of blood during the average human lifetime! Replacement of the heart has only been possible in recent years and is only minimally successful still. Without surgical or medicinal help, however, it alters its structure to compensate for various kinds of damage to its parts and it effects minor repairs upon itself.

5.2 THE HEART: GENERAL CONSIDERATIONS

The mammalian heart is a double pump, each half of it serving a separate portion of the circulation (Fig. 5-1). The right half receives blood from the general systemic circuit and pumps it to the lungs. The left half accepts blood returning from the lungs and pumps it into the general systemic circuit again. The two halves of the heart are each, in reality, separate pumps but they operate in unison with each other.

Blood from the systemic circuit enters the right atrium of the heart and blood from the lungs enters the left atrium at the same time (Fig. 5-2). From the atria the blood flows into the corresponding ventricles on the two sides of the heart. The ventricles contract in unison expelling the blood into the arteries; blood from the right ventricle flows into the pulmonary artery leading to the lungs and blood from the left ventricle enters the aorta. Blood from the lungs returns to the left atrium through the pulmonary veins and blood from the general systemic circuit returns to the right atrium by way of the venae cavae. Valves within the heart prevent the blood from flowing through it in the opposite direction.

Necessarily, the volume of blood pumped by the right side of the

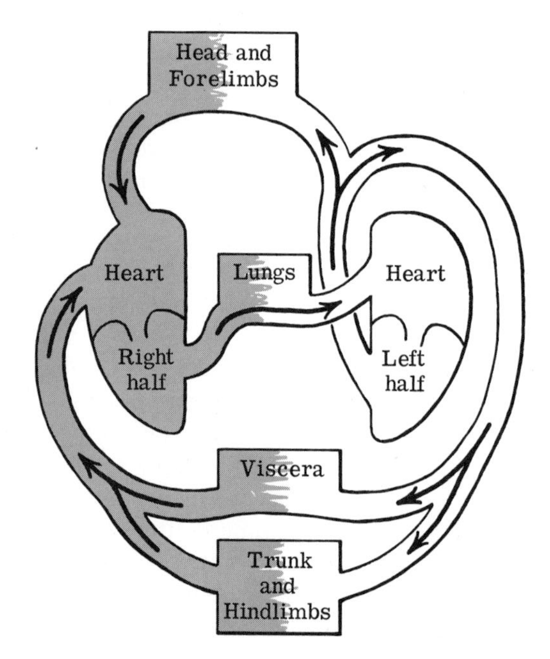

Figure 5-1 The mammalian heart is actually a double pump. The right side receives blood from the general systemic circuit and pumps it into the pulmonary circuit. The left side receives it from the pulmonary circuit and pumps it into the general systemic circuit.

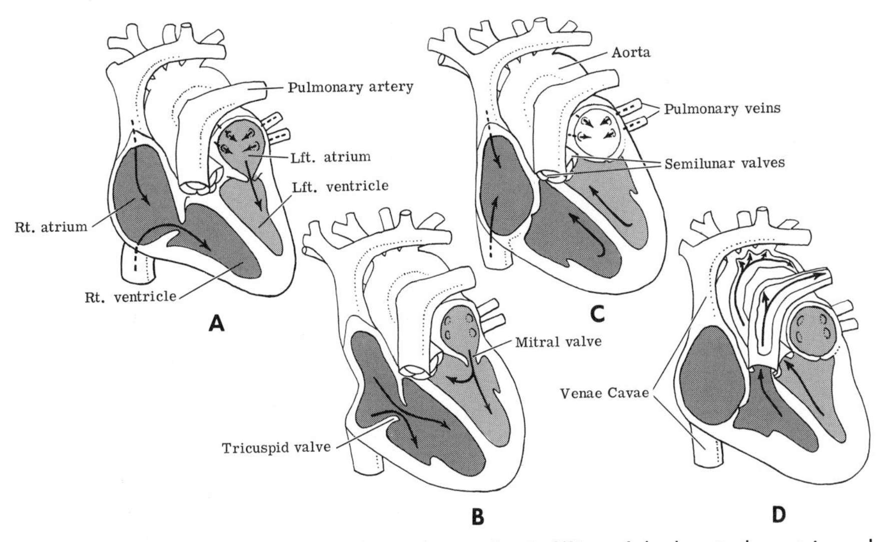

Figure 5–2 Stages in the human cardiac cycle. *A,* filling of the heart when atria and ventricles are both in diastole; *B,* atrial systole; *C,* beginning of ventricular systole (meanwhile the atria are again in diastole and filling); *D,* continuing ventricular systole and emptying. (After Cockrum and McCauley: *Zoology.* Philadelphia, W. B. Saunders Company, 1965.)

heart must equal, over any short period of time, the volume pumped by the left side of the heart. This is necessary if both the pulmonary and the systemic circuits are to remain sufficiently but not overly filled. If the right side of the heart pumped a larger volume than the left side, the pulmonary circuit would become overfilled and the systemic circuit would be depleted of its normal content of blood. This would result in increased pulmonary blood pressure and decreased systemic blood pressure. As we shall see, such changes would have drastic effects upon the physiological functioning of both circuits.

Since the pulmonary circuit is much shorter than the systemic circuit, its normal content of blood is smaller in volume. Moreover, since this imposes less frictional resistance to blood flow, there is less pressure drop through the pulmonary circuit than through the systemic circuit. This means that the right side of the heart need not be as muscular or develop as much pumping force as the left side. This difference is reflected in the comparative thickness of the right and left ventricular walls and in other details of their structure (Fig. 5-12).

The contraction of the chambers of the heart is called systole and relaxation is called diastole. During systole of the ventricles, blood is ejected into the pulmonary artery and the aorta, and the blood pressure in those arteries rises to a maximum. The maximum arterial pressure is therefore spoken of as the systolic pressure (Fig. 5-3). While the ventricles are relaxed and are being refilled with blood from the atria, the pressure in the arteries falls as a result of the run-off of blood into the capillaries and veins. Arterial blood pressure thus decreases to a minimum just before the next ventricular systole. This minimum arterial pressure is called the diastolic pressure because it coincides with ven-

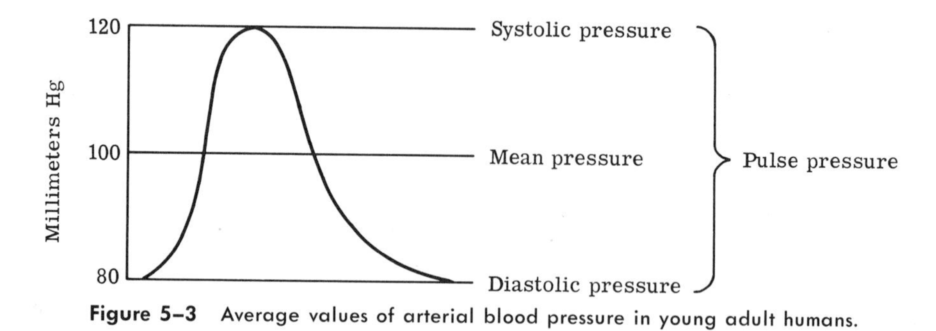

Figure 5–3 Average values of arterial blood pressure in young adult humans.

tricular diastole. In young adult humans the aortic blood pressure fluctuates, on the average, between 120 millimeters of mercury (systolic) and 80 millimeters of mercury (diastolic). Rather wide variations from the average occur in normal individuals, depending upon the degree of physical training and other factors. Generally these pressures are recorded as systolic/diastolic or, in the average example cited earlier, as 120/80. The change in pressure that occurs with each pulse wave is designated as the pulse pressure and averages 40 millimeters of mercury. The average of both the systolic and diastolic pressures is called the mean pressure; this would be 100 millimeters of mercury in the example given above. By comparison, the pulmonary arterial pressure amounts to only 22/8 millimeters of mercury in the average young adult human; that is, the pulmonary systolic pressure is 22 millimeters of mercury and the pulmonary diastolic pressure is only about 8 millimeters of mercury.

5.3 THE HEART VALVES

Each side of the heart has a valve between the atrium and the ventricle known as the atrioventricular valve. The atrioventricular valves of the two sides of the heart are similar in structure except that the right valve has three cusps or flaps while the left valve has only two. The right valve is, accordingly, called the tricuspid valve and the left valve is called the bicuspid valve. A more commonly used name for the left atrioventricular valve is the mitral valve. This name comes from the supposed resemblance of this valve to the miter (hat) worn by the bishops of liturgical Christian churches. Both atrioventricular valves are somewhat funnel-shaped structures (Fig. 5-4). The free edges of their cusps hang downward into the ventricular cavities and are anchored by tendon-like strands of connective tissue known as chordae tendineae. At the point of attachment of each of the chordae tendineae to the ventricular wall there is a small hillock of muscle called a papillary muscle.

As the ventricles contract in systole, increasing pressure within them forces the cusps of the atrioventricular valves together. This closes these valves and prevents the backflow of blood into the atria. If it were not for the presence of the chordae tendineae anchoring the valve cusps to the sides of the ventricle walls, the pressures developed

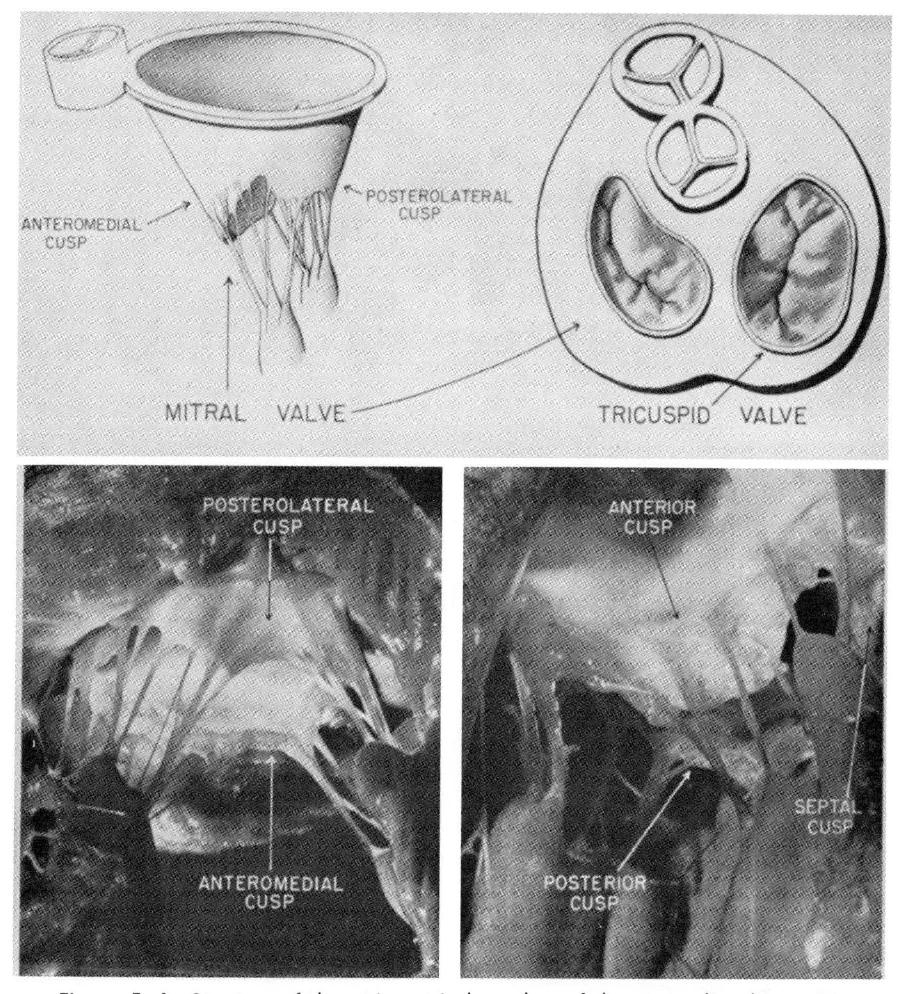

Figure 5–4 Structure of the atrioventricular valves of the mammalian heart. (From Rushmer: *Cardiovascular Dynamics,* 3rd Edition. Philadelphia, W. B. Saunders Company, 1970.)

in the ventricles would be sufficient to turn the atrioventricular valves inside out and evert them into the atrial cavities. During ventricular contraction, the ventricles shorten and their diameters are decreased. The simultaneous contraction of the papillary muscles (Fig. 5-5) prevents any slack from appearing in the chordae tendineae as a result of ventricular shortening. Otherwise such slack might be sufficient to allow the valves to be everted into the atrial chambers. When the ventricles relax, the pressure within them decreases, and the atrioventricular valves are allowed to open again. This permits blood to flow into the ventricles from the atria and from the venae cavae and pulmonary veins.

The pulmonary artery and the aorta each have a valve at the point of exit from the corresponding ventricles. These are the semilunar valves and they are quite similar in both cases, but they are totally

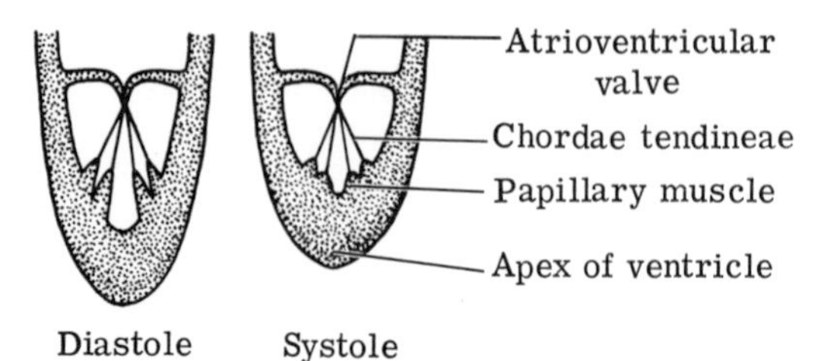

Figure 5–5 The function of the papillary muscles of the heart. See text.

unlike the atrioventricular valves in structure. Each semilunar valve is composed of three pocket-like cusps (Fig. 5-6) with the pocket opening directed away from the ventricular chamber. These pockets fold outward against the arterial walls when the valves are opened and they fit neatly together to block the lumen when the valves are closed. It is interesting to note in passing that a bicuspid valve of similar design would close just as well but could not be opened as completely. Similarly, a tetracuspid valve of the same type would only add further complexity without materially improving the opening or closure. Three cusps of this type appears to be the ideal arrangement.

When the ventricles contract in systole, the increasing pressure within them forces the cusps of the semilunar valves outward and against the arterial walls so that blood can be ejected into the pulmo-

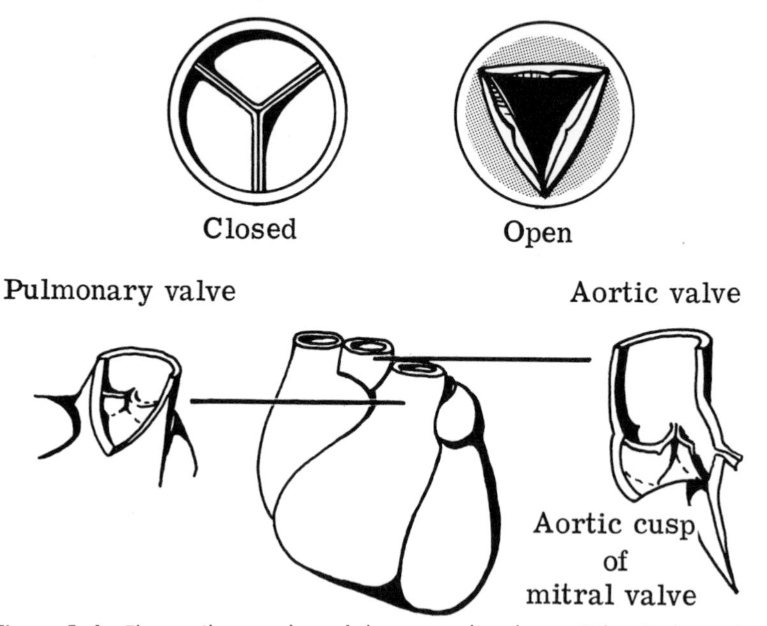

Figure 5–6 The semilunar valves of the mammalian heart. (After Rushmer: *Cardiovascular Dynamics*, 3rd Edition. Philadelphia, W. B. Saunders Company, 1970. Reprinted from McMillan and Daley: *The Action of Human Mitral and Aortic Valves Studied Postmortem by Cinematography*. Presented at the Second World Congress of Cardiology, Washington, D.C., 1954.)

nary artery and aorta. When the ventricles relax in diastole, the pressure within the ventricles falls, and as soon as it is less than the pressure within the arteries, the backwash of arterial blood attempting to reenter the ventricles catches in the pockets of the valves and snaps them shut.

It is important to recognize that all the valves of the heart are opened and closed by the differential fluid pressures on their two sides; they are not muscular structures capable of independent action.

5.4 THE CARDIAC CYCLE

Since the action of the heart is a continuous cyclic function with regularly alternating systole and diastole, we could designate any part of it as the beginning of the cycle. It is convenient, however, to begin with the ventricular systole (Fig. 5-7). As the ventricles begin to contract the pressure within them rises, and almost immediately the atrioventricular valves are closed as a result. The semilunar valves, however, remain closed for a time. Before they can open it is necessary for

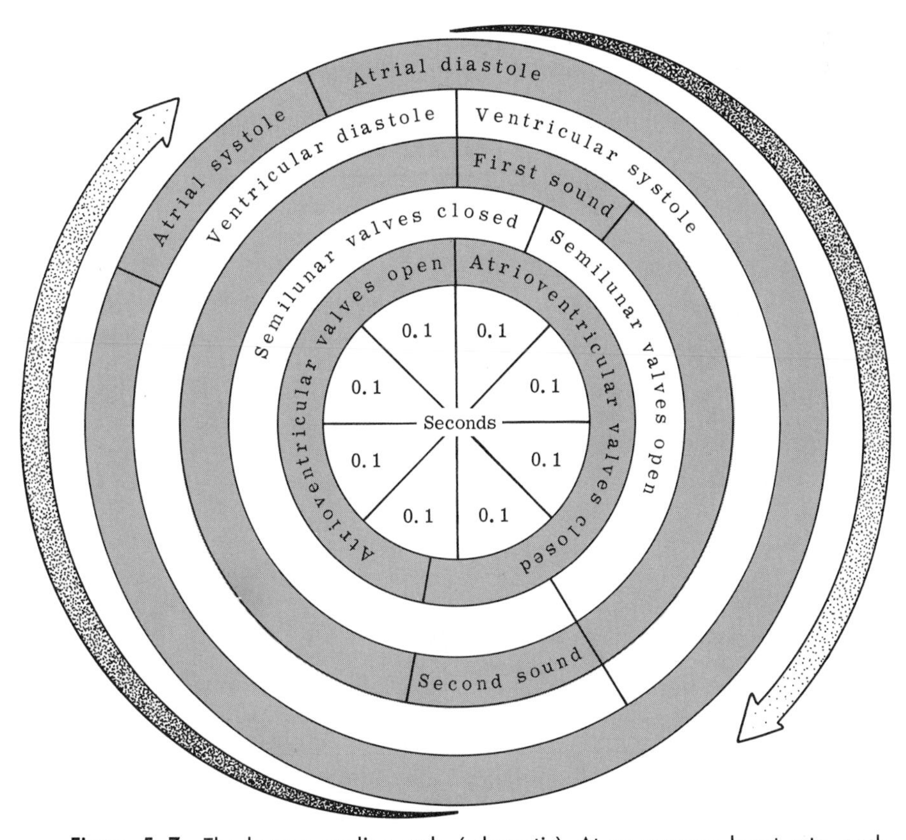

Figure 5–7 The human cardiac cycle (schematic). At an average heart rate, each beat occupies about 0.8 second, as shown in the center of the diagram. (After Cockrum and McCauley: *Zoology*. Philadelphia, W. B. Saunders Company, 1965.)

the pressure within the ventricles to rise until it exceeds the pressure in the arteries. That is, right ventricular pressure must be greater than the diastolic pressure in the pulmonary artery, and the left ventricular pressure must exceed aortic diastolic pressure before the semilunar valves can be opened. The first part of ventricular systole, therefore, produces an increase in pressure within the ventricle but no change in ventricular blood volume. This is called the isometric phase of ventricular systole (Fig. 5–8).

As soon as the semilunar valves are forced open, continued contraction of the ventricles rapidly drives the blood into the pulmonary artery and aorta. This is known as the isotonic phase of ventricular systole. As the ventricles approach the end of systole, the rate of blood ejection decreases and ventricular pressures begin to fall again. At this time the run-off of arterial blood into the capillaries and veins is also rapid, so arterial pressures also decrease. At first, the falling arterial pressure remains slightly lower than the falling ventricular pressure, and because of this the semilunar valves remain open, and blood ejection from the heart continues at a slackened pace. Ultimately, however, ventricular pressure falls below arterial pressure. When this happens arterial blood begins to move backward toward the ventricles. This backwash catches in the pockets of the semilunar valves, causing them to close, as we have already noted.

The contraction of the ventricles and the resulting turbulence in the blood produce a long, low-pitched sound detectable through a

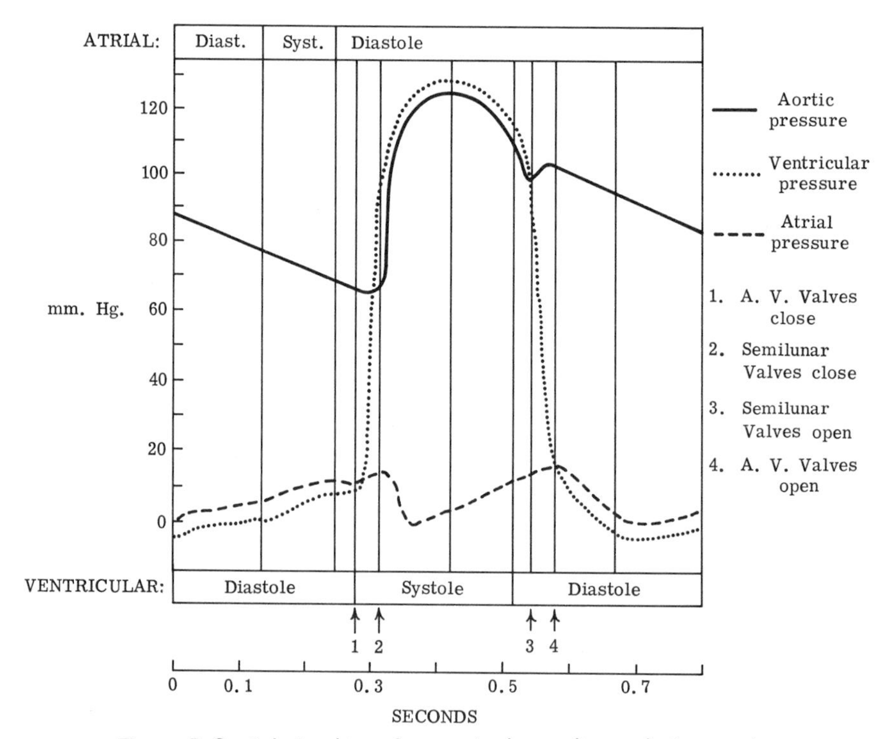

Figure 5–8 Relationships of events in the cardiac cycle (see text).

stethoscope and frequently described as sounding like the syllable "lub." This is the first heart sound. The second heart sound is a relatively short, high-pitched "dup" caused by the sudden closure of the semilunar valves. This closure also causes a small positive wave to appear in the arterial pulse. It can be recorded with appropriate equipment and is called the dicrotic wave (Fig. 5-3).

During ventricular diastole the relaxation is rapid at first, owing to the elastic nature of the heart muscles. For a time, however, the pressures in the ventricles still exceed those in the atria so that the atrioventricular valves remain closed (Fig. 5–8). Consequently, there is a period of isometric relaxation during which ventricular pressure falls but ventricular volume remains unchanged. Once ventricular pressure falls below atrial pressure, the tricuspid and mitral valves open and blood flows in from the atria and the venous drainage.

During all the preceding periods of ventricular systole and the isometric phase of ventricular diastole, venous blood has been accumulating in the atria. When the atrioventricular valves open, ventricular filling is at first very rapid. Slower ventricular filling follows as more venous blood arrives and flows directly through the atria. Finally the atria contract briefly. This atrial systole adds little or nothing to ventricular filling since the ventricle fills thoroughly anyway. It does, however, prepare the atria, by essentially emptying them, to receive blood continuously from the veins during the next period of ventricular systole and until the atrioventricular valves open again. Without the atria to receive blood during this time, venous drainage into the heart would have to stop completely, and stopping and starting with each cardiac cycle would greatly reduce the efficiency of the circulation. In the absence of a normal human atrial contraction, which sometimes occurs in certain pathological conditions, ventricular output is reduced by about 40 per cent.

5.5 CARDIAC OUTPUT

The volume of blood that is pumped by the heart each minute is called the minute-volume. This depends upon the number of beats per minute (stroke rate) and the volume of blood pumped with each beat (stroke volume). It also depends upon the volume of blood returned to the heart from the veins each minute, since, quite obviously, the heart cannot pump any more blood than comes to it. If, however, we assume that the venous return volume is adequate, then increases in stroke rate will result in increases in minute-volume. The increase in minute-volume is not, however, proportionate with the increase in stroke rate. A doubling of the stroke rate does not result in a doubling of the minute-volume of output. The reason for this is that increased stroke rates involve a decrease in the ventricular filling time, and as a result the ventricles do not fill thoroughly. Thus, the stroke volume is decreased. At very rapid heart rates the stroke volume may be decreased so much that it completely offsets the increased stroke rate. Under these conditions the minute-volume of the heart may even be reduced to values below normal.

During systole the ventricles do not completely empty themselves. A small residual amount of blood remains in them. With more complete filling of the ventricles, the succeeding beat is more forceful and blood ejection is more complete. Consequently, an increase in venous return volume may result in an increase in minute-volume of cardiac output, and this may occur even when there is no change in stroke rate. This effect of increased venous return volume is known as the Starling effect. It is probably of little value in increasing cardiac output in mammals, but there is reason to believe that it is of real significance in the hearts of lower vertebrates (Sect. 5.14). It is possible that in humans it does serve to compensate for stroke-by-stroke differences in the output of the two ventricles. As we have seen, the output of the two sides of the heart must be equal over any short period of time (Sect. 5.2).

The same conditions which usually call for increases in the minute-volume of cardiac output (*e.g.*, exercise) will also increase the rate and depth of breathing (Sect. 3.9) and the amount of muscular milking action on the veins. Both of these effects increase the amount of venous return of blood to the heart (Sect. 4.6). Also, during exercise the smooth muscles in the walls of the veins contract, thus reducing the venous capacity. This too drives more blood from the veins to the heart. At rest the veins contain about 70 per cent of the total blood volume of the body, so this response is an important one. Similar contraction of smooth muscles in the walls of arteries raises the blood pressure (Sect. 5.6) and the velocity of arterial blood flow. This is accomplished at the cost of increased resistance to arterial blood flow (Sect. 4.4) and, therefore, an increased work load on the heart. The total effect of exercise, then, is to raise the blood pressure and speed both arterial and venous flow at the very time when the increased tissue demands for oxygen and nutrients require it. Venous return of blood also increases at the very time when there is demand for increased cardiac output, thus providing the heart with the extra volume of blood required for pumping.

The rate of heart beat is under the control of the central nervous system as are the smooth muscles of the arteries and veins. This is demonstrated by the fact that adjustments similar to those discussed in the preceding paragraph may take place without any actual exercise occurring. Fear, anger, and excitement are emotions associated with possible anticipated exercise and they bring about the same responses in the respiratory and cardiovascular systems. For example, an athlete poised to begin a foot race shows an increase in rate and depth of breathing, heart rate, and blood pressure before the starting gun has been fired. The same changes in physiological function have been very marked in astronauts preceding and during the blast-off into earth orbit even though they were not engaged in any degree of exercise at the time.

5.6 ARTERIAL BLOOD PRESSURE

It is often convenient to illustrate physiological mechanisms by means of analogous mechanical models (Fig. 5-9). The pump in the model illustrated is an ordinary rubber bulb equipped with valves that permit water to pass through it in only one direction. Alternately squeezing and releasing the bulb with the hand draws water from the reservoir and forces it through any one of the four tubes we may select. If we turn the valves so that water passes through tube number one only, each squeeze of the pump will result in a spurt of water issuing from the end of that tube. When we release the bulb, however, and it begins to refill from the reservoir, water will cease to flow from the end of the tube. If the cardiovascular system operated in this way, it would mean that blood flow would occur only during systole of the heart and would cease during diastole. This, of course, is not the case.

If tube number two is used, the effect will be the same except that we shall have to squeeze harder on the bulb to force the water through the restricted jet opening. As a result the water pressure within the tube will be higher during "systole" but it will drop to zero during "diastole." Actual water flow will still be intermittent, occurring only during "systole."

The behavior of tube number three will be like that of tube number one despite the fact that it is made of thin rubber rather than of glass.

Finally, if tube number four is used, water can be squeezed out of the pump more rapidly than it leaves the system through the terminal jet opening. This is true because the thin rubber tube is elastic and can expand to accept the additional water volume. Then, while the bulb is refilling from the reservoir, the elastic return of the rubber tube to its original volume will maintain a flow of water through the jet opening. During "systole" the water pressure in the tube will rise to a maximum value since it also is being compressed by the elastic properties of the tube. During "diastole" the water pressure will fall as the tube approaches its natural diameter and its compressing force lessens. If pumping is maintained at a sufficient "stroke rate," however, the water pressure will never fall to zero as it did in the other three tubes. These

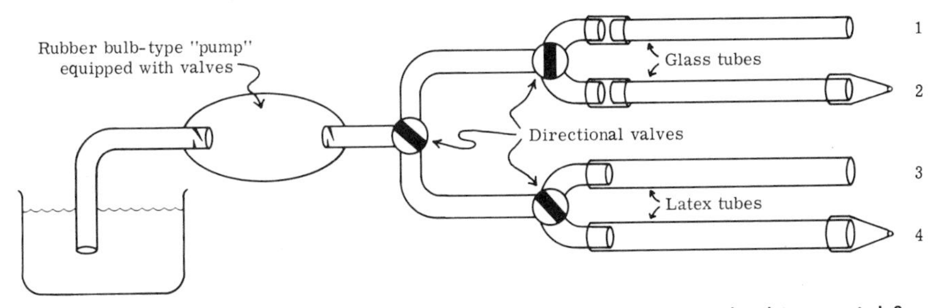

Figure 5-9 A model illustrating the importance of factors involved in arterial flow of blood. See text, Section 5.6. (From Cockrum and McCauley: *Zoology*. Philadelphia, W. B. Saunders Company, 1965.)

are the actual characteristics of the cardiovascular system. The elastic arteries expand during ventricular systole to accept blood more rapidly than it can run off through the high-resistance capillary beds (equivalent to the jet tip in our model). During ventricular diastole, then, the elastic recoil of the arteries maintains an uninterrupted flow of blood through the capillaries. The simultaneous presence of end resistance and arterial distensibility (elasticity) is necessary to produce this continuous flow as the experiments with the model demonstrate. In the absence of either of these two factors the flow of blood would be intermittent, and the pressure during diastole would fall to zero. The extra energy required to overcome the inertia of the blood flow in repeatedly starting and stopping it would be a large extra burden on the heart and would also decrease the efficiency of the circulatory system in transporting gases, nutrients, and wastes.

Arteriosclerosis is a common human pathological condition involving the deposition of materials in the walls of the arteries. As these materials become calcified they decrease the size of the lumen and also decrease the elasticity of the arterial walls. The effect of this disease is somewhat like replacing the rubber tube (number four) with the glass tube (number two) in the model (Fig. 5-9). Clinical shock, on the other hand, is a condition in which the smooth muscles in the walls of blood vessels completely relax and all of the capillaries are opened (Sects. 4.3 and 5.7). This is analogous to removing the jet tip by substituting tube number three in the model, but it is more serious since it also involves a large increase in the capacity of the cardiovascular system. As a result, circulation is decreased to very low levels.

Arterial blood pressure is dependent upon the factors just discussed (arterial elasticity and terminal resistance) as well as upon the volume of cardiac output discussed in the preceding section. Adjustments of cardiac rate and of arterial muscle contraction, then, are effective in altering arterial blood pressure. Both of these are physiological mechanisms found in vertebrate animals.

5.7 CAPILLARY CIRCULATION

We have mentioned that muscle contractions in the walls of selected arterioles will shunt blood from one place to another in the body (Sect. 4.3). A simultaneous relaxation of the arterioles in an organ or tissue can cause the shunting of blood to that area without producing any overall change in arterial blood pressure. In addition, the total number of capillaries open in any one organ or tissue is also variable. Precapillary sphincters (muscular valves) can shut off the flow to portions of a capillary bed and thus direct greater volumes of blood to other portions of the same bed (Fig. 5-10). In exercising muscle there are 20 to 50 times as many open capillaries as in resting muscle. Thus, both the route and the volume and velocity of blood flow can be varied locally or in the cardiovascular system as a whole.

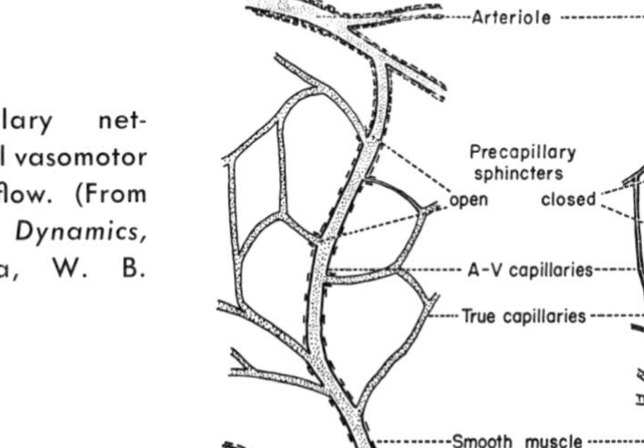

Figure 5-10 Capillary networks and the results of local vasomotor changes affecting blood flow. (From Rushmer: *Cardiovascular Dynamics,* 3rd Edition. Philadelphia, W. B. Saunders Company, 1970.)

5.8 CORONARY BLOOD FLOW

Quite obviously, the work done by the muscles of the heart is considerable, as are its demands for oxygen and nutrients. The thick myocardium (heart muscle) is supplied with its own arterial supply and venous drainage, a system called the coronary circulation. The two coronary arteries open from the aorta just beside two of the cusps of the aortic semilunar valve. These subdivide to supply all portions of the myocardium (Fig. 5-11). A system of coronary veins drains the capillary

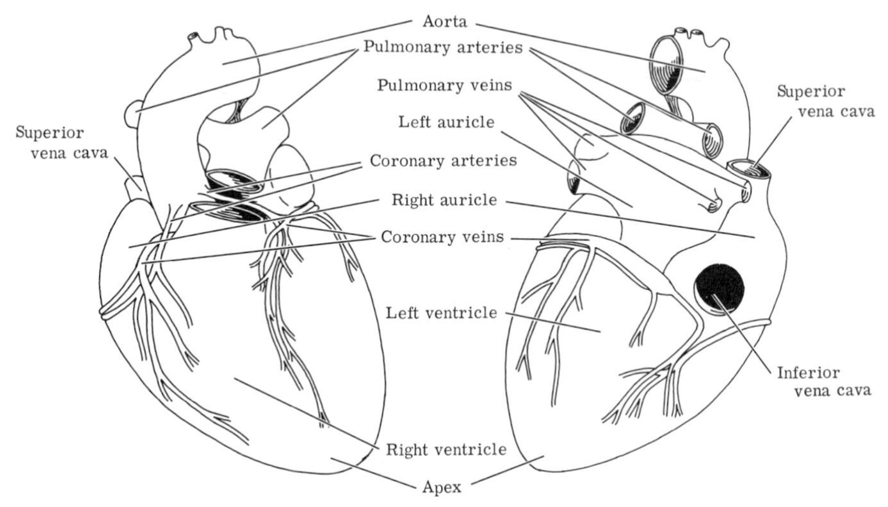

Figure 5-11 The external features of the human heart; ventral and dorsal views. The term *auricle* is synonymous with the term *atrium.* (From McCauley: *Human Anatomy: A Lecture and Laboratory Manual.* Minneapolis, Burgess Publishing Company, 1962. Reprinted in Cockrum and McCauley: *Zoology.* Philadelphia, W. B. Saunders Company, 1965.)

beds of the myocardium into an enlarged coronary venous sinus, which in turn drains into the right atrium of the heart.

The coronary arteries do not receive the full thrust of ventricular blood ejection since the openings to the arteries are sheltered behind the cusps of the semilunar valve at the time of ejection. With closure of the semilunar valve following systole, however, the backthrust of arterial blood (Sect. 5.4) fills the coronary arteries. At this time the ventricles are relaxing and the coronary arteries are easiest to fill. During systole the coronary arteries are compressed by the contracting myocardium and are harder to fill.

During ventricular systole, the contracting myocardium effectively milks the venous blood from the coronary veins, preparing them to receive more blood from the capillaries. Thus, the entire coronary system very effectively meets the enormous demands of the myocardium.

During exercise the oxygen requirements of the heart may amount to as much as 10 per cent of the total oxygen demand of the body. Under such conditions, coronary blood flow can be increased by a factor of eight or more over average resting conditions.

As a result of arteriosclerosis (Sect. 5.6), one or more branches of the coronary arteries may become partially plugged so that the areas of myocardium served by these branches are denied sufficient amounts of blood. Exercise, then, may cause pain (angina pectoris) or even a coronary heart attack. In coronary heart attacks the deprived sections of the myocardium actually suffer permanent damage through the death of cells. These damaged areas are replaced by scar tissue leaving the heart permanently weakened. Such damaged areas are known as infarctions, and if sufficiently extensive, they may cause death. Coronary heart attacks may occur suddenly and without warning, even when exercise is not involved, if sclerotic portions of coronary arteries break free in the blood stream. Such drifting materials are called thrombi, and when they reach a vessel smaller than themselves, they form an instant plug that halts blood flow through that vessel.

5.9 PULMONARY BLOOD FLOW

As we have seen, the flow of blood through the pulmonary arteries and veins equals in minute-volume the flow through the aorta and systemic circuit (Sect. 4.3). It is carried out, however, under much lower pressure (Sect. 4.4). High pressures are not necessary in such a relatively short, low-resistance system. Moreover, elevated hydrostatic pressures in the capillaries associated with alveoli of the lungs would have adverse effects. It will be recalled that a delicate balance between hydrostatic and osmotic pressures results in fluids actually leaving the capillaries to enter the interstitial matrix of most tissues (Sect. 4.5). If such fluids left the pulmonary capillaries, they would enter the alveoli of the lungs. The addition of even a little water there would decrease the alveolar air capacity and would also increase the diffusion distance between that air and the blood. Both effects would seriously reduce the efficiency of respiratory gas exchange. Since osmotic pressure of blood is the same in the pulmonary capillaries as in the arterial ends of other capillaries of the body, the hydrostatic pressure must be lower to prevent such fluid loss. Instead, the existing balance between hydrostatic and osmotic pressures in pulmonary capillaries is such that any water present in the alveoli would tend to be drawn into the capillary blood.

In cases of drowning in fresh water, water enters the lungs but is quickly drawn into the bloodstream as a result of osmotic attraction. This results in decreasing, by dilution, the osmotic pressure of the blood, which in turn causes all kinds of osmotic tissue damage throughout the body. Drowning in salt water has somewhat different effects. Because salt water has a considerable osmotic pressure of its own, it is not drawn into the blood but remains in the alveoli and passages of the lungs. If sufficiently salty, it may even result in drawing blood plasma out from the capillaries and into the alveoli.

In addition to the blood passing through capillaries associated with the alveoli—blood which enters the lungs in the pulmonary arteries—the lungs also receive another blood supply. Branches from the aorta enter the lungs and supply lung tissues which are more remote from the alveoli. These vessels carry blood at normal systemic circuit pressures so that their capillaries exchange fluids with the interstitial matrix as do capillaries of other portions of the body. These capillaries, then,

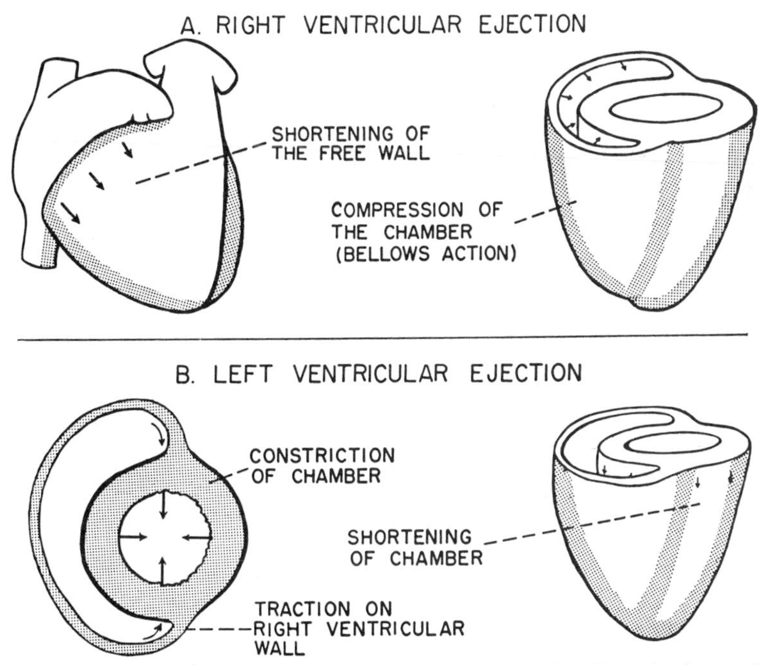

Figure 5–12 Movements of the ventricles during systole. (From Rushmer: *Cardiovascular Dynamics,* 3rd Edition. Philadelphia, W. B. Saunders Company, 1970.)

are the ones responsible for delivering oxygen and nutrients to lung cells which are further removed from the alveoli.

Pulmonary vessels are especially distensible; this is a factor in maintaining low blood pressures. It also provides for a greater minute-volume flow of blood through the lungs during periods of exercise. Pulmonary capillaries are also larger in diameter than those of other portions of the body. This provides a greater area for diffusion exchange of gases with the alveoli. Note, also, that larger capillaries are possible only because of lower hydrostatic pressures in these vessels. The law of Laplace explains why this is so (Sect. 4.2). If the hydrostatic pressure were not low, then enlarging the radius of the capillary would necessitate thicker walls to withstand the greater wall tension that would result. Larger capillaries also lower the resistance to flow and, therefore, the pressure drop in the pulmonary system. This is another important factor in maintaining low blood pressure in that system.

Correlated with the difference between pulmonary and systemic resistance is the difference in structure of the right and left ventricles of the heart (Fig. 5-12). The right ventricle is relatively thin-walled, and its cavity is long and narrow. Contraction results in a bellows-like action that displaces a large volume of blood but at low pressure. The left ventricle, on the other hand, has a cylindrical shape and relatively thick, muscular walls. Contraction of this chamber is a squeezing or wringing action resulting in high-pressure ejection of the blood.

5.10 ELECTRICAL ACTIVITY OF THE HEART

One side of each plate of an automobile battery has more negative charges (negatively charged ions) than the other side. Consequently, if one side of that plate is connected to the other side by a conducting wire, an electrical current will flow through that wire. Biological structures such as cell membranes are characteristically charged like the plates of a battery. Commonly, the inside of a cell membrane is more negative than the outside because of the differential distribution of charged ions on the two sides of the membrane (Fig. 5-13).

When a living muscle or nerve cell is stimulated in some appropriate way, there is a momentary reversal of the charge across the membrane. Permeability changes in the membrane allow charged ions to pass through it, and the change becomes positive on the inside with respect to the outside of the membrane. Immediately thereafter, the charged ions are actively "pumped" back to the original sites and the resting situation is reestablished, *i.e.*, the membrane becomes negative on the inside again. This alteration of membrane charge (membrane potential) is called an action potential and is described in greater detail elsewhere in connection with the nervous impulse (Sect. 11.3).

The physical activities of heart muscle are preceded by electrical changes (action potentials) of the same kind, and it is possible to record these changes with sensitive instruments. It is not necessary, however, to attach electrodes to the heart itself in order to record these electrical

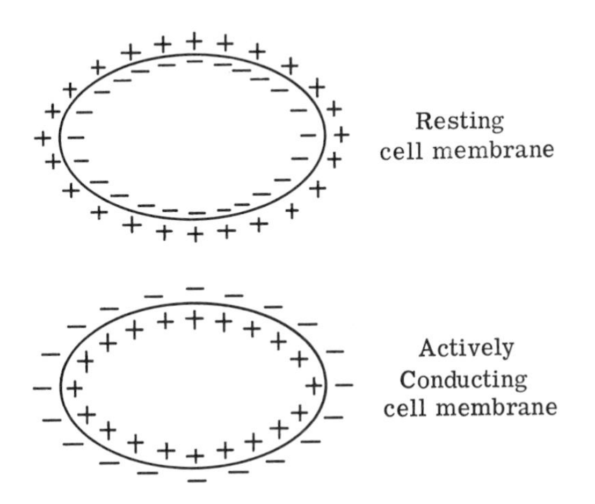

Figure 5–13 Electrical charges on a cell membrane. See text.

changes. All living tissues are able to conduct electrical currents since they are all bathed by and filled with conductive salt solutions. Consequently, the tissues lying between the heart and the body surface are conductors just as wires attached to the heart would be. It is only necessary, therefore, to attach wires to the body surface in order to pick up and record the action potentials of the heart. These are very small potentials, of course, so the recording apparatus must be sensitive and include appropriate amplifying circuits. Since skeletal muscles also produce action potentials, it is also necessary that the subject lie quietly while the recording is done. Collectively the record of action potentials in the heart is known as an electrocardiogram. Frequently this is abbreviated as ECG in the literature.* Interpretation of the electrocardiogram requires a considerable amount of experience, but the technique has proved to be of great value in clinical medicine as well as in pure research.

The myocardial cells are in intimate contact with each other (Sect. 16.6); thus, the occurrence of an action potential at one point is, in itself, an adequate stimulus to produce action potentials in adjacent cells. Stimulus at any point on the atria, therefore, results in an action potential that spreads throughout the entire atrial myocardium. Similarly, stimulus to any part of the ventricular myocardium results in excitation of both ventricles. However, atrial excitation neither spreads directly to the ventricles nor vice versa.

Electrocardiographic studies of the mammalian heart reveal that an action potential appears in the wall of the right atrium preceding each atrial contraction. It is initiated there by a small body of specialized "neuromuscular" tissue known as the sinoatrial node (SA node). This action potential spreads in a wavelike fashion over the entire atrial myocardium and each wave is followed by atrial systole. It is clear that this electrical excitation directly causes the muscle contraction (Chap-

---

*EKG is also used since the original reports of the method were in German, and in that language the prefix cardio is spelled with a "k."

ter 16). The rhythmically repeated initiation of action potentials by the sinoatrial node determines, then, the rate of the atrial beat. For this reason the sinoatrial node is often spoken of as the pacemaker of the heart (Fig. 5-14).

Although the action potential spreading over the atria does not directly excite the ventricular myocardium, it is transmitted to the ventricles indirectly. A second bit of specialized neuromuscular tissue is located just to the right of the interventricular septum and just below the atrioventricular boundary. This is known as the atrioventricular node (AV node). Extending from the atrioventricular node is a bundle of neuromuscular cells, the atrioventricular bundle. This bundle enters the wall that separates one ventricular chamber from the other (the interventricular septum) and divides there into two bundle branches. The bundle branches descend in the interventricular septum toward the apex of the heart and give off numerous small branches that end in various parts of the ventricular myocardium. These terminal branches are called Purkinje fibers.

The action potential spreading over the atrial myocardium is picked up by the atrioventricular node and transmitted by way of the atrioventricular bundle, its branches, and the Purkinje fibers to the myocardium of the ventricles. The atrioventricular node imposes a delay in this transmission so that the ventricular muscle is excited somewhat after the impulse is transmitted. The ventricles, of course, respond to this excitation by contracting (ventricular systole). The delay in transmission is necessary so that atrial systole and ventricular systole will alternate in time. If the atrial excitation spread directly to the ventricles, the contraction of atria and ventricles would be almost simultaneous. Simultaneous contraction of both the atria and ventricles would greatly reduce the pumping efficiency of the heart (Sect. 5.4).

A typical electrocardiogram shows the electrical changes associated

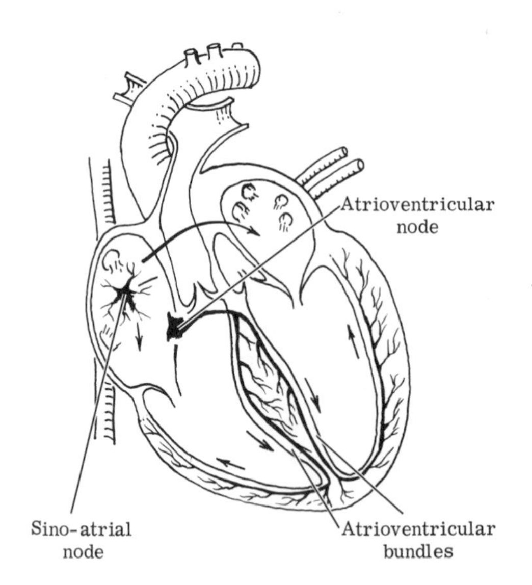

Figure 5–14 The excitation system of the human heart. (After Cockrum and McCauley: *Zoology*. Philadelphia, W. B. Saunders Company, 1965.)

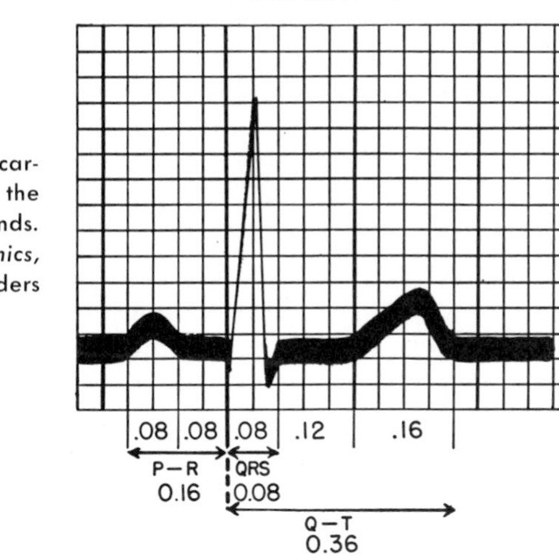

Figure 5–15 A typical electrocardiogram showing the durations of the various waves in fractions of seconds. (From Rushmer: *Cardiovascular Dynamics*, 3rd Edition. Philadelphia, W. B. Saunders Company, 1970.)

with the spreading action potentials through the heart (Fig. 5-15). The form of this record differs, of course, depending upon where the instrument's electrodes are attached to the body surface; usually several recordings are taken using different recording points on the skin.

Experimentally it can be shown that the atrioventricular node, like the sinoatrial node, is capable of initiating rhythmic action potentials but at a slower rate. The more frequently acting sinoatrial node sets the faster pace, and its presence normally prevents this same ability of the atrioventricular node from being apparent.

If the heart is completely removed from the body, the rhythmic initiation of action potentials by the sinoatrial node continues; consequently, the heart continues to beat. It is necessary, of course, to provide a suitable environment for the isolated heart if it is to continue to live and function. The experiment demonstrates dramatically, however, that the heart has intrinsic rhythmicity; it does not require any connection with the nervous system in order to beat regularly. As we shall see, however, nervous connections are necessary in order to vary the rate of beating.

5.11 CONTROL OF ARTERIAL BLOOD PRESSURE

The arterial blood pressure depends upon the resistance to flow and the minute-volume of cardiac output (Sect. 5.6). The minute-volume of cardiac output is, in turn, dependent upon the stroke rate and the stroke volume of the heart, assuming, of course, that adequate venous return is not a limiting factor (Sect. 5.5).

Even though the heart does not require innervation in order to beat, it does have a nervous supply. Cardiac muscle, like smooth muscle (but unlike skeletal muscle), has a dual innervation from the autonomic nervous system (Sect. 12.6). Parasympathetic nerves from the tenth cranial nerve (vagus) and sympathetic nerves from a ganglion of the paravertebral trunk (Fig. 12-10) both reach the heart. Experimental stimulation of the vagus nerve slows the rate of heart beat, whereas stimulation of the sympathetic nerves to the heart speeds it. The stroke rate at any given time, then, depends upon the ratio of nerve impulse frequencies along these two routes to the heart.

Since the muscles in their walls are smooth muscles, arterioles also receive autonomic innervation; at least they are supplied with sympathetic nerves. Nerve impulses along these sympathetic nerves result in contraction of the arteriole muscles and, therefore, vasoconstriction. This narrows the vessel and therefore increases the resistance to blood flow and, of course, the blood pressure (Sect. 5.6). Parasympathetic nerves, if present, could be expected to have the opposite effect, *i.e.*, vasodilation and a consequent lowering of blood pressure. Since the distending pressure of the blood within the arterioles will cause vasodilation in the absence of active muscular vasoconstriction, actual parasympathetic innervation seems unnecessary. Any decrease in the frequency of nerve impulses reaching the arterioles by way of sympathetic nerves would produce an equivalent relaxation of the arteriole muscles and therefore some vasodilation. Most authorities agree that parasympathetic innervation is absent in this case.

Autonomic innervation, then, controls arterial blood pressure by controlling the two most important factors involved: the stroke rate of the heart and the amount of vascular resistance to flow (Sect. 5.6).

Sympathetic nerve activity results in the release of the chemical substance norepinephrine at the nerve-muscle junction. This substance is much like epinephrine (Fig. 5-16) and it causes an increase in the stroke rate of the heart. Parasympathetic nerves release at their nerve-muscle junctions a different chemical substance called acetylcholine (Fig. 5-16), which causes reductions in the cardiac stroke rate. Acetylcholine is a chemical derivative of cholesterol (Fig. 10-13). The responses of the sinoatrial node in altering its rate of pacemaking are responses to these chemical substances rather than to the sympathetic

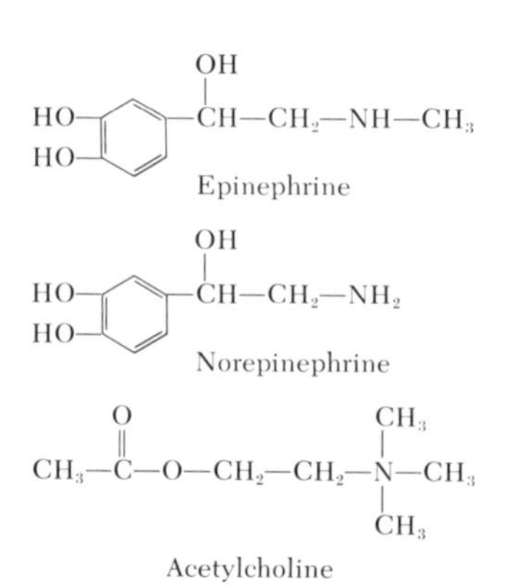

Figure 5-16 The chemical structures of epinephrine, norepinephrine, and acetylcholine.

and parasympathetic nerve impulses *per se.* Acetylcholine inhibits the node and reduces its activity rate, whereas norepinephrine stimulates it to more frequent activity.

Sympathetic activity is increased during exercise or in certain emotional states such as anger or fear. Parasympathetic activity is reduced under the same conditions. The resulting cardiac and vascular responses may be further augmented by the release of epinephrine into the blood from the medulla of the adrenal gland (Sect. 17.9). Epinephrine and norepinephrine are much alike in chemical structure and in physiological effect. In fact, both substances are released by sympathetic nerve endings and by the adrenal medulla. However, norepinephrine predominates at the sympathetic nerve endings, and epinephrine predominates in the adrenal medulla.

As in the case of the controls of respiration (Sect. 3.9), it is known that cardiovascular controls depend upon nervous centers in the medulla oblongata of the brain. Again, the exact pattern of nervous connections is not known, but it is possible to set up a model representing the probable mechanisms involved. Certainly the true mechanisms are more complex than those presented here, but they must be of the same general type.

In the medulla of the brain are groups of nerve cell bodies which initiate the nervous impulses traveling along autonomic nerves to the heart and to the smooth muscles of the arterioles. One such group of neurons is called the vasomotor center and consists of two parts, a vasodilator center and a vasoconstrictor center (Fig. 5-17). Activity of the vasoconstrictor center causes impulses to travel over sympathetic nerves to the arterioles and, therefore, causes constriction of these vessels. Activity of the vasodilator center presumably results only in inhibition of the vasoconstrictor center, since there are no parasympathetic nerves reaching the smooth muscles of the arterioles.

Similarly, in the medulla of the brain are also groups of neurons which initiate nerve impulses traveling over autonomic nerves to the heart. These include a cardioaccelerator center and a cardioinhibitory center. Impulses from the cardioaccelerator center travel along sympathetic nerves to the heart and result in an increase of its stroke rate. Impulses originating in the cardioinhibitor center reach the heart by way of parasympathetic routes and result in a slowing of its rate. The comparative activity of the two centers, then, determines the stroke rate at any given time. Similarly, the comparative activity of the two vasomotor centers determines the degree of arteriole constriction at any given time. By means of these centers in the medulla of the brain, arterial blood pressure is controlled.

Both sets of medullary centers are, themselves, controlled by the feedback (Sect. 1.6) of sensory information. Thus, they are constantly "informed" of the current state of the blood pressure and of other conditions so that appropriate adjustments in blood pressure can be made as required.

In the walls of the aortic arch and in the base of the internal carotid artery (carotid sinus) are receptors that are stimulated by the pressure of

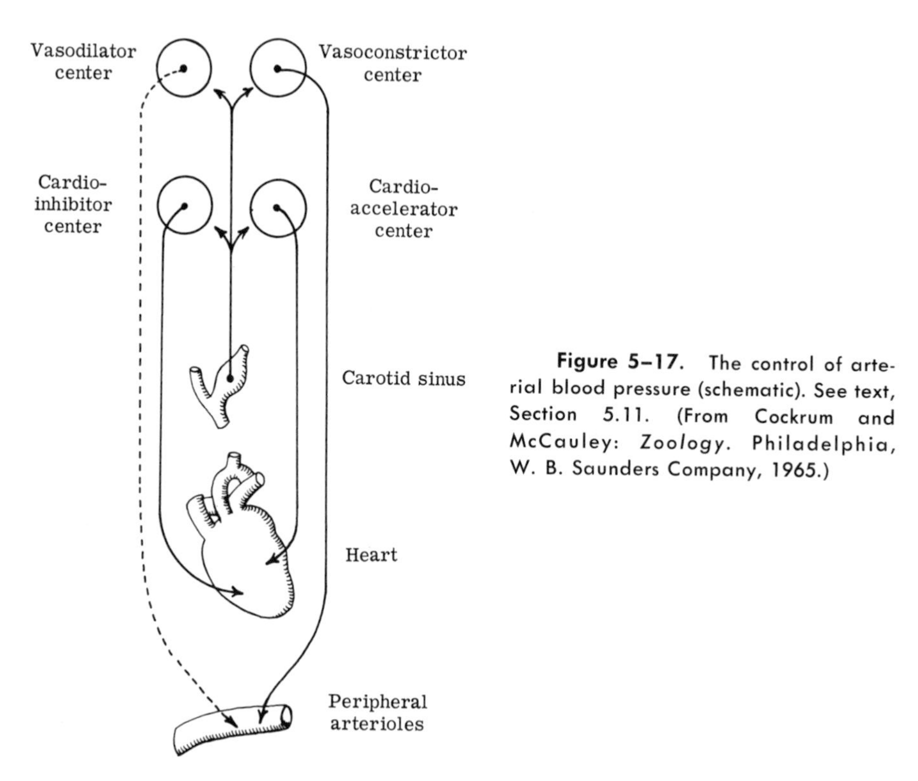

Figure 5–17. The control of arterial blood pressure (schematic). See text, Section 5.11. (From Cockrum and McCauley: *Zoology*. Philadelphia, W. B. Saunders Company, 1965.)

the blood. The greater the blood pressure, the more frequent are the impulses sent from these pressure receptors to the medulla of the brain. These impulses tend to stimulate the vasodilator and the cardioinhibitor centers, and the greater the frequency of these impulses, the greater is the degree of stimulation to these centers. Thus, if for some reason blood pressure begins to rise above appropriate levels, the increased activity of the pressure receptors leads to vasodilation (decreased vascular resistance) and decreased stroke rate of the heart. Both influences reduce the blood pressure. Conversely, a decrease in blood pressure below appropriate levels leads to decreased activity of the receptors and both of these medullary centers. As a result, the other medullary centers (vasoconstrictor and cardioaccelerator) increase their activity. This causes vasoconstriction and increased cardiac stroke rate; consequently blood pressure rises.

Under ordinary resting conditions, the arterial blood pressure is constantly monitored by the pressure receptors. These feed back the information to the medullary nervous centers, and they act to adjust the blood pressure so that it remains within appropriate limits. This feedback system is so sensitive that pressure changes coinciding with each pulse wave have momentary effects on arteriole constriction. The resulting cyclic constriction and relaxation of the arterioles helps to smooth out the pulse wave and produce a more steady flow of blood through the capillaries.

Changes in body position result in changes in blood pressure because of the altered force and direction of gravity. In a prone position the entire body is essentially on the same level as the heart, and minimum arterial pressure is required to maintain adequate circulation. When the body is erect, however, the force of gravity tends to allow blood to drain downward toward the feet, and increased vasoconstriction is necessary to offset this tendency. Increased cardiac activity is also required to overcome this increased resistance and to insure adequate blood supply to regions well above the heart, *e.g.*, the brain. If one arises quickly from a horizontal to a standing position, the necessary adjustments may not take place fast enough to avoid a momentary drop in the blood supply to the brain, and momentary dizziness or even fainting may occur. Such an event is called syncope.

Although the mechanisms discussed thus far are adequate to control arterial blood pressure at relatively constant levels under ordinary resting conditions, they are not adequate to meet the greater demands of exercise. Just as we may wish to reset the room thermostat to increase the temperature of the room, so it is sometimes desirable to reset the "pressurestat" to establish and maintain, for a time, a higher level of arterial blood pressure.

As in the case of respiratory controls, the resetting of the arterial blood pressure control system is accomplished by monitoring the blood concentration of oxygen and carbon dioxide. During exercise, of course, the oxygen concentration in the blood decreases and the concentration of carbon dioxide increases. Chemoreceptors of the carotid sinus (the same ones that operate in the control of lung ventilation) are sensitive to blood oxygen levels. The controlling cardiovascular centers of the brain are also directly sensitive to the oxygen concentration of blood and to the pH of cerebrospinal fluids (Sect. 3.9); the general mechanism of control is similar to that of respiratory control. Falling concentrations of blood oxygen result in greater activity of the vasoconstrictor and cardioaccelerator centers. Rising concentrations of carbon dioxide in the blood emphasize this effect. Similarly, increasing concentrations of carbon dioxide produce decreases in the pH of cerebrospinal fluid, and this has a similar effect on the medullary centers.

Alterations in blood pressure can also result from the anticipation of exercise due to overriding influences of higher brain centers. An athlete poised to begin a race displays elevated blood pressure, as well as increased lung ventilation even before the exercise has begun. Emotional states such as anger and fear have similar effects.

In addition to the generalized influences described thus far, there are also local effects on circulation that can be attributed to changes in the concentration of oxygen and carbon dioxide in the blood. Decreases in oxygen or increases in carbon dioxide act directly on the smooth muscles of arterioles and precapillary sphincters (Sect. 5.7) to cause relaxation (vasodilation). These effects can override a simultaneous tendency toward generalized vasoconstriction. At the very time, then, that exercise is causing a generalized vasoconstriction, the vessels in the muscles that are engaged in that exercise may be dilated instead.

This serves to shunt blood from other, less active tissues to the muscles (Sect. 4.3). The coronary arteries that supply the myocardium are particularly sensitive to such local chemical effects.

In contrast, low oxygen or high carbon dioxide concentrations in the pulmonary arterioles have the opposite effects upon local blood flow. Both cause vasoconstriction. This results in shunting blood away from poorly functioning alveoli and toward those which are better ventilated. An overall improvement in respiratory efficiency results from such local adjustments of pulmonary blood flow.

NON-MAMMALIAN CARDIOVASCULAR SYSTEMS

5.12 INTRODUCTION

Among cold-blooded vertebrates (fishes, amphibians, and reptiles) the cardiac stroke rate, the minute-volume, and the arterial blood pressure are all generally lower than among the warm-blooded vertebrates (birds and mammals) of comparable body size. The maintenance of higher body temperature allows more continuous activity and requires a more continuous oxygen and nutrient supply to the tissues. Thus, blood pressure must be higher and subject to better control in the higher vertebrates. It is recalled that the progression from aquatic to aerial respiration also necessitated alterations in cardiovascular structure and function.

5.13 CYCLOSTOMES

The hagfishes have a circulatory pattern not unlike that of many invertebrate animals. It is characterized by many open, sinusoidal spaces rather than by a system of closed vessels. Indeed, it is not like a typical vertebrate system at all.

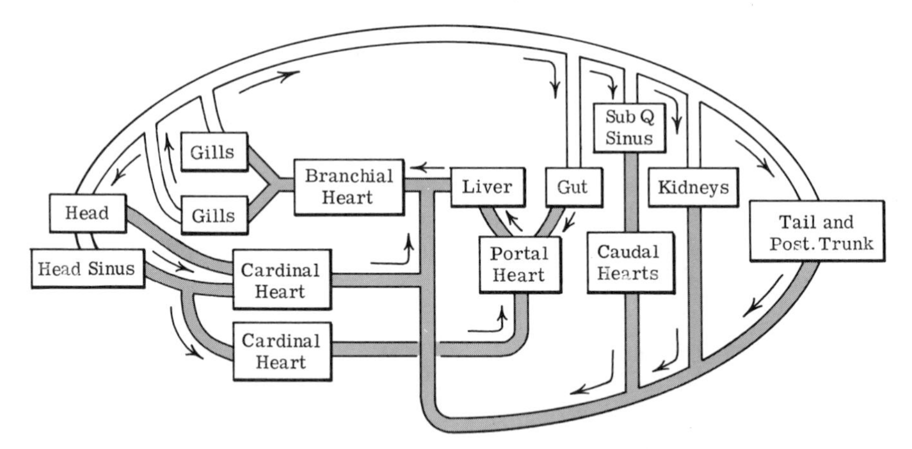

Figure 5–18 The pattern of circulation in the hagfish (schematic). See text.

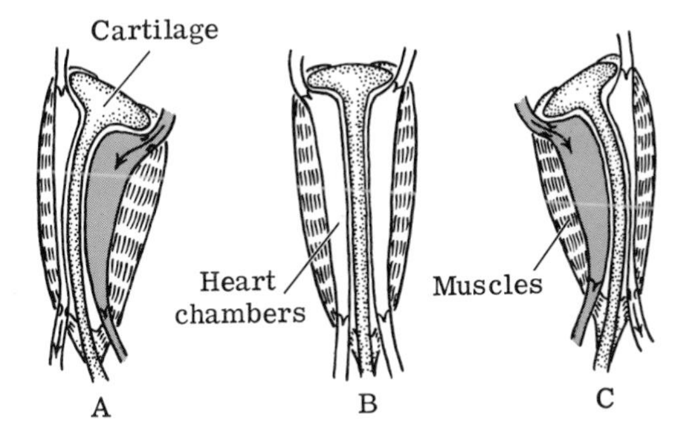

Figure 5–19 Structure and function of the hagfish caudal heart. (After Gordon, *et al.*: *Animal Function: Principles and Adaptations.* The Macmillan Company, New York, 1968.)

Under these conditions, a high-pressure, constantly flowing blood supply is impossible. At rest the hagfish's aortic blood pressure is only about 7 millimeters of mercury and it falls essentially to zero in the blood sinuses. Booster pumps, in the form of accessory hearts, exist at intervals to maintain blood flow (Fig. 5-18). Even so, circulation is sluggish and that fact greatly limits the activity of the animal.

The branchial and portal hearts of the hagfish have pacemakers (Sect. 5.10), but these are not synchronized with each other. Each of these hearts contains within its tissues some chromafin cells—cells which are much like those of the mammalian adrenal medulla (Sect. 17.9). Such cells probably release epinephrin or norepinephrin (Sect. 5-11), and if so, this stimulates the hearts to greater activity during times of exercise. There is no known opposing parasympathetic influence present, however.

With open sinuses present it is almost certain that the mechanisms controlling mammalian blood pressure would be ineffective in the hagfish. At least, constriction of the arteries would only increase the work load on the hearts without materially raising the arterial blood pressure. Exercise may, however, increase the venous return, owing to milking action (Sect. 4.6), and this may in turn produce an increase in cardiac output as a result of the Starling effect (Sect. 5.5). Probably the Starling effect is of much greater importance in cyclostomes than in mammals.

The caudal hearts of hagfishes have no cardiac muscles of their own but are operated instead by adjacent skeletal muscles (Fig. 5-19). The anatomy of these accessory hearts is completely unique among animals.

5.14 FISHES

The typical circulatory system in the fish consists of a series of closed vessels (Fig. 5-20) much like those of all higher vertebrates. Consequently, a single heart is able to maintain circulation at increased pressures, and thus greater activity of the animal is made possible.

The fish heart has a single, undivided atrium and a single, undivided ventricle (Fig. 5-21). In addition, it has two other chambers

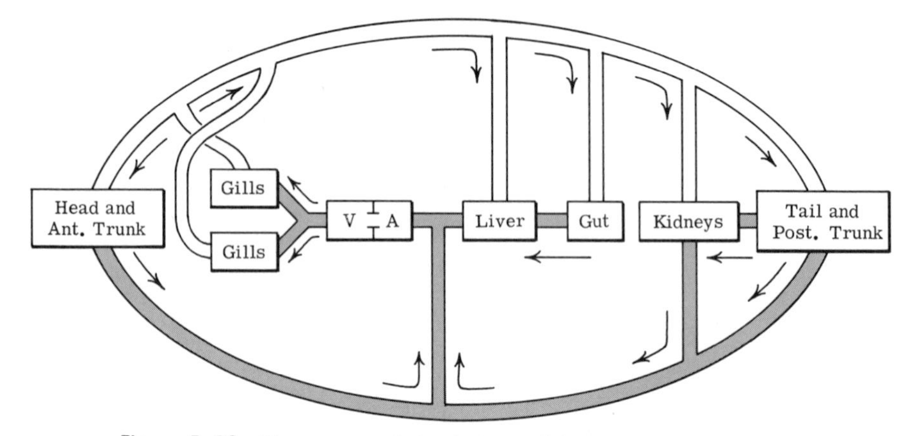

Figure 5–20 The pattern of circulation in fishes (schematic). See text.

arranged in series with the atrium and ventricle. These chambers are lacking in mammalian hearts.

Venous blood is collected in a sinus venosus, which contains pacemaker tissue. This chamber empties into the single atrium and probably plays a role similar to that played by the mammalian atrium; that is, it guarantees a continuous, uninterrupted influx of venous blood to the heart (Sect. 5.4). Blood passes from the atrium into the single ventricle and then into a final chamber, the bulbus arteriosus (or conus arteriosus). This chamber probably corresponds to the first portion of the mammalian aorta. It is equipped with several valves and apparently serves to smooth out the arterial pressure waves. Such an initial dampen-

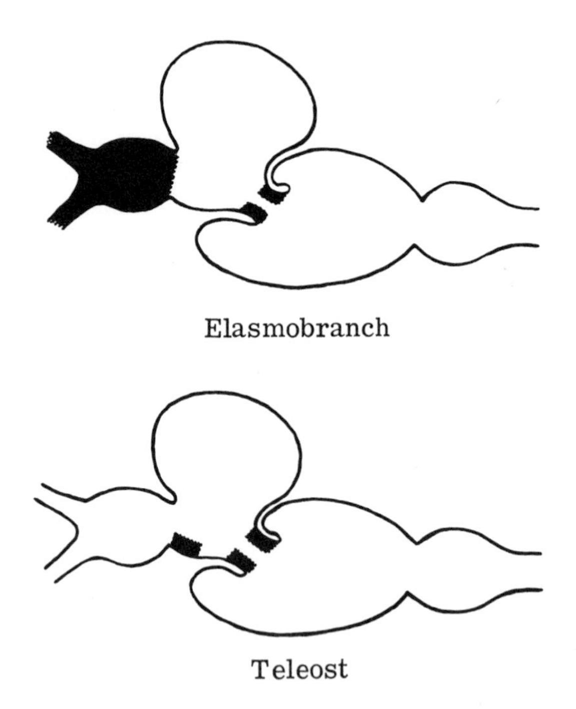

Figure 5–21 The hearts of fishes (diagrammatic). The shaded areas indicate the presence of pacemaker tissues. (After Brown: *The Physiology of Fishes*. New York, Academic Press, 1957.)

ing of pulse waves produces a more uniform, continuous flow of blood through the gills.

It has been suggested that the very low pressures of venous blood would be insufficient to inflate a heart chamber unless that chamber had very thin walls like the sinus venosus. The sinus venosus may also serve, then, to raise the blood pressure sufficiently to inflate the thicker-walled atrium. Similarly, the atrium can inflate the even thicker-walled ventricle. Each chamber may act in sequence to boost the blood pressure by degrees from venous to arterial levels.

Fish hearts receive parasympathetic nerves from the tenth cranial nerve (vagus), but there is no evidence of any corresponding sympathetic innervation. The stroke rate is not increased by increasing physical activity. Activity does increase the venous return by milking action, however, and this results in increased cardiac output due to the Starling effect (Sect. 5.5). Also, the stroke rate varies to some extent with changes in the environmental temperature. Since the metabolic demand for oxygen and nutrients also increases with temperature in cold-blooded animals, this response of the heart is of importance.

An interesting adaptation appears in elasmobranch fishes (sharks, rays), in the electric eel (*Electrophorus*), and perhaps in some other fishes as well. During inflation of the pharynx with water (Sect. 3.3), pressure receptors in the walls of the pharynx are stimulated. Nerve impulses from these receptors to controlling centers of the brain produce an inhibition of the heart rate. As a result, cardiac filling is prolonged and increased. The next heart beat, because of the Starling effect, is a more forceful one and the stroke volume of that beat is increased. Water and blood then pass over the gills simultaneously and the efficiency of gas exchange is improved.

A limiting factor in blood circulation in fish is the fact that the blood goes from the heart to the gills and then on into the systemic arteries. The pressure of blood entering the gills ranges from 15 to 70 millimeters of mercury in different fishes and may be as high as 120 millimeters of mercury in particularly active species such as salmon. The pressure drop imposed by the gill capillaries, however, may amount to as much as 20 to 40 millimeters of mercury. Thus, while the gills receive efficient, high-pressure circulation, the rest of the animal must get along on reduced blood pressures. If there were some sort of booster pump following the gills, this limiting factor would be removed and the entire body would receive a high-pressure circulation. Although no such accessory heart ever appeared in fishes, the further evolution of the heart in higher vertebrates has accomplished the same thing, as we shall see.

The pressure drop across the gills is probably of much less importance in the Dipnoi (lungfishes) since some of the branchial arches of these fishes lack gills. Thus, the total gill resistance is not as great as in other fishes. The major evolutionary advance seen in the lungfishes, however, is the addition of lungs as accessory respiratory organs (Sect. 3.2). These lungs receive blood that has already passed through the gills (Fig. 5-22) so they can only add to whatever oxygenation of the

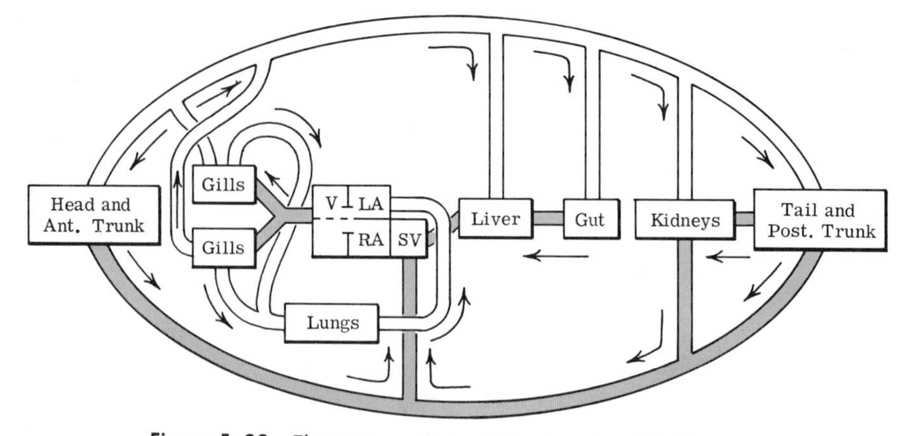

Figure 5–22 The pattern of circulation in a lungfish. See text.

blood has already taken place in the gills. Nevertheless, this gives the animal the opportunity to supplement the oxygen available from a stagnant, aquatic environment. It also increases the ability of the animal to choose between an aquatic or aerial form of respiration, at least for limited periods of time.

The atrium of the lungfish heart is divided into right and left chambers and the ventricle is also partially divided. Blood from the lungs enters the left atrium and then the left half of the ventricle. Despite the fact that the ventricle is incompletely divided, the partial division tends to keep this blood on the left side so that it is ejected into the aorta on its left side. An incompletely developed spiral valve (partition) of the aorta then shunts this oxygenated blood, which came from the lungs, into the first two branchial (gill) arches. These two arches carry blood mainly toward the head. Thus, the brain is guaranteed a sufficient oxygen supply.

Similarly, blood from the general body circulation enters the right atrium of the lungfish. This relatively unoxygenated blood passes into the right half of the ventricle and is shunted by the spiral valve of the aorta into more posterior gill arches. From these it passes mainly to the lungs. Thus, the unoxygenated blood is guaranteed good aeration in both gills and lungs.

The anatomical and functional differences in the lungfish that have just been described lay the groundwork for the evolutionary development of the cardiovascular systems of more advanced, terrestrial vertebrates.

5.15 AMPHIBIANS

The circulatory systems of adult amphibians are somewhat variable since some species are completely aquatic and have gills while others are largely terrestrial and more dependent upon lungs. The larvae of all amphibians are aquatic, of course, and they have circulatory systems similar to those of fishes. The Anura (frogs and toads) lose the gills at the time of metamorphosis and replace them with lungs. The adult circulatory patterns of frogs and toads are similar to those of lungfishes except, of course, that the gill circulation is absent (Fig. 5–23). The amphibian skin also serves as an accessory respiratory surface, and the pulmonary artery has cutaneous as well as pulmonary branches for this reason.

The atrium of the frog heart is completely divided into right and left chambers but the ventricle is undivided. The spiral valve of the aorta is better developed than in the lungfishes (Fig. 5-24), and it is probable that only minor mixing of blood from the right atrium and that from the left atrium occurs in the ventricle. At least this has been verified in those species which have been thoroughly studied. Oxygenated blood from the lungs enters the left side of the heart and goes primarily into the general systemic circuit. Unoxygenated blood returning from the general body enters the right side of the heart and is pumped primarily to the lungs and skin.

The need for a booster pump following the gills of fishes was discussed in the preceding section. The development of the lungfish and amphibian hearts has provided that booster pump, not by the development of an accessory heart, but by the division of the existing heart into right and left chambers. Blood from the right half now goes to the respiratory organs where some pressure drop occurs. Then it is returned to the other side of the heart where it is repumped to the systemic circuit. This boosts the pressure and guarantees that all tissues receive an adequate, high-pressure blood supply. This, in turn, permits the animal to engage in more activity than would otherwise be possible.

In the amphibian heart the sinus venosus is also present and serves, as in the fish heart, to receive blood from the venous drainage and to act as an initial pressure booster (Sect. 5.14). Even though the sinus venosus is not divided, only minimal mixing of oxygenated and unoxygenated blood occurs there. The amphibian heart also contains a

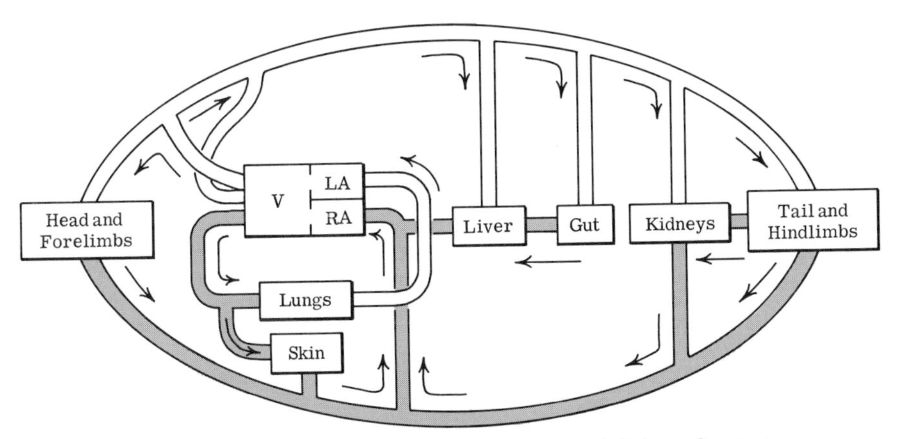

Figure 5–23 The pattern of circulation in an adult frog. See text.

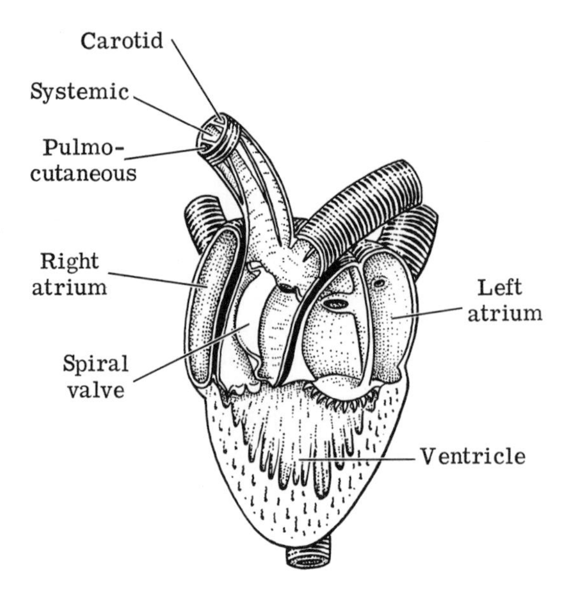

Figure 5–24 The structure of the frog heart. Oxygenated and unoxygenated blood is kept separate to a great extent by separate atria, the spiral valve of the ventral aorta, and division of the right and left aortic branches.

bulbus arteriosus (truncus arteriosus), as does the fish heart. It serves, as in fishes, to smooth out the arterial pressure wave. The heart is innervated by both sympathetic and parasympathetic nerves, which function in pressure control just as they do in mammals (Sect. 5.11).

5.16 REPTILES

The reptiles are also a diverse group. Non-crocodilian reptiles have circulatory patterns and hearts similar to those of frogs and toads except that in these reptiles the skin plays no role in respiratory gas exchange. Also, the ventricle of the heart is partially divided like that of the lungfishes (Fig. 5-25). In the crocodilian reptiles both the atrium and the ventricle are completely divided so that no mixing of oxygenated

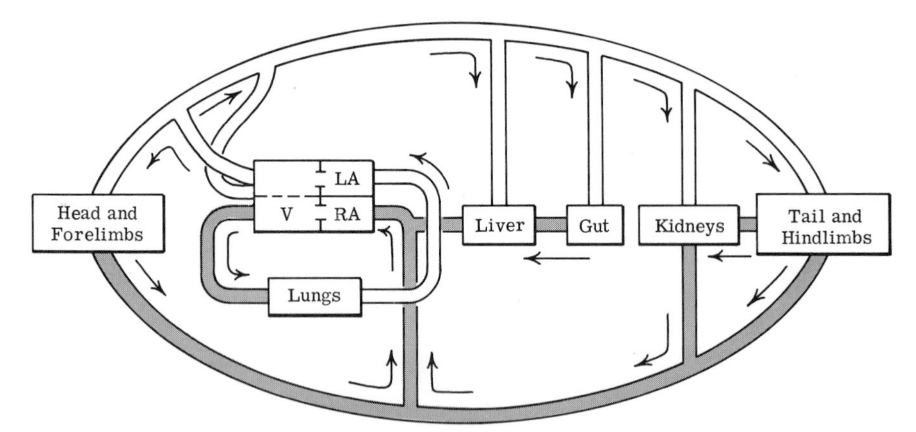

Figure 5–25 The pattern of circulation in a non-crocodilian reptile. See text.

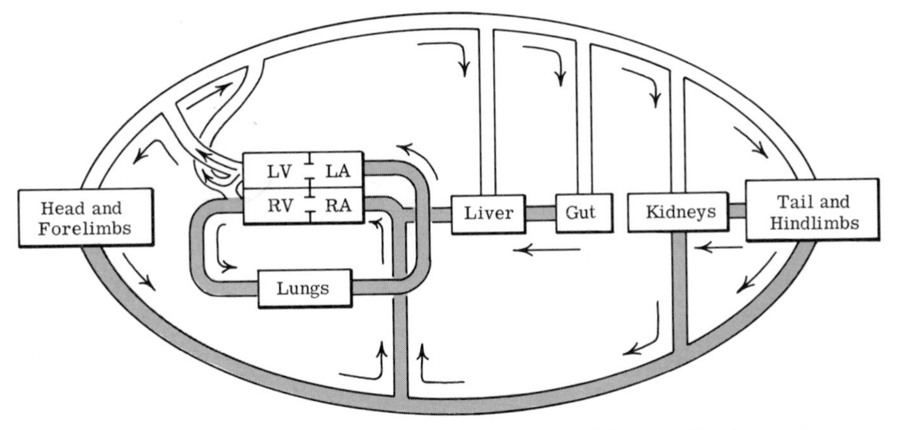

Figure 5–26 The pattern of circulation in a crocodilian reptile. See text.

and unoxygenated blood occurs. Moreover, the spiral valve of the aorta is so well developed that the aorta is divided to form essentially an independent aorta and pulmonary artery (Fig. 5-26).

5.17 BIRDS

The circulatory pattern of birds (Fig. 5-27) is essentially like that of mammals except for minor details that are of no physiological significance. For example, mammals retain the left aortic arch and lose the right one, whereas birds retain the right aortic arch and lose the left. Thus far, research has shown that the controlling mechanisms of blood pressure in birds are like those of mammals.

The heart rate of birds is greater and the arterial blood pressure is higher than in mammals of comparable size. The bird heart is proportionately larger, but the demands of flight are so great that it must also beat faster. It would be impractical to increase the minute-volume of

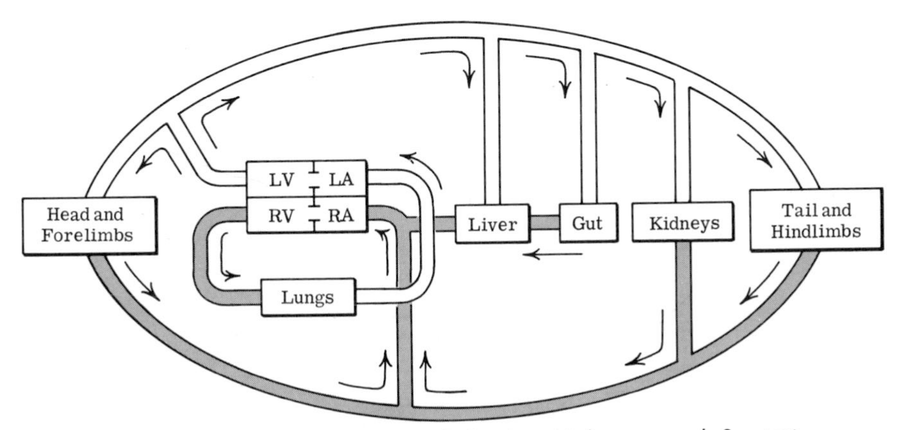

Figure 5–27 The pattern of circulation in a bird or mammal. See text.

the bird heart by developing a still larger heart since this would also increase the "payload" to be carried in flight. Yet, increased blood pressure is necessary to meet the rigorous requirements of flying. Thus, the heart and blood vessels are operating at near maximum to sustain everyday activity. Frequently normal bird blood pressures approach the maximum limits that the circulatory system can contain, and it is not unusual for the heart or aorta to rupture under conditions of great stress. Turkeys, for some reason, are especially susceptible to hypertension (abnormally high blood pressure); they sometimes have arterial pressures as high as 400 millimeters of mercury. Commercial turkey raisers have been successful in reducing hypertensive death rates in their flocks by the use of reserpine, a tranquilizing, blood pressure-reducing drug.

CHAPTER 6

RESPIRATION AND CIRCULATION: SPECIAL ADAPTATIONS

6.1 INTRODUCTION

Within each group of terrestrial vertebrates there are subgroups which have returned to aquatic environments. Evolution seldom, if ever, moves exactly in reverse, and these animals have not lost their lungs and reacquired gills. Instead, they have modified the respiratory and cardiovascular equipment typical of other vertebrates to function in slightly different ways—ways that adapt them to the aquatic environment. They are, thus, not primitive members of their group but rather highly specialized animals. Similarly, special adaptations to extremes of temperature, atmospheric pressure, and other environmental variables have developed in selected vertebrate species.

In his exploitation of the seas and of space, man has not waited for evolutionary change to adapt him. Indeed, no conceivable anatomical or physiological changes could possibly adapt him to such extremes. Instead, man has designed equipment to compensate for the intolerable aspects of these environments. Adaptive mechanisms created in vertebrate evolution are often of special interest since they demonstrate principles that can be applied to the invention of artificial adaptive mechanisms. More frequently, it is only after the development of such equipment that we learn of some animal that has been applying the same principle for thousands of years. This is true, for example, of many of the principles of aerodynamics and aircraft design. Birds have made use of various types of airfoils, wing slots, flaps, and air-speed indicators long before man developed similar equipment (Sect. 16.10). Similarly, streamlining and the use of horizontal and vertical planes to achieve stability in motion appeared in fishes thousands of years before the same ideas occurred to man (Sect. 16.8).

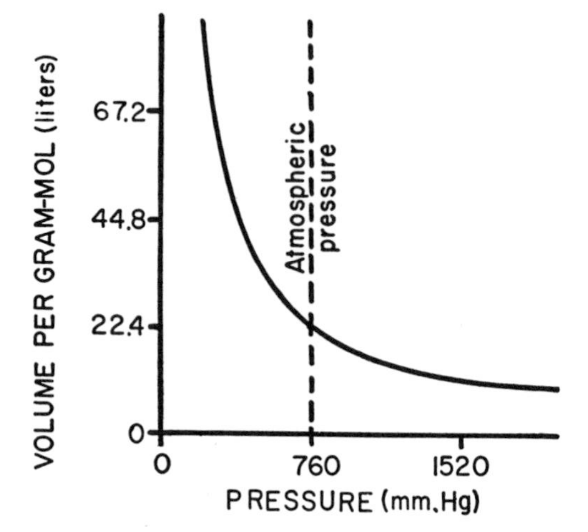

Figure 6–1 A graphic representation of Boyle's law. (From Guyton: *Textbook of Medical Physiology*, 3rd Edition. Philadelphia, W. B. Saunders Company, 1966.)

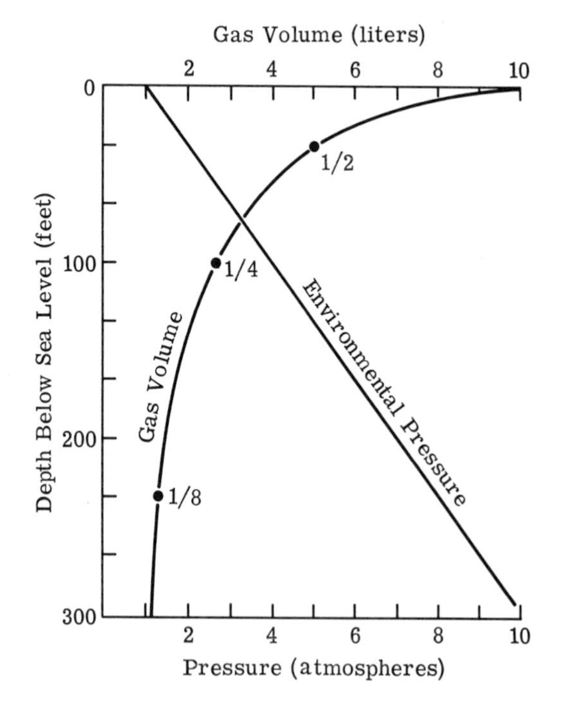

Figure 6–2 The volume of a gas decreases with increased environmental pressure beneath the surface of the sea. See text.

6.2 PHYSICAL PRINCIPLES IN DIVING

Boyle's law states that the volume of a given quantity (number of molecules) of any gas will vary inversely as the pressure upon it providing that the temperature remains constant (Fig. 6-1). Standard barometric pressure at sea level is 760 millimeters of mercury; this figure is designated as one atmosphere of pressure. It is, literally, the weight of the atmosphere above sea level. A column of sea water 33 feet deep also has a weight of 760 millimeters of mercury, so at this depth in the sea the total pressure amounts to two atmospheres. At a depth of 300 feet the pressure is equal to 10 atmospheres (Fig. 6-2). If 10 liters of air taken at sea level is lowered to 33 feet beneath the sea, its volume is compressed to 5 liters by the additional one atmosphere of pressure. If it is lowered to a 300-foot depth, its volume is further compressed to 1.25 liters or one eighth of its original sea level volume. If this air is returned to sea level, it regains, of course, its original volume as the total pressure is lowered. Thus, we can say that at any depth the compressed gas exerts an expanding pressure equal to the compressing pressure surrounding it.

In a mixture of different gases, each gas behaves as it would if it were present alone. For example, if a mixture of 20 per cent oxygen and 80 per cent nitrogen were lowered to 300 feet below the surface of the sea, the mixture would be compressed by a force of 10 atmospheres, as we have already seen. Also, the gas mixture would exert an equal 10 atmospheres of pressure against this compressing force. Under these conditions, 80 per cent of the total gas pressure (8 atmospheres) would be due to the nitrogen present, and the remaining 20 per cent of the total gas pressure (2 atmospheres) would be due to oxygen, since the gases are present in the mixture in that proportion. By convention, these partial pressures (portions of the total pressure) would be expressed as:

$$pN_2 = 8 \text{ atmospheres}$$
$$pO_2 = 2 \text{ atmospheres}$$

Similarly, the partial pressure of other gases, such as carbon dioxide and ammonia, which might be present in such a mixture would be expressed as pCO_2 and pNH_3. Since water vapor is a gas, any pressure due to its presence would be expressed as pH_2O.

6.3 GAS TOXICITY AT HIGH PRESSURE

The effects of various substances on living cells depends upon the concentrations of these substances in the tissues. This is true of gases as well as of drugs and other substances. The concentration of each atmospheric gas present in the tissues depends upon the product of two factors: (1) the intrinsic solubility (solubility coefficient) of the gas in tissue fluids and (2) the partial pressure of the gas in the environment

(Fig. 6-3). The concentration of any given gas, then, will rise in the tissues as its partial pressure rises in the environment. The concentrations present at sea level pressures are, of course, harmless, but at greater pressures the increased concentrations in the tissues may have deleterious physiological effects.

Nitrogen makes up 79 per cent of the earth's atmosphere but it is not very soluble in tissue fluids. At sea level, therefore, only small amounts are present in the tissues, and this amount is apparently without physiological effect. At higher tissue concentrations, however, it acts as an anesthetic. Humans engaged in diving who are breathing air under high pressure acquire high tissue concentrations of nitrogen and may be narcotized by it. The symptoms of nitrogen narcosis resemble those of alcohol intoxication, and for this reason the euphoria which

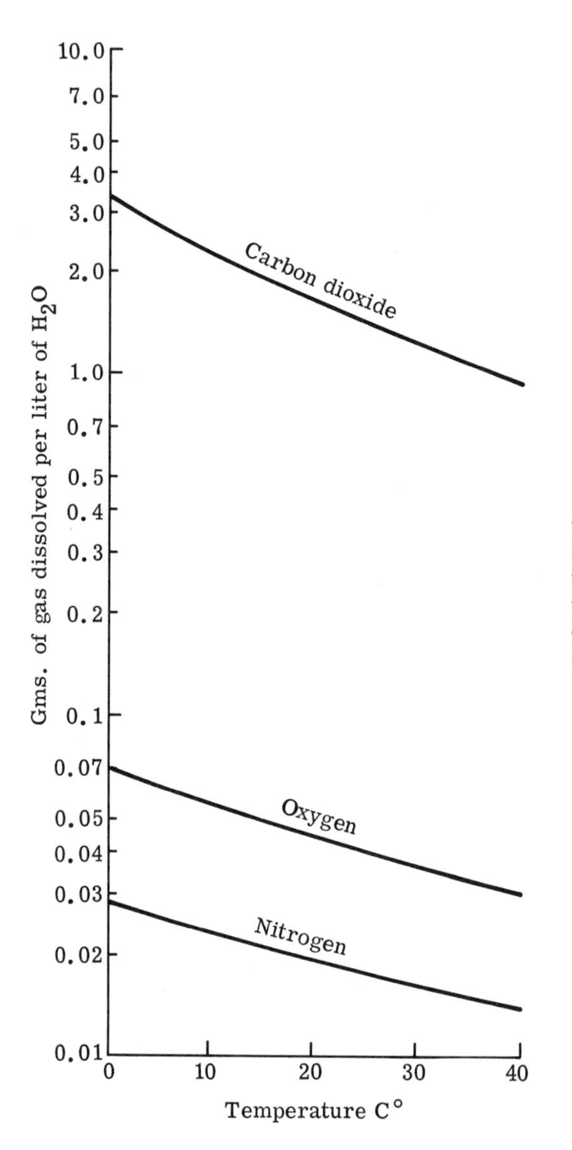

Figure 6–3 The concentration of each gas present in tissues depends upon its solubility and the partial pressure of the gas in the environment. The amount present as dissolved gas also depends upon the temperature.

first appears has been called the "rapture of the depths" or "depth drunkenness." Such symptoms may appear after an hour spent at a depth of about 140 feet. At greater depths drowsiness, clumsiness, and unconsciousness appear successively. The loss of consciousness requires an hour or more of exposure at depths in excess of 300 feet.

Carbon dioxide is, of course, produced by cell metabolism in the body and is normally washed out through the lungs during respiration. Some types of diving gear require the rebreathing of the same air supply, and when this kind of apparatus is used, the concentration of carbon dioxide continuously increases in the air supply. The increasing pCO_2 causes increased ventilation of the lungs (Sect. 3.9) up to a certain point; beyond that, increasing pCO_2 depresses the respiratory centers of the brain and causes a reduction in ventilation rate and depth. Ultimately, this reduced ventilation leads to an insufficient intake of oxygen even though there may be plenty of oxygen available in the breathing mixture. Unconsciousness and death may result. Most underwater breathing apparatus, therefore, contains a soda-lime canister that removes the carbon dioxide from the exhaled air so that it cannot be reinhaled. This kind of apparatus has been improved but is still not perfect since some amount of carbon dioxide is necessary to give normal stimulus to the respiratory centers of the brain.

Oxygen in high concentrations is tolerated well within certain limits. At sea level it is possible to breathe 100 per cent oxygen for a few days without apparent harm. Eventually, however, prolonged breathing of pure oxygen leads to necrosis (cell death) in the lining tissues of the respiratory tract. When supplied under higher pressures to a diver, pure oxygen has more serious effects. Symptoms include nausea, muscular twitching, dizziness, epileptic convulsions, unconsciousness, and even death. Apparently several mechanisms are involved.

Excess oxygen (1) increases the rate of oxidative chemical processes within the living cells and, as a result, depletes the oxidative enzymes located there; (2) increases the concentration of various chemical oxidizing agents in the tissues with the result that these substances literally burn up tissue components of many kinds; and (3) acts directly on the brain arterioles to cause vasoconstriction, which may be sufficient to seriously limit the rates at which nutrients are supplied to the brain and the rates at which carbon dioxide and other metabolic wastes are carried away from it.

In deep-diving apparatus helium is sometimes used to replace the nitrogen of normal air since helium has only about one fourth as much narcotic effect. At depths of less than 500 feet below sea level the undesirable effects of narcosis are virtually absent when helium is used. Helium is also advantageous because its solubility in tissue fluids is very low (Fig. 6-3). A disadvantage lies in the fact that the voice is rendered high in pitch and conversation is difficult to understand. Research on diving mixtures of gases is currently very active in connection with the aquanaut programs, and undoubtedly ways will be found to circumvent the various difficulties due to gas toxicities at high pressures.

6.4 DESCENT AND ASCENT IN DIVING

When a gas is in contact with water the gas molecules tend to enter the water until the limits of its solubility are reached. At equilibrium, then, gas molecules both enter and leave the water at equal rates. The tendency for gas molecules to leave the solution is called the gas tension of that solution, and it is equal to the gas pressure (or partial pressure) in the gaseous environment. If the environmental gas pressure is then reduced, more molecules of gas leave the water than enter it until a new equilibrium is reached. The emergence of carbon dioxide bubbles from a bottle of soda when the pressure is reduced by opening the bottle is a familiar example.

When a diver remains at a given depth breathing air under pressure, the gas tensions of his tissue fluids increase until they are equal to the partial pressures of the gases he is breathing. If he then ascends to a more shallow depth, the partial pressures of gases in the breathing mixture are reduced. As a result, gases in his tissues tend to come out of solution. If the ascent is gradual, the gases evolving from the tissues can leave along the same route by which they entered, *i.e.*, the lungs. If the environmental pressure is reduced rapidly by fast ascent, however, gas bubbles may appear in the tissues as they do in a suddenly opened soda bottle. These bubbles can cause severe pain, paralysis, mental disturbances, or even death, depending upon their location and size. Bubbles in the blood stream can act as emboli (Sect. 5.8) and plug smaller vessels so that blood circulation is completely cut off from certain tissues.

Since nitrogen is the gas present in the greatest volume in ordinary compressed air, it is the gas of major importance in this regard. Because the symptoms were first observed among men who worked in caissons (pressurized compartments used in the construction of underwater tunnels), the syndrome is often called caisson disease. Among deep-sea divers the same syndrome is known as divers' paralysis or the bends. A more correct term is dysbarism.

The amount of nitrogen dissolved in tissue fluids depends upon the depth (partial pressure) and, within limits, the time spent at that depth. Since nitrogen diffuses slowly through tissue fluids, some time is required for the excess nitrogen to be expelled during the ascent. The total ascent time required for safe elimination of excess nitrogen depends, of course, on the amount of it present in the tissues (Fig. 6-4). When nitrogen is replaced by helium in the breathing mixture, the situation is improved because helium has a lower solubility and because it diffuses more rapidly through the tissue fluids. Tables have been constructed showing the necessary times to be spent in decompression at each level of ascent following various lengths of time spent at various depths below sea level. If these decompression schedules are followed, the occurrence of dysbarism is unlikely.

Gases in body cavities are compressed during descent and they expand during ascent from a dive (Fig. 6-2). The amount of such

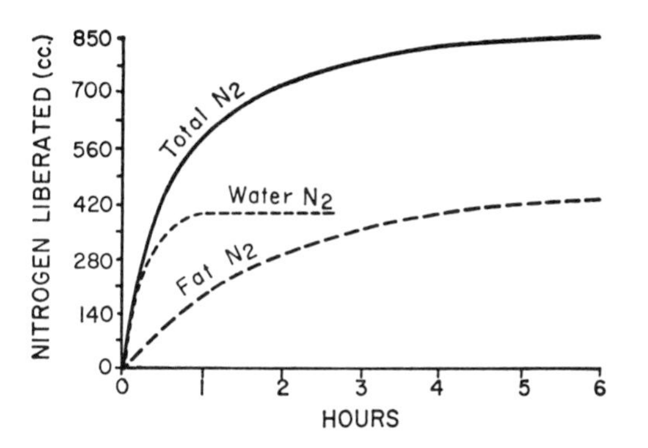

Figure 6-4 The rate of nitrogen liberation from the whole body, from body water, and from body fat of a person who has returned to sea level after prolonged breathing of compressed air at a depth of 33 feet. (From Guyton: *Textbook of Medical Physiology,* 4th Edition. Philadelphia, W. B. Saunders Company, 1971. Reprinted from Armstrong: *Principles and Practice of Aviation Medicine.* The Williams and Wilkins Co.)

expansion and contraction is sufficient to produce physical damage even during a comparatively shallow dive. If a diver inhales a maximum amount of air at the surface and then, while holding his breath, dives to a depth of 100 feet, his lungs and thorax will be compressed maximally. With any further depth his chest will begin to cave in. This, of course, is avoided by supplying the diver with air under pressure. The pressure of the air supplied is continuously adjusted as he descends so that it always equals the compressing pressure of the surrounding water. The internal air pressure in the lungs then equals the water pressure outside, and no physical damage to the thorax occurs. The familiar aqualung is equipped with a "demand valve" which automatically adjusts the air pressure to the surrounding water pressure.

Similarly, if a diver ascends while holding his breath, the air in his lungs expands and may rupture the alveoli. This can introduce air bubbles into the pulmonary blood capillaries, which may act as emboli and plug small vessels of the brain or other tissues. An ascending diver, therefore, must continuously allow air to escape from his lungs to prevent such physical damage. Intestinal gases formed while a diver is submerged can become very painful and do a great deal of physical damage during a rapid ascent for the same reasons.

6.5 DIVING VERTEBRATES

A number of vertebrate species are able to dive for prolonged periods of time and sometimes reach extreme depths (Fig. 6-5). Among these are mammals (seals, whales), birds (ducks, penguins), and reptiles (turtles, crocodiles). Quite obviously they do not suffer from any of the diving injuries discussed in the preceding section as does man. Consequently, a considerable amount of research has been aimed at discovering the physiological adaptations that make these animals such successful divers. One difference is immediately clear: they do not continue to breath during the dive as humans must. This observation, however, only raises another question: how can they survive such long periods without breathing? A human, a cat, or a dog dies of asphyxia within 7 to 8 minutes without air, but ducks regularly submerge for 10 to 20

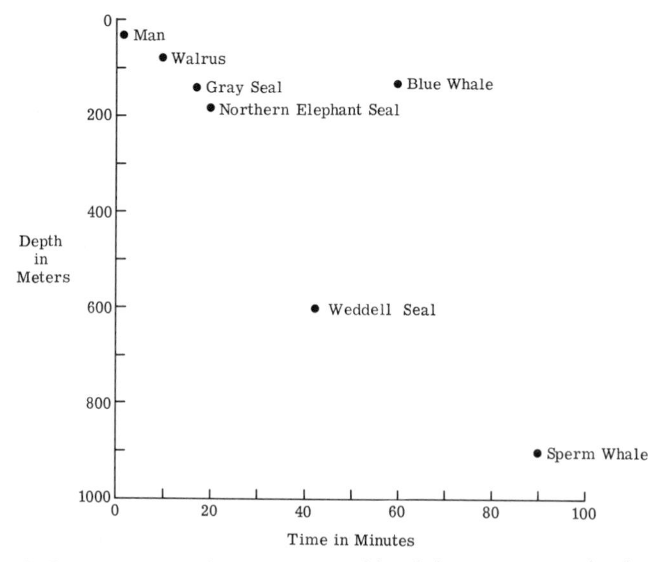

Figure 6–5 Various vertebrates are capable of diving to great depths. These dives may be prolonged for considerable periods of time.

minutes, some seals for a half hour, and certain whales for as long as two hours!

One might well suppose that filling the lungs with air before diving would enable the animal to take a supply of air with him. If the lungs were especially large, this might be a significant factor. This, however, is not the case. In the first place, such a load of air would provide unwanted buoyancy that would make diving more difficult to achieve. Second, this quantity of air would be composed of about 79 per cent nitrogen, more than enough to cause nitrogen narcosis (Sect. 6.3) or, during the ascent, dysbarism (Sect. 6.4). Instead, all diving vertebrates that have been studied empty the lungs as much as possible by forced expiration immediately before diving.

When whales and seals dive deeply, they descend as rapidly as possible and soon reach depths sufficient to compress the thorax and collapse the lungs. The thoracic cage is constructed so that no physical damage to it occurs. The compression of the lungs empties the alveoli so that nitrogen present in the residual air and the dead air space (Sect. 3.8) cannot diffuse into the pulmonary blood. Thus, none of the nitrogen remaining in the lungs can reach the tissues and dissolve in them. The dive to this depth is rapid because until there is sufficient pressure to collapse the lungs, nitrogen can diffuse into the blood. The ascent from such a dive is also rapid for the same reasons. Significantly, shallow dives, during which such thoracic compression does not occur, are much shorter in duration.

In diving reptiles a different adaptation accomplishes the same ends. Compression of the thorax increases the resistance to pulmonary blood flow, and blood is shunted away from the pulmonary circuit and into the systemic circuit through a special connecting vessel. Since pulmonary circulation is stopped by this means, there is no opportunity for nitrogen to be dissolved in the blood and carried to the rest of the animal body.

Another way in which a diving animal might carry oxygen reserves is in the blood. The blood volume of diving vertebrates is greater and the blood's capacity for oxygen is also greater than in their non-diving relatives. Larger quantities of myoglobin, which stores oxygen in the tissues (Sect. 7.3), are also present in diving forms. The quantities of oxygen that can be stored by these means, however, are not nearly sufficient to meet the requirements of an actively swimming animal for prolonged periods of time.

From the foregoing discussion it is clear that diving vertebrates do not carry any large accessory supply of oxygen with them during a dive. In the absence of such oxygen reserves, the animal must economize on oxygen utilization in every way possible. Reduced oxygen utilization is accomplished in two ways.

1. The circulation of blood is limited to those tissues which are most sensitive to decreased oxygen supply and it is denied to those tissues which are more resistant. Blood supply is maintained to the heart and brain but it is completely cut off from the muscles and skin during the dive. The oxygen stores of the blood are sufficient to supply the needs of the heart and brain; the other tissues can function for this period of time without oxygen. The decreased circulation also reduces the load and oxygen demands of the heart so that it can actually decrease its stroke rate and still maintain normal blood pressure.

2. Skeletal muscles can derive some energy from the incomplete metabolism of sugar (anaerobic glycolysis) in the total absence of oxygen (Sect. 10.5). This process results in the accumulation of an end product, lactic acid, in the tissues. In the presence of oxygen, lactic acid is further oxidized to yield much more energy plus the end products carbon dioxide and water (aerobic glycolysis). The muscles of diving vertebrates derive the necessary energy for swimming by means of anaerobic glycolysis, and as a result lactic acid accumulates in them. Since the circulation of blood to these muscles is cut off during the dive, the lactic acid remains in the muscle tissue. Immediately following the dive, however, circulation is reestablished, and the accumulated lactic acid appears in the general circulation. A part of this lactic acid is then oxidized (aerobic glycolysis), and the energy obtained from this process is used by the liver to reconvert the remaining lactic acid to sugar again. This, of course, requires oxygen, which is supplied by hyperventilation for a period of time following the dive. In a sense the animal has acquired an "oxygen debt" during the dive, and this "debt" is paid through extra respiration following the dive. Non-diving vertebrates are also capable of incurring some oxygen debt but not nearly to the same extent since their muscles cannot tolerate large concentrations of lactic acid.

It will be recalled that localized decreases in oxygen concentration or increases of carbon dioxide concentration produce localized vasodilation in mammals (Sect. 5.11). This local vasodilation overrides the generalized vasoconstriction caused by the same concentration changes operating through the carotid sinus receptors. Thus, in most mammals exhaustion of local oxygen supplies and the accumulation of carbon dioxide would result in shunting of blood *into* the exercising muscles. In contrast, as we have seen, in specially adapted mammals blood is shunted *away from* the skeletal muscles during diving. The pattern of vasoconstriction in diving vertebrates, therefore, must be different from that occurring in non-divers. It is interesting to note that typical diving responses (decreased heart rate and vasoconstriction of vessels leading to skeletal muscles) also occur in humans when the face is immersed in water. In a short time during an actual dive, however, the general and local effects of changing oxygen and carbon dioxide concentrations reverse these responses. Strangely enough, merely holding the breath does not evoke the same responses in humans. Holding the head of a seal under water does not evoke them either. Apparently there is a psychological factor that is also required. The human face must be actually under water and the seal must actually intend to dive in order that the physiological adjustments take place.

Similar adaptations appear to be present in all diving vertebrates. Seal, porpoise, hippopotamus, dugong, beaver, duck, penguin, auk, crocodile, and turtle systems all display cardiac slowing and vasoconstriction of vessels leading to skeletal muscles. In all of these, moreover, increased concentrations of lactic acid appear in the general circulation immediately following the dive. Also, a period of hyperventilation accompanied by decreasing lactic acid concentrations follows.

Similar responses occur in aquatic vertebrates that "dive in reverse," *i.e.*, those which leave the water temporarily. The situation is actually similar since they too are leaving a normal habitat and entering one in which respiration is impossible for them. The grunion, for example, is a marine fish that lays its eggs above the tidal line on sand beaches. Another is the flying fish, which sails through the air for short periods of time. In both animals the heart rate is reduced, the blood supply is cut off to skeletal muscles, and lactic acid accumulates while the animal is out of its normal environment. On the other hand, the mudskipper, a fish which spends most of the time out of the water, displays diving responses like other air breathers when it is placed in water.

6.6 PHYSICAL PRINCIPLES AT HIGH ALTITUDES

Standard barometric pressure at sea level is 760 millimeters of mercury. This pressure, of course, decreases as elevations increase above sea level (Fig. 6-6), eventually approaching zero in the upper atmosphere. Completely dry air (containing no water vapor) is composed of 79 per cent nitrogen, 21 per cent oxygen, 0.04 per cent carbon dioxide, and traces of other gases. Thus, in dry air at sea level $pO_2 = 21$ per cent of 760 millimeters of mercury or 160 millimeters of mercury.

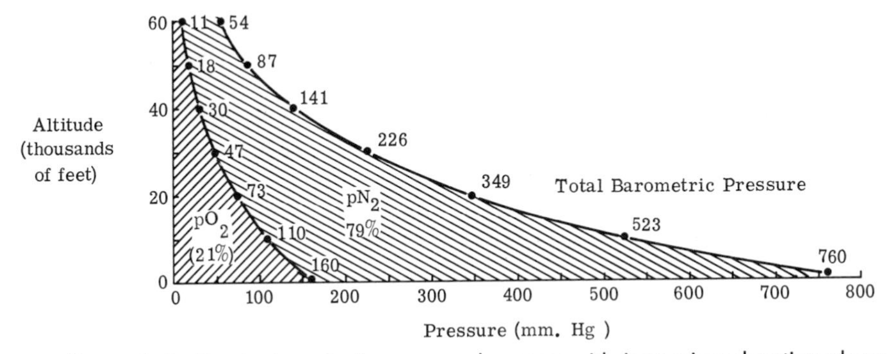

Figure 6–6 Total atmospheric pressure decreases with increasing elevation above sea level. At each elevation oxygen makes up 21 per cent and nitrogen makes up 79 per cent of the total pressure. Carbon dioxide and other gases are present in quantities that are too small to be shown in this graph.

Similarly, pN_2 = 79 per cent of 760 or 600 millimeters of mercury, and pCO_2 = 0.04 per cent of 760 or 0.3 millimeters of mercury.

However, atmospheric air is never completely dry. It always contains some water vapor, and this also exerts a partial pressure (pH_2O). If the atmosphere at sea level were composed of water vapor to the extent of 5 per cent, then pH_2O would be equal to 5 per cent of 760 or 38 millimeters of mercury. Deducting this from the total 760 leaves only 722 millimeters of mercury pressure to be divided among the three other gases. Under these conditions then pO_2 = 21 per cent of 722 or only 152 millimeters of mercury (Fig. 6–7). Similar calculations will show that the pN_2 and pCO_2 are similarly reduced to 570 and 0.29 millimeters of mercury respectively.

The actual amount of water vapor in the atmosphere depends upon the availability of water and upon the temperature of the air. Air in direct contact with bodies of open water contains a maximum amount of water vapor, *i.e.*, it is saturated. More water is required to saturate warm

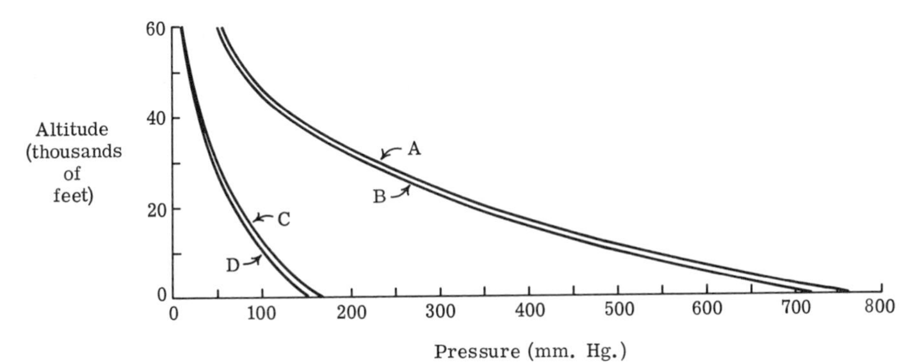

Figure 6–7 The effect of water vapor on the partial pressure of oxygen. *A*, total atmospheric pressure at various altitudes; *B*, atmospheric pressure less 5 per cent water vapor; *C*, the partial pressure of oxygen in dry air; *D*, the partial pressure of oxygen in air that contains 5 per cent water vapor.

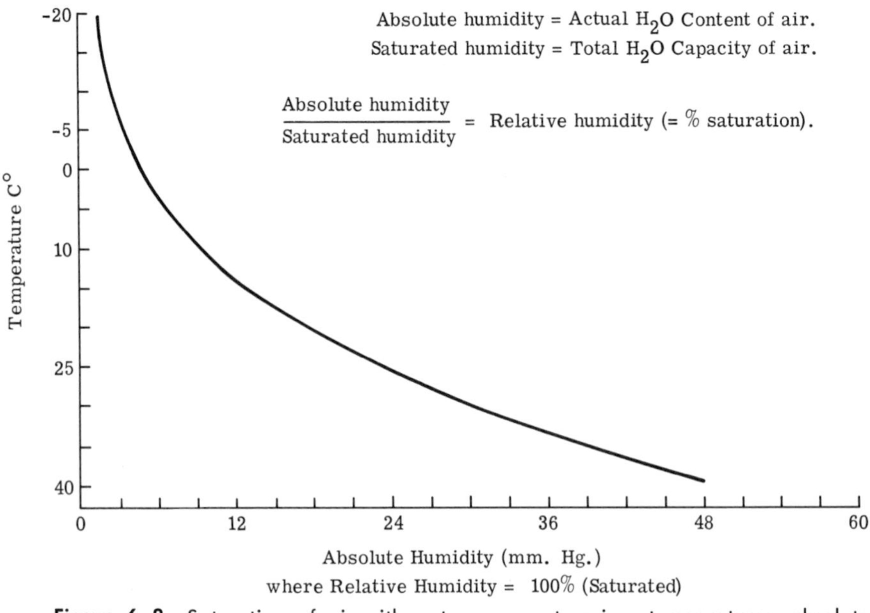

Figure 6–8 Saturation of air with water vapor at various temperatures; absolute versus relative humidity.

air than for cooler air (Fig. 6-8). Note that the partial pressure of water vapor in the atmosphere expressed as pH_2O is not the same thing as the per cent saturation of the air. The pH_2O is the absolute humidity of the air, whereas the per cent saturation is the relative humidity, which is the figure commonly published in the daily newspaper. The absolute humidity expresses the actual amount of water present in the air, and the relative humidity expresses what percentage is present of the amount which would saturate the air at the existing temperature.

Of greater importance to physiology of animals is the alveolar air. This is the air that is in contact with the respiratory membranes of the lungs, and its percentage composition and temperature are generally quite different from atmospheric air. There are two reasons for these differences: (1) In warm-blooded vertebrates the alveolar pH_2O is a constant 47 millimeters of mercury at all elevations regardless of the pH_2O of the atmosphere. This is true because the alveolar air is always at body temperature (38° C in humans) and it is in direct contact with the moist membranes of the lungs. (2) Alveolar air always contains less oxygen and more carbon dioxide than atmospheric air because oxygen is constantly being carried away by the blood and carbon dioxide is constantly being evolved from the blood.

An alveolar pH_2O of 47 millimeters of mercury amounts to only 6 per cent of the total 760 millimeters of mercury pressure at sea level, so it does not seriously interfere with the percentage composition of the other gases present (Fig. 6-9). At 20,000 feet elevation, however, the total atmospheric pressure is only 349 millimeters of mercury; thus the 47 millimeters of mercury pressure exerted by water vapor occupies 13

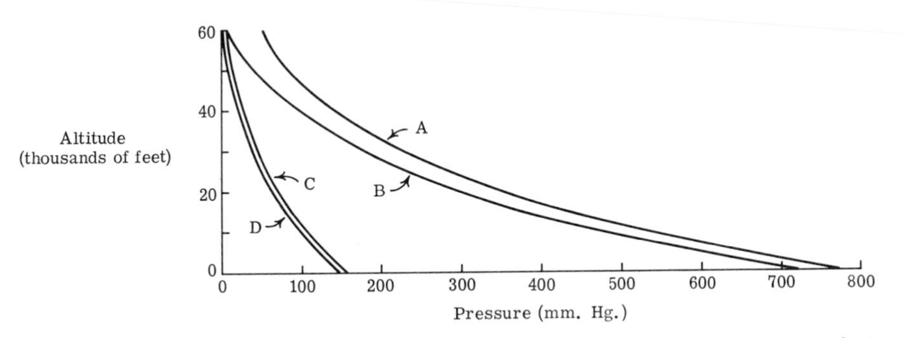

Figure 6–9 The effect of a constant pH_2O of 47 mm. Hg on the composition of air. *A*, total atmospheric pressure of dry air; *B*, total atmospheric pressure less 47 mm. Hg due to water vapor; *C*, partial pressure of oxygen in dry air; *D*, partial pressure of oxygen in air with a pH_2O of 47 mm. Hg (see text).

per cent of the total. At an altitude of 60,000 feet the total atmospheric pressure is 54 millimeters of mercury; therefore the pH_2O accounts for 87 per cent of the total. This leaves almost no room for other gases. Thus, with ascent to higher and higher elevations the constant pH_2O leaves less and less alveolar volume that can be occupied by other gases, including oxygen. Above 60,000 feet the atmospheric pressure falls below the vapor pressure for water, and all of the water in the body would boil away if a man were exposed to it.

At lower elevations the pCO_2 of alveolar air is about the same as that of pulmonary blood, *i.e.*, about 40 millimeters of mercury in humans. At higher elevations, however, the decreasing availability of oxygen leads to increases in the ventilation of the lungs (Sect. 3.9), and this lowers the alveolar pCO_2 to about 24 millimeters of mercury. Like the pH_2O discussed in the preceding paragraph, the pCO_2 occupies a disproportionate amount of the alveolar air and further restricts the amount of oxygen which can be present there.

In summary, because of the constant pH_2O and the relatively constant pCO_2 in the alveoli, the availability of oxygen decreases with increasing elevation more rapidly than it does in the atmosphere.

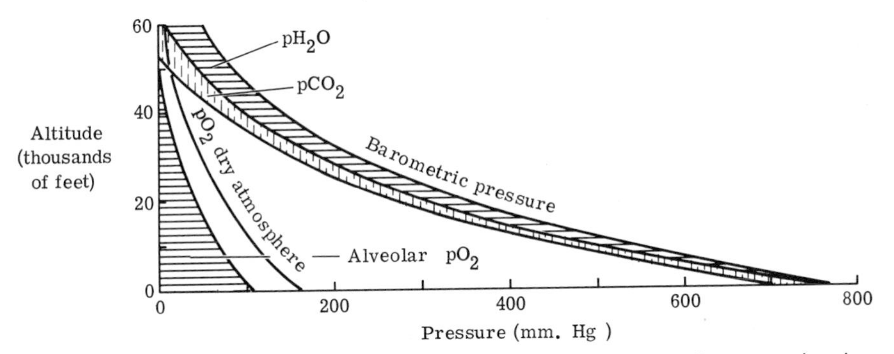

Figure 6–10 In the alveoli of the lungs the partial pressures of water and carbon dioxide increase with increasing elevation. See text.

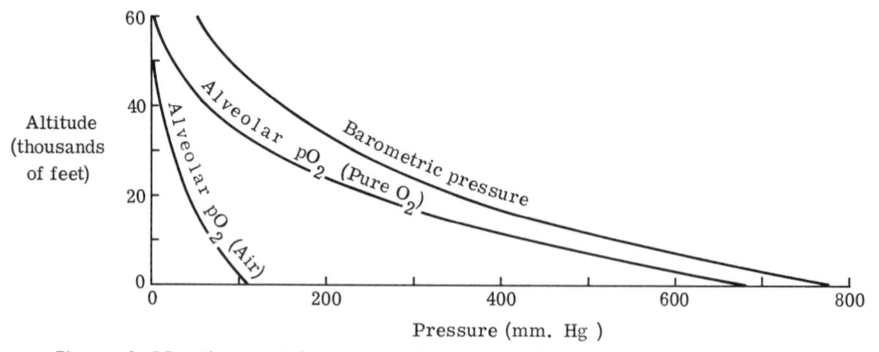

Figure 6–11 The partial pressure of oxygen in the alveoli of the lungs is improved by breathing pure oxygen. See text.

As ascent from sea level begins, the decreasing availability of oxygen is largely compensated for by increased lung ventilation. Consequently, alveolar pO_2 does not decrease as rapidly as atmospheric pO_2 does. At higher elevations, however, the alveolar pO_2 decreases more rapidly than the atmospheric pO_2, as we have just seen. The combined effect of all factors on alveolar air composition is very great (Fig. 6–10). Breathing pure oxygen instead of normal atmospheric air at the same pressure improves the situation markedly, as would be expected (Fig. 6–11).

6.7 PHYSIOLOGICAL RESPONSES TO HIGH ALTITUDE

As man ascends above sea level and the atmospheric and alveolar pO_2 begins to decrease, a slight decrease in blood pO_2 also appears. This has no appreciable effect at first except for a reduction in night vision ability. At an elevation of about 5000 feet the minimum light adequate for night vision must be increased by nearly 25 per cent above that required at sea level. At about 16,000 feet the light must be increased by 140 per cent above that required at sea level. Apparently the rods of the retina (Sect. 15.9) are extremely sensitive to oxygen deprivation.

At an elevation of 8000 feet the per cent saturation of blood hemoglobin with oxygen (Sect. 7.4) has fallen from a sea level norm of 97 to 98 per cent to about 93 per cent. This is sufficient decrease of oxygen to activate the respiratory centers of the medulla and bring about an increase in lung ventilation (Sect. 3.9). This trend continues as elevations increase until at 16,000 to 20,000 feet ventilation has increased by 65 per cent above that which occurs normally at sea level.

Beginning at about 12,000 feet and becoming more marked at higher elevations, other symptoms of anoxia appear. These include drowsiness, headache, mental fatigue, and sometimes euphoria. At about 23,000 feet more serious symptoms including muscular twitching, convulsions, and unconsciousness are present. The most serious symptom before actual loss of consciousness is the impairment of normal judgement and a decrease in ability to perform manual tasks. After an hour spent at 15,000 feet, the mental ability of a person who has ascended from sea level may be lowered by 50 per cent. Several hours spent at an elevation as low as 11,000 feet may lower by as much as 20 per cent the mental ability of a person acclimated to sea level.

After a period of several days spent at high altitude a certain amount of acclimation takes place; the individual makes physiological adjustments which compensate, in part, for the lowered availability of oxygen. At first, an increase of 65 per cent in the minute-volume of lung ventilation is maximum, as we have seen. This increase actually lowers the blood concentration of carbon dioxide, and this in turn prevents further increases in ventilation. The lowered blood carbon dioxide concentration causes an increase in the pH of cerebrospinal fluid, and apparently this is the mechanism by which further increase in ventilation is prevented (Sect. 3.9). Within a few days, however, altered kidney function corrects the pH to normal levels (Sect. 8.8). Thereafter, ventilation can be further increased to an upper limit of about six times normal sea level values.

Low blood oxygen concentration is the specific stimulus for increased hemoglobin production and the production of red blood corpuscles. There is also an increase in the total blood volume at higher elevations. A combination of these responses raises the total circulating hemoglobin by as much as 50 to 90 per cent above sea level values. The efficiency of the blood in extracting oxygen from the air and transporting it to the tissues is, thus, increased by a large factor.

Other responses to high altitude include an increase in the number of capillaries in all tissues. This decreases the average diffusion distance between the blood capillaries and tissue cells and, thus, improves the efficiency of gas exchange between the blood and tissues. The pulmonary capillaries expand, thereby increasing the total exchange area available and the total volume of blood exposed at any given moment. Finally, the metabolizing tissue cells of the body acquire larger quantities of the oxidative enzyme systems, thus increasing the efficiency of oxygen utilization at that level. This last response takes place only over long time periods and may even require genetic selection through several generations.

Populations which have existed at very high elevations for long periods of time are composed of individuals with massive chests and great ventilation capacities. Undoubtedly this and perhaps other long-term genetic adaptations have depended upon selection. The upper limit for permanent human habitation appears to be an elevation of about 17,500 feet. This is the elevation of the highest permanent town, which is located in the Peruvian Andes. Miners living in that town regularly climb an additional 1500 feet to work in a mine at an elevation of 19,000 feet. When quarters for living were erected for them at 18,500 feet, however, the men suffered loss of appetite, loss of weight, and insomnia.

One final physiological response to high altitude needs to be mentioned. Just as dysbarism (Sect. 6.4) can occur during rapid underwater ascent, so it can also occur during rapid ascent above sea level. Extremely rapid ascent over great elevational differences is required, of course, but such ascents are possible with modern jet aircraft. The same effect is also produced by sudden decompression resulting from a serious leak in a pressurized aircraft cabin or space suit.

6.8 VERTEBRATES ADAPTED TO HIGH ELEVATIONS

Birds are obviously capable of high altitude flight, an activity that makes serious demands upon respiration. Not only is the availability of respiratory oxygen decreased, but the total decrease in atmospheric pressure makes flight more difficult because the wings must work harder to obtain the same amount of life (Sect. 16.10). Certainly the very efficient respiratory apparatus of birds (Sect. 3.7) is an important factor, but little is known of other possible adaptive mechanisms. Similarly, lower vertebrates, including fishes, amphibians, and reptiles, occur at fairly high elevations, but their physiological adaptations have not been studied to any great extent.

By far the greatest amount of research in the field of high altitude physiology has been directed towards the mammals. Besides humans, considerable information has been compiled for the New World camellids of South America, *i.e.*, the vicuna, llama, and alpaca. These animals all possess increased amounts of hemoglobin in each red blood corpuscle, and this hemoglobin has an unusually high affinity for oxygen even at these lowered levels of atmospheric pO_2 (Sect. 7.4). The number of red blood cells per unit volume of blood is not markedly higher than in mammals native to lower elevations, however. Thus, these camellids achieve improved oxygen extraction from the air and transport in the blood without the disadvantage of an increase in blood viscosity.

Studies of short- and long-term acclimation to high elevations have been carried out with laboratory animals, notably the white rat. Physiological adjustments similar to those observed in humans take place in rats. In addition, rats show a lowered viability of developing fetuses and new-born young at high elevations. Undoubtedly this is simply a further manifestation of anoxia.

6.9 PHYSIOLOGICAL ADAPTATIONS OF THE GIRAFFE

The body proportions of the giraffe are unique and they require some special adaptations of both the respiratory and the cardiovascular systems even though this animal is not exposed to any extremes of atmospheric pressure. Much remains to be learned regarding these adaptations since the giraffe is not an easy animal to study. Some aspects of its physiology have been learned, however, and others can only be speculated upon.

In most quadrupeds the heart lies at a level almost exactly midway

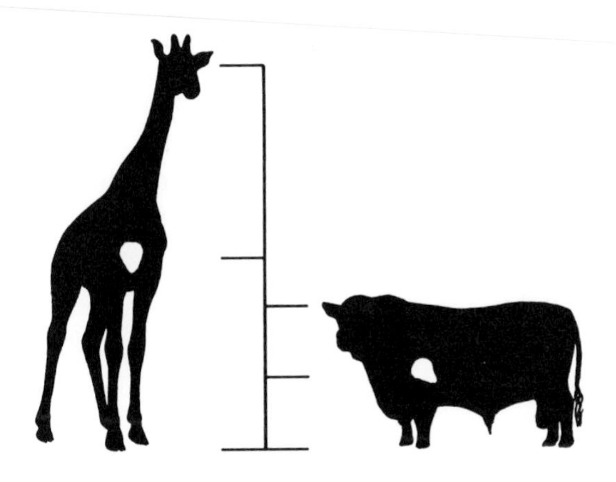

Figure 6–12 The hearts of vertebrates tend to lie about midway between the ground and the highest part of the body. (After Patterson, J. L.: Cardiorespiratory dynamics in the ox and giraffe with comparative observations on man and other mammals. Ann. N.Y. Acad. Sci., *127*:393, 1965.)

between the lowest and the highest portions of the body (Fig. 6-12). This means that the problem of venous return from the lowest parts of the body and the problem of maintaining adequate blood supplies to the highest portions are about equally balanced. Still, the heart-to-brain distance in a giraffe is such that the left ventricle must pump blood at a minimal pressure of 208 millimeters of mercury. Actual measurements in the field gave values of about 220 millimeters of mercury, but these animals were, understandably, excited by the presence of the investigators and by the procedures used in making observations. To produce such an elevated arterial blood pressure requires that the interventricular pressure actually reach 260 to 285 millimeters of mercury. Under these conditions, the brain actually receives blood at a pressure of about 90 millimeters of mercury, or at about the same pressure as in other mammals.

While the left ventricle ejects blood under these very high pressures, the right ventricle maintains a pulmonary blood pressure that is about the same as in other mammals. As we have seen, low pulmonary pressures are necessary in any animal (Sect. 5.9).

All tissues below the level of the heart are, of course, subjected to high blood pressure, which is necessary to supply the head. This is enhanced by the increased gravity effect in long vessels which carry blood into the extremities. Thus, arteries below the level of the heart have very thick walls and a smaller lumen, so they impose a larger pressure drop than the arteries of other mammals. Higher resistance is necessary to offset the high pressure before the blood reaches capillary beds since, as we have seen, the hydrostatic and osmotic pressures must be closely balanced in these areas (Sect. 4.5). For similar reasons there must also be a sharp reduction in blood pressure before it reaches the kidneys (Sect. 8.5).

When a giraffe lowers its head to the ground to drink, there must be a rapid reflex adjustment to prevent a sudden high-pressure surge of blood to the brain. Conversely, when it lifts its head, arterial pressure must be quickly readjusted to prevent syncope (Sect. 5.11). Probably the mechanisms that make these adjustments in blood pressure are similar to those which operate in other mammals, but they must be of much greater scope since there is a difference of some thirty feet between the raised and lowered positions of the head.

When the giraffe raises the head to its normal position above the body, there must also be some mechanism to check the sudden rush of venous blood from the head and neck toward the heart; otherwise the right atrium would become greatly distended. It has been reported that there is a muscular sphincter located in the superior vena cava which, presumably, performs this function.

The elongated neck contains, of course, an elongated trachea which undoubtedly contributes a greatly increased volume of dead air space (Sect. 3.7). This must be offset by enlargements of the lungs to maintain normal gas exchanges with the atmosphere. This is particularly true since the giraffe is a ruminant, and ruminant digestion (Sect. 9.14) contaminates the dead air space with additional carbon dioxide as well as with other gases such as methane.

Although some thorough studies have been carried out, the details of circulatory and respiratory adaptations of giraffes are still incompletely understood.

CHAPTER 7

BLOOD

7.1 INTRODUCTION

The first requirement for circulating blood in animal evolution was to provide transport for nutritional substances and respiratory gases between the organism surface and the metabolizing cells within the animal (Sect. 3.1). As is also true in most other organ systems of animals, additional functions have been acquired during the course of subsequent evolution. Circulating blood has also become important in the transport of hormones and many other substances and in the equal distribution of heat. Additional involvement in immunity reactions and other protective functions have also been evolved. The extreme importance of the system has made necessary the development of mechanisms to prevent the loss of large volumes of blood from injuries or other causes. Therefore, a mechanism of blood coagulation has evolved.

Both transport and non-transport functions of blood will be discussed in this chapter.

TRANSPORT FUNCTIONS

7.2 RESPIRATORY PIGMENTS

At mammalian body temperature (37 to 38° C), water will dissolve only about 0.46 volume per cent of oxygen from the air; blood plasma, because of its salinity, will dissolve even less. Such small amounts of oxygen may be sufficient to meet the needs of small animals with very low metabolic rates, and, indeed, these amounts do suffice for certain fishes of the Antarctic, for eel larvae, and for some other vertebrates under special conditions. It has been shown, for example, that carp (goldfish) can survive on the oxygen dissolved in the plasma if they are allowed to remain quietly at rest. Most vertebrates, however, require much greater amounts of oxygen than can be delivered in plasma solution. With no other oxygen transport mechanism in humans it would be necessary for the heart to pump more than 80 liters of blood per minute even if the tissues could extract 100 per cent of the oxygen from the blood. Actually, the human heart pumps only about 5 liters of blood per minute and the tissues extract only about 30 per cent of the oxygen from it. Quite obviously, then, mechanisms other than solubility in plasma must be present and these must greatly increase the capacity of the blood to carry oxygen.

A variety of colored organic substances, known as respiratory pigments, occur in animal fluids. Each of these substances has an affinity for oxygen and increases the blood capacity for that gas. A discussion of all such pigments may be found in any textbook dealing with the comparative physiology of invertebrate animals. Since only a single type of respiratory pigment occurs in vertebrates, however, we shall limit our disucssion to that one, namely hemoglobin.

7.3 HEMOGLOBIN

It is quite apparent that hemoglobin has appeared repeatedly in animal evolution since it occurs in some members of almost every animal phylum. This is not really surprising since its chemical structure closely resembles that of cytochrome oxidase (cytochrome C_3). Cytochrome oxidase is universally present in all animal cells. It and other related cytochromes are cellular enzymes that function in the oxidation of organic substances releasing their chemical energy for use by the cell (Sect. 10.7).

The hemoglobin molecule is composed of a multiple-ring compound called heme linked to a protein molecule of the globulin type (Fig. 7-1). Wherever it occurs the heme portion is the same and contains an atom of iron. The globulin portion varies, however. It is likely to be different in different species, especially if they are not closely related species. It may also vary with the age of the individual, the globulin in the fetus being different than in the adult. An animal may even have more than a single kind of hemoglobin present in its blood stream at the same time.

Variations in the oxygen-carrying capacity of the hemoglobin are generally due to differences in the globulin portion of the molecule, although factors other than the hemoglobin structure may also be involved. The combination of two or more hemoglobin molecular units to form larger molecules can also influence the oxygen affinity in certain ways, as we shall see.

Each molecule of hemoglobin, under appropriate conditions, is capable of carrying one molecule of oxygen (O_2) in a loose combination. The formation of this combination is called oxygenation of the hemoglobin.

Whether the respiratory pigment in the blood of an animal is

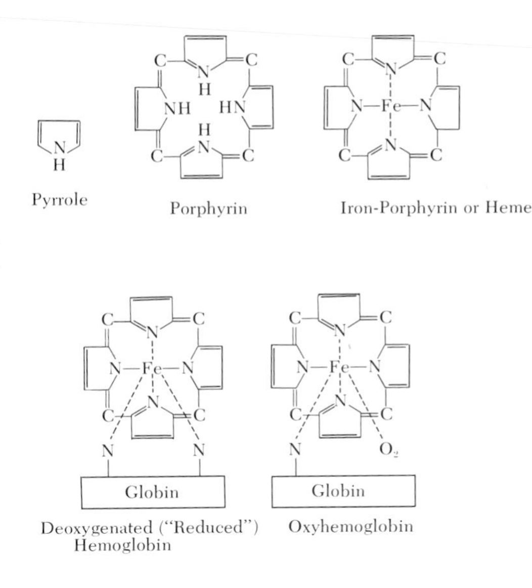

Figure 7–1 The chemical structure of hemoglobin and its component chemical groups.

hemoglobin or another type, the amount of oxygen that can be carried depends, in part, on the amount of that pigment present in a given volume of blood. It is frequently advantageous, therefore, for an animal to acquire increased concentrations of its respiratory pigment. Increasing the number of molecules of respiratory pigment in solution in the blood will, of course, also increase the osmotic pressure of the blood. Vertebrate tissues cannot tolerate much change in blood osmotic pressure. Many invertebrate animals acquire increased amounts of hemoglobin without increasing the number of hemoglobin molecules by combining two or more hemoglobin units together as a single molecule. It will be recalled that osmotic pressure depends upon the *number* of molecules, not upon their type or size. A single unit (molecule) of hemoglobin has a molecular weight of 16,500 to 17,000, but the hemoglobin molecules of earthworms have molecular weights of up to three million. This indicates that some 176 hemoglobin units are combined in each molecule of hemoglobin in the earthworm. While such combination avoids the problem of increased osmotic pressure, it must add significantly to the blood viscosity. Moreover, since all units in such large molecules are not easily oxygenated, the increased total number of hemoglobin units does not bring about a proportionate increase in oxygen-carrying capacity.

Another means by which additional hemoglobin can be present without upsetting the osmotic balance is by enclosing it within blood cells. In this way it is isolated by the enclosing cell membranes and cannot dissolve in the plasma. The presence of cells adds somewhat to the resistance to flow through capillaries, since the cells are large enough to actually rub against the capillary walls as they pass through. This disadvantage is more than offset by the advantages, however. Hemoglobin is enclosed in red blood corpuscles or cells (erythrocytes) of all vertebrate animals.

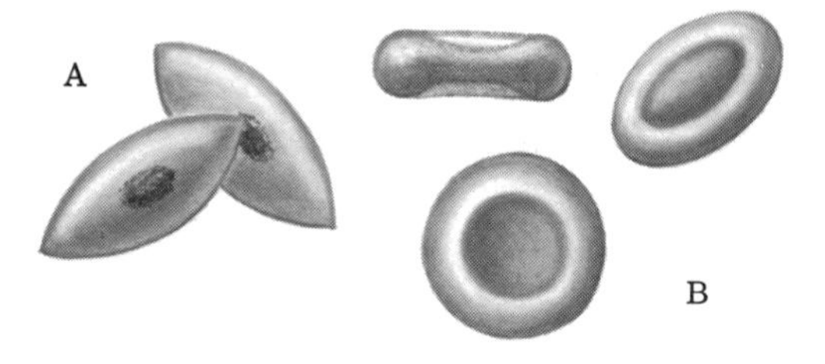

Figure 7–2 Vertebrate erythrocytes. *A*, non-mammalian; *B*, mammalian.

In all non-mammalian vertebrates and in some vertebrates (*e.g.*, the camellids), the hemoglobin is enclosed in nucleated red blood cells which are football-shaped (Fig. 7-2). In most mammals, however, the nucleus of the red cell is lost during its development before it enters the general blood circulation. It then assumes the shape of a biconcave disc. Lacking a nucleus or other normal cell organelles, it can no longer be thought of as a cell in the usual sense. For this reason, it is more correctly referred to as a red blood corpuscle. It is not capable of engaging in normal cell activities, such as synthesis of various proteins and other substances, and therefore it cannot repair itself or engage in cell division. Indeed, it is little more than a membranous sac filled with hemoglobin and stroma. The stroma is the matrix in which the hemoglobin molecules are embedded within the corpuscle.

The change, during vertebrate evolution, from nucleated cell to enucleated corpuscle was a distinct advancement that increased the efficiency of gas transport in mammals. Respiratory gases can only enter or leave the erythrocyte through its surface membrane. Consequently, gas diffusion in and out can take place most rapidly where the membrane surface area is great in proportion to the volume of enclosed hemoglobin. A thin disc has more surface area per unit of volume than any other geometric solid shape. Moreover, as we have seen, diffusion of gases will take place most rapidly where the diffusion distances are short (Sect. 1.2). In a thin disc-shaped corpuscle, every hemoglobin molecule lies very close to the surface, much closer than it would in a more spherical cell. One other advantage is conferred by the corpuscle shape: erythrocytes have a slightly greater volume when in the venous blood and a slightly smaller volume when in the arterial blood (Sect. 7.5). The concave surfaces of the corpuscle can move inward and outward like the bottom of an oil can to accommodate these changes in volume. Changes in the volume of a spherical cell would produce changes in the tension or stretch of the cell membrane, and this would contribute to the wear and tear on the cell. In a biconcave disc, however, no changes in membrane tension occur. This undoubtedly prolongs the useful life of the corpuscle, a life which is already limited by its inability to repair itself.

In the blood of most vertebrates heme units are combined in groups of four to form single, larger molecules with molecular weights of about 67,000. It will be recalled that a single hemoglobin unit has a molecular weight of only about 16,500 to 17,000. Within each of the four-unit hemoglobin molecules, each heme combines with a molecule of oxygen; therefore the whole molecule can carry four O_2 molecules. The oxygenation of each heme unit is independent if the units are separate, but when combined in this fashion there is an interaction between heme units. The oxygenation of one of the four heme units greatly accelerates the oxygenation of the other three. This so-called heme-heme effect increases the rate and efficiency of oxygen uptake as the blood passes through capillaries in the lungs.

An evolutionary progression toward the four-heme molecules in vertebrates is suggested by the fact that the lamprey has only one-unit hemoglobin, the Pacific hagfish (*Myxine*) has a mixture of one- and two-unit hemoglobins, and all other adult vertebrates have four-unit hemoglobin.

The iron atom in the heme is always in the ferrous state and does not change valence (become ferric iron) by combining with oxygen. In a true chemical oxidation reaction (Sect. 10.2) such a change in valence would occur as a result of the transfer of an electron. Clearly, then, the hemoglobin-oxygen combination is not a chemical oxidation; rather, it is referred to as an oxygenation. The combination is a loose one. The hemoglobin in the gills or lungs readily accepts the oxygen and just as readily releases it to the tissues elsewhere in the body. The formation and dissociation of the oxyhemoglobin, as the oxygenated molecule is called, depends upon the partial pressure of oxygen (pO_2) in the environment to which the hemoglobin is exposed. In the gills or lungs the pO_2 is relatively high, and oxyhemoglobin is readily formed. In other tissues the pO_2 is lower, as a result of the constant oxygen consumption by cells of those tissues. Under these conditions, the oxyhemoglobin dissociates to release the oxygen. A better understanding of these relationships can be obtained by a study of oxyhemoglobin dissociation curves.

7.4 OXYHEMOGLOBIN DISSOCIATION

The oxyhemoglobin dissociation characteristics may be determined in the following way. A sample of hemoglobin is exposed to an atmosphere of pure oxygen, and the amount of oxygen taken up by that hemoglobin is measured; several techniques for making this measurement are available. The amount of oxygen taken up under these conditions represents 100 per cent saturation of the hemoglobin. That is, every heme unit is assumed to be combined with a molecule of oxygen. The same sample is then exposed to an atmosphere containing somewhat less than 100 per cent oxygen. Under these conditions, the hemoglobin sample will release a volume of oxygen, which can be measured. The remaining oxygen still tied to the hemoglobin is recorded as a percentage of the amount required to saturate the hemoglobin. This procedure is repeated at various concentrations of oxygen (pO_2 levels) in the atmosphere and the results are plotted as a graph (Fig. 7-3). The graph is called the oxyhemoglobin dissociation curve, and it expresses

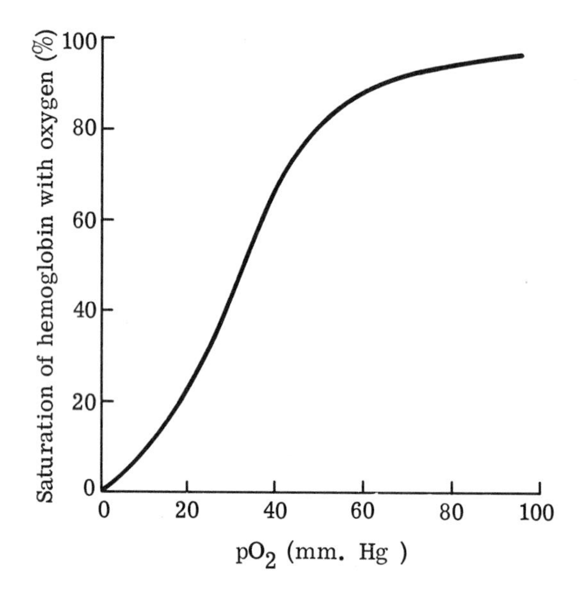

Figure 7–3 A typical oxyhemoglobin dissociation curve (see text).

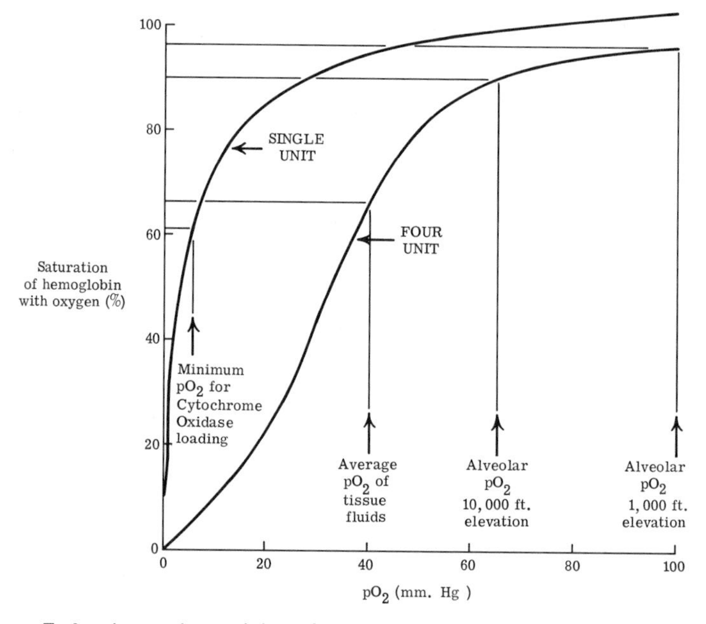

Figure 7–4 The oxyhemoglobin dissociation curve (four-unit) compared with the oxymyoglobin dissociation curve (single unit). See text.

the affinity of hemoglobin for oxygen at various levels of oxygen partial pressure.

When single unit hemoglobin is used, a hyperbolic oxyhemoglobin dissociation curve results (Fig. 7-4), but when the typical four-unit hemoglobin is used, the oxyhemoglobin dissociation curve is sigmoid in shape. The heme-heme effect of the four-unit hemoglobin molecule accounts for the difference in the shape of the curve.

The hemoglobin in the blood of most mammals characteristically loads oxygen to the extent of 96 to 98 per cent of total saturation when it is exposed to an atmosphere in which the pO_2 is 100 millimeters of mercury. This is the pO_2 of alveolar air at an elevation of about 1000 feet above sea level (Fig. 6-10). At a pO_2 of 67 millimeters of mercury, blood hemoglobin still retains about 90 per cent saturation, and this corresponds with the alveolar pO_2 at an elevation of about 10,000 feet. Thus, blood hemoglobin is able to load oxygen efficiently over the usual elevational range of mammals.

The pO_2 of dissolved oxygen in tissue fluids is usually about 40 millimeters of mercury or less. This, of course, is due to the fact that tissue cells are constantly removing the oxygen from the tissue fluids and utilizing it in their metabolism. At this pO_2 level blood hemoglobin retains only about 65 per cent saturation; the remainder of the oxygen taken up in the lungs is released into the tissue fluids. When tissue demands for oxygen are increased, as during exercise, the local tissue fluid pO_2 drops to lower levels, and even greater quantities of oxygen are released from the hemoglobin.

Muscles contain a one-unit hemoglobin molecule called myoglobin. Myoglobin loads oxygen to more than 90 per cent of its saturation capacity at the pO_2 levels commonly encountered in tissue fluids. Thus, oxygen released from blood hemoglobin can easily become associated with myoglobin. It can be either stored in that form or passed on to the tissue cells as it is needed. The oxidative enzyme cytochrome oxidase (cytochrome C_3) within each metabolizing cell ultimately utilizes the oxygen (Sect. 10.7). It can accept oxygen at a pO_2 level as low as 5 millimeters of mercury, so it can easily take it from the myoglobin as it is needed in the cell. While the combination of oxygen with hemoglobin or with myoglobin is an oxygenation, the combination of oxygen with hydrogen, which is catalyzed by cytochrome oxidase, is a true chemical oxidation. This last reaction is not reversible in animal metabolism.

The unborn mammalian fetus must be able to obtain oxygen from the maternal bloodstream by means of diffusion across the placenta. Because the fetal blood is constantly being deoxygenated in the fetal tissues, a concentration gradient exists across the placental membranes so that diffusion of oxygen from mother to fetus is favored. In many cases this diffusion is further favored by the fact that the fetal hemoglobin has a greater affinity for oxygen than the maternal hemoglobin at the same pO_2 level. The fetal oxyhemoglobin dissociation curve, therefore, lies to the left of the maternal oxyhemoglobin dissociation curve

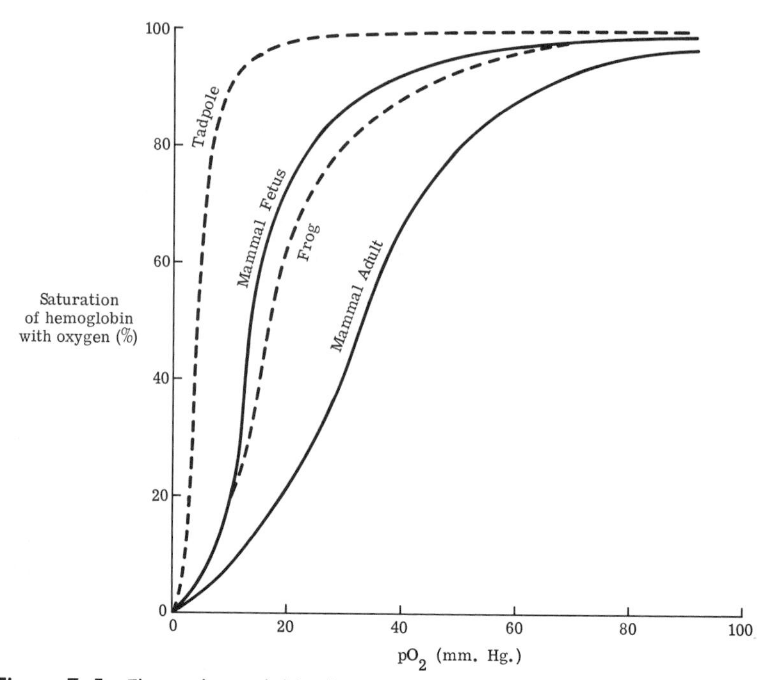

Figure 7–5 The oxyhemoglobin dissociation curves of adults are generally plotted to the right of the curves for the larva or fetus of the same species. Note also that the curve for the tadpole is hyperbolic rather than sigmoid in shape (see text).

(Fig. 7-5). The change in an individual from fetal to adult-type hemoglobin commences at birth, indicating that a different globulin comes into use at that time. Other live-bearing vertebrates, such as the ovoviviparous dogfish shark, show similar shifts from fetal to adult oxyhemoglobin characteristics at birth. Also, comparable shifts occur in amphibian hemoglobin at the time of metamorphosis (Fig. 7-5) and in various egg-laying species at the time of hatching.

The position and shape of the oxyhemoglobin dissociation curve is also influenced by the concentration of carbon dioxide (pCO_2) in the environment. Increasing the pCO_2 typically shifts the curve to the right (Fig. 7-6). Decreasing the pH (increasing acidity) has the same effect as increasing pCO_2. Both of these conditions are precisely those that occur in tissues during periods of increased metabolic activity. During exercise, for example, both carbon dioxide and lactic acid concentrations (Sect. 10.5) are increased in skeletal muscles. At any given pO_2 level, then, blood hemoglobin retains less and less of its oxygen as the pCO_2 increases or as the pH falls. Under the very conditions that increase tissue demands for oxygen, hemoglobin in the blood releases more of that gas. The effect of pCO_2 or of pH on the form and position of the oxyhemoglobin dissociation curve is known as the Bohr effect.

It should be noted (Fig. 7-6) that the Bohr effect produces large changes in the affinity of hemoglobin for oxygen at low pO_2 levels but that the effect is minimal at higher pO_2 levels. Thus, the loading of

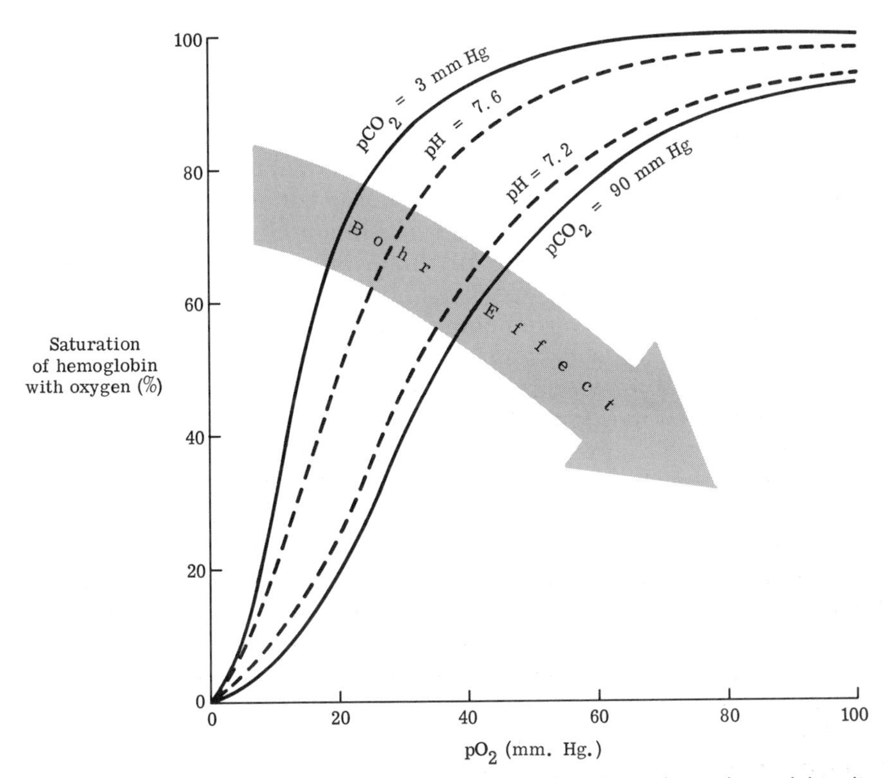

Figure 7–6 The Bohr effect: the effect of pH or of pCO_2 on the oxyhemoglobin dissociation curve (see text).

oxygen by the hemoglobin in the lungs is little changed, but the unloading of oxygen in the tissues is increased greatly.

The form and position of the oxyhemoglobin dissociation curve also varies from species to species. Fishes living in waters that contain little dissolved oxygen have oxyhemoglobin dissociation curves which lie to the left of those for fishes living in better aerated waters. This characteristic extends the upper portion of the curve to lower pO_2 levels. In other words, these fishes can efficiently load oxygen in the gills even though the pO_2 of the environment is lower than usual. Unavoidably, this means that the pO_2 of the tissues must also fall lower than usual in order that the hemoglobin give up a comparable quantity of oxygen in the tissues. Thus, the animal is able to survive the oxygen-poor environment at the cost of slightly lowered tissue pO_2. Undoubtedly, the lower availability of oxygen in the tissues necessitates curtailed physical activity on the part of the animal.

Waters having a low pO_2 generally also have increased pCO_2 levels, so an active Bohr effect would be a liability to fishes such as those discussed in the preceding paragraph. Characteristically the Bohr effect is minimal in such stagnant water fishes. Actually, it would be better if a "reverse Bohr effect" occurred, *i.e.*, a shift of the oxyhemoglobin curve toward the left resulting from increased pCO_2. In some invertebrate animals living in oxygen-poor waters such a reverse Bohr effect is present, but it does not occur in those vertebrates that have been studied.

In some fishes of normal, well-oxygenated waters, increasing the environmental pCO_2 not only shifts the oxyhemoglobin dissociation curve to the right (typical Bohr effect) but also flattens it (Fig. 7-7). This indicates that hemoglobin is not nearly saturated even at the higher pO_2 levels if the pCO_2 is sufficiently high. This response to high pCO_2 is called the Root effect and may be of importance in the inflation of the swim bladder (Sect. 3.4). Hemoglobin showing the Root effect is not present in fishes that lack swim bladders. Presumably, the Root effect would operate in this way: In the gills the hemoglobin is exposed to a high pO_2 but low environmental pCO_2, the oxyhemoglobin dissociation characteristics are therefore normal, and the hemoglobin loads oxygen to near saturation. In the gas gland of the swim bladder, however, the metabolic rate of surrounding tissues is high, so pCO_2 is also high. For the same reasons, lactic acid content of these tissues is high, and thus the pH is lowered. Both conditions produce a Root effect, and the hemoglobin readily unloads oxygen even though the pO_2 level is also high. In tissues other than those of the gas gland, the pO_2 levels are low, so that the effect of pCO_2 is the same as for other vertebrates, *i.e.*, a typical Bohr effect.

The oxyhemoglobin dissociation curves of amphibians, which are primarily aquatic, lies to the left of those for primarily terrestrial animals. These differences reflect the fact that the pO_2 of water is lower than that of air. Similarly, the oxyhemoglobin dissociation curves for animals native to high elevations (another low pO_2 environment) generally lie to the left of those for animals living at lower elevations (Sect. 6.6). Also, as we have seen, the change from oxygen-poor environment to oxygen-rich environment at the time of amphibian metamorphosis or egg-hatching in birds is also accompanied by a change in oxyhemoglob-

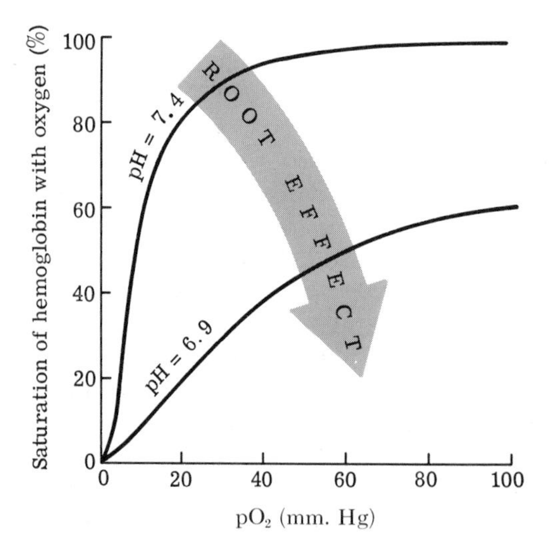

Figure 7–7 The Root effect occurs in the blood of some fishes. See text.

in dissociation characteristics. The curves for larval amphibians and unhatched birds lie to the left of those for adults of the same species (Fig. 7-5). The change in frog hemoglobin at the time of metamorphosis is also shown, and it should be noted that the larval oxyhemoglobin dissociation curve is more hyperbolic and lacks a Bohr effect, whereas that of the adult is typically sigmoid in shape and has a pronounced Bohr effect. This indicates that the four-unit combination of hemoglobin molecules and the consequent heme-heme effect (Sect. 7.3) occur only in adult blood.

The variations in the form and position of the oxyhemoglobin dissociation curve that have been discussed in preceding paragraphs are largely due to small differences in the globulin portion of the hemoglobin molecule. For example, there is a difference of only one amino acid in the globulin of sickle cell anemia victims as compared with normal human hemoglobin. Similarly, the Bohr effect depends upon the presence of one or more sulfhydryl groups (—SH) in the globulin portion of the molecule. The amino acid cysteine (Fig. 9–9) has such a sulfhydryl group. It is known that single mutations can result in the substitution of single amino acids in a given protein molecule (Sect. 9.4). Such single mutations, then, can result in striking changes in the oxyhemoglobin dissociation characteristics of hemoglobin.

Because proteins are charged molecules, they migrate at different rates when exposed to an electrical field. This characteristic is frequently used to separate the various proteins present in a mixture. A sample of the mixture is placed on a strip of paper or other material, and an electrical current is passed through the strip in a moist chamber. After allowing time for protein migration, the strip is removed and treated chemically to demonstrate the positions of the various proteins in the mixture. Since some will have migrated further than others, the proteins will be serially separated along the strip. This technique is called electrophoresis.

Electrophoretic studies indicate that many vertebrates have more than a single type of hemoglobin present at the same time. For example, many birds have one to three (usually two); horses, cattle, and sheep have two; and amphibians and reptiles contain several different hemoglobins simultaneously. The differences that can be demonstrated by electrophoretic separation are often sufficient to identify genera or even species. Of course, the oxyhemoglobin dissociation curves plotted for such mixtures of hemoglobins are the sums of the oxyhemoglobin dissociation characteristics of the different hemoglobins present. Undoubtedly, if all the different hemoglobins could be separated in sufficient quantities to permit separate oxyhemoglobin dissociation curves to be plotted, these curves would also be different. Such mixtures may well extend the ranges of some animals. For example, if an animal has two hemoglobins which show different oxyhemoglobin dissociation curves (one to the left of the other), that animal can range over wider differences in environmental pO_2 values. Acclimation to a higher elevation may involve, in part, a change in the proportions of the two hemoglobins so that more of the hemoglobin with the curve to the left is produced. This would acclimate the animal to the reduced oxygen availability of the higher elevation.

7.5 CARBON DIOXIDE TRANSPORT

While the solubility of carbon dioxide is greater than that of oxygen (Fig. 6-3), the quantities transported by the blood of vertebrates is still too great to be accounted for in this way alone. Indeed, mammalian blood actually transports at least ten times as much carbon dioxide as would be possible in solution. Quite clearly, then, mechanisms must be present for carrying carbon dioxide in other ways in the blood.

Carbon dioxide is transported in the blood primarily in the form of bicarbonate ions (HCO_3^-) in the plasma. Smaller amounts are carried in other forms as well (Fig. 7-8), and each is separately discussed in the following paragraphs. Carbon dioxide diffuses in solution from the metabolizing tissue cells that produce it. It is carried into the blood of the capillaries by fluid exchange (Sect. 4.5), after which the following series of events takes place.

1. A small amount of the carbon dioxide reacts with water in the plasma to form carbonic acid (H_2CO_3). The rate of carbonic acid formation is low; thus, very little carbon dioxide can be carried in this form. The carbonic acid that is formed reacts with various buffer systems in the blood so that it does not upset the pH of the blood plasma. For example:

$$Na_2HPO_4 + H_2CO_3 \longrightarrow NaH_2PO_4 + NaHCO_3$$

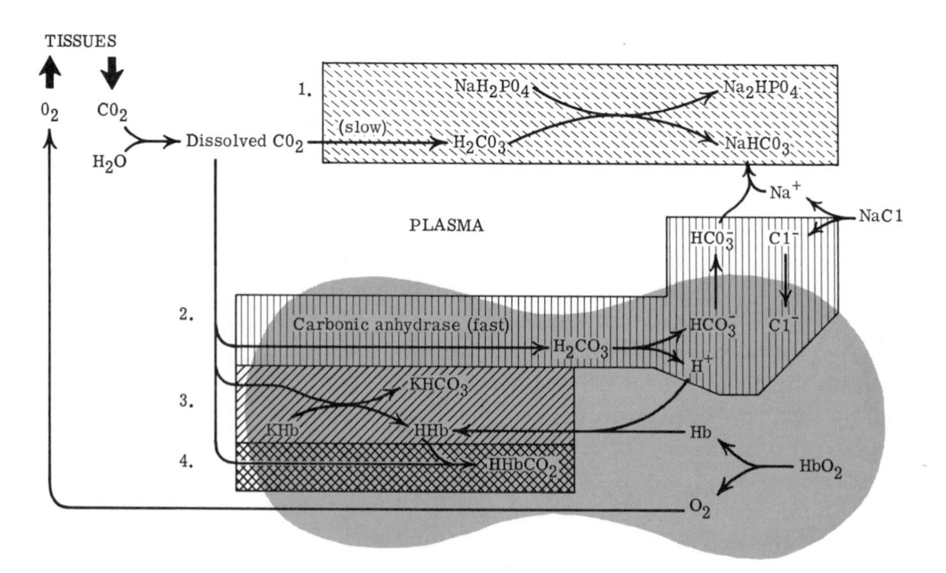

Figure 7–8 Carbon dioxide transport mechanisms (see text).

2. Most of the dissolved carbon dioxide diffuses from the plasma into the red blood corpuscles where the formation of carbonic acid is greatly accelerated by the presence of the enzyme carbonic anhydrase. The carbonic acid formed there is then ionized to yield bicarbonate and hydrogen ions:

$$CO_2 + H_2O \longrightarrow H_2CO_3 \longrightarrow H^+ + HCO_3^-$$

Most of the bicarbonate ion (HCO_3^-) then diffuses back into the plasma in which it is carried in that form. To maintain electrical balance across the corpuscle membrane, there is a simultaneous inward diffusion of chloride ions (Cl^-) from the plasma. Because of this movement of chloride ions, the process is sometimes spoken of as the chloride shift mechanism. Alternatively, it is sometimes called the Hamberger phenomenon in honor of its discoverer.

3. Some of the carbonic acid formed within the red blood corpuscle, as outlined in the preceding paragraph, remains there and reacts with hemoglobin. Oxygenated and deoxygenated hemoglobin together form an effective buffer system which prevents this carbonic acid from upsetting the pH within the corpuscle (Hb = hemoglobin):

$$KHb + H_2CO_3 \longrightarrow KHCO_3 + HHb$$

4. Finally, some of the dissolved carbon dioxide entering the red blood corpuscle reacts directly with the amino group ($-NH_2$) of deoxygenated hemoglobin to form a carbamino compound called carbhemoglobin:

$$Hb{-}N\begin{matrix}\diagup H\\ \diagdown H\end{matrix} + CO_2 \longrightarrow Hb{-}N\begin{matrix}\diagup H\\ \diagdown COOH\end{matrix}$$

The reactions discussed in the preceding paragraphs show the various ways in which carbon dioxide is loaded by the blood as it passes through the general body tissues. Of course, when the blood containing the carbon dioxide reaches the capillaries of the lungs, carbon dioxide is unloaded by reversing the direction of each of these reactions. It then diffuses from the blood into the alveoli of the lungs and is expired.

Of the four carbon dioxide transport methods discussed, the Hamberger phenomenon (2) is the most important, since it accounts for the transport of 85 to 90 per cent of the carbon dioxide present in venous blood. Because the red corpuscle membrane is an important part of the mechanism, this constitutes another evolutionary advancement attained by the enclosure of hemoglobin within blood cells or corpuscles (Sect. 7.3).

Water enters the red blood corpuscle along with the carbon dioxide, and thus increases the erythrocyte volume. For this reason the corpuscles are slightly larger in venous blood than in arterial blood.

The student should not be misled into believing that hemoglobin unloads all its oxygen in the tissues or all its carbon dioxide in the lungs. The arterial-venous pO_2 difference in a human at rest is about 30 per cent, indicating that under these conditions the blood unloads only about 30 per cent of its oxygen content to the tissues. Similar arterial-venous differences in pCO_2 indicate that only about 30 per cent of the carbon dioxide is unloaded in the lungs under the same conditions. The Bohr effect can increase oxygen utilization during physical activity, and this, of course, results in a larger arterial-venous difference in pO_2. Increased ventilation of the lungs during exercise lowers the pCO_2 of alveolar air, and this in turn removes more carbon dioxide from the blood. Thus, a greater arterial-venous difference in pCO_2 also occurs at this time. Oxygen utilization, and therefore arterial-venous pO_2 difference, is greater in birds than in mammals and greater in active vertebrates than in more sedentary vertebrates.

NON-TRANSPORT FUNCTIONS

7.6 BLOOD COAGULATION

As the various transport functions of blood became more complex during the evolution of vertebrates and as the delicate balances between hydrostatic and osmotic pressures of blood (Sect. 4.5) developed, it became more essential to prevent the loss of blood through injury. It is not surprising, then, that special mechanisms for preventing such loss also appeared and gradually increased in complexity during the evolutionary process.

In invertebrate animals it is usually sufficient to plug leaks caused by injury with cellular elements or to reduce fluid loss by contraction of the tissues surrounding the wound. Hydrostatic pressures in the circulatory system are low enough that such mechanisms are sufficient to stop the leak. In some invertebrates a gelation of protein elements of the blood occurs upon contact with the surrounding environmental medium. In the vertebrates, however, much more complex chemical processes lead to the formation of a coagulum, or blood clot, and subsequent retraction of that clot aids in pulling the edges of the wound together. The coagulum is formed by the precipitation of a special blood protein called fibrin, but this occurs only as the last step of a series of chemical reactions.

The series of chemical reactions that results in blood coagulation is complex and not completely understood. It involves the release of a substance called thromboplastin (Fig. 7-9) from the injured tissue cells and from blood platelets, or thrombocytes (Sect. 7.10). Also present in the circulating blood is another substance called prothrombin; this is combined with a third substance called antiprothrombin. Antiprothrombin masks the prothrombin and prevents it from being reactive. However, thromboplastin, released from injured tissue cells and thrombo-

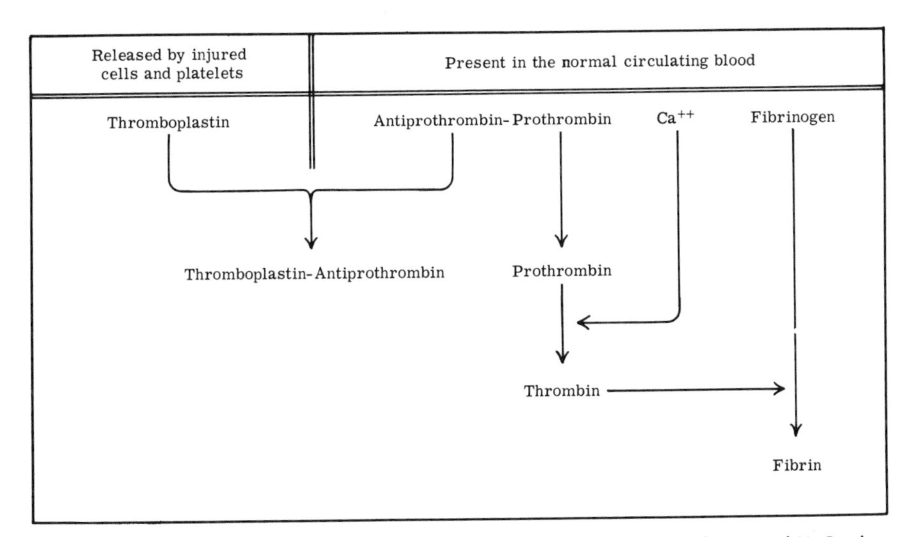

Figure 7–9 A simplified scheme of blood coagulation. (From Cockrum and McCauley: *Zoology*. Philadelphia, W. B. Saunders Company, 1965.)

cytes, acts to remove the antiprothrombin mask and, as a result, prothrombin is freed. Prothrombin is activated in the presence of calcium ions (Ca^{++}), and the active form is called thrombin. Thrombin acts in some way to convert the soluble blood protein fibrinogen to an insoluble form called fibrin. As a result, fibrin precipitates in the form of threadlike fibers. Undoubtedly, the chemical reactions leading to this final precipitation of fibrin are more complex and numerous than indicated here. Not all of the details are known, but this simplified scheme gives an accurate idea of the general kinds of reactions involved.

The precipitated strands of fibrin form a meshwork that blocks the leak or wound as piles of branches and sticks might block a flowing stream. Plasma passing between the fibrin strands carries blood cells with it, which tend to plug the openings between the strands of fibrin. Platelets (thrombocytes) carried into the fibrin meshwork in this way break down to release more thromboplastin, thus accelerating the process of fibrin precipitation. In this way a solid clot is formed that completely blocks the wound opening so that no further blood loss occurs.

Platelets within the clot then undergo what has been called the spreading reaction. They send out long, threadlike processes in all directions, and these become firmly attached to the strands of fibrin. Finally, the platelets contract again, drawing these processes inward. Because the processes are attached to fibrin strands, they draw the meshwork tighter (retraction of the clot) and squeeze out the fluids remaining in the clot. These fluids are called serum, which is essentially blood plasma minus those components that have been precipitated (fibrin). Drawing the clot together in this way also draws the edges of the wound together so that the processes of healing can commence. Squeezing out the serum also allows the surface of the coagulum, which is exposed to the air, to dry and form a hard protective scab.

Almost everyone has, as a child, fallen and skinned his knees. Commonly such wounds, while they may cover a considerable area, bleed very little. This is because a lot of tissue was damaged and, therefore, a lot of thromboplastin was immediately released. Coagulation, then, was fast and efficient. On the other hand, most men are aware that a razor cut inflicted while shaving continues to bleed for a long time. Since a razor cut damages relatively few cells, a minimal amount of thromboplastin is released and coagulation is more prolonged. The process can be hastened, as every man soon learns, by placing a small piece of paper on the wound, which provides a foreign surface not unlike the fibrin meshwork. When platelets come in contact with it, they break down to release thromboplastin and thus accelerate clot formation. Before the days of tissue paper, people commonly used spider webs for the same pupose. Unfortunately, the clot forms in intimate association with the paper, and if the paper is removed too soon, the clot is removed with it and the wound begins to bleed again.

Blood also contains a number of counterbalancing factors that prevent coagulation from occurring, normally, within uninjured blood vessels. Among these is the substance heparin which is commonly used in the laboratory to prevent blood coagulation. It is also widely used following surgical procedures to prevent the formation of internal blood clots.

The blood stored in blood banks contains a small amount of potassium (or sodium) citrate. This substance ties up all of the calcium ions in the blood and thus prevents the conversion of prothrombin to thrombin. Potassium or sodium oxylate can be used in the same way to arrest coagulation at that stage.

Many blood sucking invertebrates have evolved chemical constituents in the saliva that prevent the blood of the host from coagulating. In some cases these substances will even reliquefy blood that has already coagulated. In early America, "medicinal" leeches were commonly used to reduce black eyes. The bruised appearance of a black eye is caused by subcutaneous pools of coagulated blood, and the leech, applied to the site, is able to reliquefy the blood and feed upon it until the normal color of the skin is reestablished. On the other hand, certain snake venoms kill the snake's prey by causing abnormal coagulation of the circulating blood within the vessels.

7.7 OSMOTIC REGULATION

Soluble proteins of the blood include fibrinogen, globulins, and albumins. Fibrinogen, as we have seen, plays a most important role in blood coagulation. Globulins are involved in the formation of hemoglobin (Sect. 7.3), immune substances, or antibodies (Sect. 7.8), and in still other ways. The albumins, however, account for more than half of the total dissolved blood protein in most vertebrates. They are absent entirely only in elasmobranchs and primitive adult teleosts. They are absent from the blood of amphibian larvae but make up about half of the dissolved protein of adult amphibians. Albumins are relatively small molecules and because of their large numbers, they contribute more than any other dissolved substance to blood osmotic pressure.

The evolutionary and developmental appearance of blood albumins corresponds with the appearance of a high-pressure circulatory system. In such a high-pressure system, as we have seen (Sect. 4.5), a higher osmotic pressure is required as well; otherwise, the delicate balance controlling the exchange of fluids between the blood and the tissue matrix would be upset. Increased blood osmotic pressure probably also helps to protect marine vertebrates from osmotic desiccation (Sect. 8.11).

7.8 IMMUNITY

Another specialization of the blood (and associated lymphatic tissues) of vertebrate animals is protective in nature. This protective function depends upon the ability of certain lymphatic cells, called plasma cells, to produce substances that render the animal immune to pathogenic organisms (bacteria, viruses) or to the toxic substances which these microorganisms produce. The immunizing substances produced by the plasma cells are called antibodies or immune bodies.

In some cases immunity is naturally present. Man is completely immune to some but not all of the diseases that attack other vertebrate animals. For example, while humans are naturally immune to the causative virus of dog distemper, they have no such immunity to the rabies virus. Natural immunities probably never depend upon the production of antibodies; rather, they merely reflect the fact that most pathogenic organisms can live and reproduce only under rather specific conditions, which are found in not more than a few host species.

More frequently, an initial exposure to the infective microorganism is required, and this brings about the production of specific immunity. Subsequent exposures to the same microorganism, then, are without pathological effect for periods which may last for a few weeks to many years.

Immune bodies (antibodies) are produced by plasma cells of the lymphatic tissues, and blood globulins are the materials from which the antibodies are made. Globulins comprise a large portion of the dissolved blood proteins and are separable into three general types. These are designated as alpha (α), beta (β), and gamma (γ) globulins. The gamma globulins are the molecules usually associated with antibody production.

In the presence of a "foreign" protein, one which does not occur normally in the animal in question, the plasma cells produce modified gamma globulins that are specific for the foreign protein. Such foreign proteins are called antigens. Subsequent exposure to the same antigen that initially caused specific antibody production will then result in an antibody-antigen reaction. A number of different kinds of antibody-antigen reactions are known, but in all cases the reaction results in the destruction or elimination of the antigen. The exact nature of antibody formation and of the antibody-antigen reaction is not known with certainty, but it is thought to involve complementary molecular shapes (Fig. 7-10).

The antigen may be a part of the structural protein on the surface of a bacterium or virus or it may be a metabolic product of the microorganism. In contact with such a protein, the plasma cell is thought to modify the shape of a gamma globulin molecule so that it "fits" the shape of the antigen molecule. Once the plasma cells have produced the specific antibody, they retain for a time the ability to continue producing the same antibody even in the absence of the specific antigen. This period of time is variable and accounts for the fact that some immunities last for only a short period, whereas others may persist for many years or even for the life of the organism.

The transplanting of tissues from one animal to another involves similar foreign protein responses, since the tissue proteins of one individual may not be identical to those of another. A good deal of experimental transplantation of tissues has been carried out in lower vertebrates, and some inferences regarding the possible evolution of immune responses have been forthcoming. In lower vertebrates transplants between individuals of the same species are usually successful and transplants between individuals of closely related genera are often accomplished without any indication of antibody-antigen reactions. Antibody-antigen responses, when they do occur, result in rejection of the transplanted tissue; that is, the transplant fails to live and become a permanent part of the animal receiving it. The relative ease of transplanting tissues among lower vertebrates suggests that immune responses are relatively insensitive in these animals. It might also suggest that proteins do not differ markedly between the individuals involved, but this is a less likely possibility.

Among mammals, tissue transplants are ordinarily possible only

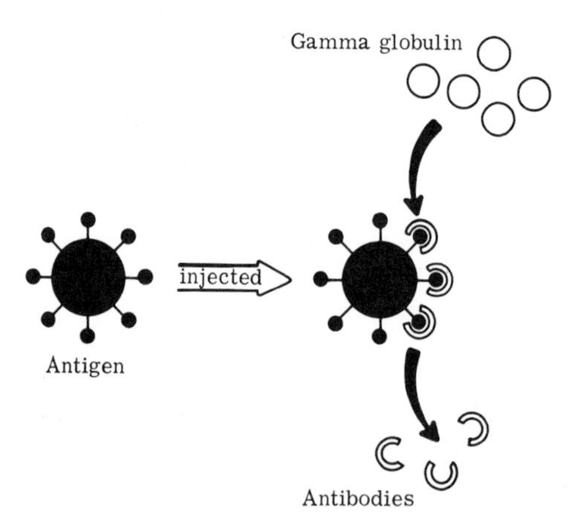

Figure 7–10 The formation of specific antibodies from gamma globulin molecules in the presence of antigen (schematic). (From Cockrum and McCauley: *Zoology*. Philadelphia, W. B. Saunders Company, 1965.)

between individuals of a closely inbred strain or between identical twins. Inbreeding of animals produces similar genetic makeup and, therefore, similar protein content. The same is true, of course, for identical twins. Attempts in recent years to transplant organs from one human to another have been completely successful only when the donor and the recipient were identical twins. Satisfactory results have been achieved in other cases by the continued use of immunosuppressive drugs or radiation treatments. Radiation treatments also inhibit the immune responses, but such treatments, thus far, are non-specific and deprive the individual of much of the normal protective function of his immune mechanisms. He may then succumb more easily to other diseases.

Since, among higher vertebrates, each individual is probably unique in his complete array of tissue proteins, some sort of "inventory" must be taken early in life. Otherwise, it would be impossible for the plasma cells to distinguish between normal and foreign proteins. It would, of course, be fatal for an individual to produce antibodies against proteins normally present in his own makeup. All evidence to date suggests that this "inventory" takes place shortly following the birth of mammals, but the nature of the inventory is not yet known. Prior to birth the immune system is completely inoperative.

7.9 BLOOD TYPES

A single mammalian red corpuscle may have several different proteins simultaneously present as parts of its surface membrane. Each of these proteins is potentially an antigen if it is introduced into the blood of another individual for whom that protein is "foreign." Each such protein is present or absent in a given individual because of the genetic makeup inherited from the parents of that individual.

It will be recalled that the genetic code occurs in the DNA of the chromosomes within the nucleus of each cell. Chromosomes are present in pairs, and the members of any given chromosome pair are spoken of as homologous chromosomes. Each member of a pair of homologous chromosomes contains sites that code similar characteristics. Thus, if one human chromosome carries a code dictating the production of protein "A" in the red corpuscle membrane, its homologous partner will carry at the same site a code dictating the production of the same protein "A," or a related protein "B," or no protein at all, "O." These three forms of the code are mutant variables and, since there are two chromosomes involved, there can be six different combinations of the three forms. These are AA, BB, OO, AB, AO, and BO. These six genetic combinations (genotypes) can form four different types of red blood corpuscles (phenotypes). These are A, B, AB, and O (Fig. 7-11). The genotype AA and the genotype AO both result in the production of protein "A" only, so the phenotype is indistinguishable. Similarly, the genotype BB and the genotype BO both result in the production of protein "B" only and, again, the phenotype is indistin-

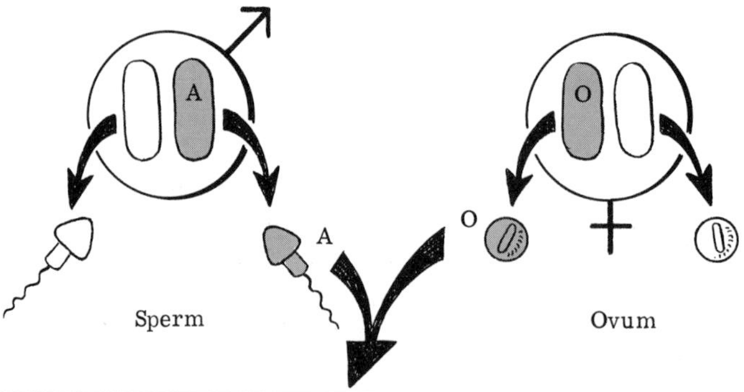

Genotype	AA	AO	BB	BO	AB	OO
Phenotype	A		B		AB	O
Agglutinogens (antigens) on red cells	A		B		A and B	none
Agglutinins (antibodies) in plasma	Anti-B		Anti-A		none	Anti-A and Anti-B

Figure 7–11 Landsteiner blood types. See text.

guishable. The genotype AB results in the production of both protein "A" and protein "B" in the same individual, and the genotype OO results in the production of neither protein "A" nor protein "B" in that individual.

The four phenotypic blood types based upon the presence or absence of the "A" and "B" proteins in the red corpuscle membrane are called the Landsteiner blood types in honor of the discoverer. In some literature these proteins are designated as "L_A" and "L_B" for the same reason. Other sets of similar proteins also occur in red corpuscle membranes and result from the action of genetic codes at different chromosome sites. For example, a protein designated as "M" and a mutant form designated as "N" form a similar family of genotypes and phenotypes in humans.

Within a short time after birth, persons who are themselves phenotypically type "A" (genotype AA or AO) acquire anti-B antibodies in their blood plasma. Similarly, phenotypically type "B" persons acquire anti-A antibodies. Type AB individuals acquire neither of these, but type O individuals acquire both anti-A and anti-B antibodies. This occurs even though no "foreign" bloods are introduced. It may be that proteins similar to both "A" and "B" occur naturally in foods, are somehow absorped intact, and lead to the usual form of antibody production, or it may be that some other mechanism is involved. In any case, the specific antibodies are soon present. Thus, there is a potential antibody-antigen reaction in any instance in which bloods from different type individuals are mixed. (Fig. 7-12), and the type of antibody-

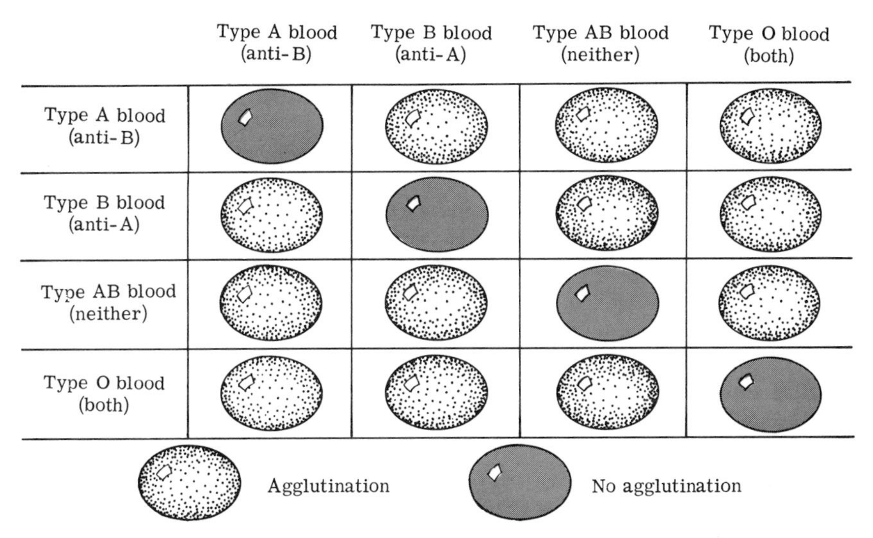

Figure 7–12 Results of cross-matching human blood types. (After Cockrum and McCauley: *Zoology*. Philadelphia, W. B. Saunders Company, 1965.)

antigen reaction that occurs is called an agglutination, the clumping together of red corpuscles in large masses. Accordingly, the antigen is often given the special name agglutinogen and the antibody is called the agglutinin. The agglutinin can be thought of as a sort of "glue" in the plasma which sticks red corpuscles together (agglutination) if the specific agglutinogen is present in the membranes of the corpuscle. The agglutination reaction is easy to observe with the naked eye. If two drops of blood of the same type are mixed on a microscope slide, the combined drops retain the normal, homogenous red appearance. If, however, the two drops of blood are of different types so that an agglutination reaction occurs, the combined drops appear like water containing large flecks of material resembling red pepper.

Quite obviously, blood types must be considered before giving blood transfusions to one person from another. The clumped red blood corpuscles that result from agglutination will act as plugs in small blood vessels and disrupt circulation to vital tissues. Because of the dilution factor involved in transfusing a small amount of donor's blood into the relatively large amount of recipient's blood, type O blood (containing no antigens) could theoretically be given to a person of any other Landsteiner blood type. Type O blood is, therefore, called the universal donor type. Similarly, type AB persons are known as universal recipients since their blood contains no antibodies against either protein "A" or protein "B." They could theoretically receive blood from a person of any other Landsteiner type. In practice, however, such transfusions are considered to be dangerous, and if at all possible, only type-specific transfusions are given.

The frequency of occurrence of different Landsteiner types in human populations varies. For example, European caucasians have the following distribution: 50 per cent O, 40 per cent A, eight per cent B, and two per cent AB; on the other hand, North American Indians are so seldom of the B type that AB is unknown among them. Attempts have been made by anthropologists to use these differences to trace the origins of various human races.

Another antigenic protein sometimes present in the membrane of the human red blood corpuscle is designated as Rh. It is so named because of the fact that it was first discovered in the red blood corpuscles of Rhesus monkeys. Several subtypes representing various mutant forms are known.

About 85 per cent of the American population has some form of Rh protein present, and these are known as Rh positive individuals (Rh+). The Rh+ phenotype may result from either the genotype Rh+Rh+ or the genotype Rh+Rh–, since Rh– is the equivalent of O in the Landsteiner types. It merely indicates that no Rh protein is encoded in that chromosome. The remaining 15 per cent of Americans are of the Rh– phenotype indicating that their genotypes are Rh–Rh–.

No anti-Rh antibodies are naturally present in any individual. They are produced only by Rh– individuals who have had Rh+ blood actually introduced into their blood streams in some way. Once this has occurred, however, the plasma cells continue to produce anti-Rh antibodies for the rest of the individual's life. Any subsequent reintroduction of Rh+ blood into this Rh– individual, then, results in an antibody-antigen reaction. In this case, the reaction takes the form of corpuscle lysis; that is, the Rh+ red blood corpuscles burst. Agglutination may also occur (Fig. 7–13). The introduction of Rh– blood into an Rh+ individual is, of course, without effect since Rh– blood, by definition, contains no Rh antigen. Quite clearly, then, the Rh characteristics of the donor and recipient must also be known before any blood transfusion is attempted.

Because damage to the placenta during pregnancy or, more frequently, at the time of childbirth may occur, it is possible for an Rh– mother to receive a small transfusion of blood from an Rh+ child (Fig. 7-14). In such cases, the mother's immune system responds by producing anti-Rh antibodies. During a subsequent pregnancy, these antibodies can diffuse across the placenta, enter the fetal blood, and there cause destruction of the child's red blood corpuscles. If uncontrolled by medical techniques, the resulting condition, known as erythroblastosis

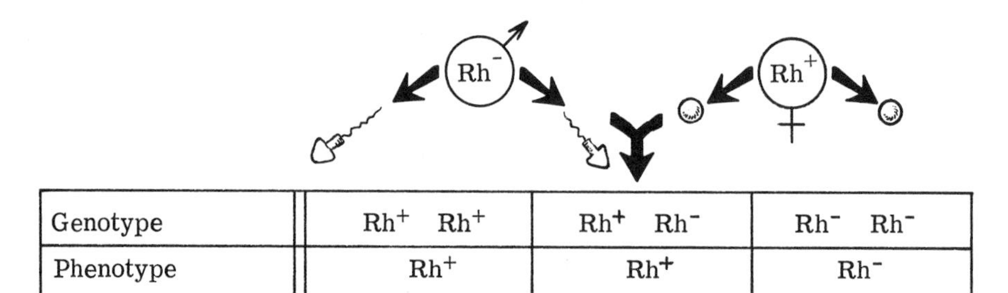

Genotype	Rh^+ Rh^+	Rh^+ Rh^-	Rh^- Rh^-
Phenotype	Rh^+	Rh^+	Rh^-

Figure 7–13 Rh blood types. See text.

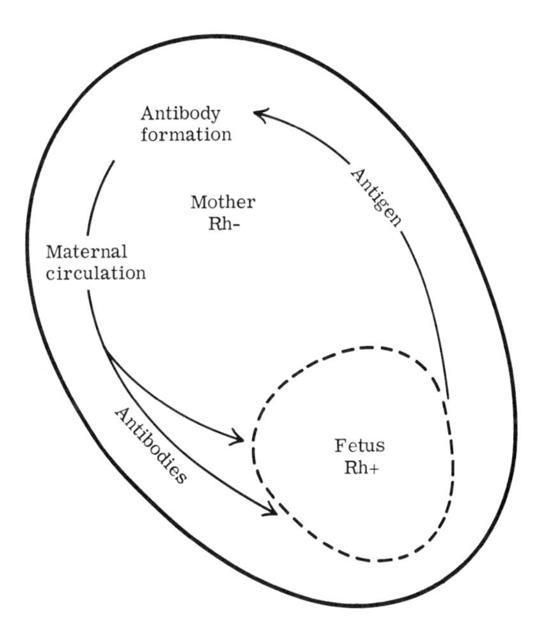

Figure 7–14 The development of erythroblastosis fetalis in the unborn human child (schematic). (From Cockrum and McCauley: *Zoology*. Philadelphia, W. B. Saunders Company, 1965.)

fetalis, may even result in the death of the unborn child. The development by the mother of anti-Rh antibodies is a slow process because of the small amount of antigen introduced. Consequently, the danger of fetal damage is small during a first or second pregnancy but increases with repeated pregnancies of this type. The inheritance of the Rh+ condition is such that this situation can only occur where the mother is Rh– and the father is Rh+. If the father has the genotype Rh+Rh–, then it is likely to occur in only half of the pregnancies, since half of the children of this couple statistically will be Rh–.

Research to date indicates that different blood types occur in most, if not all, vertebrates. For example, even salmon are known to have at least eight different blood types.

The number of different proteins that are potentially antigenic in nature is very large. Occasionally, however, the same protein (or a very similar one) may occur in more than a single species. The spiny lobster (*Panulirus*) has a naturally occurring substance that agglutinates the red blood corpuscles of certain whales. Similarly, beef heart muscle contains a protein that is sufficiently like the surface protein of *Treponema pallidum*, the causative organism of syphilis, that it will react with antisyphilis antibodies. In the Wasserman test for syphilis, beef heart antigen is used. A reaction with the blood plasma of an individual with this antigen indicates that the syphilis organism is, or has been, present in that individual and has caused him to produce antisyphilis antibodies. Such cases of unrelated antigens and antibodies are obviously not the result of acquired immunities but have come about merely by chance. The occurrence in some invertebrate animals of substances in the seminal fluids which specifically agglutinate the spermatozoa of related species may, however, serve to prevent hybridization between the species.

7.10 FORMED ELEMENTS OF THE BLOOD

Besides the liquid portion (plasma), blood contains several kinds of corpuscles, cells, and cell fragments. These are called, collectively, the formed elements of the blood. They include the erythrocytes (red cells or corpuscles), leukocytes (white cells) of several kinds, and the platelets (thrombocytes), which are really only cell fragments (Fig. 7-15).

The structure and function of erythrocytes has already been discussed (Sect. 7.3). They are produced by the red bone marrow of adults and, in young animals, by the liver. The specific stimulus to the production of hemoglobin and of erythrocytes is a lowered oxygen concen-

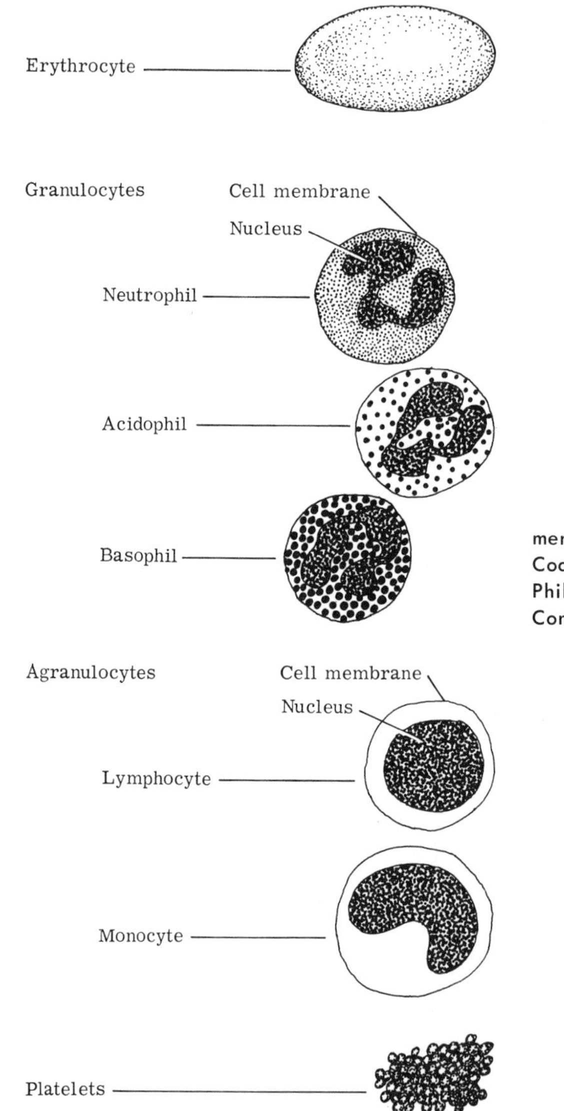

Figure 7–15 The formed elements of human blood. (From Cockrum and McCauley: *Zoology*. Philadelphia, W. B. Saunders Company, 1965.)

tration in the blood. Thus, either a reduction in circulating hemoglobin-filled corpuscles or a prolonged reduction in atmospheric pO_2 (as at high elevation) has the effect of increasing erythrocyte production. Certain special dietary factors are required for normal red corpuscle production as well (Sect. 9.5).

A human erythrocyte remains in the circulation for about four months during which time it may travel more than 200 miles through the blood vessels. An average "trip" takes it from the heart to the lungs and back and then to some part of the systemic circulation and back again—a distance of perhaps 5 to 6 feet in all. This is accomplished in about 40 seconds, and during that time the corpuscle is subject to turbulence and passes through capillaries at least twice. In the capillaries it is deformed and abraded because, as will be recalled, it is actually larger than the capillary lumen. Obviously, the wear and tear on the corpuscle is great. Lacking normal cell organelles it cannot repair itself, and ultimately the membrane ruptures and the hemoglobin spills out and becomes dissolved in the blood plasma. Both the hemoglobin and the ruptured corpuscle membrane are removed by phagocytic cells that break them down into their component parts.

The breakdown of hemoglobin yields three products: globulin, heme, and iron. The globulin and the iron are retained by the body for use in the formation of new hemoglobin. The heme is excreted by the liver through the bile ducts in the form of bile pigment (Sect. 9.18). The iron is attached to a plasma protein to form a compound called transferrin, and in this form it is transported to the bone marrow for reuse. It may also be used to supplement the iron stores of the body by being converted into a form called ferritin (Sect. 9.19).

The leukocytes (white blood cells) are of several kinds and are classified in two major groups (Fig. 7-15). Those which contain prominent cytoplasmic granules are called granulocytes, and those which lack such granules are called agranulocytes. The granulocytes are also similar in that they possess a peculiar, lobated nucleus.

The three types of granulocytes are differentiated on the basis of different staining characteristics of their cytoplasmic granules. When stained in the most commonly used blood stain, Wright's stain, the granules of each acquire a characteristic color. Those of acidophils take up the acid component of Wright's stain and are colored, therefore, a bright red. Since this red component of the stain is a substance called eosin, these cells are sometimes referred to as eosinophils. The granules of basophils are stained dark blue by the basic (alkaline) methylene blue in Wright's stain. The third type of granulocyte is given the name neutrophil because it takes up both the acid and the alkaline dyes. Consequently, its granules acquire a lavender color.

Agranulocytes all have comparatively large nuclei but no prominent cytoplasmic granules. They are differentiated on the basis of the shape of the nucleus. The nuclei of lymphocytes are spherical, whereas those of monocytes are indented or even horseshoe-shaped.

Granuloctyes are produced, like erythrocytes, in the red bone marrow, while the agranulocytes develop in lymph nodes and other lymphoid tissues such as the thymus, tonsils, and spleen. The life spans of leukocytes are not known with certainty, but they probably remain in the circulating blood for only a few hours. All leukocytes eventually leave the blood and enter the tissues; there probably all except the neutrophils and acidophils undergo further changes. Lymphocytes become large phagocytic cells known as macrophages, and basophils are probably converted into a different cell type known as a mast cell. A discussion of the identity of these cell types and their significance can be found in any textbook of histology, and the interested student should pursue the matter there.

Acidophils appear in large numbers at sites of antibody-antigen reactions. The role played in the immune response is not known, but the relationship is inescapable. Basophils frequently increase in numbers during the healing phases following tissue inflammation, but again the exact function is not known. The roles of neutrophils, lymphocytes, and monocytes are more clear-cut: they serve as the major lines of defense during inflammatory reactions. Neutrophils, in particular, are highly motile (ameboid) cells that are attracted in large numbers to sites of inflammatory tissue destruction. There they engulf (phagocytize) dead tissue cells, invading bacteria, and other detritus. The engulfed materials are digested within the neutrophil in large quantities until it, too, dies. The pus commonly associated with inflammatory sites is largely composed of dead neutrophils and other phagocytic cells of related types.

Platelets (thrombocytes) are also produced in the red bone marrow. Large cells known as megalokaryocytes in the marrow develop budlike appendages which break off and are carried away by the circulating blood. These cell fragments are platelets. Nothing is known of their ultimate fate, but it appears that in humans they are replaced about once in four days. The function of the platelets has already been discussed in connection with blood coagulation (Sect. 7.6).

Similar blood cell types occur generally in all vertebrates though they do not always fall into such distinct groups. Intergradations of cell types as well as immature forms of each are common in the circulation in many vertebrates, including mammals. Normal rabbit blood, for example, contains large numbers of such "abnormal" cells.

CHAPTER 8

KIDNEY FUNCTION: WATER AND SALT BALANCE

8.1 WATER AS A SUBSTANCE

Water was undoubtedly the medium in which life first appeared on this planet and it remains the medium containing the greatest abundance and diversity of living forms today. Of those animals that have adopted a terrestrial habitat, only some of the insects have completely escaped aquatic life. All of the others, including the mammals, must spend at least the larval period immersed in water.

Animal bodies are composed of water to the extent of 60 to 70 per cent by volume. Consequently, all the biochemical reactions of life occur in an aqueous solution. Even the respiratory gases must be dissolved in water as a first step in their transport within the animal body. Water is often a reactant or a product of biochemical reactions as well as the solvent in which these reactions take place.

Water has a number of unique physical properties that are of great importance to life:

1. Only liquid ammonia dissolves more substances than does water. Water, then, is the best possible biochemical solvent.

2. Water has a high surface tension because its molecules are more closely packed together at the interfaces between water and other media. For that reason, dissolved substances tend also to be concentrated at the surface. Under such conditions, chemical reactions between various dissolved substances occur more rapidly there. This effect is much like the action of inorganic catalysts in promoting chemical reactions by holding the reactants together.

3. Water has low compressibility but it can be deformed easily. Thus it serves well as a shock absorber and can serve as a medium for the transmission of forces from one place to another (*i.e.*, hydraulic power transmission).

4. Water is transparent to light but has a greater optical density than air (Sect. 15.2) so air-water interfaces can refract light as well. Optical systems involving aqueous lenses serve in the visual systems of animals.

5. Water has low viscosity so it serves as an efficient vehicle for the transport of substances through the animal body. With certain solutes (dissolved substances) present, however, it forms solutions of high viscosity. Such solutions form good lubricants and are present in joints and other areas involving mechanical friction.

6. Water has a high specific heat. This means that it can gain or lose a large amount of heat with only a small change in its own temperature. It acts, therefore, as a temperature buffer for the animal body. Moreover, the oceans and lakes of the world act as temperature buffers for the atmosphere. If the surface of the planet were composed of larger areas of land and smaller areas of water, the temperature extremes of the air would be much greater than they are.

7. Water has a high heat of evaporation. This means that the evaporation of a small amount of water is accompanied by the loss of a relatively large amount of heat from the water. Water evaporation is therefore an efficient cooling mechanism and is widely used in the regulation of animal body temperature. This property of water is, of course, related to the foregoing property, high specific heat.

8. Water can be held by other substances in fairly large amounts as "bound" water or water of hydration. In this form the water molecules are bound tightly to molecules of other substances but do not react chemically with them. In the bound form, water does not act as a solvent and it is not easily evaporated or frozen; its physical properties are quite different from those of "free" water. It has been shown that in certain organisms from extremely dry environments (*e.g.*, grain beetles) and in certain winter-hardy plants (*e.g.*, winter wheat), unusually large amounts of their contained water are in the bound form.

9. When water freezes, it expands and, therefore, floats. Even if no air bubbles are incorporated in the ice, it has a lower specific gravity than water in the liquid state. If water, like most other liquids, decreased in volume with lowered temperatures, ice would sink to the bottoms of lakes and seas each winter. Since the temperatures in the bottom do not rise very much in the summer, large bodies of water would be partially filled with permanent ice. This would greatly restrict the available space for living organisms and would lower the average temperatures of aquatic environments. It is difficult to say what changes such conditions would have imposed on the course of animal evolution but they would certainly have been considerable.

8.2 OSMOTIC SOLUTIONS

In solutions, the effective total concentration of dissolved substances is expressed in units called osmols. The osmolality of a solution equals its molal concentration.* For the concentrations usually encountered in biological systems, however, little error is introduced by using *molar* concentration instead of *molal* concentration. Molar concentration is expressed, of course, as *molarity* instead of *molality*. For ordinary biological purposes, then, the number of mols of solute per liter of solution (molarity) equals the number of particles of solute per liter of solution (molality). Osmotic pressure is proportional to the number of particles of solute rather than to the total amount of solute, but the two are very nearly the same for biological solutions.

The number of particles of solute per liter of solution also determines certain other physical properties of the solution. These include the amounts by which the freezing point is lowered, the boiling point is raised, and the vapor pressure is lowered. If any one of these so-called colligative properties is known, the others can be calculated mathematically. They all bear constant relationships to each other. Since osmotic pressure is difficult to measure directly, it is more commonly determined by measuring the freezing point or the vapor pressure. Techniques for making these measurements are fairly simple.

Two solutions with the same osmotic pressure are said to be isoosmotic even though the kinds of solutes in each may be different. Such isoosmotic solutions are not necessarily isotonic, however. Isotonicity requires, by definition, that there be no osmotic effect on a cell surrounded by a semipermeable membrane when that cell contains one of the solutions in question and is immersed in the other solution. Isotonicity depends, therefore, on the *kinds* of solutes, whereas isoosmotic situations depend only upon the numbers of solute molecules irrespective of their nature. A solution that has a higher osmotic pressure than another is said to be relatively hyperosmotic, while one with a lower osmotic pressure is hypoosmotic by comparison. A solution that is hypertonic to the contents of a cell which is immersed in it will cause osmotic withdrawal of water from the cell and consequent cell shrinkage. A hypotonic solution, conversely, will cause a cell immersed in it to take up water and swell.

Ocean water averages 3.5 per cent salt, and this fact may also be expressed by saying that it has a salinity of 35 parts per thousand (35‰). This gives it an osmolarity of 1.01 and a freezing point depression of −1.88° C. The solute concentration of fresh waters, by comparison, is often expressed as a percentage of sea water concentration. It may also be expressed directly in parts per thousand, of course.

We shall be less concerned in this chapter with specific solute concentrations than with relative osmotic pressures of solutions. In particular, we shall be interested in the relative osmotic pressures of the environmental medium and the internal tissue fluids of aquatic animals.

---

*A 1-molal solution contains 1 gram-mol of solute per liter of solution. A 1-molar solution contains 1 gram-mol of solute per 1000 grams of solution. (The two are nearly the same for biological solutions since a liter of pure water weighs 1000 grams.) One gram-mol is the weight in grams numerically equal to the molecular weight of the substance in question.

8.3 BODY WATER COMPARTMENTS

Water in the vertebrate body is distributed among three major compartments (Fig. 8–1). The substances dissolved in the water of each compartment differ in both kind and amount, and the movements of water from one compartment to another are subject to special controls and limitations.

The most obvious fluid compartment, but the smallest of the three in volume, is the plasma compartment. This amounts to about 3 liters in an average human. The second compartment, about 14 liters in volume in man, is the interstitial compartment. Exchanges between this and the plasma compartment occur regularly at the capillary level and have been discussed elsewhere (Sect. 4.5). The third and largest compartment is the intracellular one. The cells of the body contain approximately 29 liters of fluid. There is a constant exchange of fluids through the cell membranes between this and the interstitial compartment, but this exchange is a complex and varying one. It depends upon moment-by-moment changes in the semipermeability characteristics of the cell membrane and moment-by-moment changes in the kinds and quantities of various solutes both inside and outside the cells. The metabolic activity of the cells is responsible, in turn, for both kinds of changes.

The most labile of the three water compartments is the plasma for it is the one most nearly in contact with the environmental medium. The gills, lungs, intestine, skin, and kidneys are all organ surfaces through which exchanges with the environmental medium can take

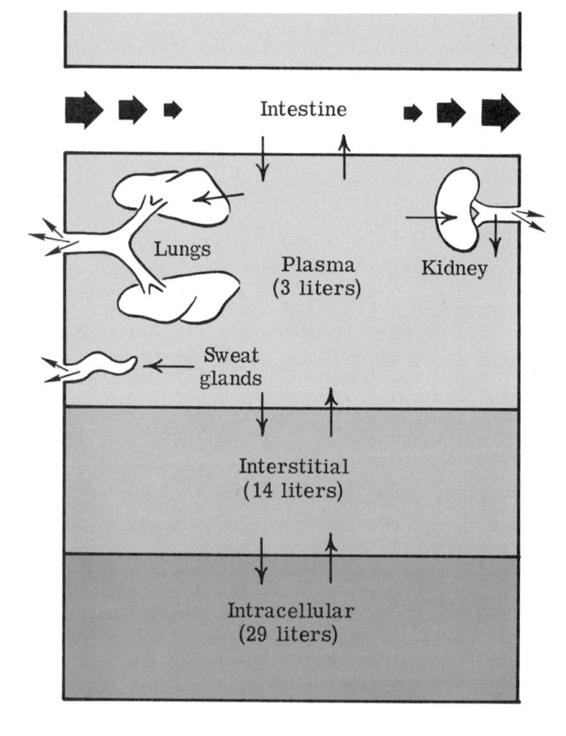

Figure 8–1 Body water compartments and the exchange of water with the environment. See text.

place (Fig. 8–2). Water can move in or out of an aminal in response to the differences in osmotic pressure found in aquatic environments. It can be lost to the air in a terrestrial environment as well. Similarly, salts (solutes) can diffuse in or out of aquatic animals in response to diffusion gradients (Sect. 1.2) and can be lost from terrestrial species by way of the kidneys, intestine, and skin. Because dissolved salts exert osmotic pressure, movements of salts have secondary effects on movements of water as well. The reverse is also true, of course. Both water and salts, finally, can be ingested as part of the diet by either aquatic or terrestrial animals.

The passive movement of water or salts through the gills or the skin of aquatic species is dictated by differences in osmolality of the plasma fluids and environmental medium. It is, therefore, not subject to any physiological control. The loss of water through the lungs of air-breathing species is determined solely by the absolute humidity of the atmosphere (Sect. 6.6), since air leaving the lungs is always saturated with water. Thus, lung-water loss is also independent of any physiological control mechanism. The loss of water or of salts with the feces is minimized in ecological situations where it is desirable to do so, but

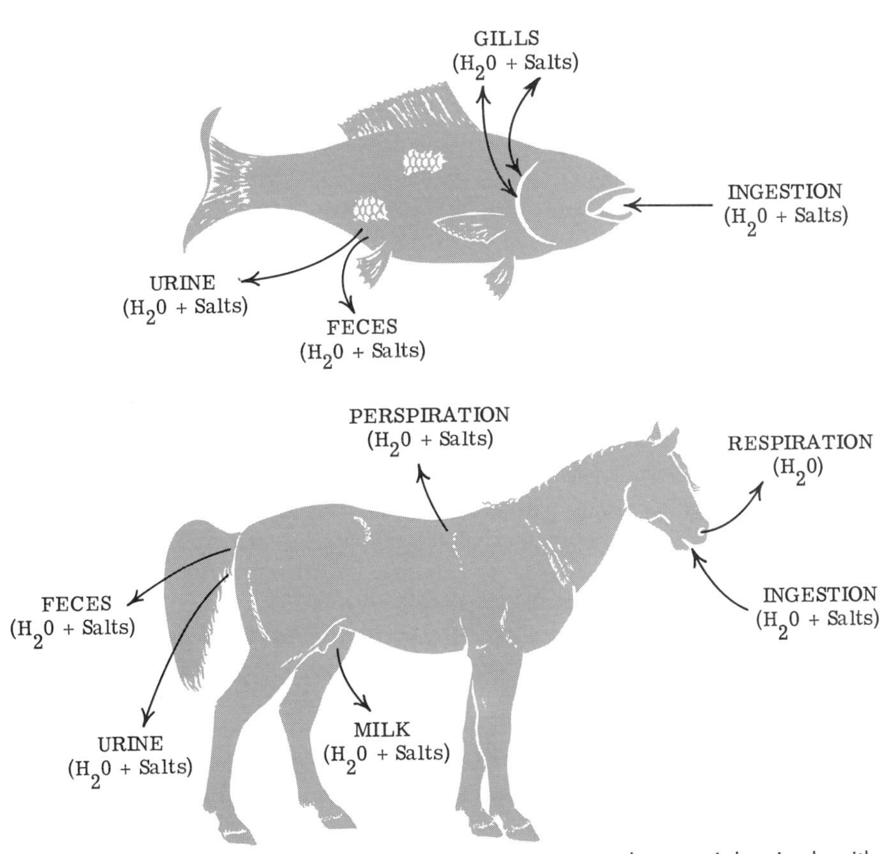

Figure 8–2 Water and salt exchanges of aquatic and terrestrial animals with their environments.

fecal water loss can never be entirely avoided. Some water loss is necessary since the feces must be wet. The amount lost is relatively constant for any given species and cannot easily be varied from time to time by physiological controls. Water and salt loss through the skin of terrestrial species is also unavoidable and is especially great in those animals which thermoregulate (control body temperature) by sweating or in those with thin, permeable skins such as the amphibians have. The amount of such skin loss is determined by the air temperature and humidity and by the rate of body heat production. Finally, some water and salt loss through the kidneys always occurs since nitrogenous wastes must be flushed out in the urine. Thus, most movements of water and salts between the environmental medium and the fluid compartments of the animal body (especially the plasma compartment) are subject to little or no physiological control.

The kidney is the outstanding exception to the general rule outlined in the preceding paragraph. It is the site of greatest importance, therefore, in the maintenance of body water and salt balance. When the intake of water exceeds the water loss from an animal, that animal is said to be in positive water balance. Conversely, when the loss exceeds the intake, the animal is in negative water balance. An equal balance (intake = loss) must be maintained over any reasonable time period in all vertebrate animals so that body water is neither depleted nor increased beyond certain rather narrow limits. The consequences of greater changes in body water content would have obvious effects on the osmotic pressures of the fluids in the three body water compartments, and, as we have seen, osmotic pressures must remain within narrow limits. Equal water and salt balances are achieved by physiological mechanisms that compensate for these unavoidable exchanges.

THE MAMMALIAN KIDNEY

8.4 RENAL STRUCTURE

The mammalian kidney is a bean-shaped organ (botanists generally describe beans as kidney-shaped seeds!) composed of a superficial cortical layer and a deeper medullary layer, both surrounding a central cavity (Fig. 8–3). The whole is enclosed by a capsule of connective tissue. At the hilum, which corresponds to the "eye" of a bean, are the communicating attachments of the renal artery and vein and of the ureter leading to the urinary bladder. Within the central cavity of the kidney, the ureter expands to form a larger chamber, called the renal pelvis. The medulla of the kidney is divided into lobes, which are called pyramids because of the roughly conical shape of each. The points of the pyramids are directed inward and each is covered by a bell-shaped extension of the renal pelvis. This extension is called a calyx. The renal artery, which is a branch of the aorta, sends branches between the renal pyramids to the cortical region and, similarly, the

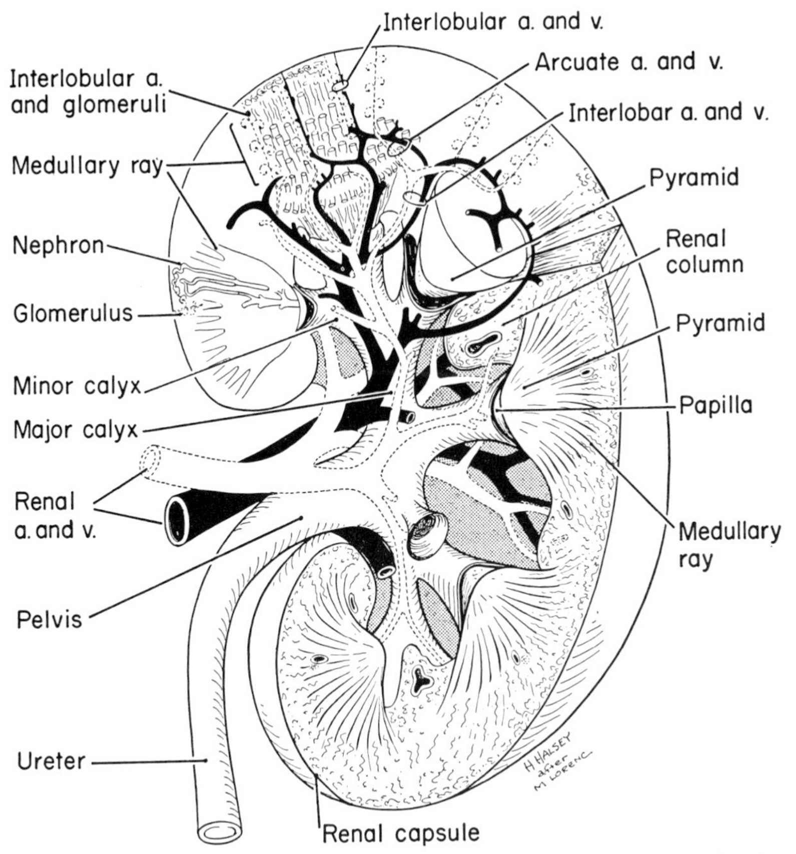

Figure 8–3 The structure of the human kidney. Only one nephron, much enlarged, is shown (*upper left*). (From Florey: *An Introduction to General and Comparative Animal Physiology*. Philadelphia, W. B. Saunders Company, 1966. After Smith: *Principles of Renal Physiology*. Oxford University Press, 1956.)

renal vein, which drains into the inferior vena cava, has interlobar tributaries.

Within the kidneys are more than a million microscopic, functional units called nephrons (Fig. 8–4). Each nephron consists of an elongate tubule closely associated at one end with a group of blood capillaries (glomerulus) and opening at the other end into a collecting tubule. The collecting tubules each receive a number of nephron tubules in this way and open, in turn, through the tips of renal pyramids into the associated calyx. The calices drain through the renal pelvis into the ureter and from there into the urinary bladder.

Each nephron tubule is divisible into four morphologically distinct portions. The first of these, Bowman's capsule, is an expanded, cup-shaped region wrapped about the glomerulus. It opens into a proximal convoluted portion. A similar but distal convoluted portion opens into the collecting tubule at the other end of the nephron. Between the proximal and distal convoluted tubules is the hairpin-shaped loop of Henle. Of the structures mentioned, only the loops of Henle (and, of

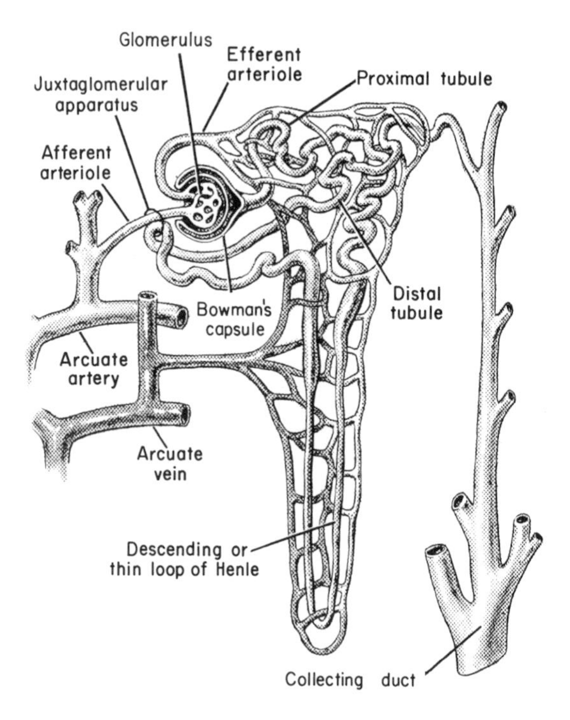

Figure 8–4 A diagram of the structure of a nephron and its blood supply. (From Guyton: *Textbook of Medical Physiology*, 4th Edition. Philadelphia, W. B. Saunders Company, 1971. Redrawn from Smith: *The Kidney*, Oxford University Press.)

course, the collecting tubules) pass into the medulla of the kidney. The rest of the nephron lies in the cortex. This fact accounts for the faintly striated appearance of the medulla as compared with the more homogeneous appearance of the cortex when the kidney is seen with the unaided eye. It also has great significance in renal physiology, as we shall see.

Arterial blood is supplied to each glomerulus through a terminal branch of the renal artery. This branch is called an afferent arteriole. It drains from the glomerulus through a smaller efferent arteriole and then passes through a plexus of vessels lying in close association with the tubules of the nephron. These vessels supply secondary capillary networks adjacent to the nephron walls. Thus, blood passing through the renal tissues passes successively through two sets of capillaries before entering the venous drainage and leaving the organ.

8.5 RENAL FUNCTION IN GENERAL

As we have seen, the glomerulus is subjected to reasonably high blood pressure through the afferent arteriole since it is almost directly supplied from the aorta with little intervening pressure drop (Fig. 8–5). The average blood pressure within the human glomerulus is about 70 millimeters of mercury. The combined membrane of the glomerular capillaries and the inner wall of Bowman's capsule is thin and permeable, and the efferent arteriole draining the glomerulus is smaller than the afferent arteriole supplying it. Consequently, a high filtration

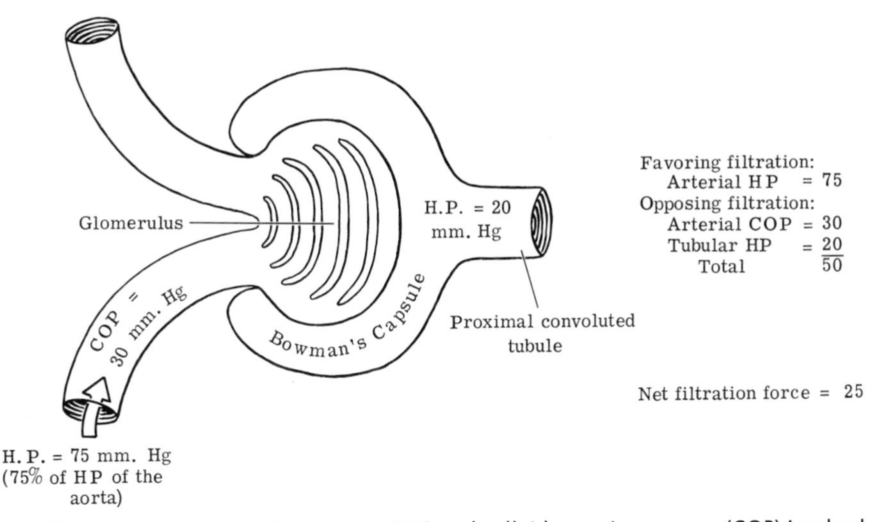

Figure 8–5 Hydrostatic pressures (HP) and colloid osmotic pressures (COP) involved in glomerular filtration. See text.

pressure forces fluids through the membrane and into the cavity of Bowman's capsule. The fluid passed through the membrane is called the glomerular filtrate and is essentially the same as blood plasma except that it does not contain blood cells or large dissolved protein molecules. All of the other normal constituents of blood are present, however. If the glomerular filtrate was allowed to flow through the nephron unchanged, all these substances would be lost to the body in the urine. Quite clearly, then, much of the content of the glomerular filtrate must be reclaimed by the body as this fluid flows through the nephron tubules.

The efferent arteriole draining the glomerulus enters a peritubular plexus of vessels supplying capillaries in close association with the nephron tubules. Reabsorption of materials from the glomerular filtrate occurs through the tubule walls, and these materials reenter the blood in the peritubular capillaries. Such reabsorption occurs largely by active transport (Sect. 1.5). This reabsorption is, of course, a selective one; materials such as sugars and amino acids, which the body cannot afford to lose, are reclaimed, whereas waste products such as urea are allowed to pass into the urine. In summary, then, glomerular filtration is a non-selective process while tubular reabsorption is highly selective.

In this manner the average human renal blood flow of some 1100 liters per day (nearly 20 per cent of the total cardiac output) yields about 180 liters of glomerular filtrate. After reabsorption from the nephrons, only about 1.5 liters remains to be passed from the body as urine.

Certain substances are very thoroughly reabsorbed from the nephrons, and these substances are called high threshold substances. The term comes from the fact that high blood levels (and, therefore, high glomerular filtrate concentrations) of these substances can be reabsorbed from the nephron tubules. In other words, the reabsorption capacity (threshold) of the kidney is very high for these substances. Sugars (*e.g.*, glucose) are high threshold substances and blood glucose levels must be very high before the renal threshold is exceeded and glucose appears in the urine (Sect. 17.8). Amino acids are also high threshold substances.

Other materials are known as low threshold substances because the kidney is capable of recovering little or none of them. These are materials such as urea which are normally present in the urine.

Of particular interest in physiology are substances for which the renal threshold is variable. These materials may be reabsorbed in greater or lesser quantities depending upon the conditions and requirements of the moment. By varying its threshold for such materials the kidney maintains appropriate balances within the body. Water and certain salts, especially NaCl and KCl, are variable threshold substances.

In addition to the filtration-reabsorption mechanisms just discussed, certain other substances are actively secreted from the peritubular capillaries into the nephron lumen and thence, are eliminated with the urine. Potassium ion is treated in this way to some extent as are certain experimentally introduced molecules such as para-aminohippuric acid (PAH).

8.6 RENAL FUNCTION IN WATER BALANCE

The interstitial fluids of the renal cortex are isoosmotic (Sect. 8.2) with the blood, but those of the renal medulla become increasingly hyperosmotic toward the tips of the pyramids (Fig. 8–6). This osmotic gradient is maintained by the kidney and makes it possible to control the threshold for water. As the glomerular filtrate passes through the nephron tubules, the following events occur:

1. The walls of the descending limb of the loop of Henle are permeable to sodium ions, and since the Na^+ concentration becomes increasingly greater in the interstitial fluids, these ions enter the glomerular filtrate, which also becomes increasingly concentrated with Na^+ as the end of the loop is approached (Fig. 8–7). Since there is some "lag" in the inward diffusion of Na^+, the concentration in the glomerular filtrate rises but does not quite reach the concentration levels that exist in the interstitial fluids adjacent to the end of the loop of Henle. Water also enters the tubule in this segment.

2. As the glomerular filtrate passes through the ascending limb of the loop of Henle, Na^+ is actively transported (Sect. 1.5) from the glomerular filtrate and back into the interstitial fluids of the kidney. In this segment of tubule, however, water is not excreted with the Na^+ because the tubule is not permeable to water. Consequently, the Na^+ concentration of the glomerular filtrate becomes progressively lower, and by the time the glomerular filtrate reaches the distal convoluted tubule, it has become hypoosmotic to the blood. In summary, the

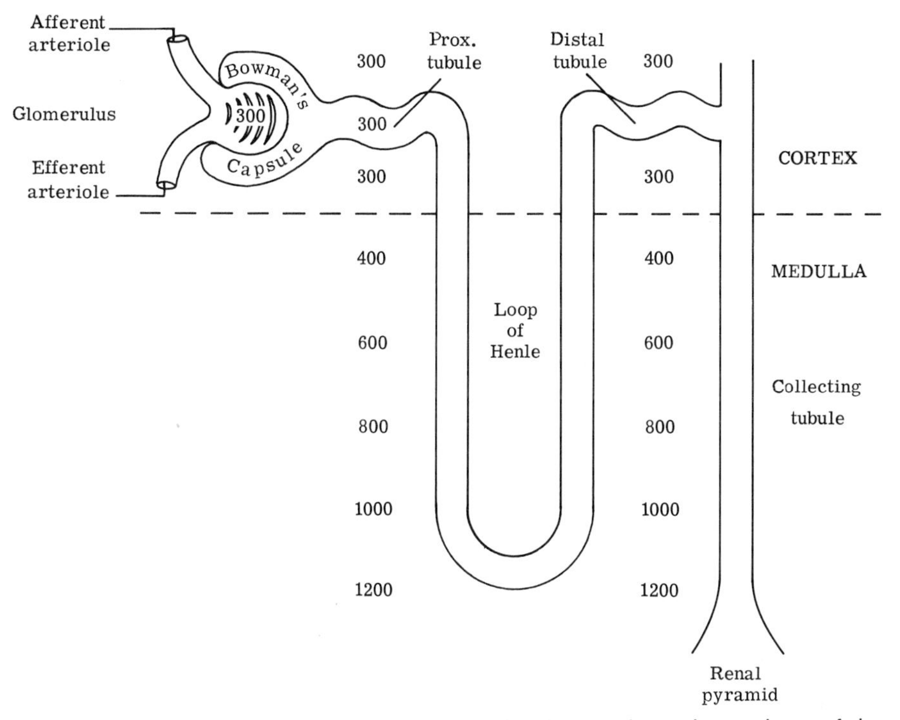

Figure 8–6 The osmotic gradient associated with juxtaglomerular nephrons of the kidney. Osmotic pressures are shown in milliosmols.

isoosmotic glomerular filtrate becomes hyperosmotic in the descending limb and then hypoosmotic in the ascending limb of the loop of Henle.

3. The distal convoluted tubule transports the hypoosmotic fluid to the collecting tubule, and in this it again descends through the renal medulla toward the renal pelvis. If, during this descent, the walls of the collecting tubule are impermeable to water or to Na^+, then no further changes in concentration occur and the hypoosmotic fluid is excreted as urine (Fig. 8–8). Under these conditions a copious, dilute urine is excreted, and this represents a net loss of water. That is to say, the threshold of the kidney for water is low. If, on the other hand, the walls of the collecting tubule are permeable to water, then water moves through them into the hyperosmotic interstitial fluids and the glomerular filtrate becomes, again, hyperosmotic. Under these conditions the urine produced is scanty and concentrated; or, we may say, the renal threshold for water is high.

In summary, the stage is set by the activity of the loop of Henle in producing a hypoosmotic filtrate. Variations in the water permeability of the collecting tubule, then, result in variations in the amount of water allowed to escape with the urine.

The permeability of the collecting tubules to water is under the control of the antidiuretic hormone (ADH), which is released by the posterior lobe of the pituitary gland (Sect. 17.4). Antidiuretic hormone increases the permeability of the collecting tubules of the kidney and

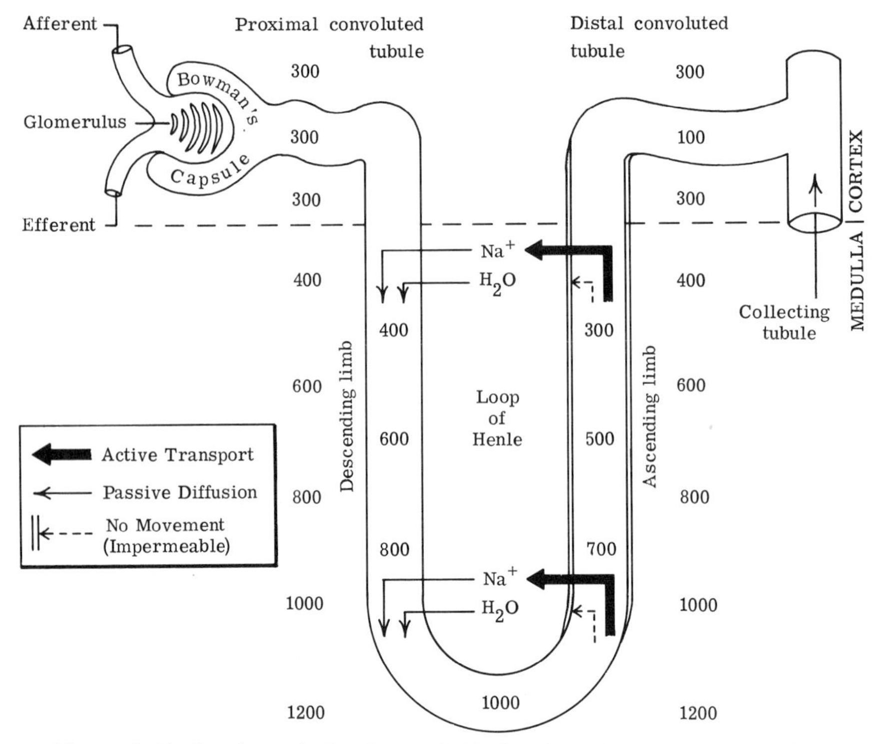

Figure 8–7 Function of the loop of Henle. Osmotic pressures are shown in milliosmols.

thus causes the retention of water by the kidney. When little antidiuretic hormone is released, the collecting tubules are relatively impermeable to water, and, as a result, urinary water loss is increased. The hormone takes its name from the word "diuresis," which means the loss of urinary water. A number of substances are known to cause increased diuresis and these are known as diuretics. Similarly, substances such as ADH decrease urinary water loss and are called, therefore, antidiuretics. Caffeine is a common diuretic substance.

The release of antidiuretic hormone by the posterior lobe of the pituitary gland is controlled, in turn, by the activity of certain osmoreceptor cells of the hypothalamus, a portion of the brain located immediately above the pituitary gland. A condition of negative water balance (Sect. 8.3) causes a rise in the osmotic pressure of the blood and, therefore, of the interstitial fluids of the hypothalamus. Under these conditions, the osmoreceptor cells there stimulate the release of larger quantities of antidiuretic hormone from the posterior lobe of the pituitary gland. The released antidiuretic hormone is carried by the blood stream to the kidneys where it increases the permeability of the collecting tubules to water. As a result, water is retained by the body and this, in turn, tends to return the body toward an equal water balance.

A condition of positive water balance, on the other hand, lowers the osmotic pressure of blood and tissue fluids, and under these condi-

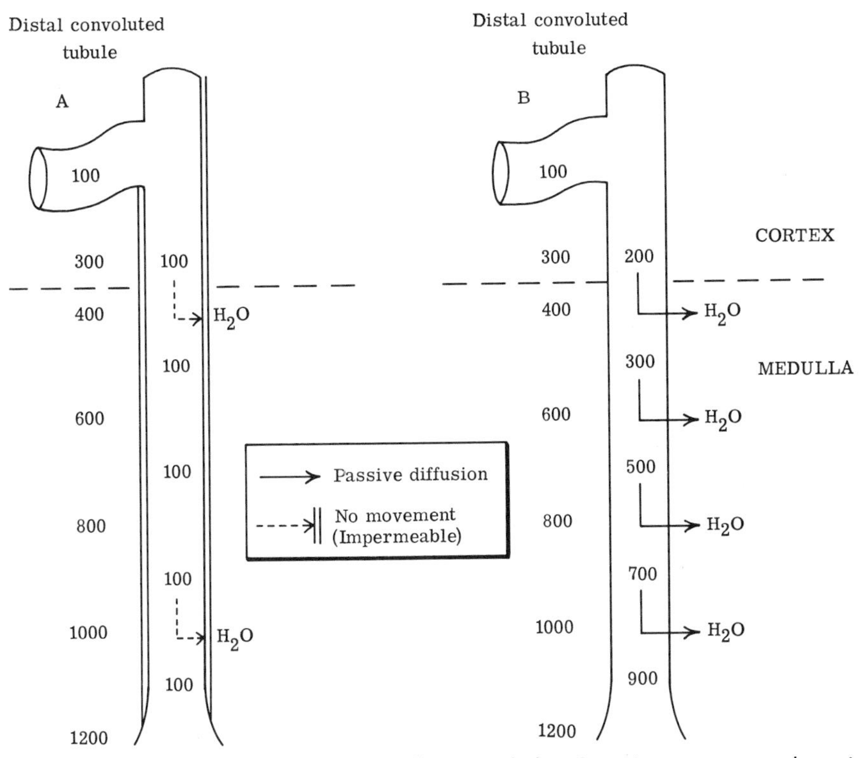

Figure 8–8 Function of the renal collecting tubules. Osmotic pressures are shown in milliosmols.

tions, the osmoreceptor cells of the hypothalamus are not stimulated. Consequently, less antidiuretic hormone is released from the pituitary gland, and the permeability of the collecting tubules is reduced. This allows more water loss in the urine and, again, a return toward an equal water balance.

Clearly, the functioning of the loops of Henle and collecting tubules in the control of water balance depends upon the maintenance of a gradient of osmotic pressure in the interstitial fluids of the renal medulla. This gradient is maintained by a specialized pattern of blood circulation in the medulla; hairpin-shaped loops of blood vessels, known as vasa recta, parallel the loops of Henle (Fig. 8–9). These carry greater or lesser amounts of sodium ion away from the renal medulla as may be required.

Both limbs of the vasa recta are permeable to Na^+, and as blood descends in these vascular loops, Na^+ diffuses from the interstitial fluids into the blood. As a result, the blood becomes more hypertonic as it approaches the end of the vascular loop. In the ascending limb of the vasa recta, Na^+ tends to diffuse out again into the interstitial fluids. There is a lag in this diffusion process, however, so that blood leaving the vasa recta is still slightly hypertonic to the blood entering them. As a result, there is a constant transport of Na^+ away from the tissues of the

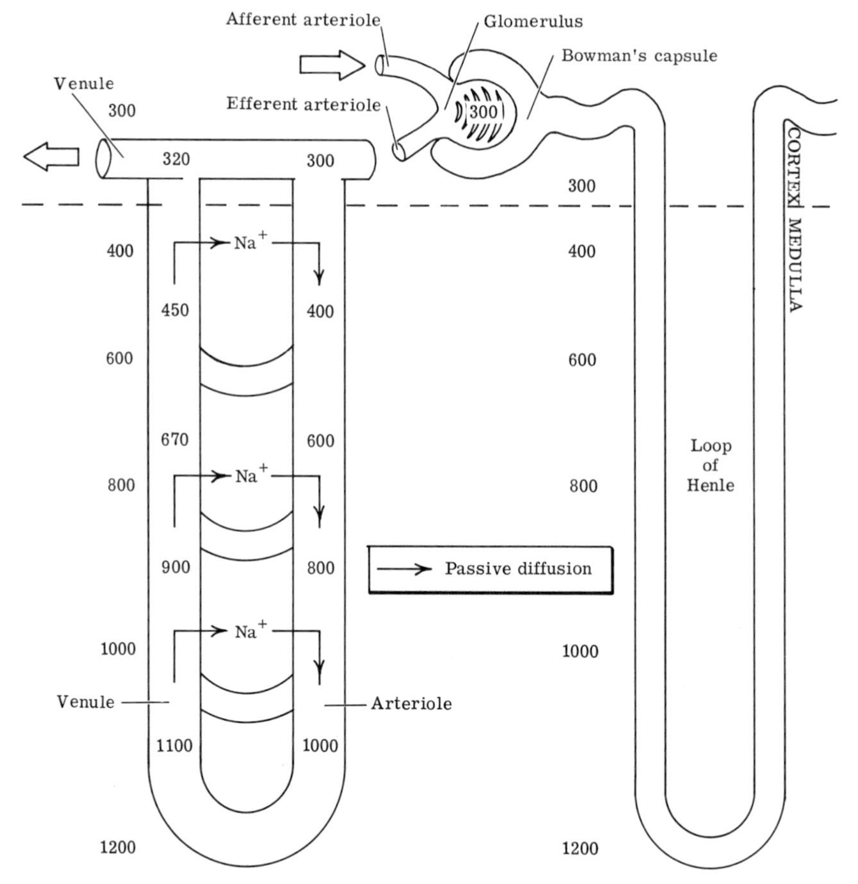

Figure 8–9 The function of the vasa recta of the kidney. Osmotic pressures are shown in milliosmols. See text.

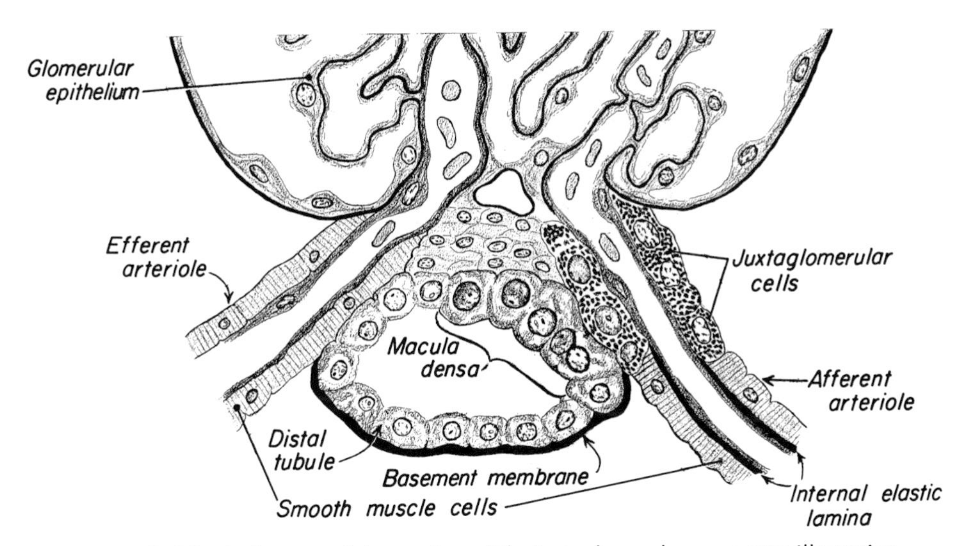

Figure 8–10 A diagram of the structure of the juxtaglomerular apparatus illustrating its possible feedback role in the control of nephron function. (From Guyton: *Textbook of Medical Physiology*, 4th Edition. Philadelphia, W. B. Saunders Company, 1971. Modified from Harn: *Histology*. J. B. Lippincott Co.)

renal medulla. This compensates for the equally continual introduction of Na^+ from the glomerular filtrate. Thus, the concentration gradient for Na^+ in the interstitial fluids of the renal medulla remains comparatively constant.

The distal convoluted tubule of each nephron passes between the afferent and efferent arterioles (Fig. 8–10). At this point the walls of the afferent arteriole and of the distal convoluted tubule are modified to form a structure known as the juxtaglomerular apparatus. The function of this structure is not completely understood, but it is thought to act in the control of the rate of glomerular filtration. Presumably the osmotic pressure of the fluids leaving the loops of Henle acts as a feedback (Sect. 1.6) stimulus in this controlling mechanism. Thus, each nephron probably regulates its own rate of glomerular filtration (probably by vasoconstriction and vasodilation of the afferent arteriole) so that fluid enters the nephron at a rate which allows it to be efficiently processed.

8.7 RENAL FUNCTION IN SALT BALANCE

The cortex of the adrenal gland (Sect. 17.9) produces steroid hormones that increase the nephron reabsorption of sodium ions and decrease the reabsorption of potassium ions. The amounts of such hormones released by the adrenal gland depend upon the actual blood concentration levels of these ions. The most important hormone in this respect is called aldosterone.

Other endocrine secretions, especially those of the parathyroid glands (Sect. 17.7), have specific effects upon the renal thresholds for calcium and phosphorus. Each such controlling mechanism operates on a feedback mechanism.

8.8 RENAL FUNCTION IN ACID-BASE CONTROL

Several renal mechanisms play roles in the physiological control of blood pH. All are really mechanisms for retaining bicarbonates which are so important as blood buffers. The renal threshold for bicarbonates is high so that this ion (HCO_3^-) is maintained in relatively high concentration in the blood (Sect. 7.5).

The metabolism of proteins and phospholipids (Chapter 10) involves the detachment from them of sulfate and phosphate groups. These groups react with water to form free sulfuric and phosphoric acids, which must be neutralized by combining with bicarbonate; thus:

$$H_2SO_4 + 2\ NaHCO_3 \longrightarrow Na_2SO_4 + 2\ H_2O + 2\ CO_2$$
$$H_3PO_4 + 2\ NaHCO_3 \longrightarrow Na_2HPO_4 + 2\ H_2O + 2\ CO_2$$

The loss of the resulting neutral salts (Na_2SO_4 and Na_2HPO_4) in the urine would constitute a serious loss of the bicarbonate used in forming them. This urinary loss of bicarbonate is minimized in two ways: (1) The neutral salt, when possible, is converted to an acid salt by reacting with carbonic acid; thus:

$$Na_2HPO_4 + H_2CO_3 \longrightarrow NaH_2PO_4 + 2\ NaHCO_3$$

The resulting acid salt (NaH_2PO_4) is excreted with the urine and the bicarbonate salt ($NaHCO_3$) is retained by the body. (2) The neutral salt is reconverted to a free acid and that acid is then combined with an ammonium ion; thus:

$$Na_2SO_4 + H_2CO_3^- + NH_3^- \longrightarrow (NH_4)_2SO_4 + 2\ NaHCO_3$$

The ammonium salt ($(NH_4)_2SO_4$) is then excreted with the urine and the bicarbonate is retained by the body. The ammonium ion is derived from the deamination of amino acids during protein catabolism (Sect. 10.13).

NEPHRON EVOLUTION

8.9 INTRODUCTION

The vertebrate nephron apparently did not evolve directly from any known excretory organ found in modern invertebrates. However, similar chemical and physical principles operate in the excretory organs of both vertebrates and invertebrates. Adult *Myxine* hagfishes have an excretory system that superficially resembles the metanephridia of invertebrates, *i.e.*, simple tubules draining celomic fluids through the body wall to the outside (Fig. 8–11). Nearby, in the celomic wall, there is a glomerulus-like capillary bed that, by forced filtration, increases the rate of formation of celomic fluid. The nephron tubule, then, reabsorbs some substances from the fluid as it drains toward the outside of the animal body. In some sharks and amphibians the communication of the nephron tubule with the celomic cavity persists, but there is a more intimate connection between the tubule and the glomerulus—a connection resembling Bowman's capsule. In higher vertebrates the direct connection with the celomic cavity is lost.

The tissue fluids of some marine invertebrates have the same total osmotic concentration as that of the surrounding sea water and there is no tendency to gain or to lose water by osmotic flow in these animals. The specific solutes responsible for internal osmotic pressure are never of the same concentration as those outside the animal, however. Therefore, mechanisms to prevent solute diffusion through the body surface (or to compensate for such diffusion as does occur) must be present. Moreover, every organism must be able to accumulate specific molecules (nutrients) necessary for its metabolism and to retain them in the body in greater concentration than in the surrounding medium. The accumulation of such materials and the excretion of wastes demands that the animal have surface membranes that are permeable to solutes, and the movement of such solutes through these membranes results in the simultaneous movement of water as well. Finally, all known animal cells maintain electrical charge differences on the two sides of the cell membrane by maintaining differential concentrations of various ions on the two sides. With all of these facts in mind, the inescapable conclusion is that active transport of substances through membranes between

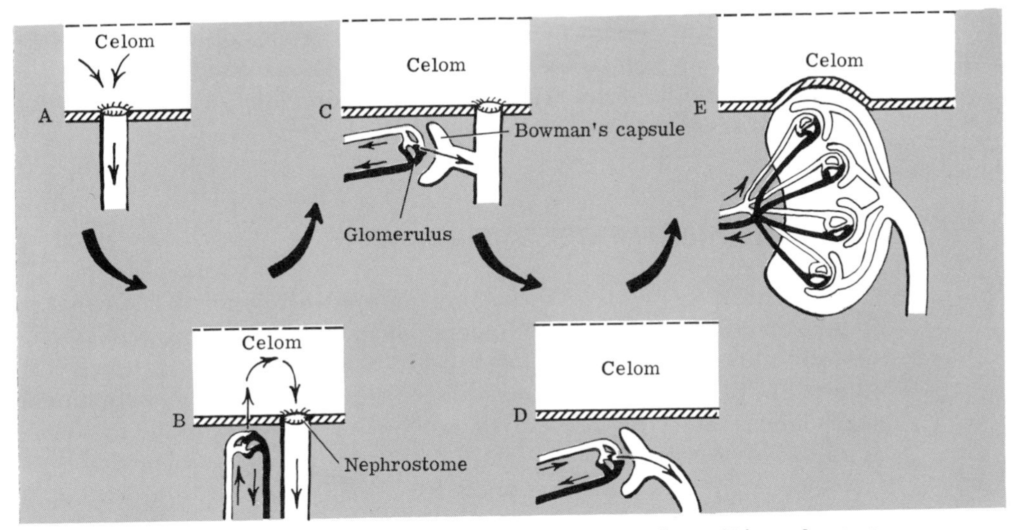

Figure 8–11 Probable stages in the evolution of the vertebrate kidney. See text, Section 9.9. (After Cockrum and McCauley: *Zoology*. Philadelphia, W. B. Saunders Company, 1965.)

internal fluids (both of cells and of whole organisms) and the surrounding medium must have appeared as early as life itself.

The salinity of sea water (Sect. 8.2) is not everywhere the same. It is higher in enclosed seas, such as the Mediterranean, where evaporation exceeds the inflow of diluting fresh water. In estuaries and other areas where fresh-water inflow exceeds evaporation, however, salinity is reduced. Such partially fresh bodies of water are said to be brackish. Some animals tolerate fairly wide ranges of salinity and are spoken of as euryhaline forms. Others are stenohaline, meaning that they can withstand only small differences in salinity.

Stenohaline animals lack physiological mechanisms for maintaining internal osmotic pressures in the face of varying external salinity. They also have low tolerance for internal changes in osmotic concentration. Euryhaline forms, on the other hand, are of two types: (1) those capable of maintaining internal osmotic concentrations even though external salinities vary widely (osmoregulation) or (2) those which simply tolerate changes in internal osmotic concentration (osmoconformity). In this book, of course, we are primarily concerned with the vertebrates, and they are generally osmoregulators.

Until recent years zoologists generally agreed that the vertebrates probably first appeared in brackish waters, since all evidence seemed to point in that direction. On the basis of certain recent evidence, some authorities now believe that the vertebrates arose in the sea. It is not possible at this time to finally resolve the question. Since the physiological evidence from kidney function supports a brackish-water origin, we will assume that point of view for the purposes of discussion. Where there is appropriate evidence pointing toward a possible marine origin, it will be noted. In any case, several evolutionary migrations between fresh and marine waters have subsequently occurred in both directions. We must, therefore, consider adaptations to both environments, *i.e.*, animals living in hyperosmotic (marine) and hypoosmotic (fresh-water) environments as compared with their own tissue fluids. Finally, we must also consider terrestrial adaptations for which the surrounding medium is air.

8.10 THE FRESH-WATER ENVIRONMENT

A fresh-water vertebrate has an internal salt concentration that is higher than that of the environment. Consequently, there is a tendency to lose salts by diffusion and to gain water by osmotic uptake. If internal salt concentrations and osmotic pressures are to be maintained, mechanisms must be present to prevent or to compensate for these tendencies (Fig. 8–12).

One might suppose that such an animal could simply develop a surface covering that would be impermeable to either water or salts. Such impermeability, however, would also prevent the vital exchange of respiratory gases and the accumulation of other specific metabolic requirements. The problem can be minimized, of course, by limiting

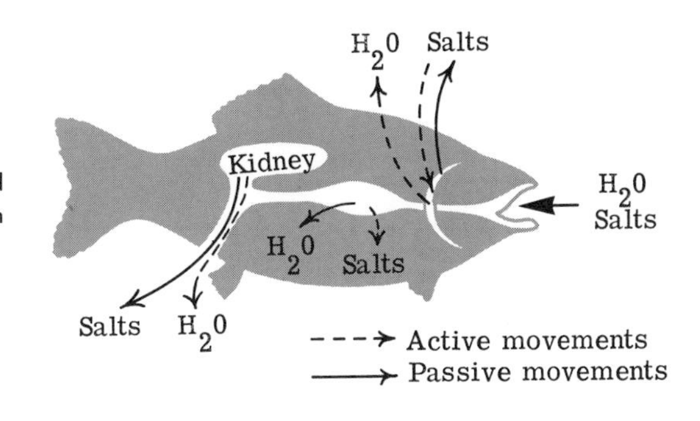

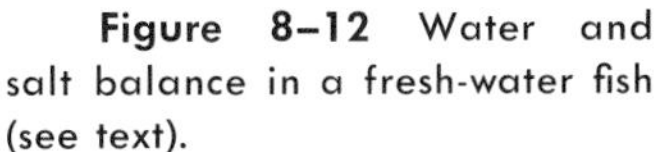
Figure 8–12 Water and salt balance in a fresh-water fish (see text).

the extent of such permeable surface membranes, and such is the case for most vertebrate animals.

Minimizing a problem does not solve it, so there is still an unavoidable salt loss and water gain through intestinal, renal, and respiratory membranes, and these losses and gains must be compensated for by other mechanisms. The greatest single factor in this compensation is the presence of a kidney which is capable of retaining salts and excreting water. The fact that all vertebrates have kidneys which are fundamentally designed to accomplish these ends argues strongly for a fresh or brackish-water origin. It would be most difficult to explain the appearance of such a kidney in a hyperosmotic environment.

The renal retention of salts is not, of course, 100 per cent efficient. Moreover, salt loss still occurs through intestinal and gill membranes as well. Some fresh-water fishes obtain sufficient quantities of salt in their diets to offset the unavoidable salt losses; others are capable of active, inward salt transport through the gill membranes. Even the freshest of natural waters contain some salts in solution, and by active transport fishes can selectively absorb them from the environment.

Fresh-water fishes drink little or no water, since the osmotic uptake through the gill membranes is already in excess of their requirements. The problem is, rather, to get rid of the excess. The production of urine that is more dilute (hypotonic) than the blood adequately compensates for the excess water uptake and maintains the animals in proper water balance (Sect. 8.6).

8.11 THE MARINE ENVIRONMENT

Marine teleost (bony) fishes have an internal salt concentration that is less than that of the surrounding environment so they tend to lose water by osmosis and to gain salts by diffusion through the permeable surface membranes. Thus, the problems are the reverse of those for fishes living in a fresh-water environment discussed in the preceding section. Impermeability of the greater part of the body surface, again, minimizes but does not solve the problem. Moreover, the possession of kidneys primarily designed to retain salts and to excrete water is a liability under these conditions (Fig. 8–13).

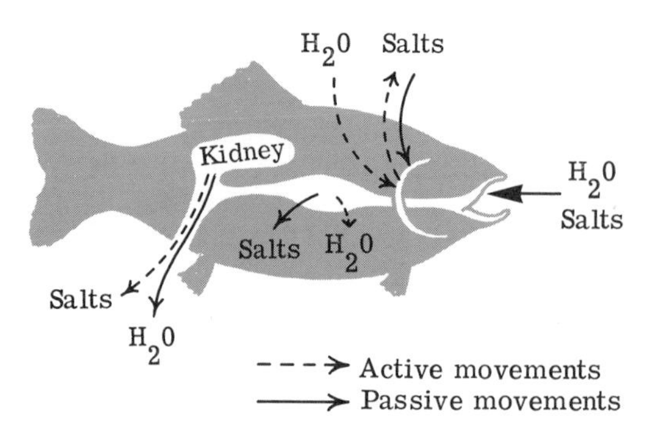

Figure 8–13 Water and salt balance in a marine fish (see text).

The number and size of the functioning nephrons and the minute-volume of blood supply to the nephrons are all reduced in marine teleost fishes. In a few cases glomeruli are completely lacking in the adults. The embryos of all fishes have glomeruli present at some stage of development, however, and this indicates probable descent from forms with typical vertebrate nephrons.

It is interesting to note that in fishes such as salmon that migrate between fresh and marine waters, both kinds of salt and water problems must be dealt with. In salmon the blood supply to the glomeruli is reduced when the animal enters sea water and is increased when it reenters fresh waters. Reduction of kidney function reduces the urinary water loss but it does not, of course, completely eliminate it. Moreover, substantial water loss also occurs through the gill membranes. Physiological mechanisms must be present, therefore, to compensate for these losses.

Unlike fresh-water fishes, teleosts in the sea drink large quantities of water. The intestinal membranes absorb much of the water as well as the monovalent salts (Na^+, K^+) contained in it. Divalent salts (Ca^{++}, Mg^{++}, $SO_4^=$, $PO_4^=$) are absorbed in only very small quantities; most of them are passed, instead, with the feces. Those divalent salts which are absorbed are actively excreted by the kidneys. The absorbed monovalent salts are eliminated by active transport through the gill membranes. This active transport through gill membranes is the same as that encountered in fresh-water fishes, which use it to obtain salts from the environment (Sect. 8.10), but here it operates in the opposite direction.

Fishes that migrate between fresh and marine waters (*e.g.*, salmon) alter the degree and direction of active salt transport through the gill membranes as the salinity of the environment changes. In fresh water salts are accumulated from the environment and in marine waters salts are actively excreted through the same membranes. Thus, these "salt pumps" are constantly moving salts against the concentration gradient regardless of the direction of that gradient.

The marine elasmobranch fishes have developed mechanisms somewhat different from those just described for teleosts. These sharks, rays, and related forms also have typical vertebrate nephrons, indicating their probable descent from fresh or brackish-water vertebrates. Thus, they have kidneys which are primarily designed to retain salts and eliminate water, *i.e.*, kidneys which are a liability in the marine environment. In addition, however, in the elasmobranch kidney there is a special segment of the nephron tubule which reabsorbs urea from the glomerular filtrate. Since urea is an osmotically active solute, the retention of urea in the blood effectively raises osmotic pressure of blood and tissue fluid. Elasmobranchs maintain an internal osmotic pressure that is higher than that of the surrounding sea water (Fig. 8–14). This has required that elasmobranch tissues develop a tolerance to urea greater than that of other vertebrate tissues. The result is that marine elasmobranchs, like fresh-water fishes, actually live in a hypoosmotic environment, in which the typical vertebrate nephron is appropriate.

The internal osmotic pressure of elasmobranchs is due, in part, to urea rather than to other salts, so there is still a tendency toward the uptake of salts from the sea by diffusion through surface membranes. Salts accumulated in this way are actively excreted by a special cloacal gland, which does not occur in other vertebrates.

Most modern sharks and rays are marine, but the fossil record indicates that they were once more common in fresh or brackish waters. A few fresh-water sharks still exist but they are probably descended from marine ancestors since they too retain urea to some extent. In the fresh-water environment such urea retention is a liability because it raises blood osmotic pressure and, thus, magnifies the salt and water balance problems of a fresh-water vertebrate (Sect. 8.10). The tendency in these fresh-water elasmobranchs to lose salts by diffusion and to gain water by osmotic uptake is even greater than in the fresh-water teleost. Yet, the tolerance to urea has become, during evolution, a requirement. The elasmobranch heart, for example, will not beat normally in the absence of elevated urea levels. Thus, it has apparently been impossible for modern fresh-water sharks to completely eliminate urea retention. The amount of such retention has been greatly reduced, however, so this additional problem has been minimized.

The compensatory mechanisms of fresh-water elasmobranchs are generally the same as those of fresh-water teleosts. The kidneys produce an abundant, dilute urine while retaining salts. The special cloacal salt-excreting gland of marine elasmobranchs is, of course, absent in the fresh-water forms.

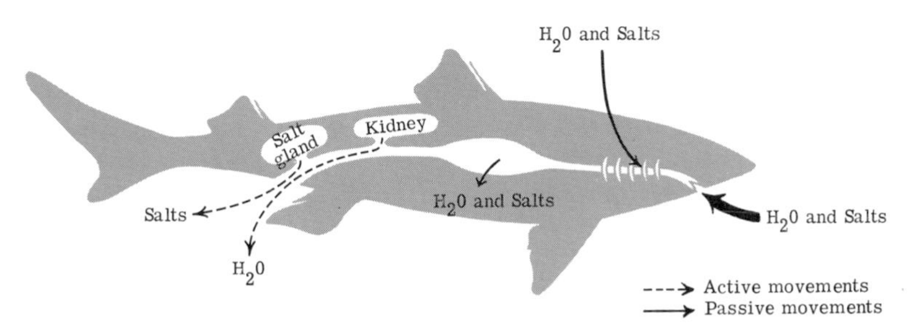

Figure 8–14 Water and salt balance in a marine elasmobranch. See text.

Certain teleost fishes living in very cold surface waters of the Antarctic have blood osmotic pressures that are above those of the surrounding sea water. This is achieved by the retention of certain unidentified organic substances in a manner analogous to the retention of urea by elasmobranchs. However, these teleosts derive an additional advantage from the depression of the freezing point of the body fluids (Sect. 8.2).

Among vertebrates, only the myxinoid cyclostomes (hagfishes) are, in a sense, osmoconformers. They have a total internal salt concentration equal to that of the surrounding sea water. Therefore, they are in osmotic balance with their environment. The kinds and concentrations of individual specific ions are different, of course, so individual ion regulation is still necessary. Selective intestinal absorption and renal retention accomplishes these requirements. This condition, much like that found in marine invertebrates, has been used to support the argument for a marine origin of the vertebrates. Certain amphibians in brackish waters also tolerate increased internal salt concentrations, however, so the argument is not a conclusive one. *Bufo viridis*, the green toad of Europe and the Middle East, and *Rana cancrivora*, the crab-eating frog of Southeast Asia, both retain urea (like elasmobranchs) and salts (like myxinoids). Salt retention is the major factor in osmoregulation of the green toad, whereas urea retention is more important in the crab-eating frog. Also, it is more likely that the myxinoid fishes represent degenerate forms rather than ancestral species.

The eggs of most aquatic vertebrates tend to conform with the environment, shrinking and swelling in response to changing osmotic forces of the medium until fertilization occurs. A decrease in the permeability of the egg membrane following fertilization isolates the internal fluids from the environment during development. The newly hatched larvae have operating regulatory mechanisms which become more efficient as further development takes place.

8.12 THE TERRESTRIAL ENVIRONMENT

In one way the terrestrial environment resembles the marine: an organism in this environment tends to lose water. Of course, the problem is one of desiccation rather than osmotic loss and it does not involve a simultaneous gain of salts by diffusion. In fact, especially where sweating is a method of thermoregulation, there may be a loss of salts as well. Nevertheless, the typical vertebrate kidney, designed as it is to excrete water, is something of a liability in the terrestrial environment and compensatory mechanisms are both physiological and behavioral.

The Amphibia live at the junction between the fresh-water and terrestrial habitat. The larval period is always aquatic, but the adults of many species are largely terrestrial. Consequently, adaptations to both habitats appear in this group of vertebrates.

The water and salt balance mechanisms of tadpoles and aquatic adult amphibians resemble those of fresh-water teleosts. Water enters the body osmotically, but reduced skin permeability keeps this at a minimum, and it is compensated for by the production of dilute urine. Salts are retained by the kidneys but some unavoidable loss occurs. This is made up by dietary salt intake and by active absorption through the gills of tadpoles and the skin of adults.

Some amphibians, especially those of desert regions, are extremely tolerant of desiccation; they can lose up to 50 per cent of their body water without suffering lasting damage. Such animals can generally absorb large quantities of water when the opportunity arises and, thus, make up their losses quickly. In desert forms the urinary bladder absorbs water from the urine and, interestingly, this reabsorption is also under the control of the antidiuretic hormone (Sect. 8.6).

Desert amphibians of the American Southwest estivate during the dry seasons and become active only during the annual summer rains when desert ponds may exist for a few days or weeks. A shortened larval period prepares the new generation for estivating before the rains cease. Estivation occurs underground where the animal can retain body water for long periods of time.

Among some amphibians, reptiles, and birds there is a reduction in the number and size of glomeruli. This is especially true of the reptiles and birds of desert areas where water retention is most critical. Reduced glomerular blood circulation also occurs in certain reptiles; it is more pronounced in lizards and snakes than in the more aquatic forms such as crocodiles and most turtles. These adaptations to a desiccating environment are reminiscent of those which have occurred in marine teleosts (Sect. 8.11).

The urine volume of reptiles, especially snakes and lizards, and that of birds is also reduced by the excretion of nitrogenous wastes in the form of uric acid (Sect. 9.14). Uric acid is quite insoluble and can be excreted with little water loss as a semisolid paste. The excretion of such wastes in the form of soluble urea in other animals involves a much greater water loss.

Birds and mammals have acquired additional control over water loss by the addition of the loop of Henle to the nephron tubules. Lower vertebrates lack this segment. The collecting tubules of birds and mammals respond to the antidiuretic hormone by increased water reabsorption as does the amphibian urinary bladder. This is an example of the older hormonal mechanism acting upon a newly evolved structure. The total length of the loop of Henle generally corresponds to the water retention requirements of the animal. Desert rodents, such as the kangaroo rat *Dipodomys* and the fish-eating bat *Pizonyx*, both of the American Southwest and the western coast of Mexico, show this adaptation to desert conditions. The kidneys of such animals have a single renal pyramid that extends through the renal pelvis and into the ureter. This long pyramid contains greatly elongated loops of Henle and similarly elongated collecting tubules. Thus, the osmotic gradient of the renal medulla is also longer, and this enables the animal to produce an even more hypertonic urine (Sect. 8.6). Both *Dipodomys* and *Pizonyx* obtain all of their water with the foods they eat; neither drinks water in the

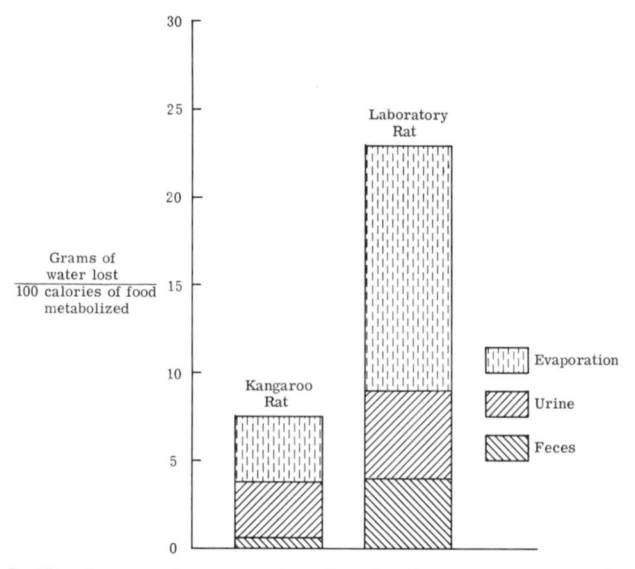

Figure 8–15 Comparative water loss for the kangaroo rat *Dipodomys* and the laboratory rat.

normal habitat. Foods provide a fair water source for the fish-eating bat, but the diet of the kangaroo rat consists entirely of dry seeds. Experimentally it is possible to induce kangaroo rats to drink water by feeding them a high protein diet. Extra protein metabolism results in extra nitrogenous waste to be excreted through the kidneys (Sect. 9.14). This, in turn, demands extra urinary water loss. Under such conditions the kangaroo rat will accept and can survive on sea water (Fig. 8–15).

The kangaroo rat's diet of seeds is high in fats, and during the metabolism of fats more metabolic water is produced than from the metabolism of other foodstuffs (Fig. 8–16). Metabolic water is the water produced as an end product of oxidation (Chapter 10). Under normal conditions this metabolic water meets the requirements of *Dipodomys* so that drinking water is not necessary.

Bird embryos enclosed within the egg shell also have limited water resources. The stored food of the yolk is high in fats and, again, this is an important source of metabolic water. Conversely, frog embryos, which experience unavoidable water uptake by osmosis from the fresh-water environment, live on high-protein food stores. The metabolism of these proteins favors increased renal water loss and thus helps the animal to maintain equal water balance.

Sometimes desertlike water problems occur even in the presence of an apparently adequate water supply. The blood-feeding vampire bat (*Desmodus*) of Mexico lives where water is generally available. Moreover, the animal obtains large amounts of water with its liquid diet. Nevertheless, *Desmodus* has a kidney structure like that of the kangaroo rat. It has been reported that the vampire bat is capable of greater urine concentration than any other mammal investigated to date. The water balance problem in *Desmodus* comes about in the following way.

To obtain sufficient amounts of metabolizable food, the animal

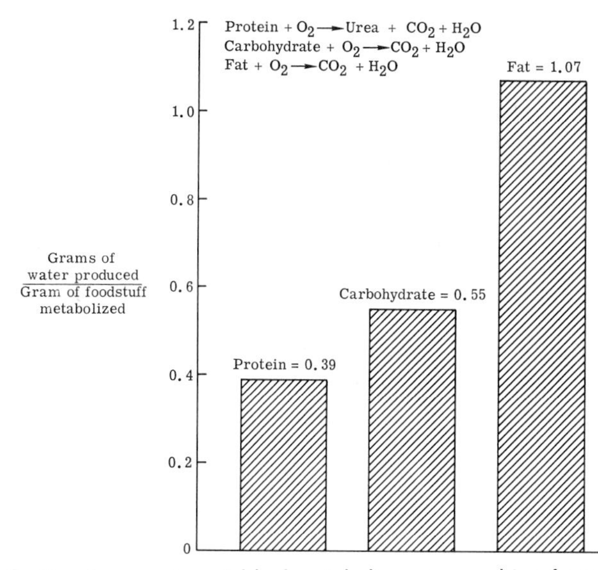

Figure 8–16 Comparative yield of metabolic water resulting from the oxidation of various foodstuffs. See text.

must drink very large quantities of blood. The water content of that blood represents additional "pay load," which limits the nutrient capacity of the stomach and inhibits flight. To permit more feeding and more efficient flight, it is imperative that the bat eliminate the water as rapidly as possible. Accordingly, the kidneys produce an abundant, dilute urine almost as rapidly as the animal drinks the blood. After feeding, the bat returns to its cave to spend the daylight hours digesting and metabolizing its blood meal. Blood has a relatively high protein content; as we have seen, this results in the production of more nitrogenous wastes (urea) which must be eliminated along with quantities of water. Thus, further water loss is necessary, yet the animal returns to its cave with a minimal amount of stored water. If it were not for the elongated renal pyramid and the elongated osmotic gradient, the animal could easily reach a state of negative water balance. Evidence suggests that *Desmodus* possesses still other physiological mechanisms for maintaining water balance, but these are not yet completely understood.

There are representatives of the reptiles, birds, and mammals that have reinvaded the sea. Since the blood of marine fishes is hypotonic to sea water, a diet of fish apparently contains enough usable water to maintain these animals in equal water balance. The unavoidable extra salt intake is excreted by some marine reptiles (turtles and iguanas) and birds (albatross) by special glands associated with the lacrimal duct or nasal passages. Such salt glands are also possessed by certain birds associated with salt marshes and estuaries.

The Egyptian camel is another animal in which water balance adaptations have been extensively studied. Most mammals cannot tolerate more than a small change in body temperature. They are, therefore, committed to relatively large expenditures of water for body cooling.

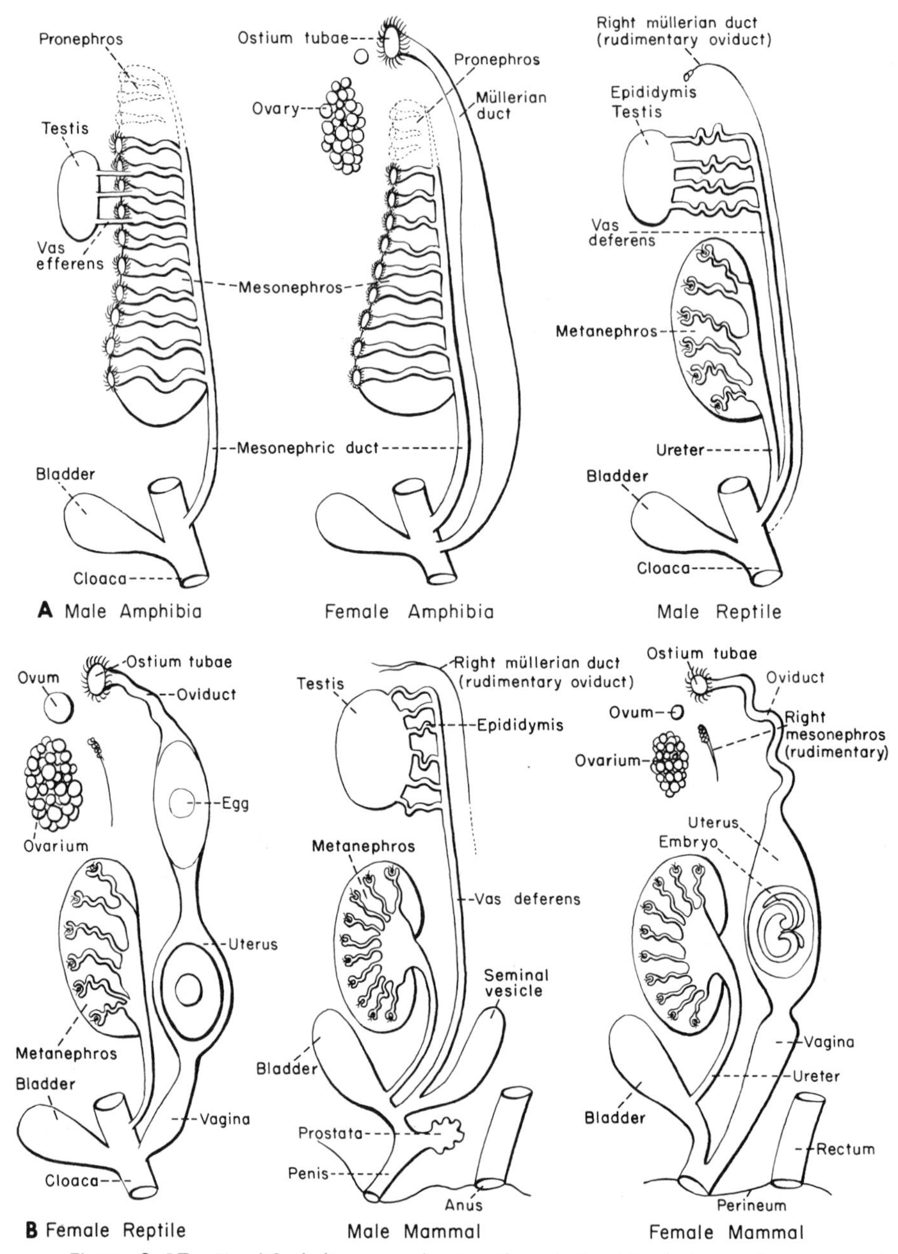

Figure 8–17 Simplified diagrams showing the relationships between the renal and reproductive organs of vertebrates. (From Florey: *An Introduction to General and Comparative Animal Physiology*. Philadelphia, W. B. Saunders Company, 1966.)

Camels, however, can tolerate a change of several degrees in body temperature. During the relatively cool desert night the body temperature falls. Then, during the first part of the hot day, it rises again to a maximum. Only after this has occurred does the animal actively thermoregulate by water evaporation. By then a part of the day has passed without such water loss, and this represents an important net saving of water.

The Egyptian camel can, in addition, tolerate a total loss of body water amounting to about 30 per cent of its weight. This compares with a maximum loss of only about 10 per cent for humans. Human water losses of this magnitude result in decreased plasma volume and consequent increases in plasma viscosity. Ultimately, the heart can no longer circulate the blood adequately and death occurs. Since plasma volume is not depleted, water lost from camels is apparently drawn from the intracellular water compartment (Sect. 8.3).

Like those amphibians which tolerate extreme desiccation, camels can replace their losses quickly when drinking water becomes available. In one reported case an animal that had been deprived of water for 16 days replaced exactly its lost weight (due almost entirely to water loss) by drinking 40 liters of water within 10 minutes time.

By comparison with many animals, man is poorly adapted to desert life. Active perspiration begins at about 89° F, and when the temperature rises to 110° F, the water loss by perspiration amounts to about one quart per hour. As the temperature rises still further, perspiration water loss increases at the rate of about 20 milliliters per hour per degree Fahrenheit. Moreover, a human cannot make good water losses of this kind by simply drinking water because perspiring also involves a considerable loss of salts. Replacement with water alone leads to osmotic problems of a serious nature. Therefore, salts as well as water must be replaced.

It is interesting to note in passing that the human kidney is perfectly capable of concentrating the urine sufficiently to permit the drinking of sea water. The fact that pure sea water is not tolerated is due, rather, to its effects on the gastrointestinal tract. There the high osmotic pressure of the sea water prevents water absorption from the gut (Sect. 9.20).

It is, perhaps, unnecessary to point out that all vertebrates minimize the problems associated with thermoregulatory water loss by seeking out the most favorable microclimate available to them. Such behavioral characteristics are of especially great importance to desert-dwelling forms. The kangaroo rat *Dipodomys,* for example, forages for food only during the dawn hours when the desert temperature is relatively low and the humidity is at a maximum. During the hot, dry part of the day, the animal remains in a cool, moist underground burrow with the entryway blocked by soil. Nearly all desert mammals and birds spend the hottest portion of the day lying quietly in some shady spot.

Reptiles depend upon behavioral controls for maintaining appropriate body temperatures since they have no well-developed physiological mechanisms for thermoregulation.

The evolution of the vertebrate kidney structure has been intimately related to that of the reproductive organs (Fig. 8–17).

CHAPTER 9

NUTRITION AND DIGESTION

NUTRITION

9.1 INTRODUCTION

In addition to oxygen and water, all animals have dietary requirements for certain additional organic substances including carbohydrates, fats, proteins, and vitamins. All of these are compounds that can be synthesized by plants and by microorganisms but not by animals. Yet, animals must have them for a variety of reasons and can obtain them only by feeding upon other organisms which contain them. Animals, like plants and microorganisms, also require an array of minerals either in the form of elements or as parts of simple compounds. These must also be present in animal diets.

The determination of specific animal requirements in the diet has not been easy. A number of complexities commonly exist which make experiments in nutrition difficult. Perhaps the greatest difficulty lies in the problem of separating a given animal from its symbiotic relationships with other organisms. Such symbioses are much more common than was recognized by the early investigators. For example, the human colon contains a variety of bacteria. These organisms not only derive their own living by feeding upon nutrients present in the gut, but they also produce substances that are required by their human hosts. Vitamin K, for example, is not present in human diets in sufficient quantities to support the production of prothrombin in the liver. We must have it since prothrombin is a necessary ingredient in the process of blood coagulation. Virtually our only source of supply is that synthesized by intestinal bacteria and absorbed from the colon. To establish experimentally, then, that vitamin K is a nutritional essential, requires prior elimination of the bacteria—a difficult thing to achieve.

In some cases an animal may require either substance A *or* substance B, but not both; this is true of certain amino acids (Sect. 9.4). Such a situation is much more difficult to prove experimentally than is the absolute requirement for a single substance. Finally, certain substances may be dietary essentials only during a particular part of the life cycle of the animal. For example, certain growth requirements may be of lesser or no importance to the mature animal.

Dietary requirements reflect inabilities of the animal in question to synthesize compounds that play essential roles in the metabolism and these inabilities are genetic in origin. It is not surprising, then, that different genetic strains of the same species may show different dietary requirements. This is particularly true of lower organisms such as bacteria and the mold *Neurospora,* both of which have been extensively used in the study of fundamental genetic principles.

Another difficulty in the establishment of dietary essentials is the fact that some of them—particularly the vitamins and certain minerals—are required only in very small quantities. It is often hard to devise diets that are completely lacking in one ingredient and yet sufficient in all other ways. To prove the specific requirement it is necessary to feed such a deficient diet to the organism and demonstrate symptoms resulting from the lack of the substance in question.

The nutritional requirements of a few mammals, insects, and microorganisms have been studied extensively, but very little research has been carried out on most vertebrates. It seems unlikely that most vertebrates differ greatly from the mammals, however, in their dietary requirements. All evidence to date indicates that most of the losses in synthetic abilities occurred rather early in animal evolution.

9.2 CARBOHYDRATES

Carbohydrates are hydrocarbons in which approximately half of the hydrogen atoms have been replaced by hydroxyl groups (–OH) and which also contain an aldehyde group ($-C\overset{\displaystyle O}{\underset{\displaystyle H}{\lessgtr}}$) or a ketone group ($-\underset{\underset{\displaystyle O}{\|}}{C}-$) on the molecule (Fig. 9–2). The central carbon chain may be three, four, or more carbon atoms in length. Consequently, carbohydrate molecules (sugars) are classified as trioses, tetroses, pentoses, and hexoses (Fig. 9–1). The six-carbon sugars (hexoses) are of primary importance in animal nutrition.

Single hexose molecules are called simple sugars or monosaccharides. Quite a number of these could theoretically exist, but only a few are of biological importance (Fig. 9–2). These include glucose, fructose, galactose, and mannose. Many of the theoretical monosaccharides never actually appear in nature at all, and many of the others are quite rare.

Monosaccharides may be linked in pairs, through the loss of a molecule of water, to form double sugars or disaccharides (Fig. 9–3). The reintroduction of water can, under certain conditions, split the disaccharide into its constituent monosaccharides, a reaction known as

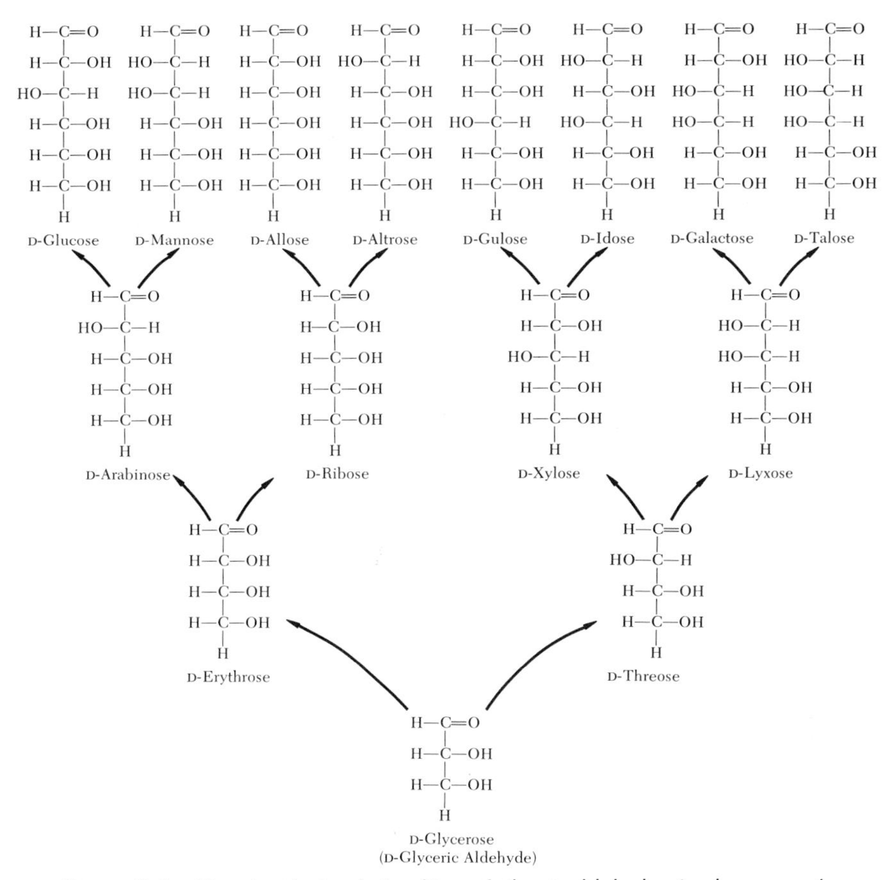

Figure 9–1 The chemical relationships of the D-aldehyde simple sugars (monosaccharides). A similar series of mirror-image L-aldehyde sugars can be drawn but none of these are of biological importance. Similarly, series of D- and L-ketone sugars can be drawn, but of these only D-fructose is of biological importance. The prefix letters "D" and "L" stand for dextrorotatory and levorotatory (right or left) and refer to certain physical properties of solutions of these sugars.

hydrolysis. Such hydrolytic cleavage occurs during the digestion of complex carbohydrates as we shall presently see. Similarly, long series of monosaccharides can be linked together to form polysaccharides, and these can be separated again by hydrolysis. Common polysaccharides include starch, glycogen, and cellulose (Fig. 9–4).

No one specific sugar molecule is essential in animal diets. Most carbohydrate sources include a mixture of the common hexoses. Hexoses are used by the animal primarily as an energy source; they supply 55 to 70 per cent of the total energy requirements of humans. Hexoses can be stored in limited quantities in the form of glycogen in the liver and muscles. Excesses beyond the glycogen capacity of these tissues are converted into fats and stored as such in the various fat depots of the body. Stored fats, then, can be reconverted and used as energy sources when needed.

2.20. FATS

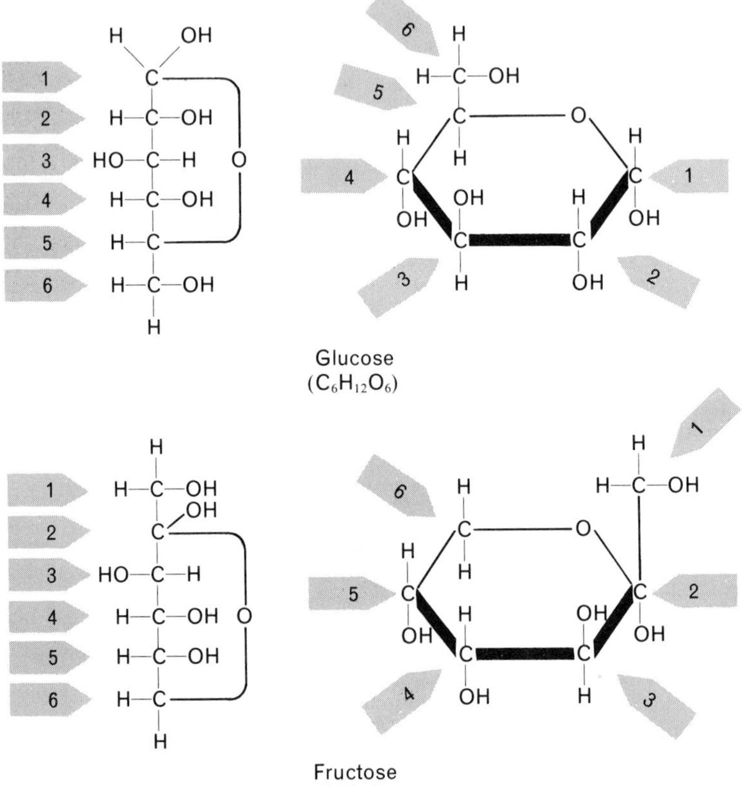

Figure 9–2 The structures of two important hexose monosaccharides. Note that they can be represented in either of two ways. Corresponding carbon atoms are labeled with corresponding numbers. From Cockrum and McCauley: *Zoology*. Philadelphia, W. B. Saunders Company, 1965.)

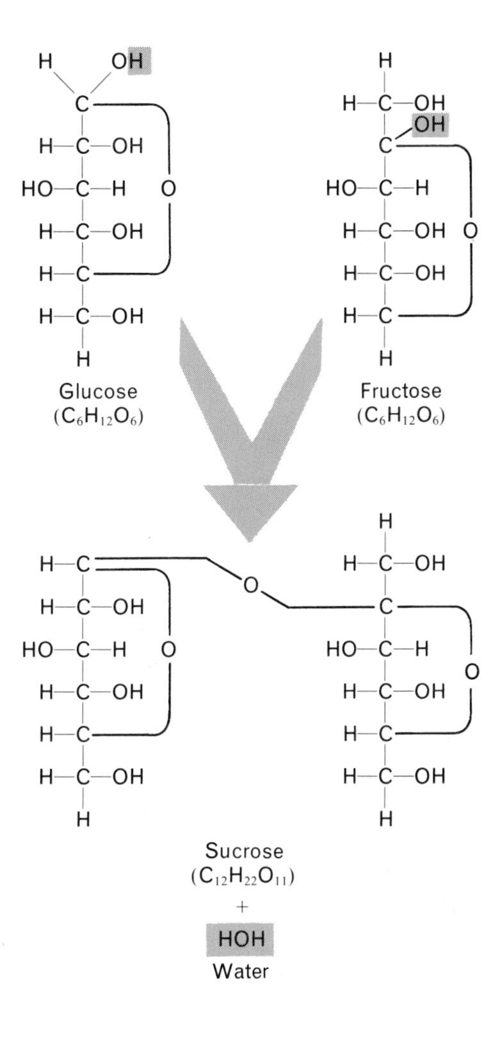

Figure 9–3 The formation of a double sugar (disaccharide) from two simple sugar (monosaccharide) molecules.

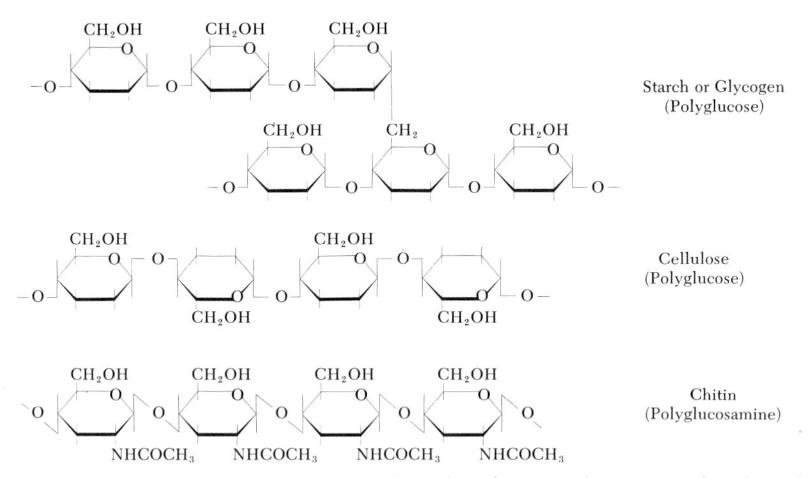

Figure 9–4 Structures of common polysaccharides. Starch consists of 250 to 300 glucose units. Glycogen is similar but contains up to 30,000 glucose units. Cellulose consists of several thousand glucose units and chitin of several thousand units of a derived glucose molecule called glucosamine. Other more complex polysaccharides such as inulin, pectin, and agar are composed of fructose, galactose, or other derived monosaccharides of various kinds.

9.3 LIPIDS

A variety of compounds are classified as lipids, but in most animal diets the simple lipids (fats) are the ones present in the greatest abundance. These simple lipids are sometimes called triglyceride fats since each molecule is composed of glycerol plus three fatty acids. The structures of these components are shown in Figure 9–5. In a triglyceride fat they are linked together by ester linkages $(\text{—C—O—}\underset{\underset{\displaystyle \text{O}}{\|}}{\text{C}}\text{—})$.

The fatty acids generally found in natural fats are straight-chain molecules composed of even numbers of carbon atoms (Fig. 9–6). They may or may not have double bonds at one or more sites in the chain; those which do have double bonds are said to be unsaturated fatty acids. The ester linkages of fats can be broken by the introduction of water (hydrolysis), which occurs during the digestion of simple fats (Fig. 9–5).

More complex lipids such as the sterols are of physiological impor-

$CH_3\text{—}COOH$	Acetic acid
$CH_3\text{—}(CH_2)_2\text{—}COOH$	Butyric acid
$CH_3\text{—}(CH_2)_{14}\text{—}COOH$	Palmitic acid
$CH_3\text{—}(CH_2)_{16}\text{—}COOH$	Stearic acid
$CH_3\text{—}(CH_2)_7\text{—}CH{=}CH\text{—}(CH_2)_7\text{—}COOH$	Oleic acid
$CH_3\text{—}(CH_2)_3\text{—}(CH_2\text{—}CH{=}CH)_2\text{—}(CH_2)_7\text{—}COOH$	Linoleic acid
$CH_3\text{—}(CH_2\text{—}CH{=}CH)_3\text{—}(CH_2)_7\text{—}COOH$	Linolenic acid
$CH_3\text{—}(CH_2)_3\text{—}(CH_2\text{—}CH{=}CH)_4\text{—}(CH_2)_3\text{—}COOH$	Arachidonic acid

Figure 9–5 The structures of some common fatty acids found in natural foods. Those with double bonds are unsaturated fatty acids. Linoleic and arachidonic acids are considered to be dietary essentials.

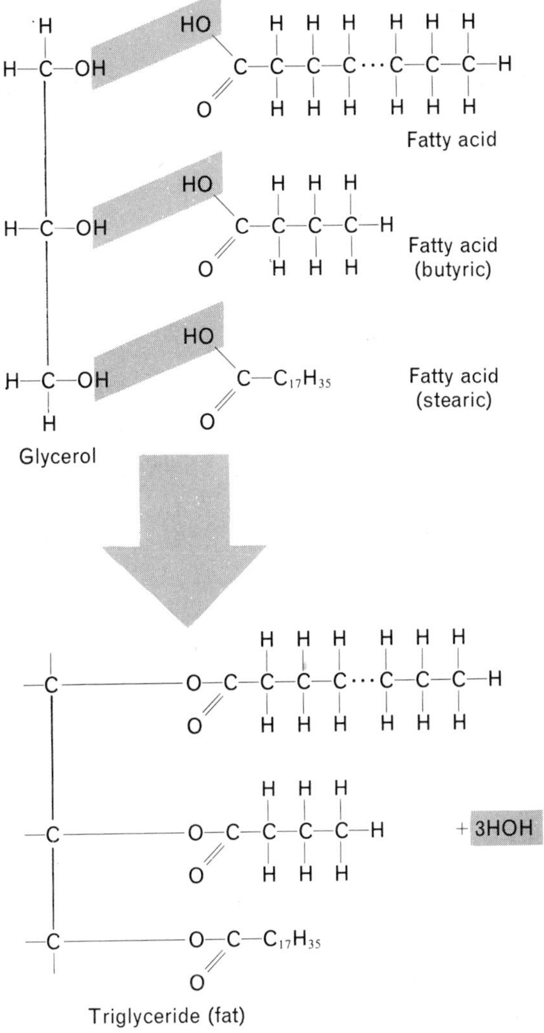

Figure 9–6 The formation of a triglyceride fat from glycerol and fatty acid. (From Cockrum and McCauley: *Zoology.* Philadelphia, W. B. Saunders Company, 1965.)

tance but they play no great part in dietary considerations. They are discussed elsewhere in this book (Sect. 10.11 and Chapter 17).

Because of their high energy value, fats are important components of vertebrate animal diets, but precise requirements are not well established. There is evidence to suggest that small amounts of linoleic and arachidonic acids (Fig. 9–5) are specifically required in the diet, and for this reason these two are sometimes called the essential fatty acids.

We have already mentioned that excess sugars are converted into simple fats for storage in the animal body and that reconversion of these fats to sugar renders them available for energy metabolism. The high energy content per gram of fat makes it an ideal form for storage. It also has high insulating properties and, for this reason, is deposited in large quantities beneath the skin of warm blooded animals living under arctic conditions.

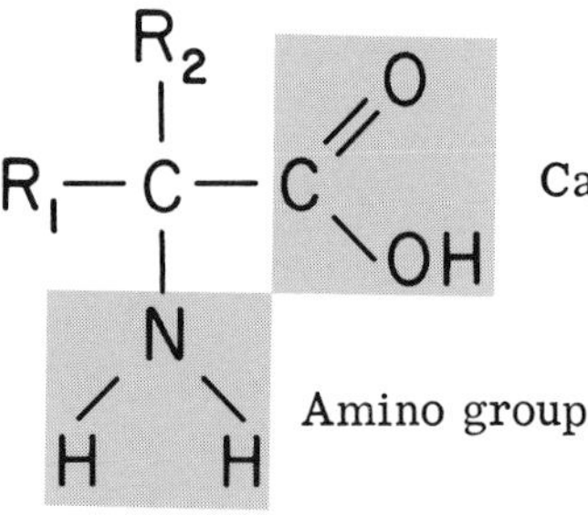

Figure 9–7 The structural relationships of the amino group and the carboxyl group in an amino acid. R indicates the portion of the molecule which varies in different amino acids. See also Figure 8–9.

9.4 PROTEINS

Proteins are long-chain molecules composed of fundamental units called amino acids. An amino acid is an organic acid in which an amino group ($-NH_2$) and a carboxyl group ($-COOH$) are present in a particular association (Fig. 9–7). Amino acids link together by forming peptide linkages (—C—N—, with O double-bonded to C and H on N) between the amino group of one amino acid and the carboxyl group of another (Fig. 9–8). A molecule of water is split out in the process. As in the case of both carbohydrates and simple fats, the molecules can be reseparated by hydrolysis. Two amino acids linked together in this fashion constitute a dipeptide, three amino acids form a tripeptide, and so forth. A series of a number of amino acids is generally called a polypeptide, and very long polypeptides are proteins.

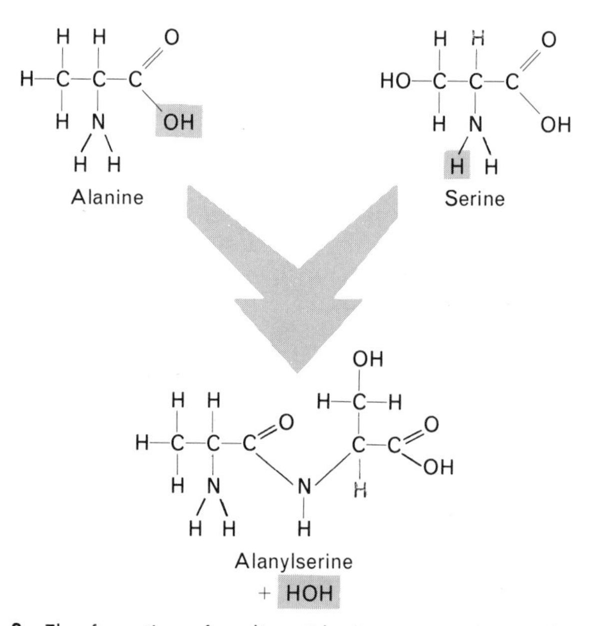

Figure 9–8 The formation of a dipeptide from two amino acids. (From Cockrum and McCauley: *Zoology*. Philadelphia, W. B. Saunders Company, 1965.)

Glycine

Alanine

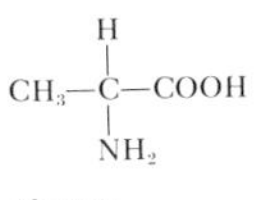

Serine

Threonine°

Valine°

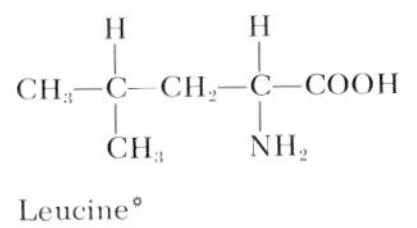

Leucine°

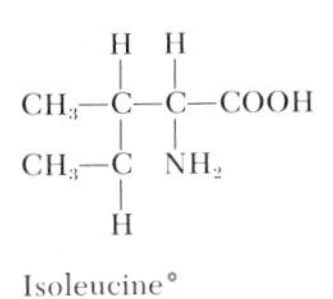

Isoleucine°

Phenylalanine°

Tyrosine

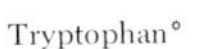

Tryptophan°

Proline

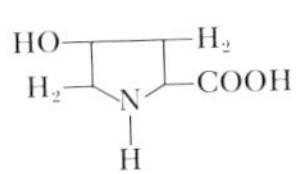

Hydroxyproline

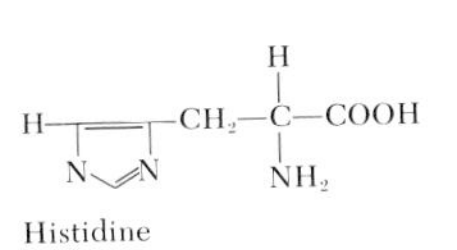

Histidine

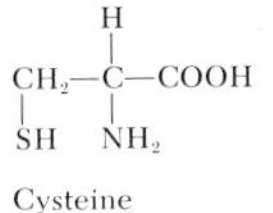

Cysteine

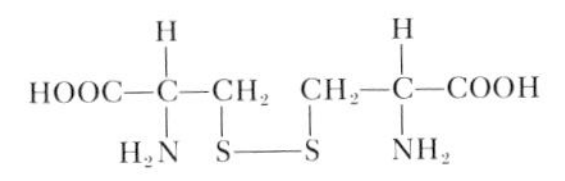

Cystine (dicysteine)

Methionine°

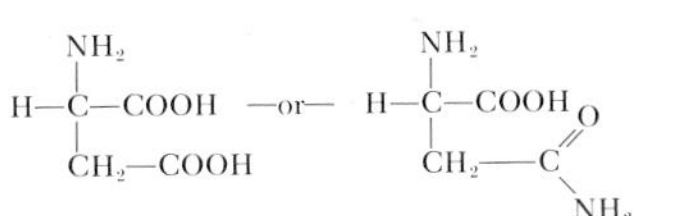

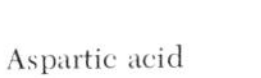

Aspartic acid —or— Asparagine

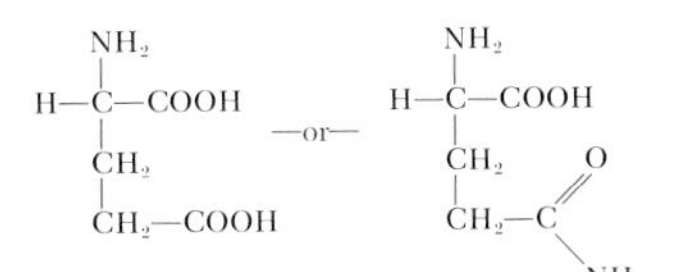

Glutamic acid —or— Glutamine

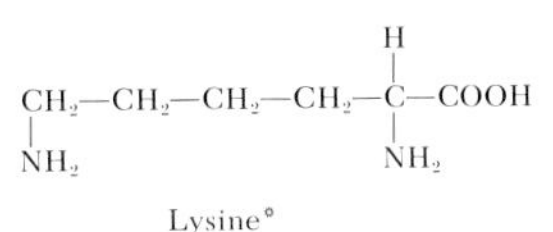

Lysine°

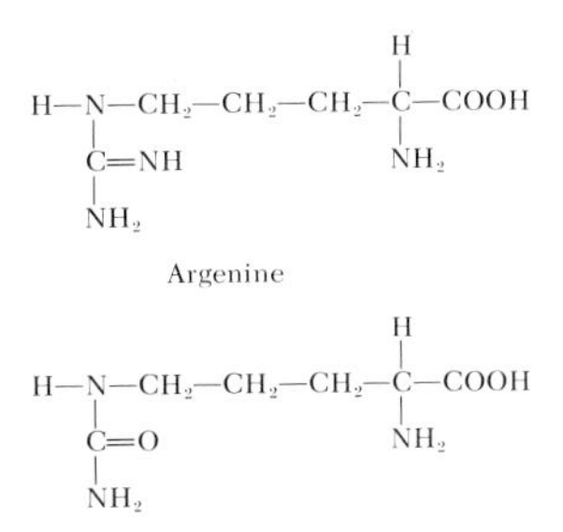

Argenine

Citrulline

Figure 9–9 Chemical structures of common amino acids. Those with rings in the molecule are called aromatic amino acids, whereas those without rings are called aliphatic amino acids. Rings containing nitrogen are called heterocyclic rings. The amino acids that are marked with an asterisk are essential in adult human diets.

The hydrolysis of natural proteins has yielded 26 different amino acids of which 20 commonly occur in animal tissues (Fig. 9–9). In some cases interconversion of specific amino acids is possible in the vertebrate body. Amino acids that can be produced in this way, then, are not specific dietary requirements so long as mixtures of other amino acids are present. Such interconversion of amino acids takes place through what is called a transamination reaction (Fig. 10–14). There are some eight to ten amino acids, however, that cannot be produced from other amino acids in this way. These are called the essential amino acids and they must be present as such in the diet. For adult humans the essential amino acids are threonine, valine, leucine, isoleucine, lysine, methionine, phenylalanine, and tryptophane. Human infants also require histidine and, perhaps, arginine. Most of the amino acids in this list are also essential for other multicellular animals which have been studied, but there are minor differences. A few microorganisms also have specific amino acid requirements, but most are capable of synthesizing all 20 of the common ones. Apparently the pattern of amino acid requirements was established quite early in animal evolution by the genetic loss of particular synthetic abilities.

In addition to small quantities of each of the essential amino acids, animal diets must contain mixtures of other amino acids in sufficient quantity to meet the requirements for normal growth, development, and reproduction. Protein synthesis from amino acids is the basis for these processes.

Amino acids cannot be stored to any significant degree in the animal body. Instead, excess amino acids are utilized as energy sources after removal of the amino group (Sect. 10.12).

9.5 VITAMINS

Vitamins are essential food factors, required in very small amounts, that aid in the maintenance of normal tissue activities. Generally they act as the functional or prosthetic groups of various enzymes (Sect. 9.8). Compounds classed as vitamins have diverse chemical structures (Figs. 9–10 and 9–11), and the list of essential vitamins is not identical for all vertebrates. For example, vitamin C (ascorbic acid) is a dietary requirement for man, anthropoid apes, and the guinea pig but apparently not for the other mammals which have been studied. Similarly, vitamin A is essential for all vertebrates and, perhaps, for certain crustaceans but not, apparently, for other animals.

The first vitamins identified were all compounds having an amino group and were called, accordingly, "vital amines." Later other essential compounds were discovered which were not amines, and the name was changed first to "vitamines" and then to "vitamins." The current list of vitamins includes compounds that have no chemical relationships to each other and they are grouped together under this name only because they are all essential dietary factors and because they apparently play similar kinds of metabolic roles. Future research may well add more compounds to the list.

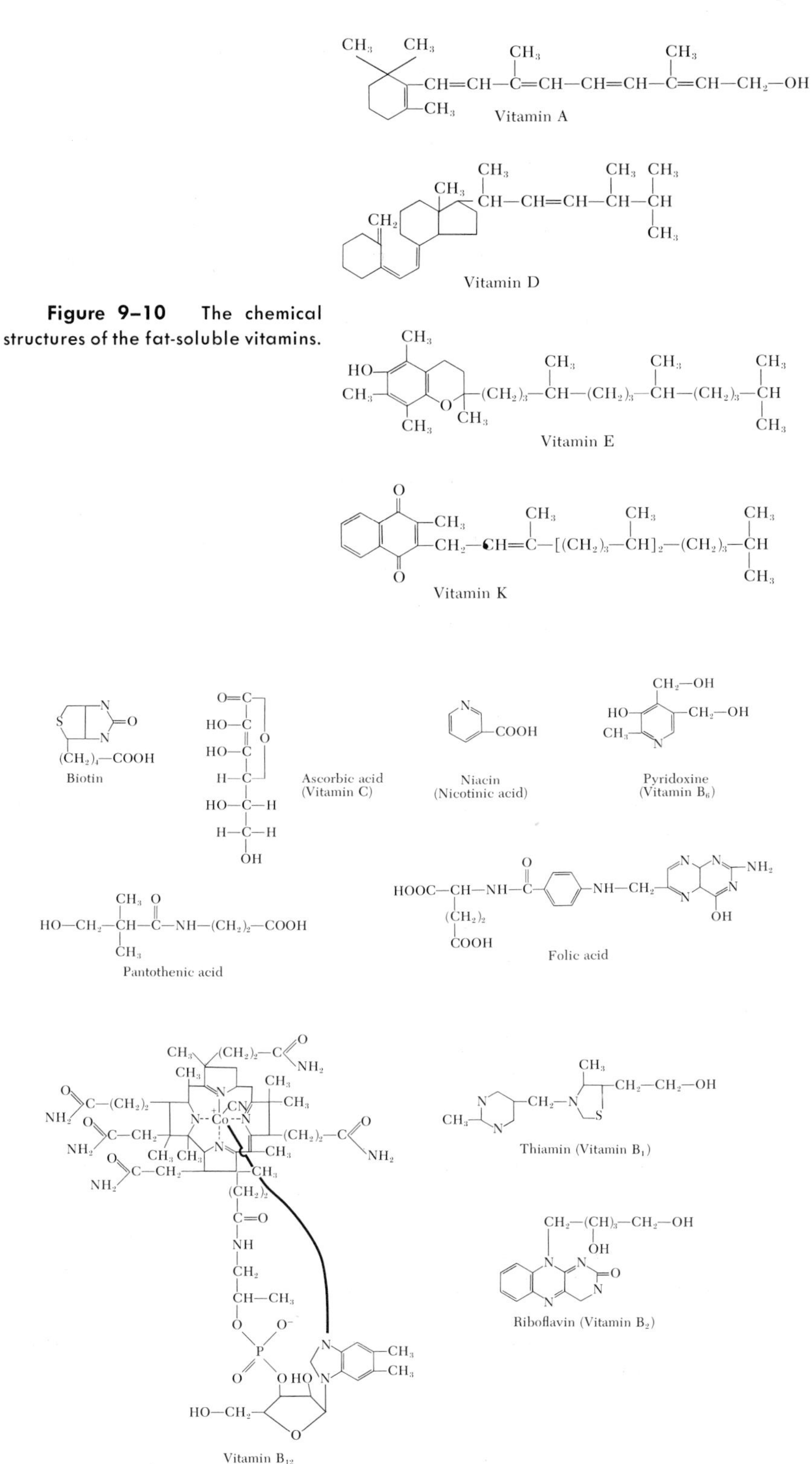

Figure 9–10 The chemical structures of the fat-soluble vitamins.

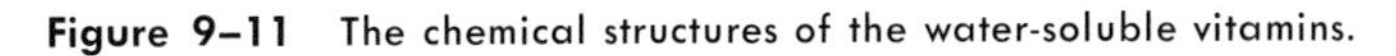

Figure 9–11 The chemical structures of the water-soluble vitamins.

The appearance of a vitamin requirement in evolution may possibly be explained in either of two ways: (1) The primitive ability (shared by most microorganisms) to synthesize these compounds was lost through genetic mutation, and the animal became dependent upon a dietary source instead; or (2) a functional requirement not present in lower forms appeared in connection with the evolution of a new physiological process. For example, vitamin A is an essential for normal vertebrate vision. It is present in the tissues of virtually all animals but does not have that function in subvertebrates. It may be that it is not an essential physiological or dietary substance except in vertebrates.

9.6 MINERALS

Certain chemical elements must be supplied in the diet in varying amounts because they are required as structural materials (*e.g.*, calcium and phosphorus in bone), as parts of various organic molecules that are synthesized in the body (*e.g.*, iron in hemoglobin), or as catalysts in metabolic reactions. An example of the latter is the requirement of calcium ions in the process of blood coagulation (Sect. 7.6).

The minerals of greatest dietary importance are calcium, phosphorus, sodium, potassium, chlorine, magnesium, iodine, iron, copper, sulfur, zinc, cobalt, bromine, and fluorine. In addition to these aluminum, arsenic, nickel, and silicon are always present in vertebrate tissues and may play vital roles that have not yet been demonstrated. The requirements for specific minerals are often difficult to assess since, quite often, they act together in complex ways. Sometimes the ratio of concentrations of two or more elements is of greater physiological importance than is the absolute quantity of either element alone. Calcium, potassium, and sodium show definite interrelationships of this kind in the maintenance of pH balance and of normal heart function. Likewise, copper catalyzes the utilization of iron in hemoglobin synthesis and cobalt influences the utilization of both copper and iron.

The element most commonly present in vertebrate tissues is calcium. Large quantities of it are deposited in bones and teeth, and it is present to the extent of about 10 milligrams per cent in blood plasma. Considerable amounts of calcium are also present in the soft tissues as well. Calcium ions catalyze many metabolic chemical reactions and play important roles in the controls of membrane permeability. The mechanisms by which it accomplishes these ends are largely theoretical and need not concern us at this time. Discussions of the subject may be found in any textbook of cellular physiology.

The deposition of calcium in bone is influenced by the hormone of the parathyroid gland and by vitamin D (Sect. 17.7). The disease called rickets may result from deficiencies of either the mineral or the vitamin.

Phosphorus, perhaps, plays more different metabolic roles than any other mineral. It is an important part of energy storing (Sect. 10.3) and transferring molecules (Sect. 10.4), and the addition or removal of phosphorus from specific molecules is often involved in biochemical reactions (Sect. 10.5). Phosphorus is also deposited along with calcium (as calcium phosphate) in bones and teeth.

Most of the magnesium in the body is deposited in bones, but it too plays a number of metabolic roles, usually as a catalyst. In some instances magnesium and calcium have opposing effects on physiological processes.

Sodium in the body is present almost entirely in solution in the extracellular body fluids, while potassium is mostly confined to the intracellular fluid compartment (Sect. 8.3). The differential distribution of these two elements is important in maintaining the resting electrical potential of cell membranes (Sect. 11.3). In addition, both ions play other physiological roles, some of which are discussed elsewhere in this book.

Manganese acts as an enzyme activator and is present in greatest concentration in the pancreas, gonads, skin, muscle, and bone. Iron is a vital part of the hemoglobin molecule, related molecules such as myoglobin, and the cytochrome enzymes (Sect. 7.3). Copper catalyzes several metabolic chemical reactions including the incorporation of iron in the hemoglobin molecule. Cobalt also catalyzes hemoglobin formation and is, in addition, a part of the vitamin B_{12} molecule (Fig. 9–11). Iodine is a necessary part of the thyroid hormone in which it is attached to the amino acid tyrosine (Sect. 17.6).

Sulfur is a part of the structure of the amino acids methionine, cystine, and cysteine as well as some of the vitamins. Zinc occurs in the enzyme carbonic anhydrase and is, thus, associated with carbon dioxide transport in the blood (Sect. 7.5). It is also an essential in certain metabolic reactions, especially in the liver. Fluorine has been shown to be a requirement in the formation of sound, cavity-resistant teeth but has no other recognized function. Virtually all of the fluorine in the diet is completely excreted by the body. Molybdenum and selenium may play roles in digestion and in facilitating the intestinal absorption of other materials. No metabolic functions within the body have been demonstrated for these substances, however.

Minerals may be supplied as such or as parts of molecules in the diet. An ordinary balanced diet contains more than enough of all of the common minerals. Tables of important sources in foods are available in textbooks of nutrition.

FEEDING AND DIGESTION

9.7 INTRODUCTION

To make appropriate use of nutrients it is necessary for a vertebrate animal to (1) separate them selectively from the general environment, (2) break them down mechanically and chemically, (3) absorb them into the body, and (4) either assimilate them as integral parts of body structure or oxidize them as sources of energy. Also, as we have seen, some nutrients are used to control the utilization of other nutrients (*e.g.*, as catalysts).

On the basis of the kinds of food selected, various animals are classified as carnivores, herbivores, omnivores, or they may be grouped even more specifically (Fig. 9–12). Often food selection is highly selective, *i.e.*, a given animal may feed upon one or only a few very specific species of animals or plants. Much more unusual (and successful!) is omnivorous feeding like that of man. An animal with specific feeding habits is totally dependent upon a continuing supply of that particular kind of food, whereas omnivorous feeding allows the animal to adapt to whatever kinds of food may be available. Both the seasonal activity and the geographic range of an animal may be increased by omnivorous feeding habits.

Correlated with feeding habits are various adaptations of the feeding apparatus. Adaptations of locomotion and of sensory systems may also be correlated with the types of foods selected. The vertebrates are generally characterized by the possession of movable jaws used in the capture and holding of animal prey or the gathering of plant materials. In a few specialized animals, such as anteaters and hummingbirds, the jaws have no such functions. Mammalian jaws are generally capable of mastication (chewing), a function not shared by other vertebrates. Various specialized vertebrate teeth are modified to suit a host of feeding habits. Some fishes have pharyngeal teeth that are movable. They can be folded back to permit the swallowing of live prey and they can be elevated to prevent the escape of the prey. The fangs of poisonous snakes can also be folded back or erected. They are modified in form and associated with special glands for the injection of poison.

A. Holophytic	Food is at least in part synthesized in the body from inorganic sources by means of photosynthesis (i.e., autotrophic).
B. Saprophytic (or Saprozoic)	Food is absorbed from dissolved nutrient materials in water (as among the protozoa in some cases).
C. Holozoic	Food consists of ingested plant or animal materials (i.e., heterotrophic).
1. Herbivorous	Food consists entirely of plant materials.
a. Frugivorous	Food consists of fruits.
b. Gramnivorous	Food consists of seeds.
c. Nectarivorous	Food consists of nectar of flowers.
d. Others	
2. Carnivorous	Food consists entirely of animal tissues.
a. Insectivorous	Food consists of insects.
b. Piscivorous	Food consists of fish.
c. Sanguivorous	Food consists of blood.
d. Others	
3. Omnivorous	Food consists of both plant and animal materials.
D. Mixotrophic	Food is secured from two or more of the major sources listed above.

Figure 9–12 The feeding habits of animals (terminology). (From Cockrum and McCauley: *Zoology*. Philadelphia, W. B. Saunders Company, 1965.)

Certain fishes are equipped with extensions of the gill apparatus known as gill rakers. They serve to strain from the water the tiny microorganisms and invertebrates upon which the animal feeds. The whalebone whales and the basking sharks are also filter-feeders, although their filtering apparatus has a different anatomical derivation.

Parrot fishes have a bill-like mouth without teeth that is equipped with a pharyngeal grinding mill composed of flattened plates. These are used to grind up corals so that the contained animal polyps can be digested. The calcareous portions of the coral are quickly expelled through the anus.

Other adaptations of the mouth include the suckers found in fishes that live in swiftly flowing waters or that attach themselves to other fishes for purposes of feeding. The remora, for example, attaches itself to sharks in order to accompany them and feed upon the leavings of the shark kills. Lampreys attach themselves to other fishes and by means of a special rasping tongue feed upon the flesh and blood of the host. The economic loss to the Great Lakes fishing industry has been severe for a number of years because of the depredations of lampreys.

In most vertebrates the tongue is used in the mixing and swallowing of foods and in some cases to capture or extract the food from the environment (*e.g.*, the rasping tongue of lampreys). Frogs have sticky tongues that can be flipped out to considerable distances to capture insects; anteaters insert their long sticky tongues into colonies of ants or termites in order to withdraw them from their burrows; and hummingbirds have tubular tongues which they use like sipping straws to obtain nectar from flowers.

In most vertebrates the esophagus serves only as a conducting pathway to the stomach, but in birds a part of the esophagus is modified as a crop in which foods can be temporarily stored. A feeding bird is itself more vulnerable to being captured and eaten. There is an advantage, therefore, in being able to feed rapidly without the necessity for mastication. Foods stored in the crop can subsequently be passed to the stomach at a more appropriate rate (where the equivalent of mastication occurs).

The stomachs of most vertebrates are characteristically composed of a single chamber in which food is stored and chemical digestion begins, but in some vertebrates the stomach is modified to contain additional chambers that perform special functions. In ruminant animals, such as cattle, a portion of the stomach contains a large population of symbiotic microorganisms that contribute to the nutrition of the host animal, as we shall see presently. In birds the stomach is characteristically divided into two chambers, one of which functions in typical digestive processes. The second chamber serves in the mechanical breakdown of foods or in the temporary storage of indigestible materials (Sect. 9.13). The Crocodilia are similarly equipped with a food-grinding compartment or gizzard.

In the small intestine further chemical breakdown of foods occurs under the influence of enzymes and other substances secreted from the intestinal walls, pancreas, and liver. Absorption of digested products also occurs primarily in this portion of the digestive tract.

The large intestine or colon functions in the reabsorption of water from the remaining intestinal content and in the preparation of these undigested materials for evacuation. Generally, microorganisms are present in the colon that aid in the formation of the fecal material. Quite commonly they also contribute to the nutrition of the host (Sect. 9.20). In some vertebrates one or more diverticula, or pouches communicating with the colon, are present and serve to contain populations of such microorganisms.

The passage of food materials through the digestive tract and the digestion, absorption, and evacuation of wastes are stages in a programmed sequence which is under nervous and hormonal control. The controlling mechanisms modify these actions in accordance with the types of foodstuffs to be dealt with and other physiological factors.

9.8 ENZYMES

Enzymes are either proteins or complex compounds made up of a protein plus a non-protein molecule. The non-protein portion of such enzymes is called the prosthetic group of the molecule, and this role is often played by a vitamin. Each enzyme is a highly specific organic catalyst, which increases the rate of a specific chemical reaction without being itself chemically altered by the process. Almost every chemical reaction in animal metabolism requires the presence of such an enzyme. Since this is also true of unicellular organisms, we must conclude that an array of enzymes was already present and available before the first appearance of life on Earth. Experimental evidence indicates that this was quite possible under the conditions existing at that time (Sect. 1.7).

An enzyme functions by combining with the substance on which it is to act (the substrate) to form a complex enzyme-substrate molecule. This event somehow alters the substrate so that some particular chemical reaction is facilitated. Once this reaction has taken place, the substrate is altered so that it no longer "fits" the enzyme molecule. As a result, the enzyme is freed and becomes available to combine with another molecule of substrate and catalyze the same reaction again.

Within limits, increased quantities of enzyme accelerate the chemical reaction even more. If twice as much enzyme is present, then twice as many molecules of substrate can react simultaneously. The theoretical limit is the situation in which there are equal numbers of enzyme and substrate molecules present; quite obviously additional enzyme molecules, having no substrate molecules to act upon, could not hasten the reaction any further. Metabolic reaction rates, thus, can be controlled by varying the amount of enzyme in accordance with the amount of substrate present, and this can be accomplished in several different ways: (1) Reserves of enzymes can be stored in inactive states so that they can be released or activated quickly as the need arises.

Activation may depend upon the simultaneous presence of a second enzyme (coenzyme) or some particular ion such as calcium or magnesium. Activation may, instead, depend upon the unmasking of the enzyme by removal of a part of the inactive molecule. (2) The synthesis of certain enzymes is dependent upon the presence of the appropriate substrate. That is to say, the enzyme is not even produced unless the substrate on which it is to act is present. Such inductive enzymes, as they are called, are produced in response to some feedback control initiated by the specific substrate. This is in contrast to the constitutive enzymes which are always present. (3) The amount of enzyme synthesis may be repressed by decreases in the amount of substrate present or by increases in the concentration of the reaction products. The mechanisms by which enzyme production is facilitated or repressed are not completely understood, but the advantages in terms of metabolic economy are obvious. Theoretical mechanisms are commonly discussed in works on cytology and genetics.

Proteins are said to be amphoteric compounds; that is, the amino and carboxyl groups on the molecule are differentially ionized at different pH values. This alters the chemical reactivity of the molecule making it act more as an acid (carboxyl ionization) or as a base (amino ionization). Consequently, protein reactions are pH-dependent. Since enzymes are fundamentally proteins, it is not surprising to note that there are optimum pH ranges in which the catalyzed reaction occurs most rapidly.

All chemical reactions proceed more rapidly at higher temperatures, but there are upper limits for enzyme-catalyzed reactions. This is true because the proteins are temperature-labile compounds; that is, they are chemically altered by high temperatures. Such an alteration is called denaturization and commonly occurs to proteins at temperatures above 40 to 41° C. Denaturization of an enzyme molecule renders it permanently ineffective as a catalyst.

Enzymes are produced by living cells, and most enzymes remain to catalyze reactions within the same cell. Even the enzymatic breakdown of foodstuffs occurs intracellularly in lower animals; each cell feeds upon and digests its own food just as though it was an independent unicellular organism. In higher animals, including vertebrates, food digestion is extracellular. The enzymes are released in solution into the lumen of the digestive tract by the cells which synthesize them. In some instances the enzyme is secreted by the cell and in others enzyme release depends upon the disintegration of the cell. Enzyme-producing cells are sloughed from the intestinal glands and are then disrupted to release the contained enzymes. It seems likely that this kind of release may have been an evolutionary intermediate between intracellular digestion and the extracellular secretion of digestive enzymes.

Especially among mammals, digestive enzyme secretion or release varies with the diet. When, for example, the proportion of carbohydrate to protein is increased in the diet, the production of carbohydrate-splitting enzymes (carbohydrases) increases and the production of protein-splitting enzymes (proteases) decreases. Digestive enzyme production is more constant in animals that spend greater lengths of time in feeding, but it is appropriately turned off and on in those animals that feed only periodically. For example, the process of digestion, as we shall see, is virtually constant in cattle.

9.9 SALIVA

Salivary glands in the oral region of most vertebrates secrete a moistening fluid containing mucin for lubrication and variable amounts of salts including bicarbonates. The pH of human saliva is near neutrality, whereas that of cattle is quite alkaline, thus reflecting differences in the kinds and quantities of dissolved substances. In some vertebrates the saliva also contains a starch-splitting enzyme called salivary amylase. Man, apes, elephants, and a few rodents produce high concentrations of this enzyme, while pigs, bears, ruminants, most rodents, fishes, and many amphibians produce lower concentrations. Still other vertebrates apparently lack salivary amylase altogether. The effectiveness of salivary amylase in humans is probably low since the food is immediately passed to the stomach where the pH is far below the effective range for this enzyme. The question is more academic than practical, however, since the pancreas produces a similar enzyme (pancreatic amylase) that digests starch when the food reaches the small intestine. The pH of the intestinal contents is within the optimum range for this enzyme.

The saliva of lampreys contains a substance that prevents the coagulation of the blood of its host or prey. Proteases and lipases (fat-splitting enzymes) may also be present in the saliva of some reptiles (*e.g.*, *Heloderma*, the Gila monster of Mexico and the Southwestern United States). This is subject to some doubt, however, because the esophagus of this animal contains glands much like those of the stomach. These enzymes may be regurgitated from the esophagus and, therefore, not really a part of the saliva as such.

Mammalian salivary glands are under nervous control and are stimulated by the thought, smell, or taste of food. The percentage composition of water, mucin, salts, and enzyme in saliva varies with the composition of the food introduced into the mouth, but little is known of the mechanisms responsible for these differential controls.

When a negative water balance exists (Sect. 8.3), the flow of saliva is reduced and the mouth and pharynx become drier. It is believed that sensory receptors located in the pharynx respond to such drying by inducing the sensation of thirst. Thus, salivary flow may play an intermediate role in the maintenance of proper water balance.

The salivary glands are stimulated by stimulation of the parasympathetic nerves and inhibited by stimulation of the sympathetic nerves leading to them (Sect. 12.6). Thus, situations that lead to sympathetic activity, such as anger or fear, result in inhibition of salivary activity and a consequent drying of the mouth.

9.10 DEGLUTITION

Swallowing or deglutition may be divided into three phases as it occurs in humans and, probably, in other mammals. In the first phase, the chewed food is rolled into a ball on the tongue and forced backward into the pharynx. The tongue follows this movement, at first slowly and then more rapidly so that the food is projected into the pharynx (Fig. 9–13). During the second phase of swallowing, the tongue remains in the same position blocking the opening from the mouth and elevating the soft palate to block the opening from the nasal cavities. At the same time, the epiglottis tips slightly and the larynx moves upward toward it so that the opening to the trachea (the glottis) is closed. The esophageal opening dilates somewhat and the walls of the pharynx contract forcing the food into the esophagus. The third phase of deglutition involves the movement of the food through the esophagus to the stomach. This is accomplished by dilation of the esophagus below and constriction above the moving bolus of food. These muscular responses result from nervous reflexes initiated by the distention of the esophagus; they proceed toward the stomach in a wavelike manner "milking" the food along with them.

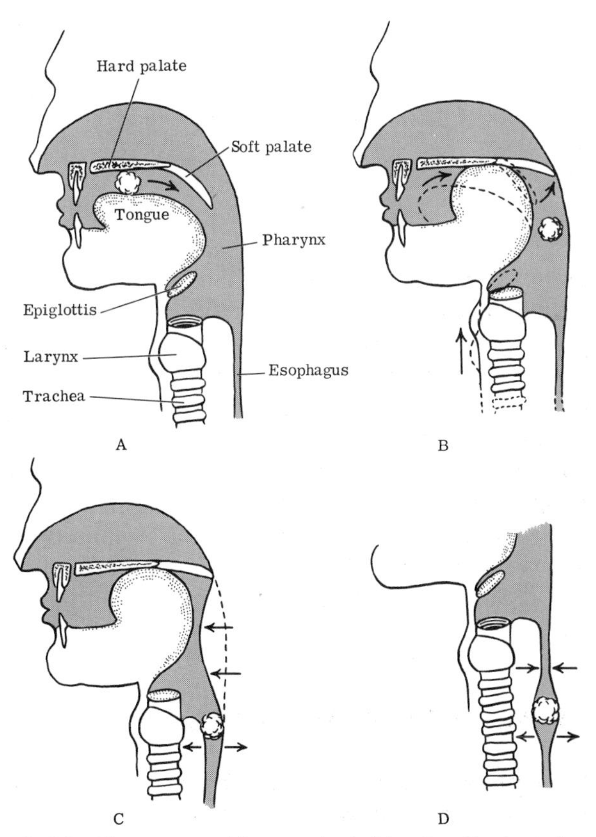

Figure 9–13 The process of human deglutition (swallowing). See text.

In some lower vertebrates (*e.g.*, frogs), the entire pharynx and esophagus are ciliated and food particles are propelled to the stomach by the coordinated action of the cilia. It is interesting to note that the esophagus of fetal mammals is similarly ciliated, but this feature disappears before birth.

9.11 THE ESOPHAGUS

In most vertebrates the esophagus is merely a pathway along which food moves from the pharynx to the stomach, but in some it may also aid in food storage. Teleost fishes have a sphincter at the end of the esophagus that prevents water from entering the stomach; terrestrial vertebrates do not possess such a valve. In animals that swallow their prey whole, the size of the meal may exceed the capacity of the stomach so that the esophagus remains distended for some time. It is not uncommon, for example, to encounter sea birds with the tail of a fish hanging from the mouth while the head of the same fish is undergoing digestion in the stomach. In many other birds a portion of the esophagus is modified as an enlarged crop for food storage. This modification has probably occurred more than once in evolution because not all bird crops appear to be homologous structures. The possession of a crop enables a bird to eat rapidly and consume large quantities of small foods (*e.g.*, insects, seeds) when they are available. Rapid feeding also reduces the period of time during which the bird is particularly exposed to predation. The crop stores the food and forces it into the stomach at an appropriate rate for the digestive processes taking place there. Probably there is no actual digestion of foodstuffs in the crop, although some investigators have reported the presence of an amylase in esophageal secretions of birds. During storage there, foods do undergo moistening and swelling, which prepares them physically for the subsequent gastric activity.

Among doves and pigeons the epithelial lining of the crop becomes thickened in both sexes while they are incubating the eggs in the nest. Just before the eggs hatch, the epithelial cells begin to slough away and disintegrate to form a liquid known as pigeon milk. This fluid, which is rich in fat and protein, is regurgitated and fed to the newly hatched offspring during their first two weeks of life. It is interesting to note that while pigeon milk and mammalian milk are totally different materials produced by unrelated glands, the same hormone, apparently, controls the production of both (Sect. 17.13) and both are used for the same purpose.

9.12 THE TYPICAL VERTEBRATE STOMACH

The vertebrate stomach is apparently not homologous with the stomach of any living invertebrate animal and there is no true stomach in the most primitive vertebrates (*e.g.*, the lamprey). It is also unique among animal organs in that high concentrations of hydrochloric acid are secreted by glands in its walls. In some vertebrates (*e.g.*, humans) the gastric juices may have a pH of less than 1.0 during the digestion of a meal because of the presence of this acid. Gastric juices also typically contain a protein-splitting enzyme called pepsin. This catalyzes the splitting of the protein molecule at particular sites, generally adjacent to an aromatic amino acid in the chain. Long-chain proteins are, thus, split into shorter polypeptides in the stomach.

The cells that produce the gastric protease are themselves composed of protein to a large extent. If such a cell produced an active protease, that enzyme might well disrupt the structure of the cell. Instead, pepsin is produced by these cells in the form of an inactive proenzyme called pepsinogen. Pepsinogen, then, enters the lumen of the stomach and is converted into the active form, pepsin, by the action of hydrochloric acid. The walls of the stomach are also, of course, composed to a large extent of protein and one might wonder why these are not digested. Sometimes this apparently occurs (ulceration), but the lining of the normal stomach is protected from protease activity by a coating of resistant mucin. This mucin is also secreted by glands in the walls of the stomach.

Probably the primitive function of the vertebrate stomach was the storage of foods and their release into the intestine at a rate which permitted efficient digestion to occur there. In addition, during evolution hydrochloric acid perhaps played a bacteriostatic role, preventing the spoilage of stored foods. It may also have served to kill living prey that had been swallowed whole. Finally, the evolution of pepsin-secreting cells made it possible to begin the digestion of proteins and thus break down the structure of tissues which had been eaten.

The stomachs of certain young mammals (*e.g.*, calves) secrete a milk-coagulating enzyme called rennin. Hydrochloric acid clots milk efficiently, so this enzyme, if present, would have little value in adult humans. It may be present in the stomachs of babies, however, since the hydrochloric acid concentration is low during infancy and milk is the sole diet for a time.

Mammalian gastric juice also contains a gelatinase and a gastric lipase. The former acts specifically on the gelatin molecule—a specific kind of protein—while the latter splits fatty acids away from the glycerol molecule. It is especially effective in splitting off butyric acid (Fig. 9–6).

When gastric secretion and digestion have progressed far enough to render the foodstuffs liquid in consistency, the pyloric valve of the stomach allows it to pass slowly into the small intestine. Continuous wavelike contractions of the stomach wall serve to mix the contents and to force it through the pyloric valve.

9.13 THE BIRD STOMACH

Bird stomachs are typically divided into two chambers, but the development and relative importance of the two parts vary. In herbivorous birds and in those carnivorous species which swallow whole mollusks and crustaceans, the muscular gizzard predominates (Fig. 9–14). The lining of this chamber secretes a keratinous material that hardens in the form of plates or ridges used for the grinding of foods. In addition, many birds swallow quantities of sand to aid in the process. This preliminary grinding is the equivalent, of course, of chewing by mammals and it achieves to a remarkable degree the mechanical disruption of food materials. The gizzard of a turkey can grind up nut shells and even metal bits. In one remarkable experiment, a tube of sheet iron which could be dented only by a force of 80 pounds, was completely flattened and partly rolled up after spending 24 hours in a turkey gizzard! In Egypt, reportedly, manufactured copies of small items such as carved stones are force fed to fowls so that the action of the gizzard can "age" them. Subsequently they can be sold as the authentic artifacts of the pharaohs.

Some carnivorous birds, such as owls, swallow their prey whole, and in these species the gizzard serves a different purpose. Indigestible materials such as bones, hair, and feathers are stored in the gizzard during the digestion of the remaining food material. These materials are rolled into soft pellets which are periodically regurgitated and disposed of through the mouth. By careful inspection of such owl pellets it is often possible to identify the exact species on which it has fed. In many cases whole, undamaged skulls of rodents, bats, or of other birds may be present in the pellet.

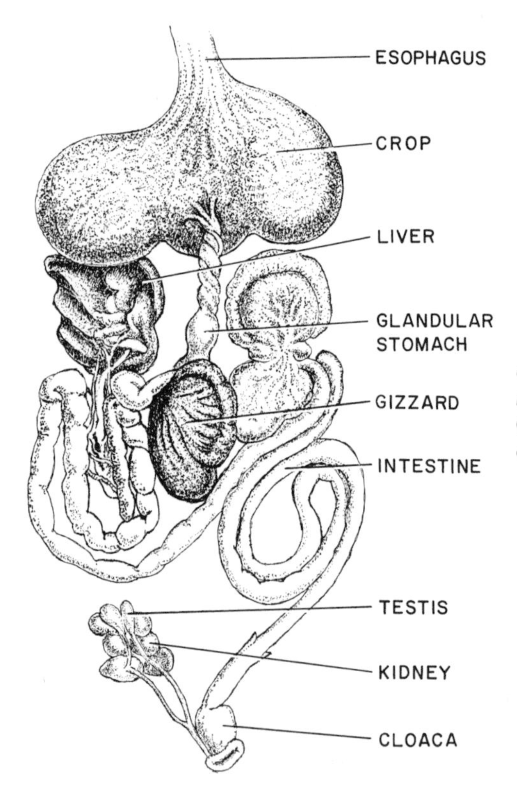

Figure 9–14 The digestive tract of a pigeon. (From Welty: *The Life of Birds*. Philadelphia, W. B. Saunders Company, 1962. After Schimkewitsch and Stresemann.)

The second portion of the typical bird stomach functions as a digestive organ like the stomachs of other vertebrates. Pepsin is present and the hydrochloric acid concentration may be sufficient, in some species, to completely decalcify and dissolve mollusk shells and even relatively large bones!

The Crocodilia have gizzards much like those of birds and they are used, similarly, in the grinding of foods. These reptiles swallow small stones to aid the process. The stomach stones of extinct archosaurs (certain dinosaurs) are sometimes found with the fossilized bones and are recognizable by their sizes and shapes.

9.14 THE RUMINANT STOMACH

The stomachs of ruminant animals, such as cattle, are highly modified structures consisting of four chambers (Fig. 9–15). One of these, the rumen, is a large chamber in which important digestive processes are carried out by symbiotic microorganisms. Much of the food value of the grasses eaten by the cattle is in the form of cellulose, which they cannot digest. They lack any specific cellulase of their own, but this enzyme is produced by the microorganisms present in the rumen. The grass is swallowed whole as the animal grazes. Then, usually while lying down and resting, the animal periodically regurgitates masses of mixed grass and microorganisms from the rumen and chews this "cud." The chewed and mixed food is returned to the rumen together with quantities of alkaline saliva for further microbial digestion. The alkaline saliva serves to neutralize acids that are produced by the microorganisms as by-products of their own metabolism. If it was not for this neutralization, the rumen pH would soon be so low that it would kill the microorganisms.

During the process of digestion in the rumen, the cellulose in the food is hydrolyzed, and the microorganisms use a part of the resulting sugar to support their own growth and multiplication. Periodically the contents of the rumen pass into a second chamber, the reticulum, and

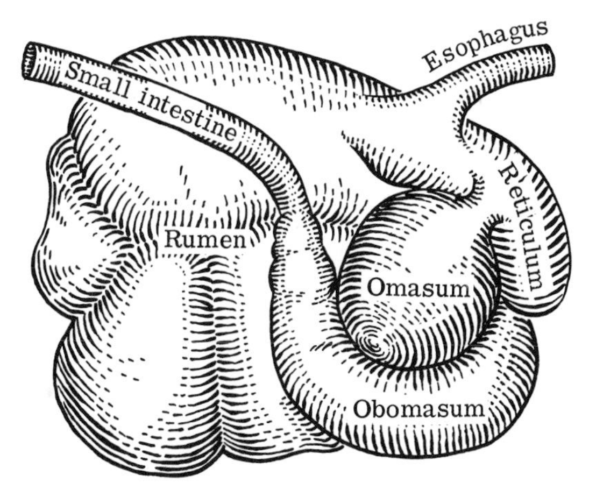

Figure 9–15 Structure of a typical ruminant stomach.

are pressed by muscular contractions. The liquid portions containing microorganisms and their digestive products are passed into a third chamber, the omasum, and the solid portions are again returned to the rumen for further digestion. In the omasum the water content is partially absorbed, and the remainder is carried on to the fourth and last chamber of the stomach. In this chamber, the obomasum, typical vertebrate gastric digestion occurs.

In summary, cattle and other ruminants feed plant materials to their colonies of microorganisms in the rumen. These microorganisms digest the cellulose and utilize most of the resulting sugars to support their own growth and multiplication. Ultimately the microorganism colony is harvested by the host animal and digested for its own nutrition.

In addition to the breakdown of cellulose, the microorganisms in the rumen contribute to the well-being of the ruminant in other ways: (1) They can utilize ammonia and urea as materials from which to synthesize amino acids. These substances are metabolic wastes of the host ruminant and are excreted by other vertebrates in the urine. Ruminants, however, secrete urea in the saliva, which is passed into the rumen with the "cud." There the microorganisms synthesize amino acids and incorporate them into their own structural proteins. Ultimately, when the host digests the microorganisms in the obomasum, they form a source of dietary protein. This mechanism for recycling amino acid nitrogen also confers an important ability in limiting water loss (Sect. 8.12) and helps to adapt ruminant animals to desert environments. (2) Most of the B vitamins are synthesized by the microorganisms, and this guarantees an adequate intake of these essential substances by the host. (3) The sugar metabolism of the microorganisms yields short-chain fatty acids such as acetic acid, proprionic acid, and butyric acid. These can be utilized directly by the host as energy sources (Sect. 10.10). In fact, cattle must synthesize needed glucose from these fatty acids rather than receive it as an end product of starch digestion as do other mammals.

Although ruminant digestion is beautifully adapted to the ecology of grazing animals, it also has serious disadvantages, and not all grazing animals by any means have this type of stomach. In the first place, the ruminant system is extraordinarily bulky and heavy. The weight of the reticulum and rumen of a cow may amount to as much as 15 per cent of the entire animal weight. Second, microbial digestion is a time-consuming process. Food remains in the cow rumen for about 60 hours, in the omasum for 8 hours, and in the obomasum for 3 hours. This contrasts markedly with the 3 to 4 hours required for digestion in the human stomach and even more markedly with the rapidity of digestion in birds. Seeds of berries eaten by birds may appear in the feces within 15 minutes, and a shrike can completely digest a whole mouse within three hours. Note that these figures refer to *complete* digestion, not just gastric digestion.

Cattle produce as much as 60 liters of saliva per day. This volume of saliva, rich in bicarbonates, is necessary to neutralize the acids produced by microbial digestion. The resulting carbon dioxide and methane gas are normally removed from the stomach by eructation (belching), but under certain conditions, gases can gather so rapidly as to cause dangerous bloating. The eructation of these gases imposes additional limitations on efficient ventilation of the lungs by contamination of the dead air space (Sects. 3.8 and 6.9).

Ruminant digestion in cattle is only one of a number of examples of symbiotic relationships with microorganisms in digestion. Certain marsupials (*e.g.*, the short-tailed wallaby of Australia) have very similar rumenlike stomachs that have arisen independently in the evolution of these animals. The New World peccary also has a stomach structure that suggests the same sort of development in a more primitive state. Most microbial populations within vertebrates are found in the large intestine, however. The kinds of microorganisms differ, of course, and are highly specific in some cases. For example, the honeyguide (*Indicator*) harbors microorganisms which can digest the beeswax on which this African bird feeds.

9.15 GASTRIC CONTROL

In mammals the secretion of gastric juices and the muscular activity of the stomach are under both nervous and hormonal control. Three separate kinds of gastric controls are well known: (1) As with the secretion of saliva, the sight, smell, or thought of food stimulates gastric activity by way of nervous routes. (2) The distention of the stomach with food is a further stimulus to activity. If a balloon is swallowed and then inflated by way of a tube leading through the esophagus, the resulting mechanical distention activates both the glands and the muscles of the stomach walls. (3) The presence of food materials (especially meats) in the stomach exerts a chemical effect which is stimulating. Probably certain nutrient molecules are actually absorbed directly from the stomach to induce the stimulation. In response to these substances, tissues near the pyloric end of the stomach produce a hormone called gastrin. This hormone, then, stimulates secretion of gastric juices, particularly hydrochloric acid. Two different gastrins are known from the hog stomach, and each is a relatively short polypeptide. Gastrin is also produced by gastric distention and by nervous stimulation of the stomach so it is apparently involved in the first two stimulating mechanisms mentioned above as well. (4) Finally, as the liquefied gastric contents, now called chyme, leave the stomach to enter the duodenum, another hormonal factor makes its appearance. Particularly in response to fats in the chyme, tissues of the duodenum (the first section of the small intestine) produce the hormone enterogastrone. This hormone inhibits the activity of the stomach. In summary, the expectation of food and the actual presence of food in the stomach both stimulate it to digestive activity. When its task is nearing completion, and materials are leaving the stomach, a final controlling mechanism inhibits or "turns off" the stomach.

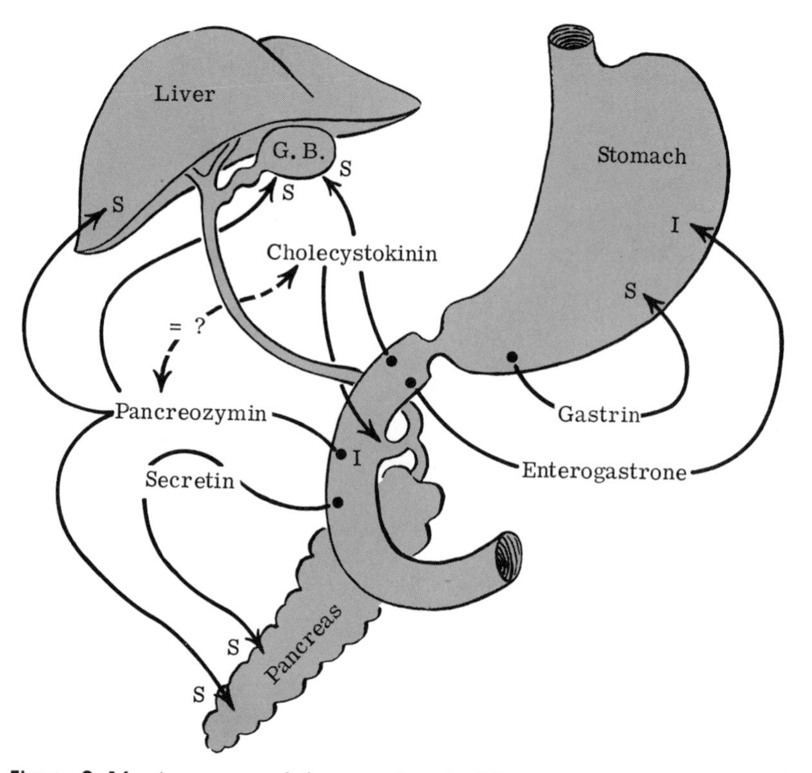

Figure 9–16 A summary of the gastrointestinal hormones, showing their places of origin and action. S indicates a stimulating effect and I indicates an inhibiting effect upon the target organ.

Both gastrin and enterogastrone are true hormones even though they are not produced by one of the classic endocrine glands such as the pituitary, thyroid, or adrenal (Chapter 17). Sometimes these two hormones together with others associated with digestion are grouped together as the gastrointestinal principles (Fig. 9–16).

Little is known of the gastric controls of non-mammalian vertebrates. There is no evidence for the natural presence of gastrin in the frog stomach but it is effectively stimulated by the experimental introduction of mammalian gastrin. Mammalian gastrin is also effective when introduced into the bird stomach.

9.16 THE SMALL INTESTINE

Glandular elements of the small intestine produce juices containing a number of digestive enzymes. Among these are a protease (or polypeptidase) called enterokinase, a lipase, an amylase, and others. Some of these are released by cellular secretion and some by actual sloughing and subsequent disintegration of glandular cells (Sect. 9.8). Secretory activity of the intestinal glands is probably under hormonal control. Incomplete evidence suggests the existence of two gastrointestinal principles, enterokinin and duocrinin, both of which are involved in such stimulation. They are thought to be produced by intestinal tissues in response to the presence of chyme within the intestinal lumen.

Several different kinds of muscular activity occur in the small intestine (Fig. 9–17). These serve to mix the chyme with digestive juices, bring all parts of the chyme into contact with the intestinal epithelium so that absorption of digestive products can proceed efficiently, and to control the rate at which the chyme moves along through the intestine. Muscular activity in the small intestine can be classified as follows: (1) Segmentation movements are constrictions occurring at intervals along the intestine. They appear briefly and then disappear without moving from the same site. New constrictions of the same type then appear at different points in a random sequence. This type of squeezing or chopping action serves to mix the chyme. (2) Pendular movements involve the contraction of longitudinal muscles first on one side of the intestine and then on the other that result in bending the tract back and forth with a snakelike motion. This also serves, mainly, to mix the chyme. (3) Peristaltic movements are muscular contractions much like segmentation except that they move along the gut for short distances before disappearing. This serves to "milk" the chyme along the tract in short stages. Occasional antiperistaltic movements of the same sort travel in the opposite direction and act to reduce the net rate

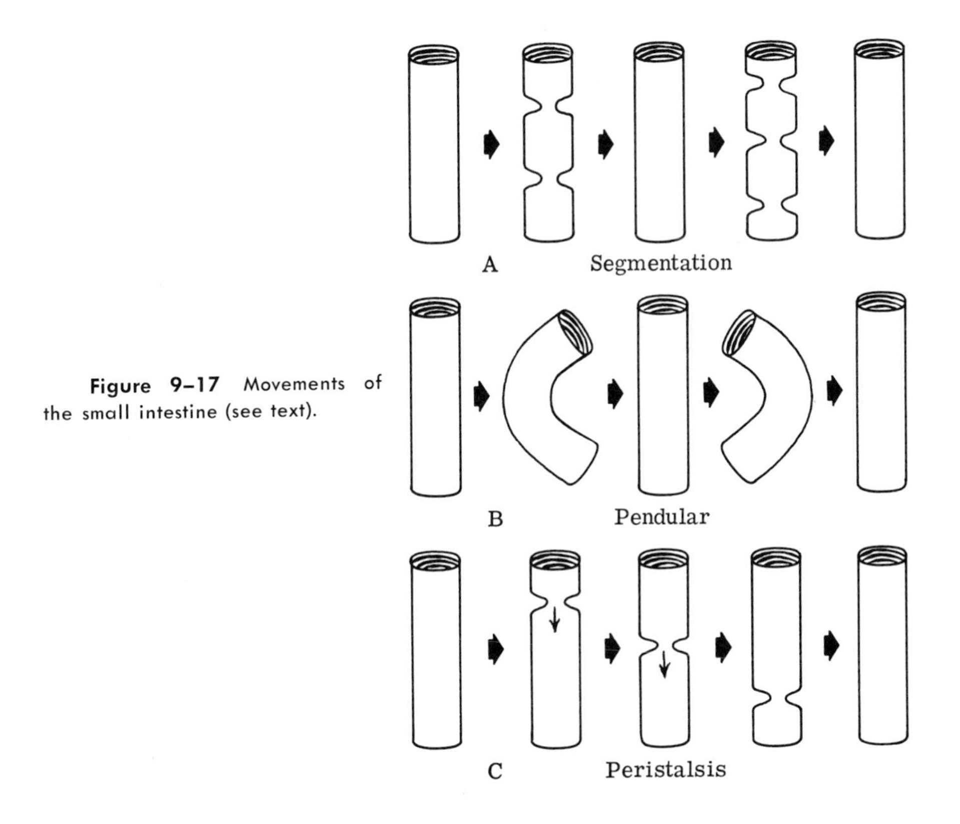

Figure 9–17 Movements of the small intestine (see text).

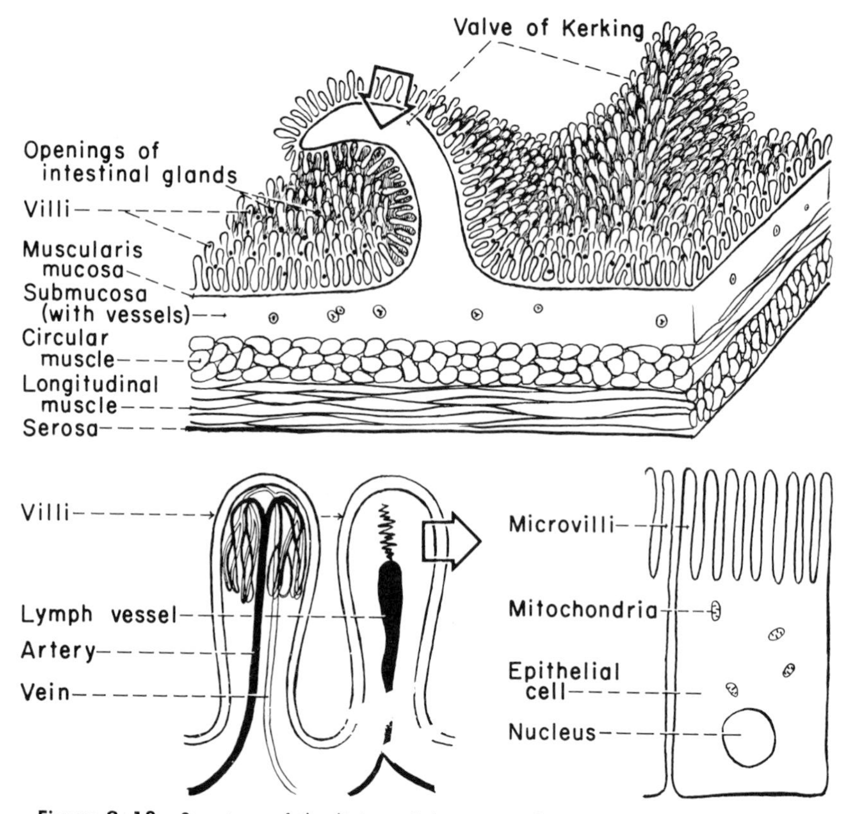

Figure 9–18 Structure of the lining of the mammalian small intestine. (From Florey: *An Introduction to General and Comparative Animal Physiology*. Philadelphia, W. B. Saunders Company, 1966.)

of chyme movement. Also, occasional peristaltic rushes sweep over considerable lengths of the intestine speeding the forward movement of chyme.

The epithelial lining of the intestine is marked by the presence of numerous fingerlike extensions known as villi (Fig. 9–18). The villi increase the surface area available for absorption of digestive products and they, too, are in constant motion. They alternately shorten and lengthen and they also sway from side to side. Some evidence exists for another gastrointestinal hormone, villikinin, that stimulates these activities of the villi.

9.17 PANCREATIC SECRETION

The pancreas of most vertebrates is a dual organ. Most of its bulk, the acinar portion of the pancreas, is devoted to the production of pancreatic juice which contains digestive enzymes. The organ also contains, however, isolated groups of endocrine cells of two kinds (Sect. 17.8). These groups of cells are known as the islets of Langerhans. Lampreys have no separate pancreas, but the walls of the small intestine contain cells like those of the typical vertebrate pancreas. Many of these cells are located in diverticula or outpocketings which contain enzymatic secretions resembling those of the pancreas of other vertebrate animals. Other similar diverticula contain cells resembling those of typical vertebrate islets of Langerhans.

In the elasmobranchs the pancreas appears as a distinct organ and contains both acinar and endocrine cells. In the teleost fishes, however, only the endocrine portion is present as a compact organ. The acinar portion is scattered as a number of diffuse glands in the mesentery of the small intestine. The pancreatic tissue of some mammals, including the rat, is also diffuse but contains both acinar and endocrine cells.

In a typical vertebrate the pancreatic secretions are carried from the acinar portions of the organ through a pancreatic duct which opens into the lumen of the duodenum. These juices contain two peptidases, both of which are released as inactive proenzymes (Sect. 9.12). The inactive forms are called trypsinogen and chymotrypsinogen. The enterokinase of intestinal juice converts trypsinogen to the active enzyme trypsin, and this enzyme, in turn, converts chymotrypsinogen to the active form chymotrypsin.

Trypsin catalyzes the hydrolysis of peptide bonds at sites adjacent to amino acids having more than a single amino group (dibasic amino acids) such as arginine and lysine (Fig. 9–9) while chymotrypsin splits bonds adjacent to amino acids with aromatic (ring) structures such as tyrosine or phenylalanine. These two enzymes, like pepsin, are endopeptidases; that is, they break protein molecules or polypeptides into shorter lengths.

Pancreatic juice also contains enzymes called exopeptidases because they remove the terminal amino acids from polypeptide chains. Aminopeptidase removes terminal amino acids that have a free amino group, while carboxypeptidase removes amino acids with a free carboxyl group from the other end of the chain. Collectively, the pancreatic enzymes mentioned thus far convert polypeptide molecules to dipeptides and free amino acids. A dipeptidase in the intestinal juice completes the process by separating the dipeptides so that only free amino acids remain.

Pancreatic juice also contains pancreatic amylase, which is much like salivary amylase in function (Sect. 9.9). Present also is a pancreatic lipase that splits fatty acids away from glycerol by hydrolysis (Sect. 9.3). Sufficient sodium bicarbonate is secreted in the pancreatic juice to make it somewhat alkaline. All of the enzymes of the pancreatic and intestinal juices have optimum activity in the pH range of 7 to 9.

There is little evidence for the existence of any nervous control of the pancreas, but at least two gastrointestinal principles are known in mammals. The presence of acid in the chyme entering the intestine from the stomach stimulates duodenal tissues to produce the hormone secretin. Secretin was the first of the gastrointestinal principles to be discovered and, indeed, the first substance to be given the name "hormone." It stimulates the release of water and sodium bicarbonate from the pancreas in order to neutralize the acids from the stomach. A second hormone produced by the duodenal tissues in response to the presence of chyme (especially fats) is called pancreozymin. This hormone specifically stimulates the secretion of digestive enzymes from the pancreas (Fig. 9–16).

9.18 BILIARY FUNCTION

Another unique feature of vertebrate digestion is the secretion of bile by the liver. Bile is a fluid containing substances of two general kinds, namely bile pigments and bile salts. The bile pigments are derived from the hemoglobin molecule when worn-out erythrocytes are broken down. The iron of the hemoglobin molecule (Fig. 7–1) is salvaged for reuse, but the porphyrin component is excreted by the liver in the bile. This pigment plays no further role in most vertebrates other than to color the feces, although in some birds it is used to color the feathers or egg shells. The bile salts are derived from the choline molecule (Fig. 10–13) and certain amino acids and they have an important function in the digestion of fats.

Bile salts are not enzymes but they assist the lipases in the intestine by aiding in the emulsification of fats. An emulsion is a stable mixture of fat and water. Homogenized milk is a common example of such a stable mixture. The milk proteins coat the surface of small fat droplets and prevent them from coalescing with each other to form larger drops. This kind of stabilization prevents the milk fat (cream) from separating and floating to the surface as it does in non-homogenized milk. Homogenization is a process of breaking the fats into very small droplets so that stabilization can occur. By similarly preventing the coalescence of fat droplets in the chyme, the bile salts provide a much greater surface area for the action of lipases. This expedites the digestion of fats in the chyme.

In some, but not all, vertebrates the bile is temporarily stored in the gallbladder. There it is concentrated by reabsorption of some of the water content. A gallbladder is present in lampreys so apparently this was an early vertebrate development. In animals having a gallbladder, bile is released into the duodenum only when fats are present there in the chyme. Again, a hormonal control is involved. In response to the presence of fat in the chyme, the duodenal tissues produce a hormone called cholecystokinin. This hormone causes contraction of the gallbladder and relaxation of the muscular sphincter or valve at the point where the bile duct enters the duodenum. This dual action results in pouring of bile into the small intestine. There is evidence to suggest that cholecystokinin and pancreozymin (Sect. 9.17) may be the same hormone acting in two different ways.

9.19 INTESTINAL ABSORPTION

Enzymatic processes described in the foregoing sections break proteins into their constituent amino acids, carbohydrates into monosaccharides, and fats into fatty acids and glycerol molecules. Other enzymes, not specifically mentioned, are also present in the intestinal juice and these act upon other substances present in the chyme. For example, there are nucleases which act upon nucleic acids. Finally, these end products of digestion are absorbed through the walls of the small intestine and pass into either the venous or the lymphatic drainage from the gut.

The surface area of the intestinal lining available for absorption is increased by several anatomical features. An elaborate series of folds projecting into the intestinal lumen forms the spiral valve seen in some cyclostomes, elasmobranchs, and teleost fishes. It probably evolved from the single, longitudinal fold known as the typhlosole found in certain protochordate animals. In addition to the spiral valve, teleost fishes also have many blind pyloric caeca—small outpocketings in the intestinal lining. These contain glandular elements and are involved in the absorption of digested materials as well. Some fishes and all of the amphibians and reptiles have intricate networks of ridges and folds, and often these border deep glandular crypts in the intestinal lining. Birds and mammals also have such folds on the intestinal lining, and in addition its surface is covered by the small, fingerlike villi extending into the intestinal lumen (Fig. 9–18). It is interesting to note in passing that the villi grow constantly from the base and slough at the tip. This releases some 250 grams of cellular debris into the human intestinal tract each day. The disrupting cells from the villi release cellular enzymes, and this may be an important source of the digestive enzymes present in intestinal juice (Sect. 9.8).

Both monosaccharides and disaccharides are absorbed into cells of the intestinal wall. Within these cells the disaccharides are split by intracellular digestion to form monosaccharides. Finally, all of the monosaccharides are passed on through these cells and into the blood of intestinal capillaries. Venous drainage from the intestine flows by way of the hepatic portal system to the liver.

In much the same way amino acids and dipeptides are absorbed by intestinal cells, where the dipeptides are split into their constituent amino acids by intracellular digestion. Finally, all of the amino acids are passed on into the blood stream and carried away by the hepatic portal vein.

Some fats may be absorbed as mono-, di-, or even triglycerides, but most are absorbed in the form of free fatty acids and glycerol. In any case, they are generally reunited to form triglycerides again before passage through the intestinal cells is completed. Some short-chain fatty acids remain free and are passed, together with the reconstituted fats, into the lymphatic drainage, for the most part, from the intestine. Some may also enter the blood. The content of intestinal lymphatic capillaries becomes milky white following a high fat content meal. This is due to the inclusion of droplets of fat called chylomicrons in the lymph. From their milky appearance, the intestinal lymphatics are given the special name lacteal vessels. These vessels drain into the lymphatic thoracic duct, which drains, in turn, into the venous system near the heart. Thus, ultimately the contained fats enter the blood stream.

Water passes easily through the intestinal epithelium to enter the blood and lymph. In addition, some water is taken up by intestinal cells by the process of pinocytosis; the lining cells of the gut form small surface vacuoles in a manner reminiscent of ameboid feeding but on a much smaller scale. Droplets of water trapped in this way are carried into and through the cells to enter the blood or lymphatic vessels. There is some evidence that fat droplets may also be taken in by this mechanism to some extent.

The permeability of the intestinal lining to monovalent ions such as sodium (Na^+) is usually great and these ions are easily absorbed. Divalent ions such as calcium (Ca^{++}) do not penetrate as easily, so absorption of these ions involves active transport by the intestinal cells. That is, the cells actively "pump" the ions through the lining rather than depend upon simple diffusion processes. A discussion of the possible mechanisms of active transport may be found in any textbook of cellular physiology.

9.20 THE COLON

By the time the chyme reaches the large intestine or colon, the useful and absorbable components have been removed from it in large measure. The remaining mass is composed of indigestible or non-absorbable materials plus water. The loss of the water content would be a serious drain for a terrestrial or marine vertebrate since, in both cases, the availability of fresh water for drinking is generally limited. One of the major functions of the colon is the recovery, by absorption, of water from the fecal material. Constipation and diarrhea are conditions often associated with excessive absorption or non-absorption respectively.

In most vertebrates the colon contains large populations of symbiotic microorganisms that play various roles. Those in humans, for example, synthesize vitamin K which is then absorbed by the host. This is virtually the only source of this vitamin in humans, and without it the liver cannot synthesize prothrombin. Prothrombin, it will be recalled, is necessary in the process of blood coagulation (Sect. 7.6). In some mammals (*e.g.*, horse, rabbit) and birds (*e.g.*, chicken, turkey) there is a large colonic diverticulum (cecum) which serves to hold the microorganism population. In some cases the vital substances produced by the microorganisms cannot be directly absorbed by the host animal. Reingestion of fecal material is then necessary. In fowls an occasional evacuation of the colon results from a partial emptying of the cecum and is particularly rich in such vital substances. The bird selects these particular feces for reingestion. Rabbits and chickens raised in cages with open wire bottoms must be given special vitamin supplements in the diet since the feces fall through the cage bottom and are not available to the animal for reingestion.

The colon undergoes accordionlike movements which aid in mixing the contents. Occasional mass movements, not unlike the peristaltic rushes of the small intestine (Sect. 9.16), also occur. These move the forming feces toward the terminal end of the colon. When sufficient material has gathered there, evacuation takes place by means of a nervous reflex action. Interestingly, distention of the stomach by a meal induces mass movements in the colon by way of nervous reflex mechanisms. Thus, when a new meal begins its digestive processing, the colon is prepared to receive the wastes resulting from that process.

CHAPTER 10

METABOLISM

10.1 INTRODUCTION

We have seen that digestive processes convert foodstuffs by hydrolysis into simple forms that are absorbable from the small intestine (Chapter 9). Monosaccharides (glucose, fructose, and galactose), amino acids, and fats are absorbed and transported by the blood or lymph to other portions of the body.

In general, sugars are used as sources of energy to power energy-consuming reactions such as muscular activity, glandular secretion, maintenance of membrane potentials, syntheses of various substances, and the active transport of materials through membranes. Sugars not immediately needed for energy production may be stored in one of two ways: (1) limited quantities of glucose can be polymerized to form large molecules of glycogen (animal starch), especially in the liver and in muscle tissue (Sect. 9.2), or (2) almost unlimited quantities of sugar can be converted into fat for storage in that form.

Amino acids are used primarily as building blocks for the synthesis of proteins. Proteins are required for growth, repair, and other structural roles as well as for use as controlling factors (enzymes, some hormones). In the absence of sufficient quantities of sugar, about 60 per cent of the amino acids normally present can be converted into sugars or related molecules. This conversion requires the removal of the amino group by a process of deamination (Sect. 10.12). The remaining 40 per cent of the amino acids have molecular structures that make deamination impossible; these, then, are not available as potential energy sources.

Fats are used as an energy source to about the same extent as are sugars. Fat is an efficient form in which to store energy because it contains more potential energy per gram than either sugars or amino acids. Fats are also used as structural components of cells. Cell membranes, for example, are composed of layers of protein and lipid (Figs. 2–2 and 2–3).

Sugars, amino acids, and fats are all used for other purposes besides those already mentioned. All three form parts of special, complex molecules of various kinds. The functions and identities of these special molecules will become apparent as we encounter them in subsequent sections of this book.

Other dietary components such as vitamins and minerals also perform vital functions in animal metabolism, and these too will be discussed in various sections of this chapter.

ENERGY METABOLISM

10.2 OXIDATION-REDUCTION

When either an electron or a hydrogen atom is transferred from one kind of molecule to another, the molecules losing the electron or hydrogen atom are said to be oxidized. Conversely, the molecules accepting the electron or hydrogen atom are said to be reduced by this reaction. Thus, the oxidation of one compound and the reduction of another occur simultaneously. The compound that is oxidized is called the reducing agent (or reductant), and the compound that is reduced is called the oxidizing agent (or oxidant).

The most common oxidation-reduction reactions in biological systems are of the hydrogen transfer type:

If molecule B in the scheme above happens to be oxygen (O_2), then molecule B—H_2 will be water (H_2O). In many instances, however, oxygen as such is not involved in oxidation-reduction reactions. Also, since the transfer of an electron is equivalent to the transfer of a hydrogen atom, many reactions do not involve hydrogen as such either. Nevertheless, such reactions are true oxidation-reduction reactions. The following reaction is an example of one in which neither oxygen nor hydrogen is involved:

In this example, an electron has been transferred from Fe^{++} to the iodine molecule causing it to ionize (the equivalent of being reduced). The simultaneous oxidation of the iron is accompanied by a change in its valence from Fe^{++} to Fe^{+++}.

The synthesis of organic molecules by green plants is essentially a series of reductions, and carbon dioxide and water are the materials from which complex organic molecules are synthesized by this process (photosynthesis). Photic energy from the sun is trapped by the chlorophyll of the plant and provides electrons which are then used to reduce other substances. The incorporation of these electrons provides the energy necessary to bind atoms together in the formation of carbohydrates, proteins, and fats (Fig. 10–1). This incorporated energy can be released again by oxidizing these complex molecules back to carbon dioxide and water. Both processes are carried out by green plants.

Animals lack the means for trapping electrons (chlorophyll) from solar radiation so they must obtain their energy by consuming the plants (or other plant-eating animals) and oxidizing the organic molecules contained in these foods. The electrons trapped during photosynthesis are removed (oxidation) and used by the animal to power its own syntheses (reduction) and other energy-consuming activities. Carbon dioxide and water are the end products of the process.

If oxidation of foodstuffs is carried out experimentally by the direct introduction of oxygen, all of the energy available from their degradation is released in a single burst of heat. Metabolizing tissue cells cannot use energy in this form nor can they handle it efficiently in such large quantities. Instead, metabolic oxidation is a stepwise process that releases energy in smaller amounts. Moreover, to a considerable extent the energy is released as chemical energy rather than as heat.

To function efficiently, it is necessary that the energy-releasing reactions be closely coupled with energy-consuming processes so that the energy can be transferred directly from the one to the other. Such coupling is accomplished by specific enzymes and energy-transferring molecules. Since the energy to be transferred may be in the form of electrons or hydrogen atoms, different kinds of energy-transferring substances are required in different instances. All of these substances are present within each metabolizing cell because it is there that energy-producing oxidations and energy-consuming syntheses (reductions) occur.

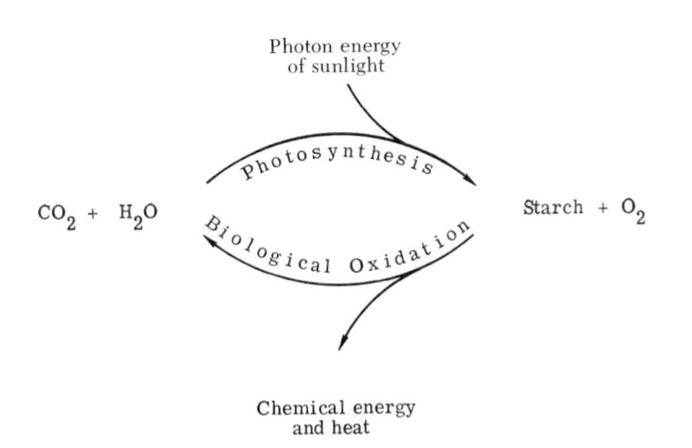

Figure 10–1 Energy flow in biological systems. Shown here are the fundamental chemical equations of photosynthesis and biological oxidation.

10.3 ADENOSINE TRIPHOSPHATE

The compound adenosine triphosphate (ATP) is present in all cells and functions as an energy-storing or energy-transferring molecule (Fig. 10–2). The last two of the three phosphate groups of the molecule are attached by bonds called high-energy bonds. These are represented in the structural formula by the symbol ~ . A large amount of energy (nearly 8000 calories per gram-mol of ATP) is required to unite each of these phosphate groups with the adenosine monophosphate (AMP) molecule. Adding one phosphate to adenosine monophosphate converts it to adenosine diphosphate (ADP) and "stores" in the energy-rich phosphate bond about 8000 calories of potential energy. Adding a second phosphate to convert the molecule to adenosine triphosphate adds another 8000 calories of potential energy. When either of these terminal phosphates is removed, that energy becomes available and can be transferred to some other molecule.

During the stepwise oxidation of foodstuffs in the body, energy is released in successive "bundles." When any bundle amounts to as much as 8000 calories, it can be used to bind an inorganic phosphate to the adenosine monophosphate or diphosphate molecule. That amount of energy is then trapped or stored in the energy-rich bond that is formed. Subsequently, when phosphate groups are again removed, the stored energy is transferred to some other energy-consuming chemical process. In some instances the phosphate group is itself transferred to another compound to raise its store of available chemical energy. This will allow the phosphorylated compound to undergo reactions which would otherwise be energetically impossible.

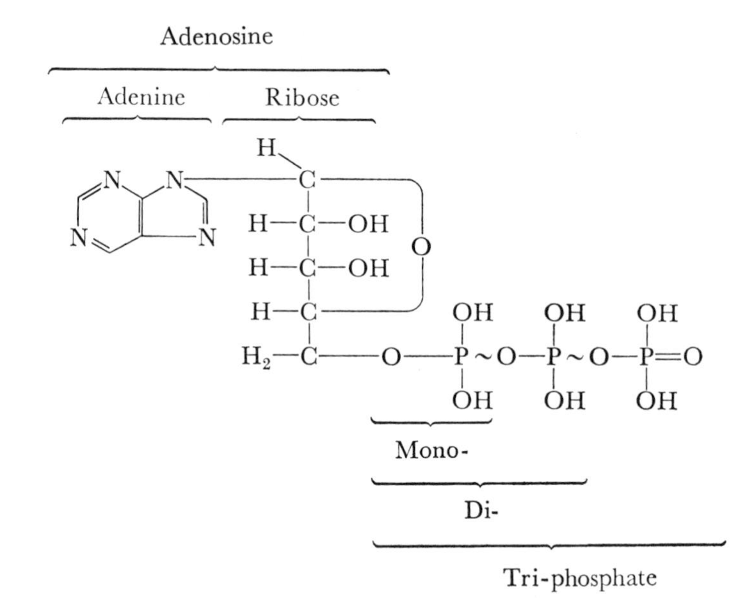

Figure 10–2 The chemical structures of adenosine monophosphate, diphosphate, and triphosphate.

Examples of the processes just described will be found in a number of the reactions discussed in subsequent sections of this chapter.

10.4 HYDROGEN ACCEPTOR MOLECULES

Many of the stepwise oxidations of foodstuffs are of the type in which hydrogen is removed from the molecule (Sect. 10.2). Each such reaction requires the presence of a specific dehydrogenase enzyme to catalyze the reaction and a hydrogen acceptor molecule which is reduced in the process. Four hydrogen acceptor substances occur commonly in animal systems:

1. Nicotinamide adenine dinucleotide (NAD) is the most common

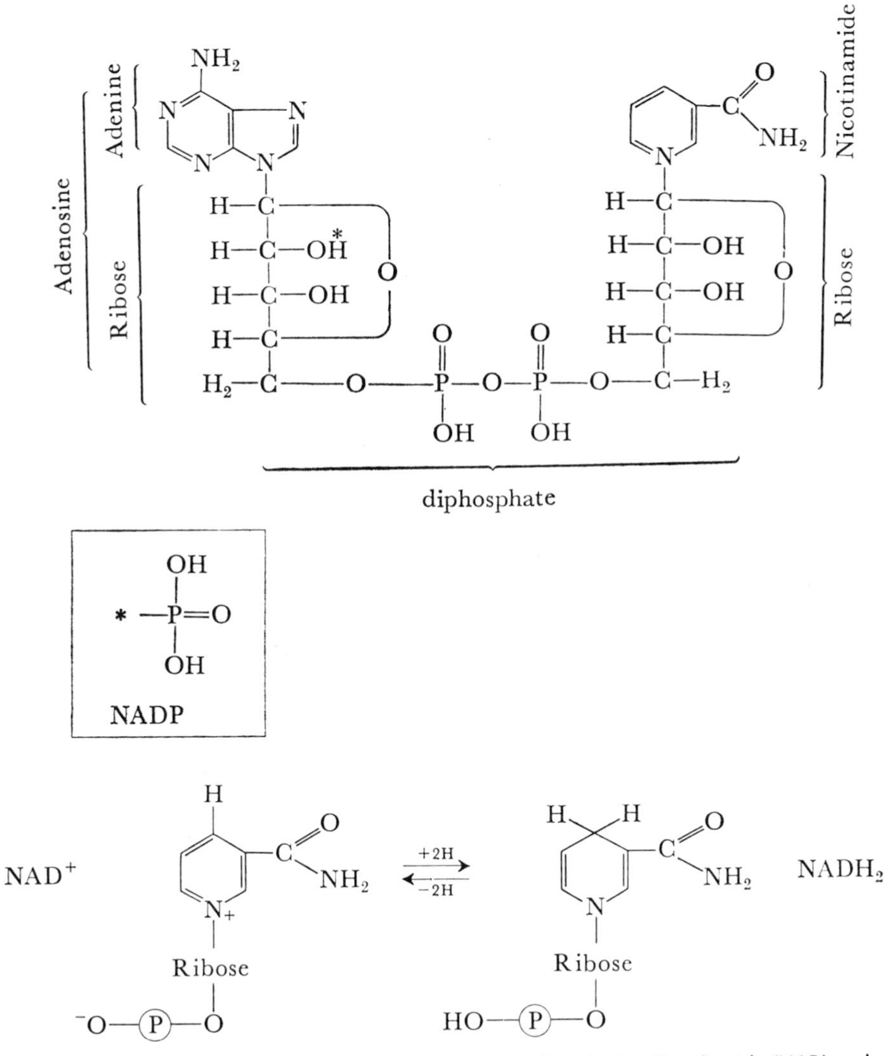

Figure 10–3 The chemical structure of nicotinamide adenine dinucleotide (NAD) and nicotinamide adenine dinucleotide phosphate (NADP). Also shown is the way in which NAD acts as a hydrogen acceptor.

of the hydrogen acceptor molecules (Fig. 10–3). It has a complex structure consisting of a purine molecule (adenine), two five-carbon sugar molecules (ribose), a molecule of nicotinamide, and two phosphate groups. Nicotinamide is the vitamin niacin amide (Fig. 9–11). A large part of the NAD molecule is similar in structure to that of adenosine monophosphate (Fig. 10–2). The compound is reduced by accepting two hydrogen atoms and can be reoxidized by subsequently passing these hydrogens to some other molecule:

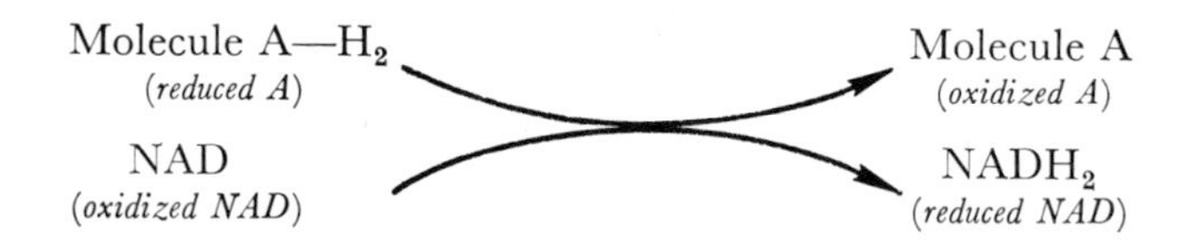

2. Nicotinamide adenine dinucleotide phosphate (NADP) is similar to NAD in structure except that an additional phosphate group is attached to one of the ribose molecules (Fig. 10–3). It functions in a reduction reaction in the same general way as does NAD:

3. Flavin adenine dinucleotide (FAD) is a similar molecule (Fig. 10–4) except that the nicotinamide group of NAD and NADP is replaced by a flavin group, which is derived from the vitamin riboflavin (Fig. 9–11). FAD functions much like NAD and NADP:

4. Coenzyme Q is a quinone derivative that is converted to a hydroquinone by accepting two hydrogen atoms (Fig. 10–5). This compound is related to vitamins E and K (Fig. 9–10). Coenzyme Q accepts hydrogen only from reduced FAD:

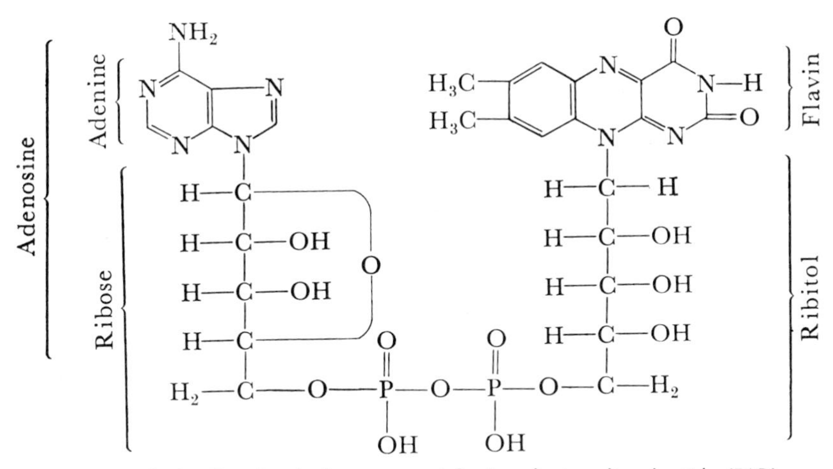

Figure 10–4 The chemical structure of flavin adenine dinucleotide (FAD).

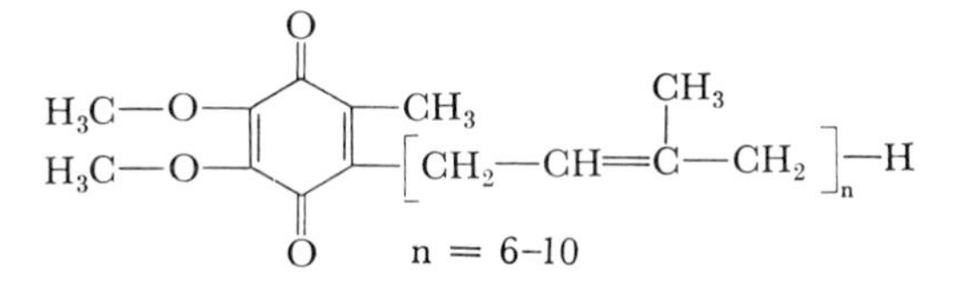

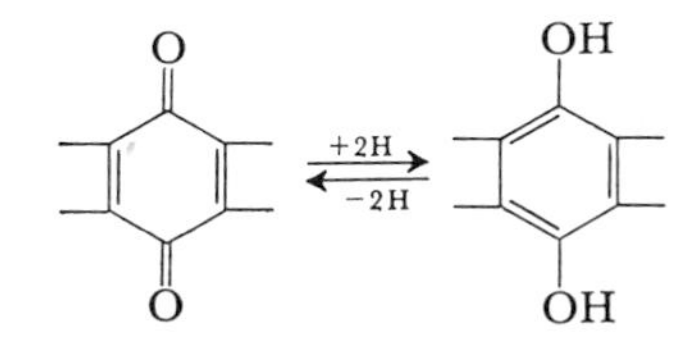

Figure 10–5 The chemical structure of coenzyme Q. Also shown is the way in which coenzyme Q acts as a hydrogen acceptor.

10.5 ANAEROBIC GLYCOLYSIS

The first several reactions in the oxidative degradation of glucose are not dependent upon the presence of oxygen, but subsequent stages in its complete oxidation are. For that reason, the initial reactions are spoken of collectively as anaerobic (without air) glycolysis (glucose breakdown).

The sequence of reactions through which at least 95 per cent of the glucose in anaerobic glycolysis passes is known as the Embden-Meyerhof pathway; its name is derived from the names of the coinvestigators who first proposed this series of reactions. Subsequent studies have proved them to be correct. The numbers that precede each paragraph in the following description correspond to the reaction numbers illustrated in Figure 10–6.

1. In the first reaction, a phosphate group is transferred from adenosine triphosphate (ATP) to the sixth carbon of the glucose molecule to yield ADP plus glucose-6-phosphate. This phosphorylation occurs as soon as a glucose molecule enters a metabolizing cell. Then, since the cell membrane is not permeable to the phosphorylated glucose molecule, the glucose is trapped within the cell and is readily available for use in subsequent reactions.

2. Glucose-6-phosphate is converted to fructose-6-phosphate by an internal molecular reorganization. Note that dietary fructose could be phosphorylated by ATP and enter the reaction sequence at this point as well.

The remaining reactions of anaerobic glycolysis occur within the mitochondria of the cell (Sect. 2.2). Each step requires a specific enzyme and it is believed that these enzymes are arranged in sequence on the inner membrane of the mitochondria so that the degrading sugar molecule can be passed easily from one to another. The speed at which anaerobic glycolysis occurs suggests that some such disassembly line procedure must be involved.

3. A second phosphate group is transferred from ATP to produce fructose-1,6-diphosphate and another molecule of ADP. The two phosphate groups obtained thus far from ATP molecules raise the energy level of the sugar molecule and enable it to undergo subsequent reactions. These reactions would be essentially impossible without this additional level of molecular energy.

4. The six-carbon fructose molecule is broken into two three-carbon fragments, each of which retains one of the phosphate groups. One of these three-carbon fragments, dihydroxyacetone phosphate, is then converted by molecular reorganization to a form that is identical with the other three-carbon fragment. Thus, the final product is two molecules of glyceraldehyde-3-phosphate for each original molecule of glucose. From this point on, each compound in the Embden-Meyerhof scheme is preceded by the number 2 to indicate this fact.

5. Each molecule of glyceraldehyde-3-phosphate gains a second phosphate group and, at the same time, is oxidized by the loss of hydrogen. In this instance the phosphate group comes from inorganic

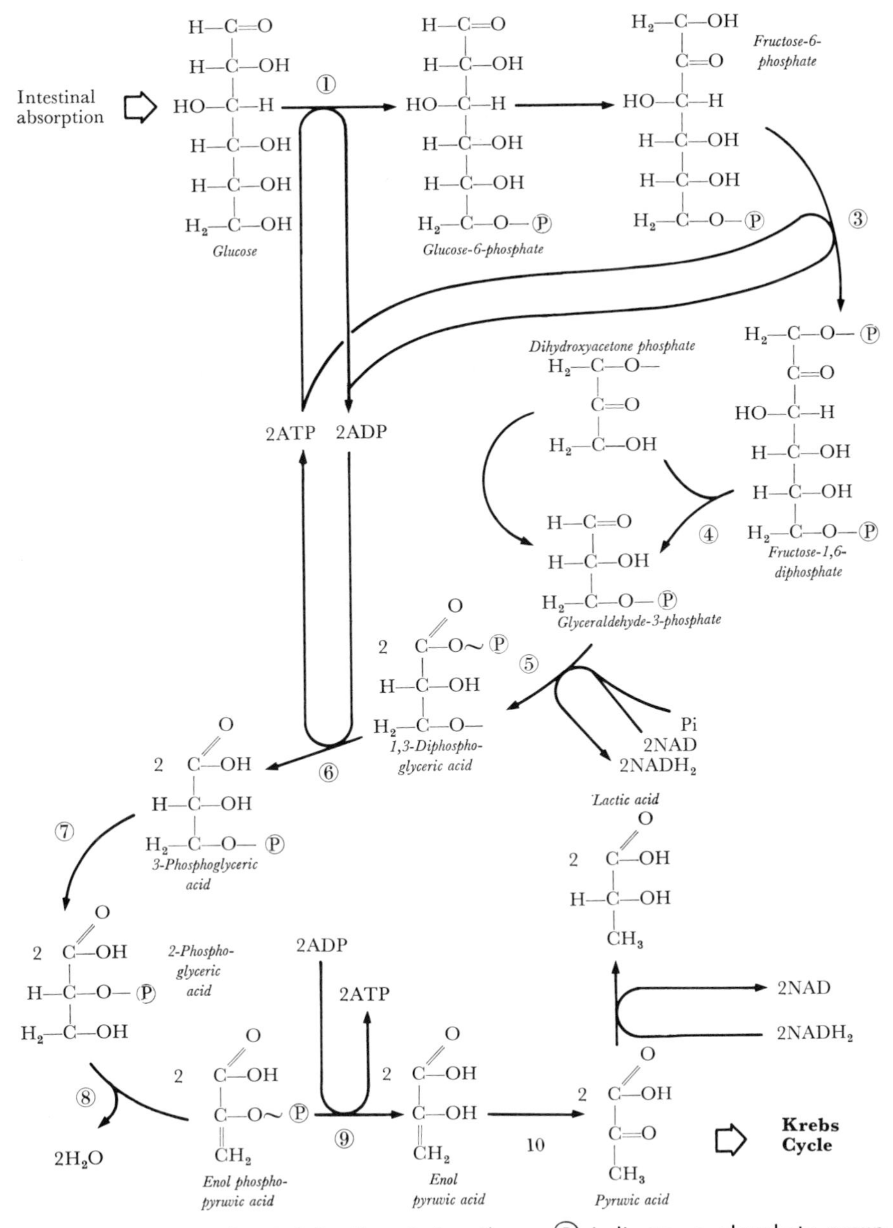

Figure 10–6 The Embden-Meyerhof pathway. Ⓟ indicates a phosphate group attached to the molecule. Pi indicates an inorganic phosphate molecule. See text.

phosphate rather than from ATP, and the hydrogen acceptor molecule is NAD (Sect. 10.4). Thus, NAD is reduced while glyceraldehyde-3-phosphate is both oxidized and phosphorylated in the same reaction.

6. Each molecule of the 1,3-diphosphoglyceric acid formed in the preceding reaction now gives up one of its two phosphate groups to become 3-phosphoglyceric acid. The phosphate group is transferred to a molecule of ADP, converting it to ATP. Note that sufficient ATP is formed in this reaction to make up for the ATP loss in reactions 1 and 3 preceding. One might say that the phosphate energy was "loaned" in reactions 1 and 3 and then repaid to the ATP in reaction 6.

7. The remaining phosphate group is shifted to a different position on the molecule, converting it to 2-phosphoglyceric acid.

8. Through the loss of a molecule of water the compound is converted to enol phosphopyruvic acid.

9. The phosphate group is transferred to a molecule of ADP to form more ATP. This converts the enol phosphopyruvic acid to enol pyruvic acid.

10. The enol pyruvic acid is spontaneously converted to keto pyruvic acid by molecular rearrangement.

For each original molecule of glucose the reactions thus far have produced a net gain of two molecules of ATP. As we shall see in subsequent sections, the further oxidation of the pyruvic acid formed results in the release of more energy and the consequent formation of much more ATP. These further oxidations require, ultimately, the presence of oxygen, so they are sometimes referred to collectively as aerobic glycolysis.

In the temporary absence of sufficient oxygen to carry out aerobic glycolysis, anaerobic reactions can continue to provide some energy in the form of ATP. In this event one would expect pyruvic acid to accumulate within the cell and, through mass action effects, slow down the reactions leading to its own formation. Such would, indeed, be the case except that under these conditions pyruvic acid is converted to lactic acid (reaction 11). Lactic acid can diffuse out from the cell and into the tissue fluids and blood. Thus, pyruvic acid, as such, does not accumulate in the cell, and the reactions of anaerobic glycolysis are not inhibited. The conversion of pyruvic acid to lactic acid is a reduction, and the reducing agent is $NADH_2$. Thus, $NADH_2$ is simultaneously oxidized to become NAD. Ultimately, the lactic acid formed is reconverted to pyruvic acid (or to glucose) so the hydrogen is returned to the NAD. Again, we might say that hydrogen is loaned by $NADH_2$ to accomplish the conversion but it is later repaid by the reversed reaction.

The $NADH_2$ formed in reaction 5 represents potential energy derived from the reactions of anaerobic glycolysis. This energy will be used, as we shall see, for the formation of more ATP during aerobic glycolysis. Thus far, then, each molecule of glucose has yielded two molecules of ATP plus two molecules of $NADH_2$, both of which represent energy taken from the glucose molecule. The reactions of anaerobic glycolysis actually release about 56,000 calories, of which only about 16,000 were captured as ATP. The $NADH_2$ formed will ultimately yield an additional amount of ATP, but most of the energy released in anaerobic glycolysis is lost as heat. The process of capturing energy as ATP is only about 40 per cent efficient.

10.6 THE KREBS CYCLE

The Krebs cycle is a series of reactions comprising the first portion of aerobic glycolysis. Since it involves citric acid in an early reaction, it is sometimes called the citric acid cycle; and because several of the intermediate compounds in the cycle have three carboxyl (COOH) groups in the molecule, it is also sometimes called the tricarboxylic acid cycle. It is a cycle because it involves a series of reactions that returns again and again to the same compound (Fig. 10–7).

The reactions in the Krebs cycle are oxidations (dehydrogenations) and decarboxylations. The oxidation reactions result in the reduction of NAD, NADP, and FAD, all of which serve as hydrogen acceptors (Sect. 10.4). The decarboxylation reactions result in the production of carbon dioxide, one of the two final end products of glucose oxidation.

The numbers preceding the paragraphs in the following description correspond to the reaction numbers illustrated in Figure 10–7.

1. Each molecule of pyruvic acid produced by anaerobic glycolysis (Fig. 10–6) is combined with a molecule of coenzyme A to form acetyl coenzyme A and carbon dioxide. At the same time, two atoms of hydrogen are transferred to NAD, reducing it to $NADH_2$. Thus, this reaction is both an oxidation and a decarboxylation. Coenzyme A (Fig. 10–8) is a complex molecule composed of adenine, ribose-3-phosphate, pantothenic acid, thioethanolamine, and two phosphate groups. The pantothenic acid is a vitamin (Fig. 9–11). In part, the coenzyme A molecule resembles NADP (Fig. 10–3).

2. Acetyl coenzyme A combines with oxaloacetic acid and a molecule of water to yield citric acid and at the same time release the coenzyme A molecule. Thus, the coenzyme A is used over and over to introduce two carbon fragments (acetyl groups) into the Krebs cycle and simultaneously produce carbon dioxide and reduced NAD.

3. Citric acid is converted to isocitric acid by an internal rearrangement of the hydrogen and hydroxyl (OH) groups.

4. Isocitric acid is oxidized (dehydrogenated) to produce oxalosuccinic acid.

5. Oxalosuccinic acid is decarboxylated to yield alpha-keto glutaric acid and carbon dioxide.

6. Alpha-keto glutaric acid combines with coenzyme A to form succinic coenzyme A and carbon dioxide. Much like reaction 1, this is both an oxidation (dehydrogenation) and a decarboxylation reaction. In this reaction, however, NADP is the hydrogen acceptor.

7. Succinic coenzyme A is converted to succinic acid, and the coenzyme A is released in the process. Thus, coenzyme A is again used over and over as in reaction 1. In addition, a molecule of guanosine diphosphate (GDP) captures sufficient energy from the reaction to bind

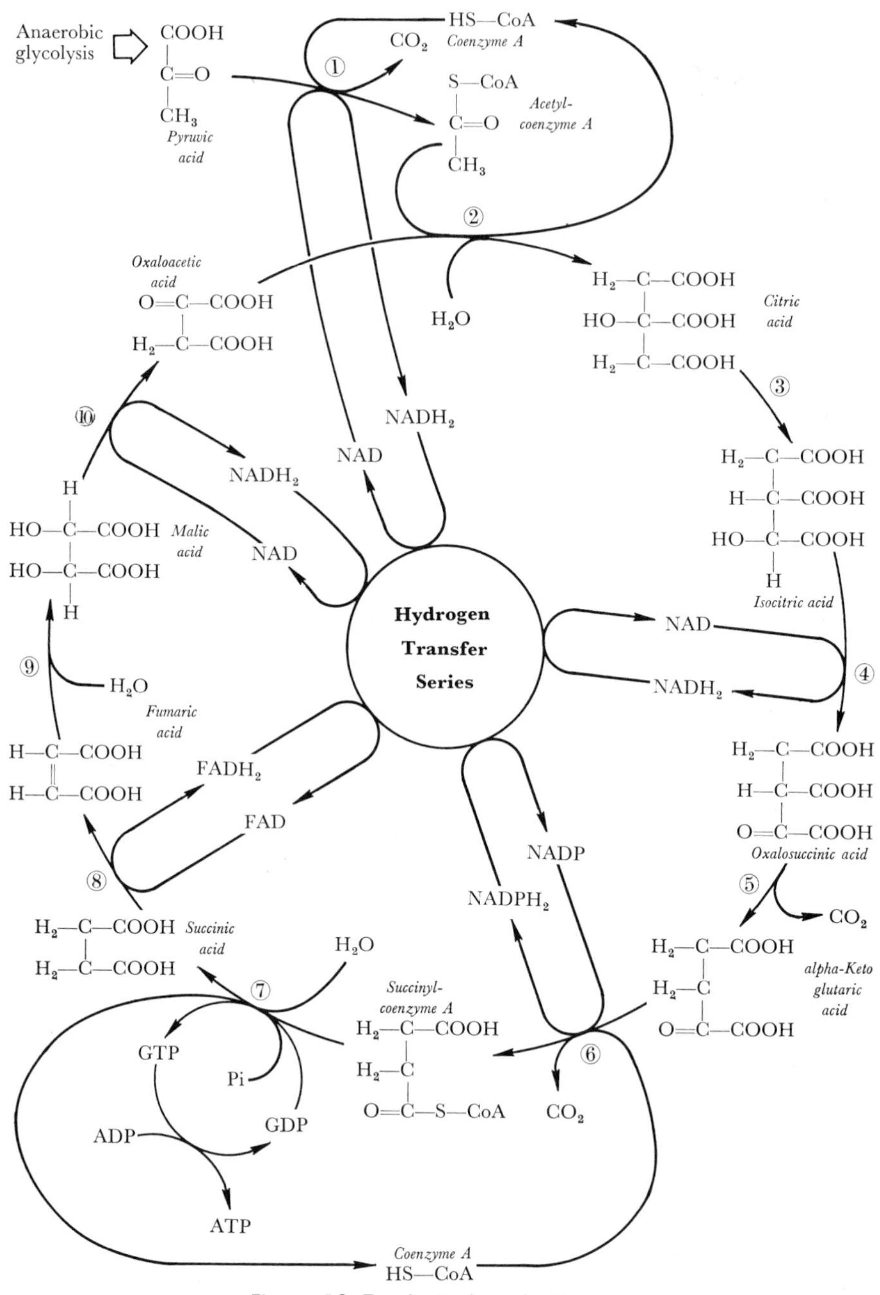

Figure 10–7 The Krebs cycle. See text.

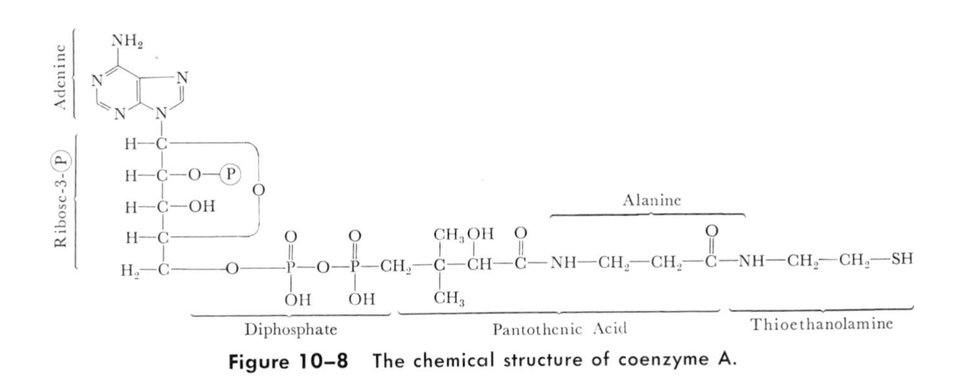

Figure 10–8 The chemical structure of coenzyme A.

to itself an inorganic phosphate group, becoming guanosine triphosphate. Guanosine is much like adenosine in structure (Fig. 10–2), so this reaction is much like the formation of ATP from ADP. The GTP formed can subsequently pass the phosphate on to a molecule of ADP to form ATP and at the same time yield GDP again. For technical reasons biochemists believe that this ATP is lost again and should not be counted as a part of the useful energy captured during glycolysis.

8. Succinic acid is oxidized to yield fumaric acid; in this reaction FAD serves as the hydrogen acceptor molecule.

9. Fumaric acid is converted to malic acid by the introduction of a molecule of water.

10. Malic acid is oxidized to form oxaloacetic acid, and here NAD serves as the hydrogen acceptor. The oxaloacetic acid formed can then combine with another molecule of acetyl coenzyme A (reaction 2), and the cycle is repeated.

For each molecule of glucose the process of anaerobic glycolysis produces two molecules of pyruvic acid (Fig. 10–6), which then enter the Krebs cycle (Fig. 10–7). For each molecule of pyruvic acid passing through the Krebs cycle the following products are formed: three molecules of $NADH_2$, one molecule of $NADPH_2$, one molecule of $FADH_2$, and three molecules of carbon dioxide. This does not count the apparently useless ATP produced in reaction 7. Double these product quantities is produced, then, for each molecule of glucose begun with (Fig. 10–10). The carbon dioxide is an end product that is removed by diffusion into the blood and ultimately from the blood by respiratory activity (Sect. 7.5). The reduced NAD, NADP, and FAD, however, enter another series of reactions known as the hydrogen transfer series (or the electron transfer series). In this final series of reactions still more energy is trapped in the form of ATP.

10.7 HYDROGEN TRANSFER SERIES

This final portion of aerobic glycolysis consists of a series of oxidation-reduction reactions that terminate with the final combination of hydrogen with oxygen to form water. Water, it will be recalled, is the other final end product, along with carbon dioxide, of complete glucose oxidation. The carbon dioxide has already been produced in the Krebs cycle reactions described in the preceding section.

At three points in the hydrogen transfer series sufficient energy is released to permit the formation of ATP from ADP and inorganic phosphate. This means that at each of these points the oxidation-reduction reaction releases at least 8000 calories of energy.

The numbers preceding the paragraphs in the following description correspond to the reaction numbers illustrated in Figure 10–9.

1. Either $NADH_2$ or $NADPH_2$ can be reoxidized (to NAD or NADP) by passing its hydrogen to FAD. In the process FAD is reduced (to $FADH_2$), and a molecule of ATP is formed from ADP plus inorganic phosphate. Note that the $FADH_2$ formed in the Krebs cycle can enter the series at this point but in doing so it will not cause the production of any ATP. Thus, $FADH_2$ contains less potential energy than either $NADH_2$ or $NADPH_2$; that is, it can produce less total ATP in the hydrogen transfer series of reactions. The NAD or NADP formed in this reaction are, of course, able to return to the Krebs cycle to be rereduced. Thus, they serve, along with FAD, as hydrogen transporting molecules between the Krebs cycle and the hydrogen transfer series.

2. The reduced FAD ($FADH_2$) passes the hydrogen on to coenzyme Q (Sect. 10.4), reducing coenzyme Q and reoxidizing itself in the process.

3. Reduced coenzyme Q is then reoxidized in a different way: an electron is removed from the hydrogen atom, converting it to a hydrogen ion (H^+), and is then passed to a molecule of cytochrome *b*. Remember that transferring an electron in this way is equivalent to transferring the hydrogen atom itself; that is, it is an oxidation-reduction reaction (Sect. 10.2). Thus, cytochrome *b* is reduced and coenzyme Q is reoxidized by this electron transfer. One can think of the hydrogen ion (H^+) as merely being set aside at this point for later use in the final reaction of the hydrogen transfer series. Cytochrome *b*, like other cytochromes to be encountered in subsequent reactions, is hemoglobin-like in structure. Since, like hemoglobin, it contains an atom of iron, this reduction changes the valence of the iron from Fe^{+++} to Fe^{++}.

4. The electron is then passed from cytochrome *b* to cytochrome c_1 in a similar fashion. The valence of iron in both cytochromes is changed in the process as cytochrome *b* is reoxidized and cytochrome c_1 is reduced. At the same time, sufficient energy is made available to form another molecule of ATP from ADP and inorganic phosphate.

5. Cytochrome c_1 is reoxidized by passing the electron to cytochrome *c*, reducing it in the process.

6. Cytochrome *c* passes the electron to cytochrome *a*. This reoxidizes cytochrome *c* and reduces cytochrome *a*.

7. Cytochrome *a* passes the electron to cytochrome a_3, and at the same time a molecule of ATP is formed in the same manner as before.

8. Cytochrome a_3 (sometimes called cytochrome oxidase) passes the electron to an atom of oxygen, converting it to a hydroxyl group

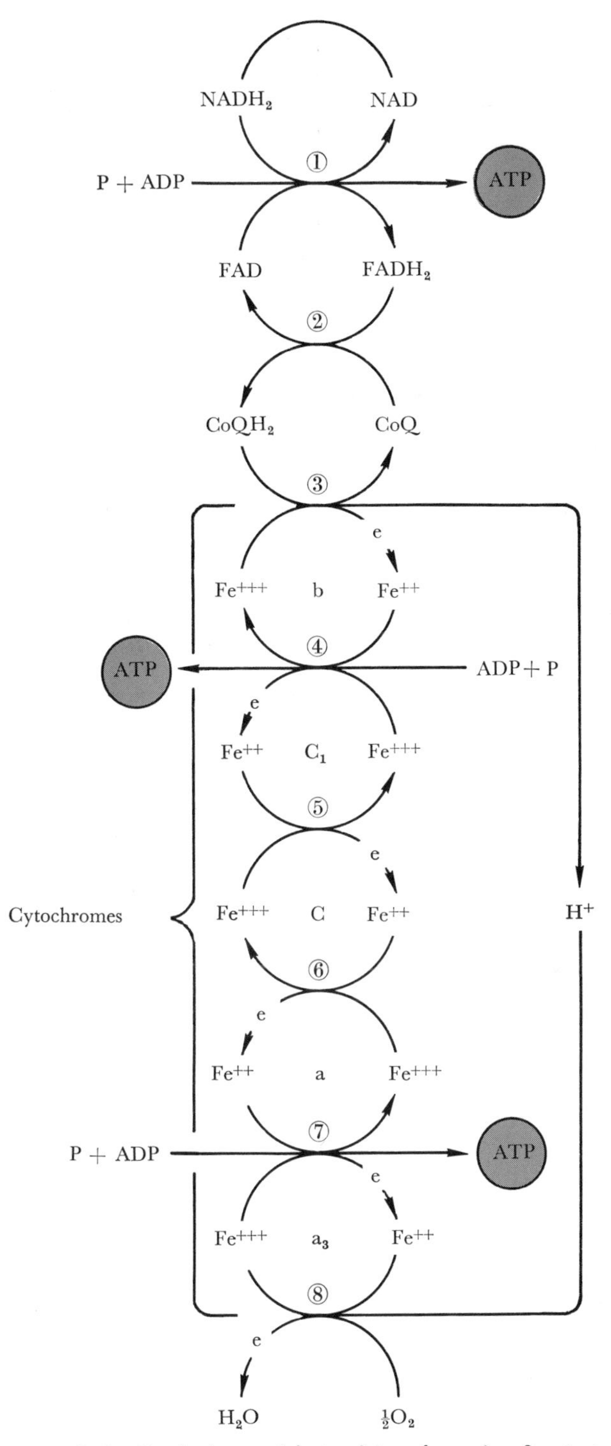

Figure 10–9 The hydrogen (electron) transfer series. See text.

	For each mol of glucose oxidized:				
	Reaction Number	$NADH_2$ (mols)	$NADPH_2$ (mols)	$FADH_2$ (mols)	ATP (mols)
Anaerobic Glycolysis (Fig. 10-6)	5	2			
	9				2
Krebs Cycle (Fig. 10-7)	1	2			
	4	2			
	6		2		
	7				2*
	8			2	
	10	2			
Total mols formed		8	2	2	2
Hydrogen Transfer Series mols ATP/mol input		3	3	2	1
		24	6	4	2
Total ATP yield		36			

*These 2 mols of ATP are not retained (See Sect. 10.6)

Figure 10–10 A summary of the ATP production accompanying the Embden-Meyerhof pathway (anaerobic glycolysis), the Krebs cycle, and the hydrogen (electron) transfer series.

(OH), which combines with the hydrogen ion produced in reaction 3 to form a molecule of water.

For each molecule of $NADH_2$ or $NADPH_2$ entering the hydrogen transfer series, three molecules of ATP are formed. Since each molecule of glucose yields a total of eight molecules of $NADH_2$ and two molecules of $NADPH_2$, this accounts for 10 times 3 or a total of 30 molecules of ATP formed. For each molecule of $FADH_2$ entering the hydrogen transfer series, only two molecules of ATP are formed. Each molecule of glucose led to the production of two molecules of $FADH_2$, so this accounts for another four molecules of ATP produced. Finally, it will be recalled that two molecules of ATP are produced directly during anaerobic glycolysis (Fig. 10–6, reaction 9). Thus, for each molecule of glucose completely oxidized to carbon dioxide and water, a total of 36 molecules of ATP are formed (Fig. 10–10). Each molecule of ATP represents about 8000 calories of trapped energy; the 36 molecules of ATP contain, therefore, about 288,000 calories obtained from each molecule of glucose. Since glucose actually releases some 673,000 calories during its oxidation, this represents an efficiency of about 42 per cent. The remainder of the energy not captured by ATP is lost as heat. Remember that "bundles" of at least 8000 calories must be provided to form ATP. Smaller "bundles" are lost completely and energy "bundles" of more than 8000 calories are partially lost in the same way.

10.8 CONTROL OF ENERGY PRODUCTION

The ATP produced by the oxidation of glucose serves as a readily available store of energy for use by the metabolizing cells of the body. The contraction of muscle, for example, draws energy from ATP and converts it again to ADP and inorganic phosphate (Sect. 16.3).

In the absence of ADP, glucose oxidation cannot proceed. Therefore, the ratio of ADP to ATP concentration in a cell determines the rate of glucose oxidation in that cell. As existing ATP stores are depleted by cell energy utilization, the process of glucose oxidation is accelerated to replace the ATP.

In addition, several endocrine substances (Sects. 17.8 and 17.9) exert controls on glucose oxidation by influencing the availability of glucose itself to the metabolizing cells.

LIPID METABOLISM

10.9 INTRODUCTION

Triglycerides (fats), sterols, and certain other compounds are all classed as lipids, but of these the triglycerides (Sect. 9.3) are the most abundant. Triglycerides are stored in the adipose tissues of the body and constitute an extremely important source of metabolic energy.

Although the average human diet probably contains larger amounts of energy in the form of carbohydrate (starches and sugars), this can be stored only to a limited extent as glycogen. Remaining sugars must be converted to triglyceride and stored in that form in the fat depots of the body. The storage of energy in the form of triglycerides is most economical since a gram of fat contains more calories than does a gram of either carbohydrate or protein.

Fats are absorbed from the small intestine after having been hydrolyzed, for the most part, to glycerol and free fatty acids (Sect. 9.19). Apparently these digestive products are reassembled to form triglycerides in the intestinal cells, however, since they appear in that form in the lymphatic drainage of the intestine. In the lymph they appear as minute droplets known as chylomicrons, which give the lymph a milky appearance. For this reason the lymphatics draining the intestinal walls are known as lacteal vessels. Ultimately, the lymphatic vessels empty their contents into the venous blood at a point near the junction of the jugular and subclavian veins. In the blood the fats are bound to plasma proteins to form molecules known as lipoproteins.

Lipoprotein lipase is a blood enzyme that catalyzes the hydrolysis of triglycerides. The component fatty acids and glycerol then become available to cells as energy sources, or else they are transported to fat depots for storage. Small amounts are also removed by the liver to be used in the synthesis of other substances such as cholesterol (Sect. 10.11).

The fat depots are the major areas of concentration of adipose tissues. In humans about 50 per cent of the fat depots are located subcutaneously, mostly in the trunk region. Another 45 per cent of the fat depots are associated with organs of the abdominal cavity. The adipose cells of these depots (Sect. 2.6) reconstitute the fatty acid and glycerol molecules to re-form triglycerides and then store them in that form. Stored triglycerides are released to the blood again when they are rehydrolyzed into their component fatty acid and glycerol molecules. The storage and release (turnover) of fats in this manner is a continuous process, and the fat content of any given adipose cell is rapidly changing at all times. The turnover rate is 100 per cent within a period of two or three weeks even in the most obese people.

At first glance it may seem that the constant repeated hydrolysis and reconstitution of triglycerides during digestion, transport, and storage is a pointless repetitive process. It must be remembered, however, that there are several fatty acids that occur commonly in animal and plant triglycerides. The three fatty acids attached to each glycerol molecule to form triglyceride may be any of these fatty acids and may be present in any combination. The characteristic fatty acids combined to form plant or animal (food) fats are probably seldom the same as those typical of human fat depots. Thus, the reconstituted stored fat may be quite different from the fat that was eaten. Moreover, the proportions of different fatty acids in human storage depots change with acclimation to different climates. In very cold climates, the fatty acids stored tend to be more unsaturated (contain more double bonds) and, consequently, have lower melting points than those stored in warmer climates. Most of the fatty acids of human depots are palmitic, stearic, and oleic acids (Fig. 9–5).

When required as an energy source, the glycerol portion of the triglyceride molecule can be phosphorylated and then enter the Embden-Meyerhof pathway (anaerobic glycolysis) as glyceraldehyde-3-phosphate (Fig. 10–6). The fatty acids can also be used as energy sources by all tissues. Indeed, there is increasing evidence to suggest that fatty acids sometimes serve as the major source of metabolic energy. The mammalian heart, for example, derives about 70 per cent of its energy from fatty acid oxidation rather than from glucose oxidation.

10.10 FATTY ACID OXIDATION

The most probable sequence of reactions whereby fatty acids are oxidized as energy sources is called the beta oxidation pathway. Two-carbon fragments are successively split off from the fatty acid molecule. This splitting requires prior activation by combination with coenzyme A (Sect. 10.6). Thus, the entire fatty acid ultimately yields a number of molecules of acetyl coenzyme A units, each of which can enter the Krebs cycle. Most fatty acids available for biological oxidation have even numbers of carbon atoms—usually 14 or more—and they generally yield only two-carbon fragments (acetyl coenzyme A) with no odd carbon atoms left over. There is advantage to considering, however, the oxidation of a fatty acid with only six carbon atoms. The process is identical regardless of the total length of the carbon chain, and the six-carbon fatty acid (hexanoic acid) provides an instructive comparison with the oxidation of a six-carbon sugar such as glucose (Sects. 10.5, 10.6, and 10.7).

The numbers preceding the paragraphs in the following description correspond to the reaction numbers illustrated in Figure 10–11.

1. Hexanoic acid is combined with a molecule of coenzyme A to yield the activated coenzyme A derivative compound. The energy of both of the two energy-rich phosphates of ATP is required to make this combination. Thus, a molecule of ATP is converted into AMP (adenosine monophosphate), and two molecules of inorganic phosphate are released. A molecule of water is also split off in this process.

2. The coenzyme A derivative compound is oxidized to form an unsaturated molecule. Since FAD serves as the hydrogen acceptor molecule, $FADH_2$ is formed.

3. A molecule of water is introduced to the double bond formed in the preceding reaction. This converts the molecule to a beta-hydroxy compound.

4. The beta-hydroxy compound is again oxidized and converted to a keto compound. This time NAD serves as the hydrogen acceptor and is reduced to $NADH_2$.

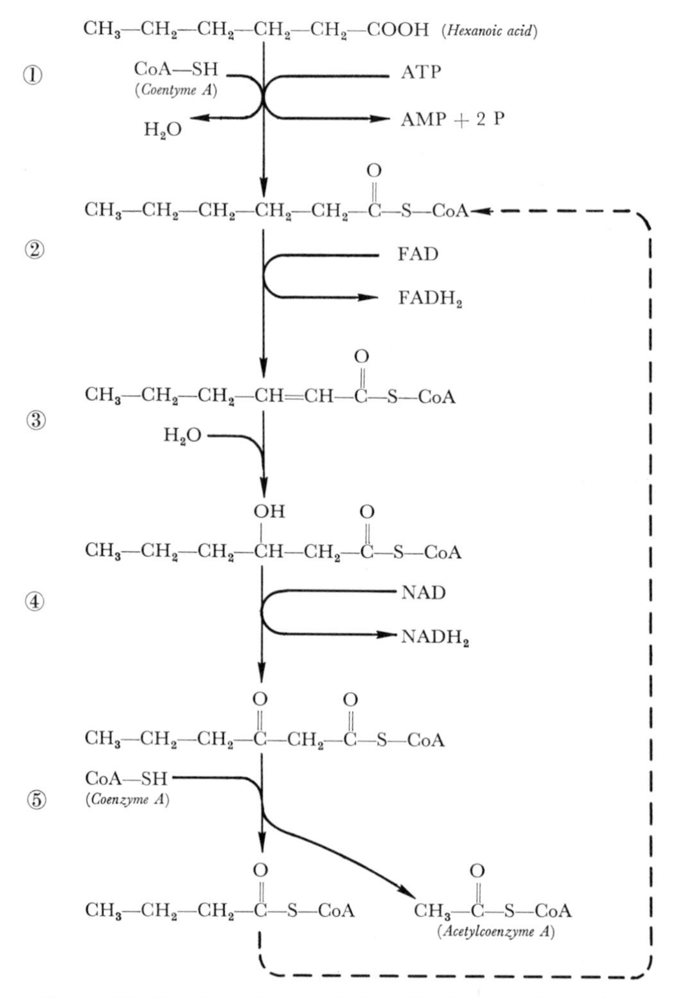

Figure 10–11 Biological oxidation of a fatty acid. See text.

5. Another molecule of coenzyme A is introduced and it attaches to the newly formed keto group. At the same time, the two terminal carbon atoms are split off together with the coenzyme A molecule that was attached to them in reaction 1. Thus, a molecule of acetyl coenzyme A is formed, and the fatty acid molecule is left with only four carbon atoms. These four carbons are already attached to coenzyme A and are therefore activated as in reaction 1. Therefore, reactions 2, 3, 4, and 5 are repeated, yielding two more molecules of acetyl coenzyme A. Thus, the hexanoic acid has been oxidized to form a total of three molecules of acetyl coenzyme A, each of which can enter the Krebs cycle (Fig. 10–7, reaction 2).

A review of Sections 10.6 and 10.7 will show that each molecule of acetyl coenzyme A entering the Krebs cycle ultimately causes the production of 12 molecules of ATP. Thus, hexanoic acid will produce 3 times 12 or 36 molecules of ATP. In addition, the oxidation of hexanoic acid yields two molecules of $FADH_2$ and two molecules of $NADH_2$, which can be introduced into the hydrogen transfer series (Sect. 10.7). Together these will result in the production of an additional 10 molecules of ATP. Thus, for each molecule of hexanoic acid completely oxidized, 46 molecules of ATP are formed. The equivalent of two molecules of ATP are used to activate the reactions (reaction 1) so there is a net gain of 44 molecules of ATP. It will be recalled that the complete oxidation of glucose (also containing six carbon atoms) yielded only 36 molecules of ATP (Fig. 10–10). Clearly, then, fats contain more potential energy per molecule than sugars.

More than half of all the beta oxidation of fats occurs in the liver, and this organ itself cannot use all of the acetyl coenzyme A that is formed. The excess acetyl coenzyme A molecules are combined in groups of two to form acetoacetic acid (Fig. 10–12) and transported by the blood in that form to other tissues. There the acetoacetic acid molecules can be reseparated into two-carbon fragments and recombined with coenzyme A in order to enter the Krebs cycle. Acetoacetic acid may also be converted into beta-hydroxy butyric acid or into acetone. All three of these compounds are known collectively as ketone bodies and they may occur in excess in the blood when the rate of fatty acid oxidation is abnormally high. Such is the case in the disease diabetes mellitus (Sect. 17.8), in which the normal metabolism of glucose is inhibited and fatty acid oxidation is increased. Ketone bodies are organic acids and, as such, they must be buffered by the blood. Excessively high ketone body concentrations, therefore, lead to a depletion of buffer systems and, if untreated, to an eventual decrease in blood pH, which may ultimately cause diabetic death.

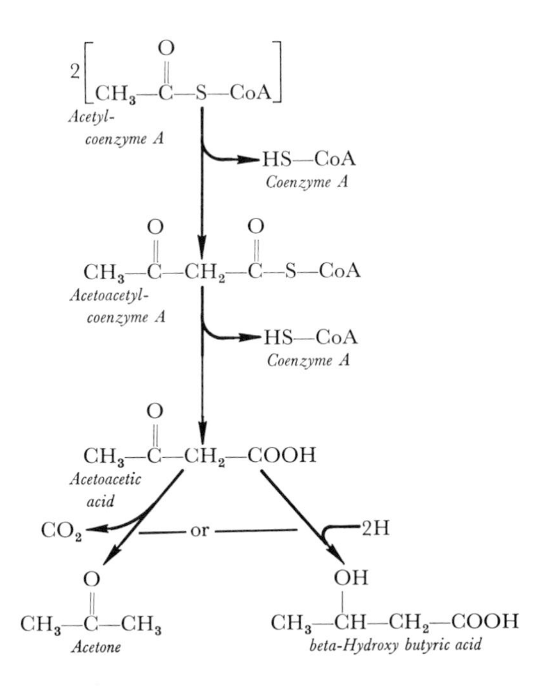

Figure 10–12 The formation of ketone bodies from acetyl coenzyme A. The ketone bodies are acetoacetic acid, acetone, and beta-hydroxy butyric acid.

The synthesis of fatty acids from glucose occurs by way of pathways which are almost the reverse of the beta oxidation pathway. Glucose is first converted to pyruvic acid by means of the Embden-Meyerhof pathway (anaerobic glycolysis). Pyruvic acid is converted into acetyl coenzyme A (Fig. 10–7, reaction 1). Acetyl coenzyme A molecules are then linked together by successive reductions to form fatty acids. Finally, the fatty acids are linked to glycerol molecules to form triglyceride fats. In this way excess sugars in the diet can be converted to fats and stored in that form in the fat depots of the body.

10.11 CHOLESTEROL METABOLISM

A certain amount of cholesterol is a requirement of the body because it is used for the synthesis of steroid hormones (Chapter 17) and

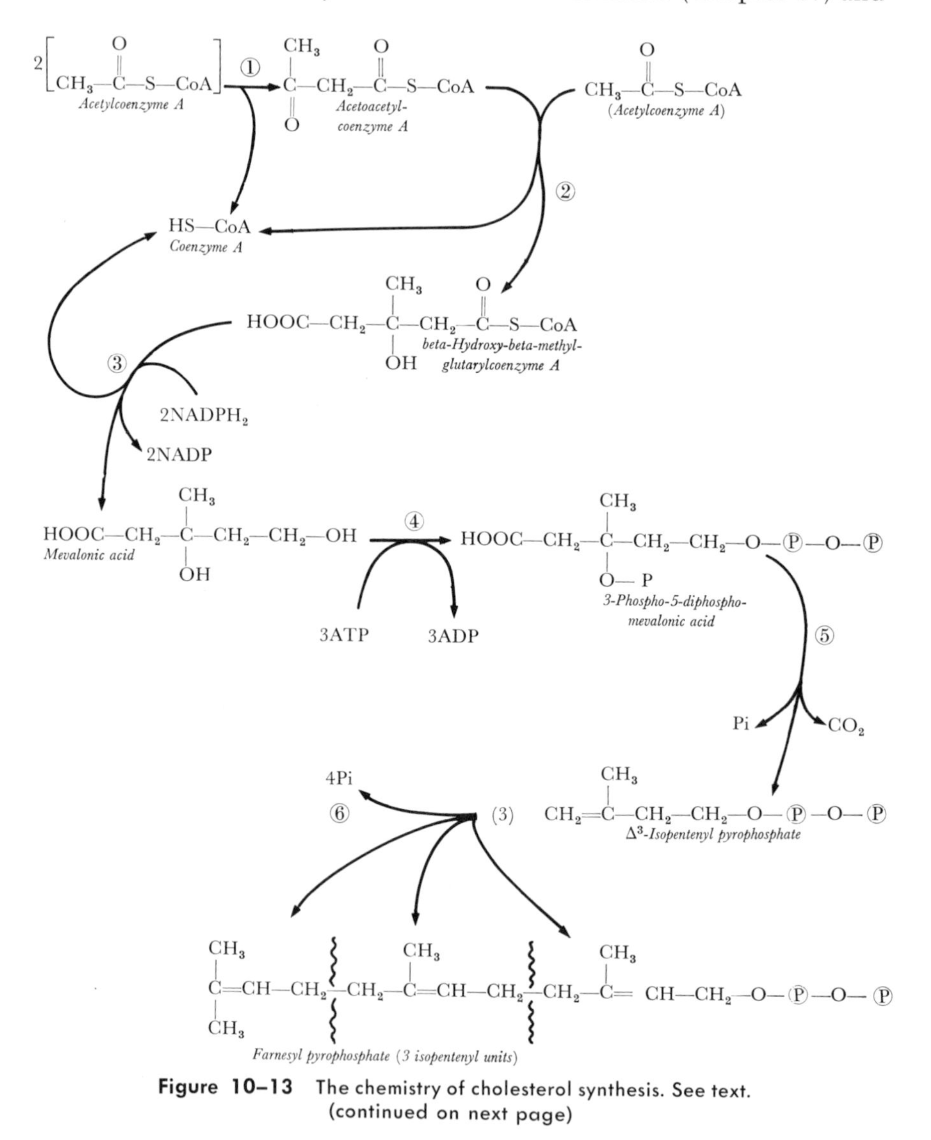

Figure 10–13 The chemistry of cholesterol synthesis. See text.
(continued on next page)

bile salts (Sect. 9.18). Excess cholesterol may be deposited in the walls of arteries, causing atherosclerosis, and this, in turn, may lead to mineral deposition (arteriosclerosis) at the same sites. Such deposition narrows the arterial lumen and leads to hypertension (Sect. 5.6); deposited particles that break free into the circulation may cause thrombosis (Sect. 5.8). Because of these pathological associations, the control of cholesterol metabolism has received a great deal of attention in recent years.

The cholesterol present in foods is readily absorbed from the small intestine; in addition, it is synthesized in the liver and elsewhere. When the dietary supply of cholesterol is low, the rate of cholesterol synthesis is increased. Thus, the avoidance of cholesterol in foods is probably of limited value in lowering the blood cholesterol levels of the body, but lowered intake of saturated fatty acids (those without double bonds) does appear to be somewhat effective in reducing blood cholesterol. Saturated fatty acids usually occur most frequently in ani-

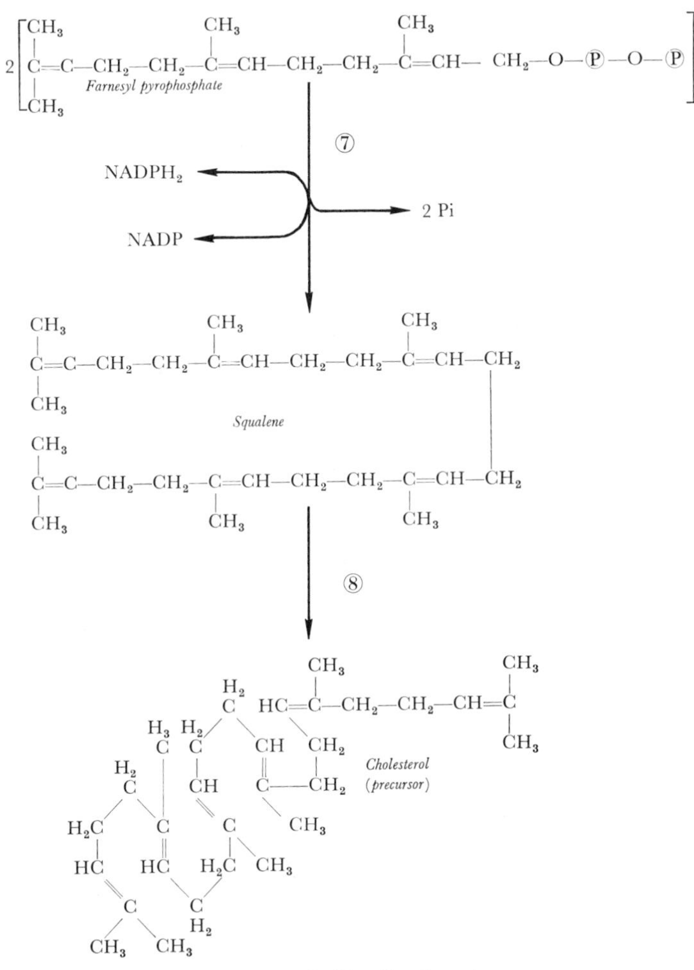

Figure 10–13 *Continued.*

mal fats, whereas unsaturated fatty acids are more characteristic of plant fats. However, there are exceptions to both of these generalities.

The synthesis of cholesterol begins with the combination of three molecules of acetyl coenzyme A (Fig. 10–13). This combined molecule is simultaneously reduced by $NADPH_2$ to form a molecule of mevalonic acid. This molecule is phosphorylated by ATP to form delta^3-isopentenyl pyrophosphate. Three molecules of delta^3-isopentenyl pyrophosphate are combined to form a single molecule of squalene. Two molecules of squalene are combined to form a very long-chain molecule, which can be written in a form that resembles the steroid molecule. Subsequent internal reorganization of this molecule, to close the rings of the steroid portion, results in a compound that is closely related to cholesterol.

AMINO ACID METABOLISM

10.12 INTRODUCTION

As we have seen (Chapter 9), proteins are generally hydrolyzed completely to their constituent amino acids during the process of gastric and intestinal digestion. The amino acids are then absorbed and transported by way of the hepatic portal vein to the general body circulation.

Amino acids are used by tissue cells of all kinds for the synthesis of structural proteins, enzymes, hormones, and so on. Alternatively, some of them can be deaminated and then used as energy sources. Very little storage of amino acids as such is possible in the body.

Proteins consist of long chains of amino acids arranged in specific sequences. Thus, a deficiency of any one particular amino acid may prevent the formation of a specific protein, and this may deprive the body of some essential enzyme, hormone, structural protein, or other protein derivative. In many cases, the supply of a given amino acid may be increased by conversion of another more abundant amino acid. This conversion is carried out by a process known as transamination (Fig. 10–14) and results in the simultaneous conversion of a keto acid to another form. The specific amino acids that cannot be derived by these means are called the essential amino acids because they must be supplied, as such, in the diet in adequate quantities (Sect. 9.4).

To be used as energy sources, amino acids must first be deaminated (Fig. 10–14). This converts them into keto acids, which can enter the Krebs cycle either directly or indirectly since many of them are identical with intermediate compounds in that cycle. Keto acids may also be converted into fatty acids by means of the acetyl coenzyme A intermediate. Only about 60 per cent of the mixture of amino acids present in the average diet can be deaminated, so the remaining 40 per cent are not available as potential energy sources.

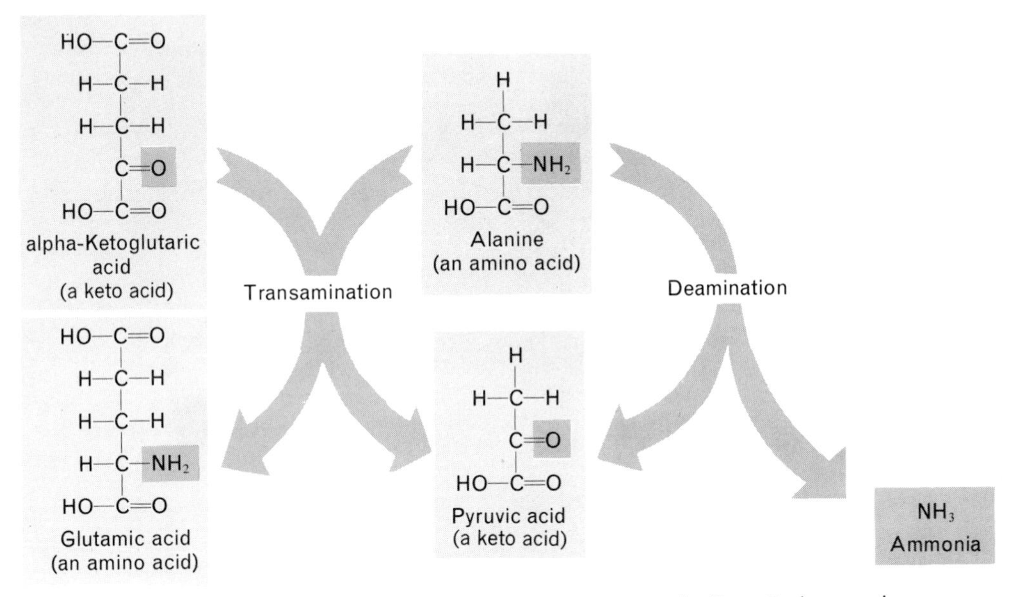

Figure 10–14 Transamination and deamination of amino acids. (From Cockrum and McCauley: *Zoology*. Philadelphia, W. B. Saunders Company, 1965.)

10.13 UREA SYNTHESIS

The deamination of amino acids produces ammonia which, because of its toxicity, must be excreted from the body. In mammals, most of the ammonia is combined with carbon dioxide to form the compound urea, which is then excreted by the kidneys. The sequence of biochemical reactions involved in urea synthesis is called the Krebs-Henseleit cycle (Fig. 10–15) and it interlocks with the Krebs cycle (Fig. 10–7) in an interesting fashion.

The numbers preceding the paragraphs in the following description correspond to the reaction numbers illustrated in Figure 10–15.

1. Carbon dioxide and ammonia are combined to form a molecule of carbamyl phosphate. This requires phosphorylation by one molecule of ATP, which transfers its terminal phosphate to the carbamyl molecule, and it also requires the energy from another terminal ATP phosphate bond. Thus, two molecules of ADP and one of inorganic phosphate are also produced.

2. The carbamyl phosphate formed in reaction 1 is combined with an amino acid called ornithine. In the process, the carbamyl phosphate is dephosphorylated and the phosphate group appears as inorganic phosphate. The carbamyl-ornithine compound produced is called citrulline.

3. Citrulline is combined with another amino acid called aspartic acid. This reaction, again, requires the input of energy from both of the energy-rich phosphate groups of a molecule of ATP. Thus, adenosine monophosphate (AMP) and two molecules of inorganic phosphate are produced. The combined citrulline-aspartic acid compound is called argininosuccinic acid.

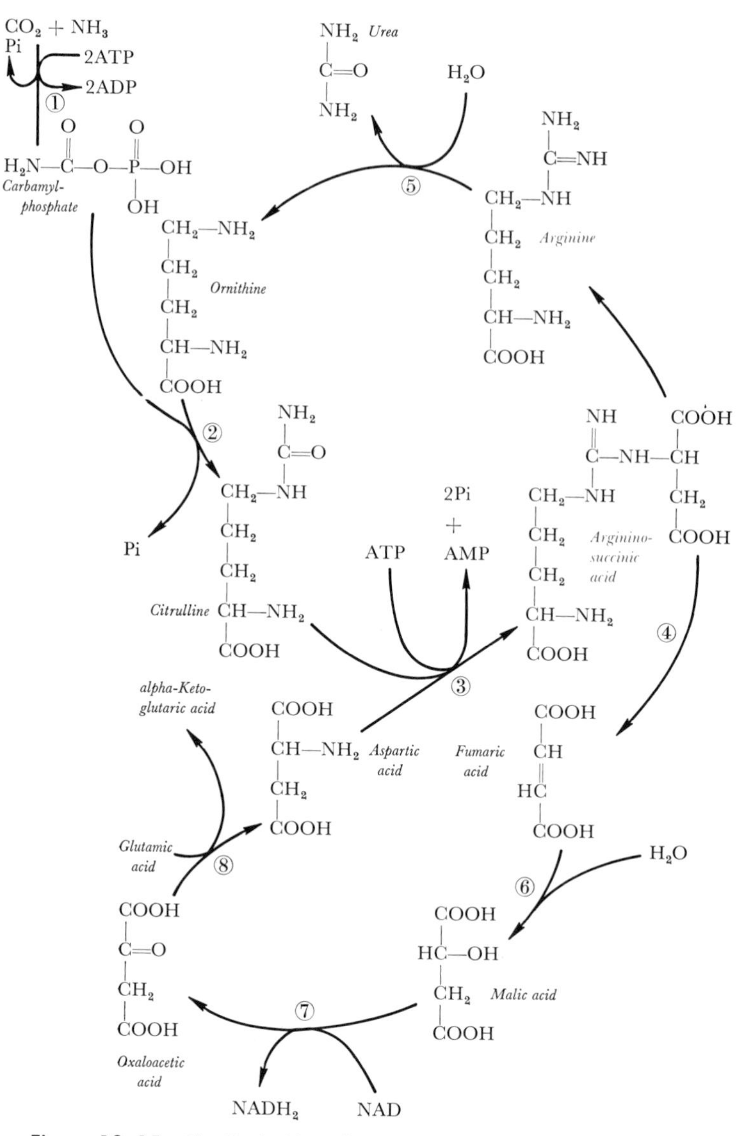

Figure 10–15 The Krebs-Henseleit cycle (urea formation). See text.

4. Argininosuccinic acid is split into two molecules. One of these is the amino acid arginine and the other is the dicarboxylic acid fumaric acid. Fumaric acid is one of the compounds that play a role in the Krebs cycle, and further reactions of this compound are described as reactions 6, 7, and 8 below. Returning now, however, to the further reactions of the other product, arginine:

5. Arginine reacts with a molecule of water to produce two products, namely, urea and ornithine. Thus, urea has been produced, and at the same time ornithine is reconstituted and ready to combine with another molecule of carbamyl phosphate and repeat the cycle (reaction 2).

6. The fumaric acid resulting from reaction 4 is, as we have seen, one of the compounds of the Krebs cycle (Fig. 10–7). As in the Krebs cycle, fumaric acid is converted to malic acid by the addition of a molecule of water.

7. Again as in the Krebs cycle, malic acid is oxidized (dehydrogenated) to oxaloacetic acid with NAD acting as the hydrogen acceptor.

8. Oxaloacetic acid enters a transamination reaction (Fig. 10–14) with the amino acid glutamic acid. In effect, oxaloacetic acid trades its keto group for the amino group of glutamic acid and, as a result, a new keto acid (alpha-keto glutaric acid) and a new amino acid (aspartic acid) are formed. Alpha-keto glutaric acid is also a compound in the Krebs cycle and can be utilized there. The other product, aspartic acid, can be combined with citrulline, as in reaction 3, to keep the Krebs-Henseleit cycle operating.

The making of a molecule of urea has required the input of energy from the equivalent of four molecules of ATP. The oxidation of malic acid in the linked portion of the Krebs cycle has recovered, as $NADH_2$, the equivalent of three molecules of ATP (Fig. 10–9) so there has been a net "cost" of one ATP to form the urea. Nature does not pay such a price unless there is ultimate benefit to the animal. In this case, the benefit comes about through the resulting economy in water loss.

The elimination of ammonia as such requires a great deal of renal water loss to wash it out from the body. Urea elimination is much less costly in terms of water because its toxicity is lower and, therefore, the blood concentrations can be higher (Sect. 10.14).

Instead of combining ammonia with carbon dioxide in the Krebs-Henseleit cycle, the ammonia can be used for reaminating keto acids, converting them again to amino acids; or, alternatively, ammonia can be used by the kidneys to conserve bicarbonates (Sect. 8.8).

About 90 per cent of the urine nitrogen excreted by most vertebrates comes from the deamination of amino acids. The remainder comes from the metabolic breakdown of other nitrogen-containing compounds namely purines and pyrimidines. The end product of pyrimidine catabolism is also ammonia and, like that formed from amino acid deamination, it is also converted into urea by mammals. The metabolic breakdown of purines, on the other hand, results in the formation of a different excretory product (Sect. 10.15).

10.14 AMINO ACID NITROGEN EXCRETION

The ammonia resulting from amino acid deamination and from pyrimidine catabolism is excreted as ammonia by fresh-water fishes. It will be recalled that these aminals produce a very copious dilute urine as a means of overcoming the continuous, unavoidable uptake of water by osmosis from the hypotonic environment (Sect. 8.10). This abundant excretion of water serves well to wash out the ammonia as rapidly as it collects. Moreover, the fresh-water environment is sufficiently extensive to dilute the excreted ammonia well below toxic levels. Animals that excrete ammonia in this way are spoken of as ammonotelic animals (Fig. 10–16).

Marine teleost fishes, on the other hand, live in an osmotically dehydrating medium and must strictly limit urinary water loss (Sect. 8.11). Ammonia cannot be washed out with sufficient rapidity to prevent toxic levels from occurring in the blood. Thus, the ammonia must first be somewhat detoxified. This is accomplished by conversion of ammonia to urea, a much less toxic compound. Still, some water must be expended to excrete urea, since in high concentration it may exert undesirable osmotic effects. Indeed, the marine elasmobranchs, certain cold-water teleosts, and a few amphibians (Sect. 8.11) have capitalized upon the osmotic activity of urea to control body water balance.

As we noted in the preceding section, energy in the form of ATP is required to form urea from ammonia (Fig. 10–15) but this is the price that such animals must pay to prevent osmotic desiccation of their bodies. Animals that excrete amino acid nitrogen in the form of urea are known as ureotelic animals (Fig. 10–16).

The ability to convert ammonia to urea appears wherever there is need to conserve urinary water but where environmental water is not entirely absent. Estivating lungfishes cease to excrete ammonia as the active, non-estivating lungfishes do. Instead, they excrete urea (or in some cases store it). When the completely aquatic amphibian *Xenopus laevis* (the South African clawed toad) is experimentally kept out of water, it ceases to excrete ammonia and instead converts it to urea. The urea is stored in the animal until it is returned to the water. Then the urea is reconverted to ammonia and excreted as such. Such evidence suggests that the ability to form urea appeared early in vertebrate evolution but that the ability of the kidney to excrete it appeared later. The development of these abilities was probably an important part of the total array of developments that permitted the first brief journeys onto land.

The appearance in amphibians of the enzymes necessary for converting ammonia to urea occurs at the time of metamorphosis. Urea conversion first appears at the time when the animal is beginning to spend greater amounts of time on land and lesser periods in the water. Some of these enzymes occur in lower vertebrates in which urea excretion has not developed; presumably these enzymes serve other metabolic functions in such animals. The key enzyme, the evolutionary appearance of which completes the picture and makes urea formation possible, is probably carbamyl phosphate sythetase. This enzyme catalyzes the first reaction in which ammonia and carbon dioxide are combined and phosphorylated by ATP (Fig. 10–15).

Certain vertebrates, notably the birds and some reptiles, further convert urea to uric acid before excreting it (Fig. 10–17). Uric acid is even less toxic than urea. Moreover, it is only slightly soluble so it can be precipitated from solution and either stored or excreted as a paste. Very little renal water loss is associated with such excretion; thus, uric acid excretion is extremely economical for animals in a desert environ-

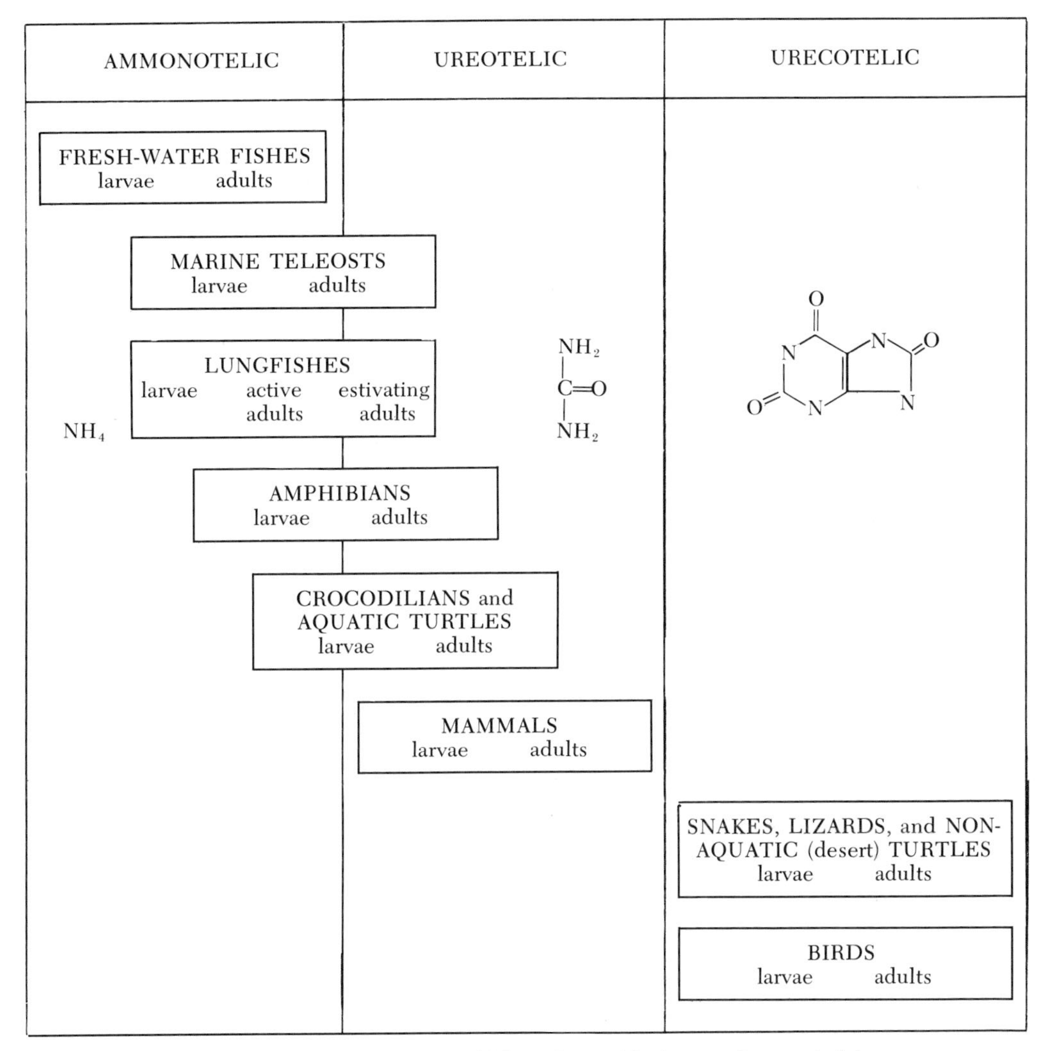

Figure 10–16 The chemical forms in which amino acid nitrogen is excreted by larvae and adults of various vertebrates.

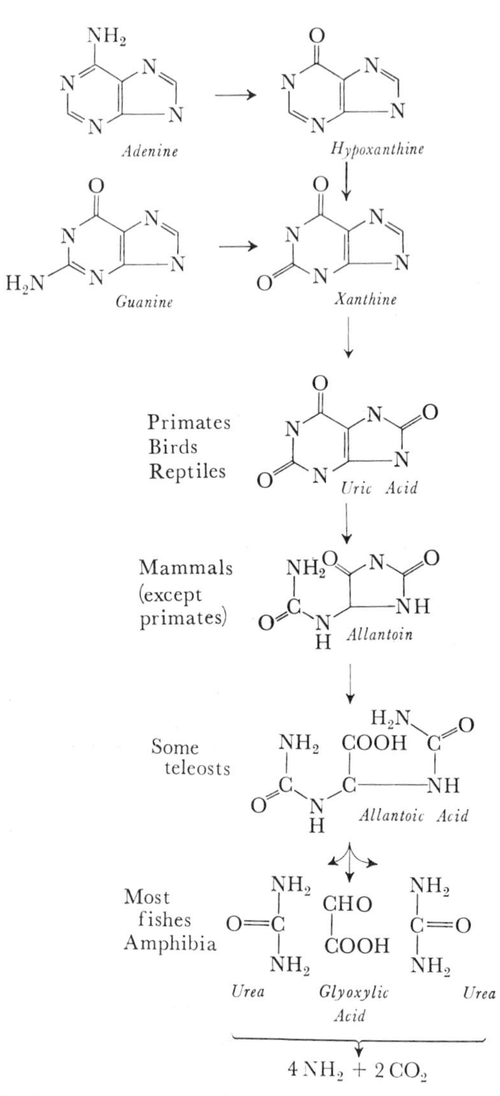

Figure 10–17 Purines are excreted in different forms by different animal groups. See text.

ment. This greater water economy is made possible at the cost of still greater energy expenditure, since the formation of uric acid from urea requires additional ATP energy input. The birds and the squamate reptiles (snakes and lizards) are preadapted to desert life by having this ability. Such animals are spoken of as urecotelic forms (Fig. 10–16).

The association between the form in which ammonia nitrogen is excreted and the habitat of the animal involved has been recognized for many years. A relationship of perhaps greater significance was pointed out by Needham in 1931, however. "The main nitrogenous excretory product of the animal depends on the conditions under which its embryos live, ammonia and urea being associated with aquatic prenatal life and uric acid being associated with terrestrial prenatal life."

The larvae of aquatic animals develop in eggs which, like the adults, are surrounded by abundant water into which ammonia can be allowed to diffuse as rapidly as it is formed by the developing embryo. Generally speaking, the eggs do not regulate internal osmotic pressure to the extent that the adults do (Chapter 8), and the conversion of ammonia to urea by marine fishes is probably an adult modification. The larvae of mammals, like those of fishes, develop in an aquatic environment. Ammonia is converted to urea, however, since it must pass through the maternal bloodstream where the toxicity of ammonia would be damaging. The larvae of amphibians and those of aquatic reptiles (crocodilians and most turtles) are similarly developed in an aquatic environment, or at the very least in extremely moist situations. Thus, ammonotelic and ureotelic excretion is also practical for these embryos. In most cases amphibians and aquatic reptiles actually excrete a mixture of ammonia and urea in varying proportions which correlate with the habitat.

The birds and the remaining more terrestrial reptiles (snakes, lizards, and some turtles) develop relatively impermeable, enclosing "cleidoic" eggs. The lack of any communication between the internal materials of the egg and the surrounding environmental medium means that everything necessary to the developing embryo must be enclosed within the egg. Moreover, it means that the developing embryo cannot easily dispose of any metabolic wastes that accumulate during the developmental period. Ammonia, as we have seen, cannot be allowed to accumulate at all because of its high toxicity. Urea cannot accumulate in large quantities because of its osmotic activity. The conversion to uric acid, however, allows nitrogenous wastes to be stored as non-toxic, non-osmotic, solid materials within the egg. Thus, the ureotelic excretion pattern probably arose in connection with the development of the cleidoic egg. Once established for this purpose, it also conferred special advantages on the adults by preadapting them for the invasion of dry habitats.

It would be a real advantage if animals that develop in cleidoic eggs could metabolize fats and carbohydrates exclusively so that no nitrogenous end products would appear and require excretion. That is, of course, impossible since protein turnover is necessarily associated with growth and development. The energy stores of cleidoic eggs are largely in the form of fats in the yolk, however. As we have seen, fats yield large amounts of energy per unit of weight (Sect. 10.10). Moreover, upon oxidation they yield larger amounts of metabolic water (Sect. 8.12). Both characteristics suit well the needs of such an isolated developing embryo. In aquatic larvae, such as frog tadpoles, where water availability is no problem, egg food stores contain larger amounts of protein and smaller amounts of fat.

10.15 PURINE NITROGEN EXCRETION

The purine molecule, unlike the pyrimidine molecule (Sect. 10.13) is catabolized to ammonia and carbon dioxide by only a few animals and these are all invertebrates. Many other invertebrates and all vertebrate animals excrete instead intermediate compounds of various sorts. The phylogenetic significance of this variability in the form of purine excretion is less clear-cut than that of amino acid and pyrimidine excretion discussed in the preceding section. There is some indication that among the vertebrates there has been a progressive loss during evolution of the enzymes required to catabolize purines completely to ammonia and carbon dioxide. Such a loss of ability would be expected in view of the advantages, for higher vertebrates, of excretion of larger molecules; these advantages were discussed in the preceding section.

CHAPTER 11

THE NERVOUS SYSTEM: PRINCIPLES OF INTEGRATION

11.1 INTRODUCTION

We have seen that physiology is the study of those mechanisms by which an animal (1) maintains relatively constant internal conditions of temperature, pressure, and so forth despite widely fluctuating environmental influences, (2) obtains various required substances from the external medium and returns waste materials to it, (3) carries out internal processes directed toward maintenance, growth, repair, and reproduction, and (4) interacts with other organisms, both of its own kind and of other species, in appropriate ways. In all cases, the kind and degree of animal response, if it is to be appropriate, must depend upon the conditions existing at that time. This means that the animal must continuously monitor all sorts of environmental factors. It must be able to sense accurately changes in temperature, pressure, gas concentrations, and other physical parameters of the external medium. If the animal is motile, as all vertebrates are to some degree, it must also be able to detect barriers, predators, food sources, and others of its own species. Likewise, to make appropriate adjustments of internal processes the animal must monitor a variety of physical conditions within itself. Such factors as blood pressure, osmotic pressure, temperature, muscle tensions, lung inflation, and blood gas concentrations are as important to the physiological well-being of the animal as are conditions of the external medium.

A variety of receptor cells or organs occur among vertebrate animals. Each is particularly sensitive to one or two specific kinds of environmental factors. Some respond easily to pressure, others to chemical changes, and still others to light. Thus, any significant change in the environment, either external or internal, is a potential stimulus to one or more kinds of receptors.

The response of a receptor to an environmental stimulus is merely an extension and specialization of a property common to all living tissue, namely that of irritability. The specificity of receptor response to a particular kind of stimulus results from the individual characteristics and physical locations of receptor cells and from the modifying influences of surrounding cells. In some cases such surrounding cells form parts of the total receptor organ. A single example will serve to illustrate the point. The receptor cells of the vertebrate eye are most sensitive to and respond most effectively to changes in light intensity. This is partly due to the intricate array of surrounding tissues that are specialized to form a cameralike organ with automatic exposure control and focusing mechanisms. It is also due to the fact that the receptor cells contain photosensitive chemical substances. These same cells, however, will also respond to other kinds of environmental stimuli. Pressure exerted with a fingertip against the side of the eyeball is an effective stimulus, but because the brain can interpret signals from the retina of the eye only in terms of light, the conscious sensation is of that type. Electrical and thermal stimuli elicit similar responses. Thus, retinal receptors retain the general irritability of all living cells but have heightened sensitivity to photic stimuli and are situated where other kinds of stimuli are unlikely to reach them.

The response of an animal to changing environmental conditions is generally complex, requiring the integration of several kinds of information from different receptors and involving, perhaps, more than a single kind of physiological adjustment. Often information stored as memory—a special input permitting the animal to learn and profit from experience—is also utilized. Such complex integration requires the presence of a computerlike central nervous system equipped with data storage or memory banks. The central nervous system must quickly integrate all of the input available to it and provide an appropriate output which, through initiating activity of effector organs, results in purposeful physiological responses. These responses and their results must be monitored by the same or other receptors, or both, and the "feedback" input (Sect. 1.6) must be integrated in the same way. The monitoring process permits the responses to be continued, modified, or stopped as required to achieve a desirable final result. An extremely mechanistic view of the vertebrate animal, then, would show it as a complex, highly automated machine that combines sensitive receptors of many kinds, a finely controlled set of effector organs, and an extremely efficient central computer. These three basic units are in constant communication through networks of peripheral nerves so that all activity can be coordinated and purposeful.

Nerve impulses—the signals that pass along nerves—are of only one kind (Sect. 11.3) whether they originate in the responses of retinal cells of the eye to light, of carotid sinus cells to changing pO_2 of the blood (Sect. 3.9), or in the activity of the brain destined to cause the contraction of skeletal muscle. The differences in the kind and degree of effect depend upon (1) the origin and destination of the impulses, (2) the number of similar conducting nerve cells (neurons), and (3) the number of impulses per unit length of time traveling along each neuron (impulse frequency). Recall, for example, that all impulses reaching the brain along nerve pathways leading from the retina of the eye are interpreted as light regardless of the true nature of the stimulus eliciting them. Similarly, greater amounts of light entering the eye may stimulate larger numbers of receptor cells, lead to greater impulse frequencies from each receptor cell, or both. Thus, a good deal of information is encoded in the impulse signals sent from the retina to the brain.

When of sufficient magnitude to stimulate receptor cells, changes in environmental conditions cause these cells to initiate nerve impulses which then travel along peripheral nerves to the central nervous system. The particular pathway that they follow is, of course, related to the kind of environmental stimulus, since the impulses originated in the specialized receptor which alone makes use of that pathway. When the stimulus is a large or abrupt one, the receptor response is greater so that more receptor cells may be involved, or each may initiate a greater number of impulses in the associated nerve cells, or both. Thus, the number of transmitting neurons and the frequency of impulses transmitted by each are factors related to the intensity of the original stimulus. The receptor "translates" actual environmental changes into coded patterns of nerve impulses. In electrical recording instruments, such receptorlike parts are called transducers; these convert specific kinds of environmental events into recordable electrical signals in an analogous way.

Within the central nervous system, the incoming coded signals may require integration with other sensory input signals and perhaps with stored memory data. As the student turns the pages of this book, signals from the eyes, fingers, and muscles of the forearm must be integrated in this way, and as he reads the book, memories of the meanings of the symbols with which the words are printed must also be brought into play. Such internal integration of sensory input in the brain requires internal controls of routing, amplifying, damping, and recoding, which we shall consider in more detail later. Much of this data processing depends upon events that occur at nerve synapses, the junctions between successive neurons (Sect. 11.4).

Ultimately, the central nervous system initiates patterns of nerve impulses which travel through other peripheral nerves (motor nerves) to various effector organs. These nerve impulses are identical in character with those of sensory nerves, but because they are transmitted to muscles, glands, and other effectors, they result in adjustments of physical position or of physiological processes. As in sensory coding, the frequency of impulse transmission and the total number of parallel neurons involved determine the degree of effector response. Each effector organ "decodes" the signals and responds in the proper way and to the appropriate degree.

In some instances the central nervous system does not directly produce the changed activity of the effector organ; instead, it may adjust the activity of an endocrine gland (Chapter 17), and this gland, in turn, exerts chemical influences on other effector organs by altering the amount of hormone secretion into the blood stream. Typically such indirect and relatively slow-acting control is used for long-lasting responses, whereas the fast-acting direct innervation of effectors is used for short-term and frequently changing responses.

11.2 REFLEX MECHANISMS

In very simple, somewhat stereotyped stimulus-response situations no great amount of central integration is necessary because there is only a single kind of appropriate response. Generally that response must occur immediately without any delay imposed by the central nervous system. Placing a finger in a flame, for example, results in the immediate withdrawal of the finger—the only reasonable response to such a painful stimulus. The withdrawal of the finger involves the activation of the flexor muscles of the arm and forearm, and the response is called, therefore, a flexor reflex. It actually takes place before the brain has consciously interpreted the sensory input as pain. This is true because the reflex mechanism takes place through a simple and direct nervous arc involving only a portion of the spinal cord (Fig. 11–1). An all-but-direct link between the sensory and motor neurons within the spinal cord results in an immediate contraction of flexor muscles, causing withdrawal of the burned finger. The more intense the stimulus, the more violent is the withdrawal by reason of the involvement of greater numbers of neurons in both the sensory and motor portions of the reflex arc. In fact, if the stimulus is intense enough, the response may radiate to other portions of the body as well so that in addition to removing the finger, one may even leap back-

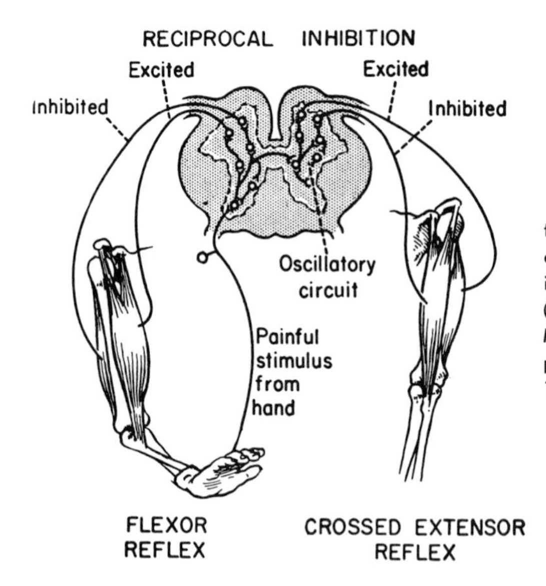

Figure 11–1 The flexor reflex, the crossed, or contralateral, extensor reflex, and the reciprocal inhibition of antagonistic muscles. (From Guyton: *Textbook of Medical Physiology*, 3rd Edition. Philadelphia, W. B. Saunders Company, 1966.)

wards away from danger almost before realizing consciously that an accident has occurred.

Whenever any muscle is at rest, it is nevertheless partly in a contracted state (Sect. 16.5). This state of partial contraction is known as tonus. Now, when a given muscle contracts, either reflexly or voluntarily, the tonus of opposing muscles must be reflexly inhibited so that they do not unnecessarily oppose the action of the contracting muscles. For example, contraction of the biceps muscle, which flexes the upper extremity at the elbow, would be opposed by the normal tonus of the triceps muscle, which extends the same limb. Therefore, whenever the biceps is stimulated to contract, the tonus of the triceps is inhibited, and this occurs by way of spinal cord level reflex connections. In the case considered in the preceding paragraph, the withdrawal of the finger from the flame occurs faster and more efficiently because of this inhibition of tonus in the antagonistic triceps muscle. Such simultaneous and opposite controls of antagonistic muscles involves the principle of reciprocal innervation and occurs reflexly in the spinal cord.

Completely unopposed muscle responses, especially when they occur suddenly and strongly as they would in a flexor reflex induced by a painful stimulus, tend to be uncontrolled and jerky. Two other spinal cord reflex mechanisms act to smooth such actions by first opposing the response slightly and then releasing it as follows: (1) The contraction of any skeletal muscle results in the stretching of an antagonistic muscle. Contraction of the biceps necessarily results in stretching the triceps muscle. Sensory receptors within the stretched muscle are stimulated and act, by way of a simple, cord level reflex arc, to cause a contraction of that muscle sufficient to resist the stretch. This is called a myotatic or stretch reflex and it acts to damp the sudden onset of any skeletal muscle activity. A sudden contraction of the biceps is, thus, momentarily resisted by contraction of the opposing triceps muscle. (2) If the initial contraction of the biceps continues, the increasing stretch of the triceps muscle then activates still other receptors within the triceps tendons, and these, also by way of a simple reflex arc, initiate a reverse myotatic reflex (or Golgi tendon reflex). This reflex results in a sudden disappearance of the myotatic response; the further contraction of the biceps is then unopposed.

Since the pull of gravity tends to cause limb flexion and, therefore, the stretching of extensor muscles, certain myotatic reflexes function also as antigravity responses. Suppose, for example, that flexion of the lower extremity begins to occur at the knee of a standing human. The force of gravity will tend to continue such flexion and, if totally unopposed, will cause the individual to fall to the ground. Flexion at the knee, however, excites the stretch receptors of the extensor muscles of the limb (quadratus femoris muscle group), and this elicits a myotatic reflex. The extensor muscles contract and resist the further flexion of the knee joint. This may stop or even reverse the flexion (extend the limb) to prevent falling, or if the force causing the flexion is too great (perhaps the individual desires to kneel or fall or the load he is bearing is too great), the reverse myotatic reflex soon takes over and continued flexion occurs unopposed. This sudden unopposed flexion is sometimes spoken of as the clasp-knife effect. The antigravity myotatic reflex just discussed can be demonstrated by tapping the patellar tendon just below the kneecap to produce the familiar knee jerk or kicking (extension) response (Fig. 11–2). Similar antigravity responses can be demonstrated at the ankle joint, and even though humans do not walk on all four limbs, antigravity responses for the elbow and wrist joints are present. Interestingly, the South American tree sloth, which hangs upside down from tree limbs, uses flexor muscles rather than extensor muscles to resist the pull of gravity. In this animal typical myotatic responses involve the contraction of flexor muscles.

The pressure of the soles of the feet against the ground stimulates receptors in the skin of the soles. By way of simple reflex arcs this input into the spinal cord results in extension of the legs. This supportive reflex can easily be demonstrated by pressing lightly against the footpad of a reclining dog or cat. Even in this position the animal will respond by extending the stimulated limb. If the limb of the animal is first flexed to a considerable degree, the effectiveness of this stimulus disappears. This reflex is of importance in walking since the limb extends against the ground while in contact with it. Flexion of the limb removes the contact and, therefore, the tendency to extend the limb. Further flexion then becomes easier.

The reflexes discussed thus far all result in muscle contraction (or inhibition) on the same side of the body on which the initiating stimu-

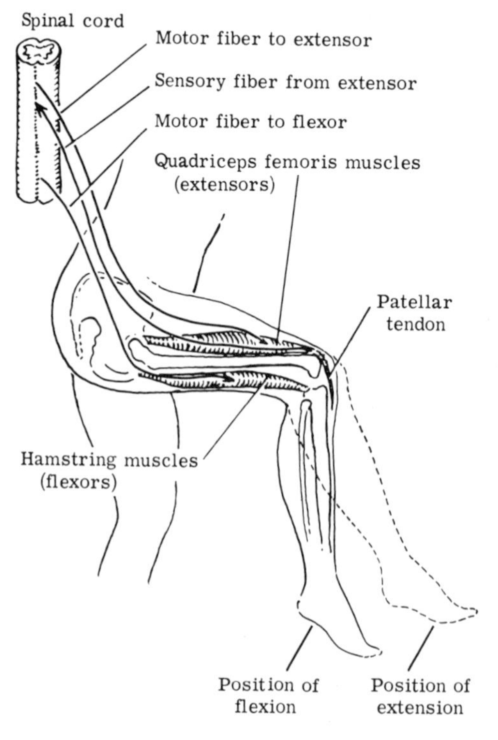

Figure 11–2 The patellar reflex (schematic). See text, Section 11.2. (From Cockrum and McCauley: *Zoology*. Philadelphia, W. B. Saunders Company, 1965.)

lus has occurred. Such reflexes are called ipsilateral reflexes. In addition, however, there are simple spinal cord reflexes that are set in motion by a stimulating event on one side of the body and result in muscle activity on the opposite side of the body. These are called contralateral reflexes. One such contralateral response results from the crossed extensor reflex (Fig. 11–1). When the right leg is flexed, the crossed extensor reflex causes an immediate extension of the left leg. The value of such a reflex when maintaining an upright position while walking or while shifting the weight of the body from one foot to the other is quite obvious.

Most spinal cord reflexes tend to fatigue and disappear rather quickly after they are evoked. The same reflex is then more difficult to elicit a second time until recovery has taken place. At the same time, however, opposing reflexes are easier to elicit. This phenomenon is known as the rebound phenomenon. For example, if a flexor reflex occurs in the right leg, a second flexor reflex is more difficult to elicit for a time thereafter; that is to say, a more intense stimulus is required to cause the same reflex to reoccur immediately. At the same time, an extensor reflex is easier to elicit in the same leg. In walking, then, the alternate flexion and extension of each leg is facilitated by this rebound phenomenon.

In quadruped vertebrates the alternate activity of the hind legs is coordinated with that of the forelegs to give the animal a characteristic four-legged pace. This coordination is accomplished by long spinal reflexes which are similar in principle but slightly more complex in anatomical linkage than the reflexes discussed previously. Some animals have more than a single pattern of four-legged coordination. Horses, for example, can be trained to utilize three or more different patterns, and this fact illustrates that a variety of long spinal reflexes are available. Biped vertebrates like humans do not use the "forelegs" for locomotion after the period of infancy, yet the long spinal reflex connections remain functional and, as a consequence, we swing our arms as we walk. This swinging is in the proper sequence with leg movements for walking (Sect. 16.11).

Movements of the head and neck are also coordinated with body and leg movements by special neck reflexes. These also operate on a spinal cord level and are little more complex than the long spinal reflexes just discussed. Neck reflexes keep the head steady despite the side-to-side movements of the body during walking. They also direct the head and limbs toward common goals in leaping and other special activities. As a quadruped vertebrate walks in a straight line, the shoulders may move from side to side. If equal and opposite bending of the neck did not also occur, the head would be swung from side to side as well. This characteristic movement of the shoulders and of the head is a consequence of the basic pattern of locomotion common to all vertebrates (Chapter 16). Even in humans, in which no lateral movements of the shoulders ordinarily accompany locomotion, one sometimes sees rhythmic side movements of the head coordinated with walking. If the head of a standing cat is lifted by placing a finger beneath its chin and raising it so that the animal is looking upward, the hind legs will flex and the fore legs will extend. This places the animal in a crouching position appropriate to leaping upward in the direction of its gaze. Conversely, if the animal's head is pressed downward, the forelegs will flex and the hind legs will extend. This again places the animal in the appropriate position for leaping downard in the direction of its gaze. Similarly, turning the cat's head toward the right will cause flexion of the right foreleg and extension of the left foreleg—movements appropriate to stepping toward the right. All of these responses are mediated by simple neck reflex connections within the spinal cord and, therefore, require no integration within the brain.

Besides the simple cord reflexes discussed thus far, there are other reflex mechanisms that require some integrative processing in lower brain levels; the controls of blood pressure (Sect. 5.11), respiration (Sect. 3.9), body balance (Chapter 14), and focusing of the eyes (Chapter 15) are examples. Still more complex integrative action by higher brain centers takes place when experience (conditioning) or perhaps even conscious judgment are required. The conditioned reflex is well-known to most people; it was demonstrated by Pavlov that ringing a bell and feeding a dog at the same time could establish such a pattern. After a time the bell alone becomes an adequate stimulus to elicit salivation in the dog by way of reflex pathways. It is really impossible to draw a sharp line between this simple kind of conditioned reflex and a variety of more complex behavior. For example, conforming to the mores, styles, and behavior patterns of our society seems to have many of the features of conditioning!

11.3 THE NERVOUS IMPULSE

The complete nature of the nervous impulse is not entirely known, but it involves both electrical changes at the surface of the neuron (nerve cell) and internal changes, which proceed in a wave-like manner at speeds up to 100 meters per second in some mammalian nerves.

Nerve cell membranes, like the membranes of other living cells, have an electrical charge when they are at rest (not conducting any impulses). Characteristically, this charge is positive on the outside and negative on the inside of the membrane (Fig. 11–3). This means that the neuron membrane is like a plate of an automobile battery—if a wire connects one side of the membrane with the other side, an electrical current will flow through the wire. When a battery is charged, it is said to have an electrical potential; that is, it contains potential energy in the form of a differential charge on the two sides of its plates. Similarly, the differential charge across the membrane of a resting nerve cell is spoken of as the resting membrane potential. It has been shown that this resting potential is due to a differential distribution of ions on the two sides of the membrane. The cell membrane actively "pumps" sodium ions (Na^+) outward and allows potassium ions (K^+) to accumulate inside

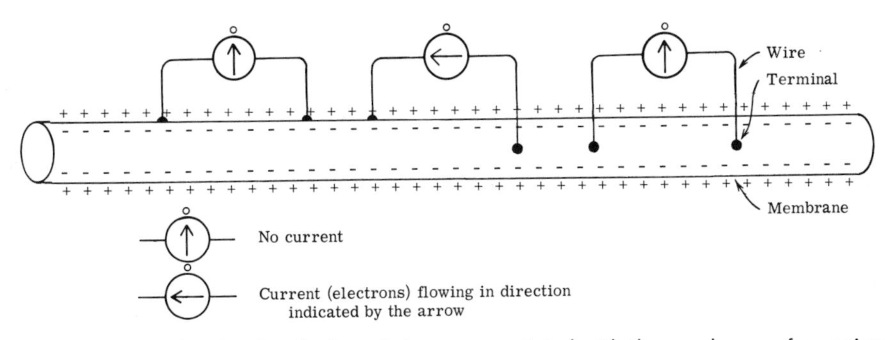

Figure 11–3 The distribution of charges associated with the membrane of a resting (non-conducting) neuron. No current flows when both terminals of a measuring device are on the same side of the membrane, but electrons flow from the inside of the membrane to the outside when one terminal is on each side. (From Cockrum and McCauley: *Zoology.* Philadelphia, W. B. Saunders Company, 1965.)

the cell. The additional presence of negatively charged organic molecules (proteins) inside the cell more than counterbalances the positively charged potassium ions and gives the inside a net negative charge. The cell membrane is not permeable to these organic molecules. It is, however, slightly permeable to both sodium and potassium. These ions, therefore, constantly "leak" through the membrane, and as a consequence, the cell must continuously operate its "sodium pump" to maintain the resting condition. The same is true of an automobile battery—it will not continuously retain its potential unless it is periodically charged.

When a neuron is stimulated, the permeability of the membrane at the point of stimulus is altered; as a result, sodium ions and potassium ions are able to diffuse readily through the membrane in accordance with the existing concentration gradients for each (Sect. 1.2). That is, sodium ions diffuse inward and potassium ions diffuse outward. The permeability becomes greatest for sodium ions, and they rush into the cell rapidly. Potassium ions leave the cell somewhat more slowly, since the permeability to these is not increased as much. The result is that momentarily there is a large combined population of sodium and potassium inside the cell—a population sufficient to counterbalance the negatively charged organic molecules there and give the inside a net positive charge. At the same time, the outside surface of the membrane has been depleted of sodium ions, and it momentarily has a negative charge with respect to the inside. Thus, the change in membrane permeability produces a momentary reversal of membrane change; the membrane becomes negative on the outside and positive on the inside. Subsequently, the continued outward diffusion of potassium ions removes positive charges from the inside and restores them to the outside. This continues until the distribution of both sodium and potassium has reached an equilibrium, and, as a result, the membrane potential begins to return toward the resting state. At the same time, however, the "sodium pump" of the membrane increases its activity and the permeability characteristics of the membrane return to normal. As a result, sodium ions are again pumped outward, potassium ions are accumulated inside, and the resting membrane potential is reestablished. The momentary reversal of membrane charge and the subsequent reestablishment of the resting potential just described are together called an action potential. The action potential coincides with and is at least a part of the nervous impulse. It sweeps over the surface of the nerve cell from the point of origin (point of stimulus) to all other parts of the cell in a wavelike fashion.

As one portion of the neuron membrane is reversed in polarity by the action potential, this is in itself an adequate stimulus to adjacent

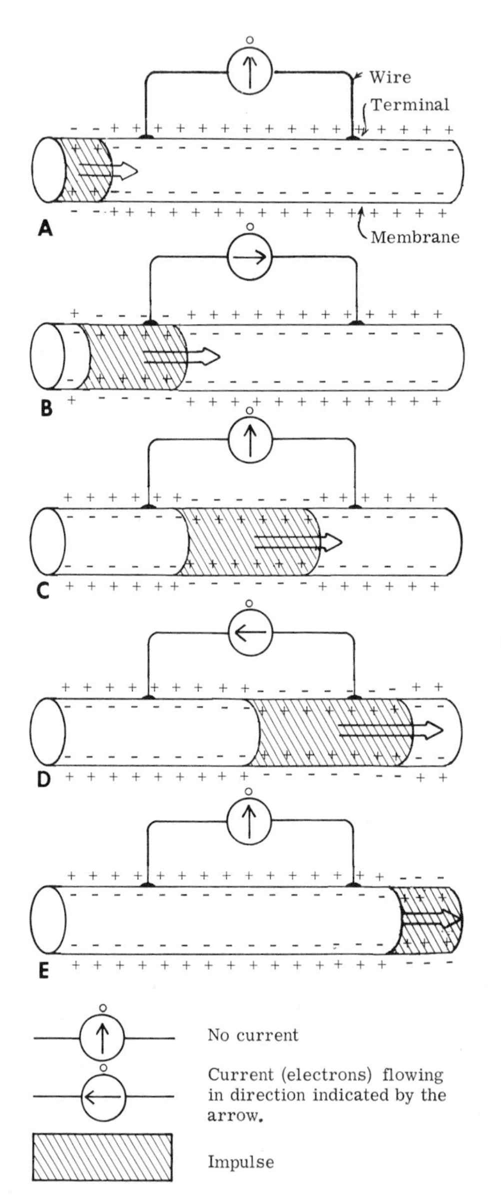

Figure 11–4 The conduction of an action potential along a neuron. The flow of electrons through a current measuring device is shown. See text, Section 12.3. (From Cockrum and McCauley: *Zoology*. Philadelphia, W. B. Saunders Company, 1965.)

portions of the membrane so that they follow suit. The action potential, therefore, spreads along the membrane with recovery of the resting membrane potential following behind (Fig. 11–4). This conductive excitatory state, as it is sometimes called, appears as a wave of negativity of the outer surface of the membrane.

Certain large, rapidly conducting neurons are covered with layers of a substance called myelin; smaller, slower-conducting neurons have no such covering. Myelin is a lipid material that provides electrical insulation and protection for the neuron (Fig. 11–5). The myelin is associated with Schwann cell membranes, which are wrapped in a spiral fashion around the nerve cell process (axon). This covering is interrupted at intervals by nodes of Ranvier, at which points the neuron membrane is exposed (Sect. 2.4). Ions cannot flow easily through the insulating myelin but pass through the neuron membrane with ease at the nodes of Ranvier. As a result, the action potential in such neurons appears to leap from node to node. Actually, the electrical disturbance spreads through the insulated internodes by electrical conduction, and it is only at the nodes where membrane polarity is momentarily reversed. This minimizes the task of the "sodium pump" and speeds the entire process of action potential transmission so that myelinated neurons often carry impulses at up to 500 times the speed achieved in non-myelinated neurons. The fastest neurons are used for reflex mechanisms (Sect. 11.2) involving skeletal muscles, whereas slower neurons act in connection with smooth muscles and in other ways.

The application of a stimulus to a sensory receptor induces in it a receptor potential or generator potential (Sect. 13.2). This potential constitutes an effective stimulus to the associated sensory neuron and elicits an action potential in it. Often the receptor potential persists for a time so that as soon as the associated neuron has reestablished its resting potential, it is restimulated. In this way a single stimulus can evoke trains of nerve impulses traveling toward the central nervous system. The numbers of such impulses and the frequency at which they occur in a sensory nerve depend upon the magnitude and the duration of the receptor potential, and this, in turn, is dependent upon the intensity of the stimulus.

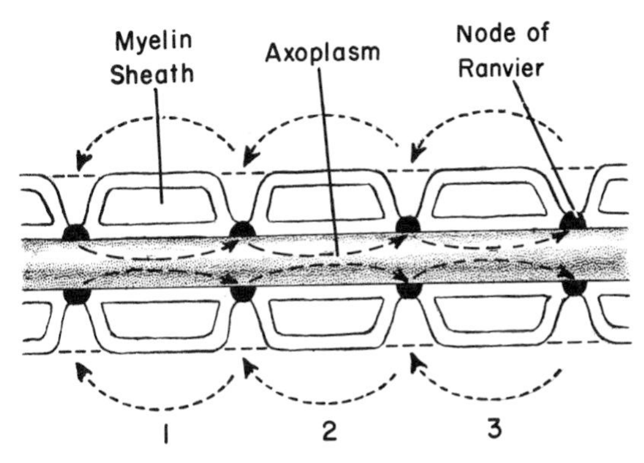

Figure 11–5 Saltatory conduction along a myelinated axon. (From Guyton: *Textbook of Medical Physiology*, 4th Edition. Philadelphia, W. B. Saunders Company, 1971.)

11.4 THE SYNAPSE

Within the central nervous system (and in a few cases outside the CNS) individual neurons end by contacting one or more other neurons. At such interneuron junctions, called synapses, nerve impulses may be transferred from one neuron to others. It is at the synapses that the real integrative functions of the central nervous system presumably occur, since at these points impulses may be routed in different directions or modified in various ways. Synapses are versatile in function, whereas individual neurons can only conduct unchanged impulses from one end of the cell to the other.

The simplest function of a synapse is to transfer unchanged the action potential from the presynaptic neuron to the postsynaptic neuron. The activity pattern of the postsynaptic neuron is then identical in every way with that of the presynaptic neuron. Such impulse transfer comes about in this way. The presynaptic neuron terminal is an enlarged structure, sometimes called an end button, that contains many tiny vesicles filled with "transmitter substance" (Fig. 11–6). The arrival of an action potential along the presynaptic neuron causes some of these vesicles to move to the surface and discharge their contents into the space (synaptic cleft) between this neuron and the postsynaptic neuron. The transmitter substance, then, alters the permeability of the postsynaptic neuron membrane so as to establish an action potential in it.

The neuromuscular junction, at which a motor neuron delivers impulses to a muscle, is in many ways similar to a neuron synapse. Transmitter substance is released in the same way and serves to establish an action potential in the muscle cell membranes. This muscle action potential sweeps over the muscle and in some way triggers muscle contraction (Sect. 16.3). The anatomy of the neuromuscular junction is slightly different but the principles of function are like those of a typical synapse.

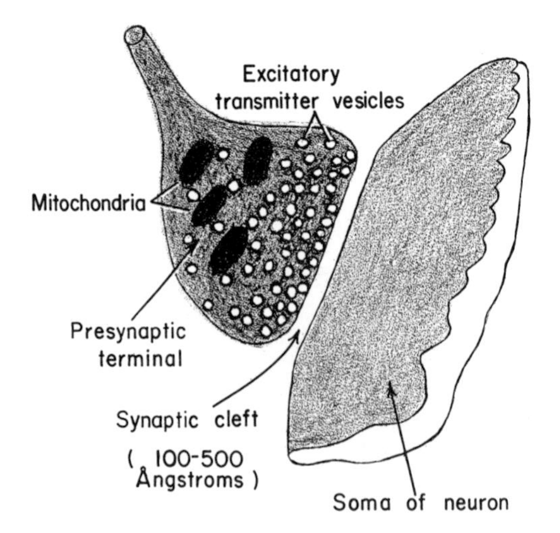

Figure 11–6 The physiological anatomy of a synapse. (From Guyton: *Textbook of Medical Physiology*, 4th Edition. Philadelphia, W. B. Saunders Company, 1971.)

If a neuron is experimentally stimulated near its midpoint, it will conduct action potentials in both directions. Only one side of a synapse has vesicles containing transmitter substance, however, so impulses can cross synapses in one direction only. Thus, synapses act as "valves" to control the direction of nerve impulses in the body.

The chemical nature of the transmitter substances within the central nervous system is not entirely known but it seems likely that there are at least three different substances present. Acetylcholine (Fig. 5–16) is known to be a transmitter substance at certain synapses outside the central nervous system (Sect. 12.6) and at most neuromuscular junctions; probably it also acts at some synapses within the central nervous system. Similarly, norepinephrine (Fig. 5–16) acts as a transmitter substance at some neuromuscular junctions (Sect. 12.6) and may also function at some central nervous system synapses. Gamma amino butyric acid (GAMA) is suspected as being a third transmitter substance within the central nervous system, and there may well be still others. Certainly inhibitory substances as well as stimulatory substances must exist since sometimes the result of nervous input to the central nervous system is inhibition of a process rather than stimulation of it. The reciprocal inhibition of antagonistic skeletal muscles is an example (Sect. 11.2).

Probably in all cases within the central nervous system many presynaptic neurons make contact with the same postsynaptic neuron. The response of the postsynaptic neuron is, then, the sum of the influences (stimulatory and inhibitory) of the presynaptic neurons. Similarly, each presynaptic neuron probably makes contact with a number of postsynaptic neurons. Such complex connections make possible a wide variety of postsynaptic responses (Fig. 11–7) as follows:

1. Postsynaptic neuron #1 receives only stimulatory influence. Assuming then, that a sufficient amount of transmitter substance is re-

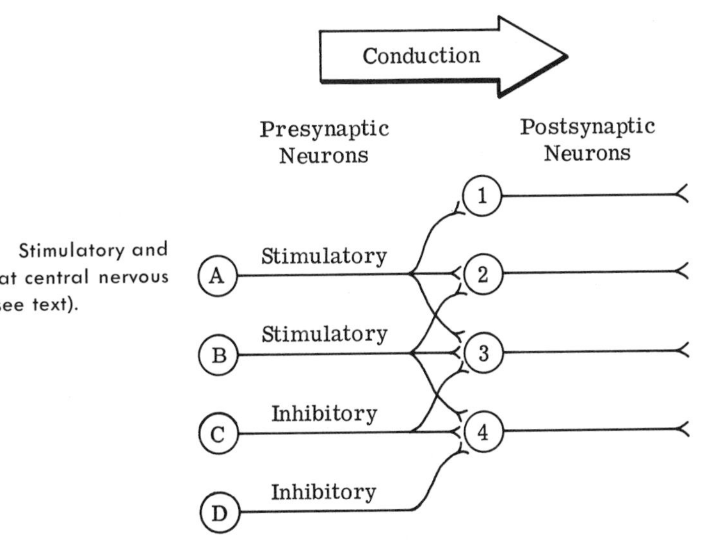

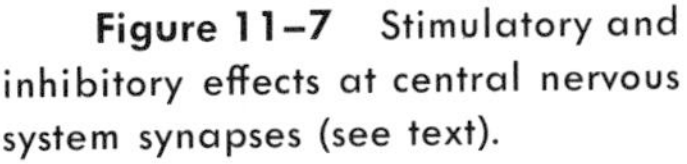
Figure 11–7 Stimulatory and inhibitory effects at central nervous system synapses (see text).

leased from presynaptic neuron "A," it will respond in a like manner. The postsynaptic neuron may, however, require a particular minimum number of action potentials in neuron "A" within a short time period (impulse frequency) to build up a sufficient concentration of transmitter substance in the synaptic cleft to accomplish this. Thus, the additive effects of a number of presynaptic action potentials may produce only a single postsynaptic impulse. This effect is called temporal (time) summation. The number of presynaptic impulses required to produce postsynaptic activity depends, presumably, upon the stimulus threshold (irritability) of the postsynaptic neuron. For example, a very light touch on the skin may evoke only a few sensory action potentials and, therefore, fail to produce any conscious awareness or response to the stimulus. A more forceful touch, however, evokes a greater frequency of sensory potentials, thus leading to temporal summation at some central synapse and, consequently, awareness and response to the stimulus. This mechanism acts to damp out small, unimportant inputs so that the central nervous system is not constantly bombarded with confusing and nearly useless information.

2. Neuron #2 receives stimulatory influences from two presynaptic neurons "A" and "B." If either of these is active alone, the amount of transmitter substance released into the synaptic cleft may be insufficient to stimulate the postsynaptic neuron. If both "A" and "B" are simultaneously active, however, the effects may be added together and produce action potentials in neuron #2. This mechanism is much like the one described in the preceding paragraph except that two presynaptic neurons are involved and they may bring sensory information from quite different portions of the body. Consequently, this kind of summation is called spatial (space) summation. For example, a slight tickling in a nostril may be insufficient to elicit a sneeze reflex. If, however, the person simultaneously looks at a bright light, spatial summation occurs at some central nervous synapse where both sensory pathways converge and a sneeze occurs. This is a familiar phenomenon among people living in bright tropical, desert, or seacoast areas where sneezing frequently occurs whenever one steps outside into the sunlight. This is especially true of blue-eyed people because their eyes are more sensitive to bright light. Similarly, simultaneous stimulation of the foot and the belly of a pithed frog will produce flexion of a hind leg even when the stimulus intensities are so low that neither stimulus alone is adequate to produce the response. This is also an example of spatial summation at some central nervous synapse.

3. Neuron #3 receives stimulatory influence from neurons "A" and "B" and inhibitory influence from neuron "C." It may be that this neuron will respond to input from either "A" or "B" alone unless there is a simultaneous input from neuron "C." Even then, of course, it might respond if both "A" and "B" are active. The sneeze reflex can be inhibited by firm pressure exerted by a finger on the upper lip; this is sufficient to offset the effects of either nostril tickling (if not too intense) or bright light alone, but it may not be sufficient to inhibit both of these stimulating influences simultaneously.

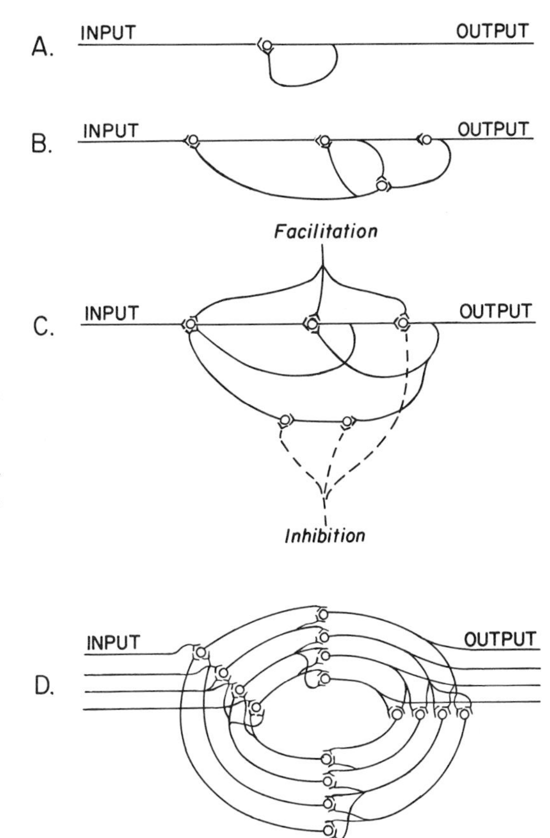

Figure 11–8 Neuron circuits within the central nervous system illustrating the mechanism of after-discharge. (From Guyton: *Textbook of Medical Physiology,* 4th Edition. Philadelphia, W. B. Saunders Company, 1971.)

4. Neuron #4 receives stimulatory influence from presynaptic neuron "B" and inhibitory influence from neurons "C" and "D." It may be that it will respond to input from "B" only when both "C" and "D" are both inactive; that is, either "C" or "D" alone may be sufficient to cancel the stimulatory effect of "B." On the other hand, it may be that the combined influence of both "C" and "D" are required to cancel the effects of "C."

Thus, a variety of responses and non-responses can be "programmed" into the central nervous system by the anatomical associations of neurons with differing thresholds (irritabilities) and transmitter substances.

Sometimes the activity of the postsynaptic neuron may outlast, in time, the activity of the presynaptic neurons. This results in a phenomenon known as after-discharge, which presumably can be explained on the basis of neuron connections of a special kind. Such connections are called reverberating or self-reactivating circuits (Fig. 11–8). In these circuits postsynaptic neurons repeatedly restimulate presynaptic neurons (and therefore themselves) until synaptic fatigue or other factors interfere. Synaptic fatigue probably results from the temporary exhaustion of the supply of transmitter substance available.

Transmission across a synapse requires more time than conduction of an action potential for a comparable distance along a neuron. The greater the number of synapses in a given nervous pathway, the more synaptic delay is introduced. Use of such synaptic delay is presumably involved in the timing of repetitive actions of certain kinds. The function of the pneumotaxic center in timing the rate of lung ventilation (Sect. 3.9) is an example.

The central nervous system has several levels of control. The cerebrum (Chapter 12) is the highest level and exerts overriding influences on all lower levels. The midbrain, medulla oblongata, cerebellum, and spinal cord are successively lower levels of the hierarchy and can exert generalized inhibitory or facilitatory influence on reflex activities at lower levels.

Repetitive stimulations of the same kind over a period of time may result in an increase in the number of presynaptic terminals at synapses along the pathway involved. As a result, it becomes progressively easier for signals to be transmitted across these synapses. It is probably this kind of development that allows for improvement of skills through practice, and it may be that similar synaptic changes are involved in the establishment of certain kinds of conditioning.

CHAPTER 12

THE NERVOUS SYSTEM: FUNCTIONAL ANATOMY

12.1 INTRODUCTION

The parts and regions of the vertebrate nervous system can be classified in a number of ways. A useful and convenient classification for our purposes (Fig. 12–1) is based partly on anatomical and partly on functional considerations. In succeeding sections of this chapter we shall discuss the human nervous system, since it is better known than other vertebrate nervous systems, in the same sequence as they are listed in our classification scheme. The student will want to refer to this scheme from time to time as he reads this chapter.

All vertebrates, with certain exceptions, have the same fundamental parts and regions of the nervous system, although they are developed to different degrees in different species. There are four major divisions of the vertebrate central nervous system: (1) the prosencephalon or forebrain, (2) the mesencephalon or midbrain, (3) the myelencephalon or hindbrain, and (4) the spinal cord (Fig. 12–2). The remaining portions of the nervous system, outside the central nervous system, are collectively referred to as the peripheral nervous system. This consists largely of nerve tracts leading to and from receptors and effector organs. The peripheral nervous system is composed, typically, of 12 pairs of cranial nerves associated with the brain and variable numbers (in different vertebrates) of spinal nerves that occur in pairs along the spinal cord. The entire central nervous system is hollow, the central cavity varying from a tiny central canal in the spinal cord to relatively large ventricular spaces in the brain. Also, the entire central nervous system is covered by layered membranes called meninges. Both the internal cavities and the spaces surrounding the central nervous system between the layers of meninges are filled with a special cerebrospinal fluid.

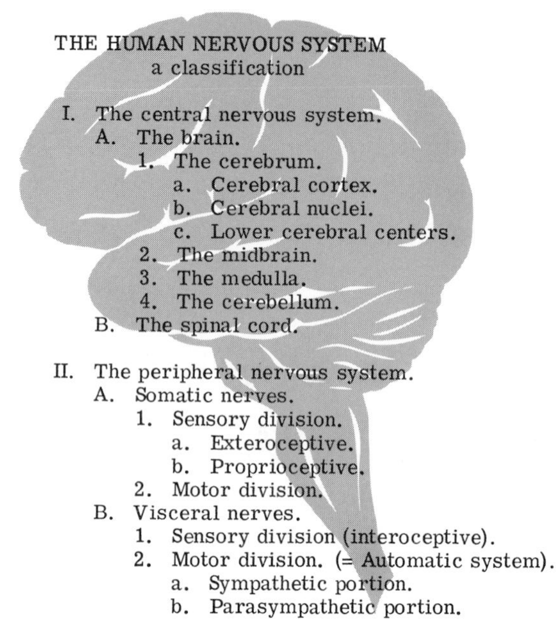

THE HUMAN NERVOUS SYSTEM
a classification
I. The central nervous system.
A. The brain.
1. The cerebrum.
a. Cerebral cortex.
b. Cerebral nuclei.
c. Lower cerebral centers.
2. The midbrain.
3. The medulla.
4. The cerebellum.
B. The spinal cord.
II. The peripheral nervous system.
A. Somatic nerves.
1. Sensory division.
a. Exteroceptive.
b. Proprioceptive.
2. Motor division.
B. Visceral nerves.
1. Sensory division (interoceptive).
2. Motor division. (= Automatic system).
a. Sympathetic portion.
b. Parasympathetic portion.

Figure 12–1 A classification of the parts of the nervous system. See text.

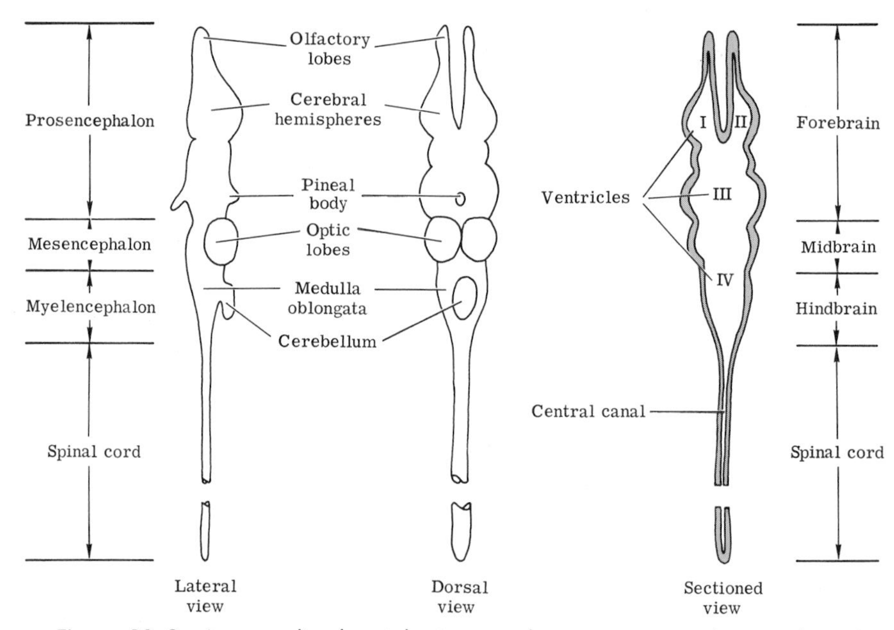

Figure 12–2 A generalized vertebrate central nervous system showing the relationships of the various parts. (After Cockrum and McCauley: *Zoology*. Philadelphia, W. B. Saunders Company, 1965.)

12.2 THE HUMAN BRAIN: STRUCTURE

The human brain consists of the same general regions as the brains of lower vertebrates. The great development of the cerebral hemispheres (part of the forebrain) plus the repositioning of parts to accommodate the upright posture of man somewhat distorts the pattern, however (Fig. 12–3). The cerebral hemispheres hide from view the other portions of the forebrain, most of the midbrain, and a part of the hindbrain when the entire brain is seen in lateral view. A part of the medulla oblongata and the cerebellum (both portions of the hindbrain), however, can be seen.

The cerebrum contains two large cavities (ventricles I and II) that lie lateral to the midline and a third cavity (ventricle III) in the midline. Much of the substance surrounding these cavities is composed of nerve tracts and collections of nerve cell bodies; the latter lie in regions called ganglia or nuclei. The exposed surface of the cerebrum consists of nerve cell bodies arranged to form the cortex of the cerebrum. This surface cortex is marked by a number of deep fissures and shallow sulci that give it a convoluted appearance. The convolutions between sulci (singular sulcus) and fissures are called gyri (singular gyrus). At first glance these appear to be distributed in a random pattern. Actually the same pattern of fissures, sulci, and gyri occurs on every human brain, and each of these structures is given a definite name by anatomists. We will be concerned here with only a few names of landmarks on the surface of the cerebrum.

The right and left cerebral hemispheres are separated from each other by a deep sagittal fissure. Each hemisphere is further divided by

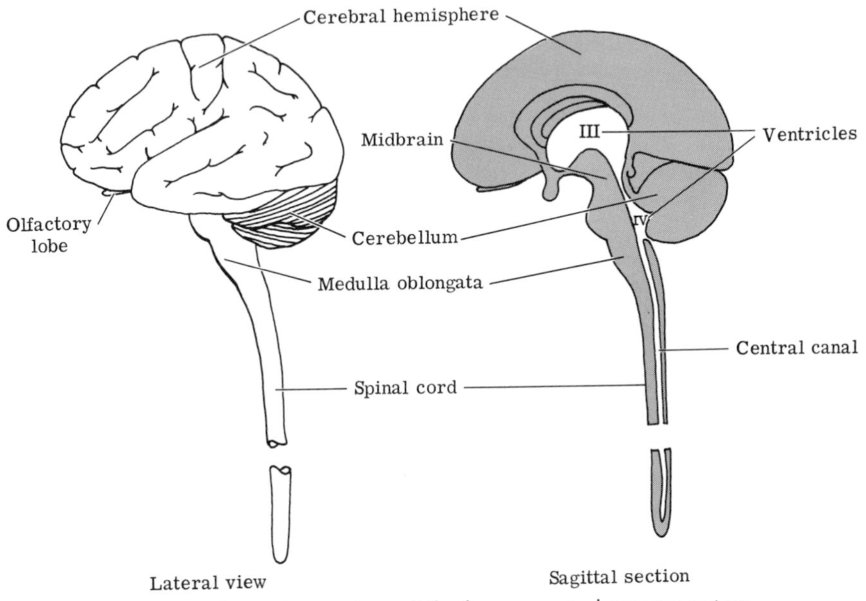

Figure 12–3 Major portions of the human central nervous system.

a vertical central fissure and a diagonal lateral fissure. These, together with other more arbitrary landmarks, divide the cerebral cortex into four immediately visible lobes (Fig. 12–3), namely the frontal lobe anteriorly, the temporal lobe laterally, and the parietal and occipital lobes posteriorly. When the margins of the lateral fissure are spread apart, it is possible to see the surface of a fifth lobe, called the central lobe, hidden beneath overlying extensions of the frontal and temporal lobes.

The cavities within the cerebral hemispheres (ventricles I and II) extend forward into the frontal lobes, backward through the parietal lobes and into the occipital lobes (Fig. 12–4). They also have extensions laterally and anteriorly into the temporal lobes. Each of these lateral ventricles communicates through a small interventricular foramen (hole) with the single median cavity (ventricle III). This third ventricle lies between and slightly below the level of the lateral cavities. From

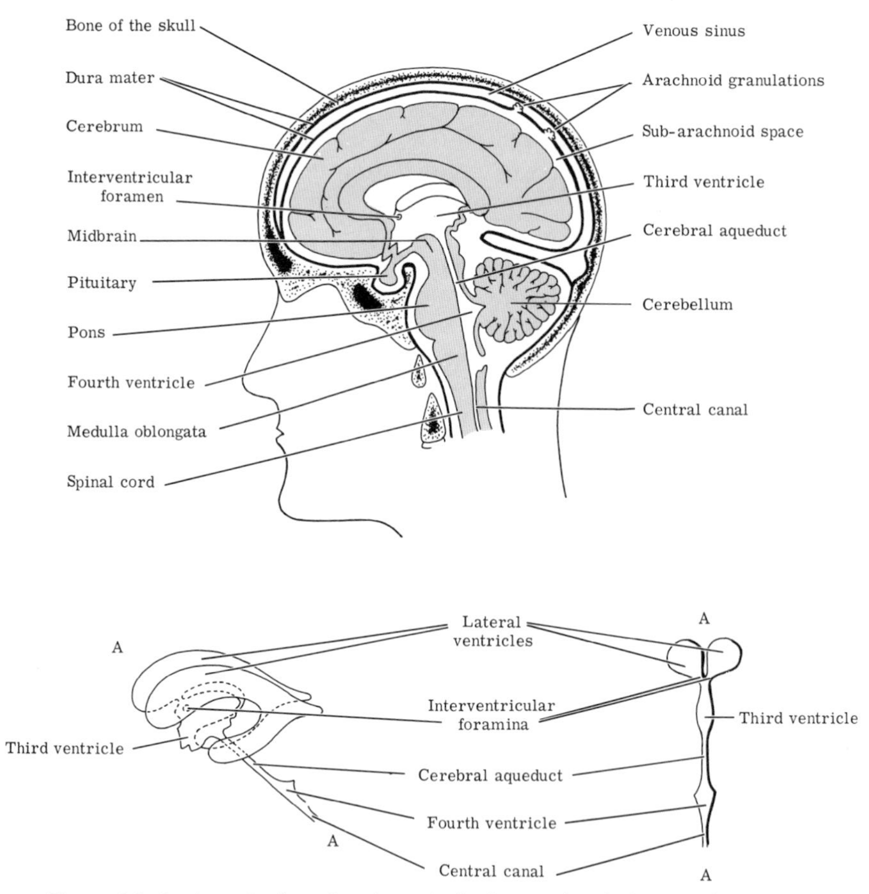

Figure 12–4 A sagittal section through the human head showing the relationships of the brain and its internal cavities to external structures and spaces. Insets show the internal cavities in lateral and frontal views. AA indicates the same axis in each inset. (From Cockrum and McCauley: *Zoology*. Philadelphia, W. B. Saunders Company, 1965.)

the third ventricle a narrow canal, the cerebral aqueduct, passes through the midbrain region to open into a fourth cavity (ventricle IV) located in the hindbrain. The fourth ventricle, in turn, communicates with the tiny central canal of the spinal cord and also opens to the surface of the brain through three small openings, namely the foramina of Luschka and Magendie.

The outer surface of the entire central nervous system is covered by a thin, closely adherent membrane called the pia mater. This membrane dips into all of the fissures and sulci following faithfully the contours of the central nervous system (Fig. 12–5). Over the pia mater is a second membrane called the arachnoid membrane. It is attached to the pia mater by delicate spider-web-like strands of tissue; the space between the arachnoid membrane and the pia mater is crossed only by these strands. This is the subarachnoid space, and like the internal ventricles and passages, it is filled with cerebrospinal fluid. The arachnoid membrane dips into the fissures and the larger sulci but does not follow as closely the contours of the central nervous system as does the pia mater. Thus, the subarachnoid space is relatively large in certain places. Finally, a third membrane, the dura mater, encloses the whole. The dura mater is a tough, thick layer which is doubled in some places. In such places the inner layer dips into the major fissures of the brain while the outer layer follows more closely the inner surface of the skull. The dura mater is, indeed, continuous with the connective tissue covering (lining) of the bones surrounding the central nervous system. This connective tissue, which covers all bones, is called the periosteum. In

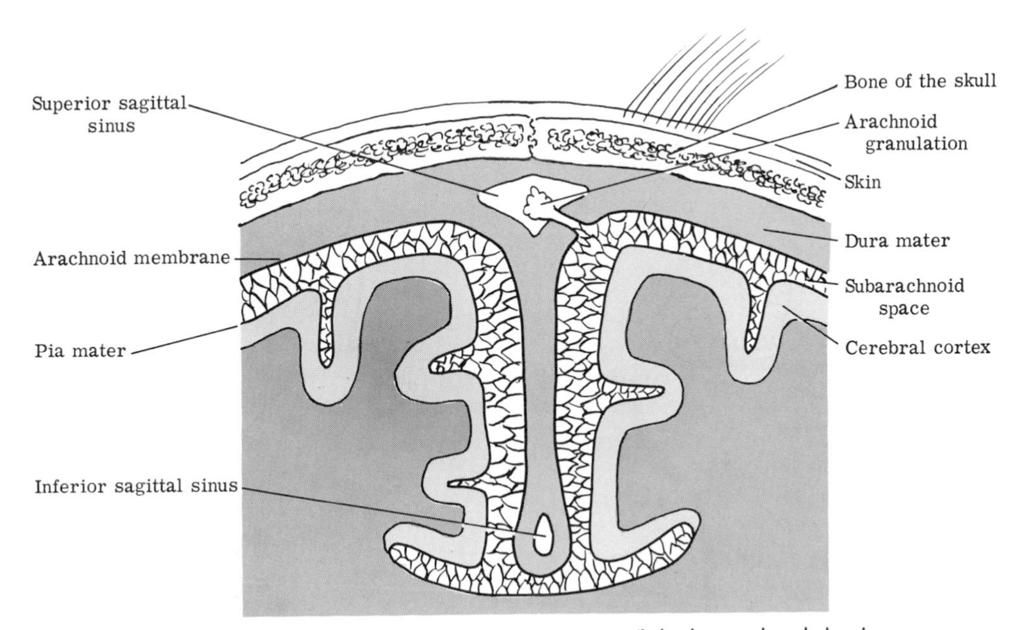

Figure 12–5 A frontal section through the upper part of the human head showing the relationships of the brain to the skull and meninges. (From Cockrum and McCauley: *Zoology*. Philadelphia, W. B. Saunders Company, 1965.)

regions where the dura mater is doubled there are relatively large venous blood sinuses between the two dural layers that serve as the major routes of venous drainage from the brain. Like the two other layers of meninges described above—the pia mater and the arachnoid membrane—the dura mater also extends downward to envelop the spinal cord. In this region it even extends laterally to enclose the first portion of each spinal nerve.

Branches of cerebral arteries are closely associated with the walls of all four of the internal ventricles of the brain. The endothelium of associated capillaries together with the lining membranes of the ventricles (ependyma) form a specialized membrane in the part of each ventricle in which the cerebrospinal fluid, a special filtrate of blood plasma, is formed. This specialized membrane is called the choroid plexus, and the cerebrospinal fluid formed by it is continuously secreted into the cavities of the ventricles. As the fluid is produced it flows through the communicating passages between the ventricles and escapes through the openings from the fourth ventricle (foramina of Luschka and Magendie) into the subarachnoid spaces surrounding the central nervous system. In the regions of the venous blood sinuses the arachnoid membrane here and there penetrates through the inner layer of dura mater and into the blood sinuses. At these points, called arachnoid granulations, cerebrospinal fluid passes from the subarachnoid spaces into the venous blood and is carried away with the venous drainage. Thus, cerebrospinal fluid, derived from arterial blood through the choroid plexus, circulates through the interior of the central nervous system, bathes the exterior of the central nervous system, and is ultimately returned to the blood through the arachnoid granulations. This fluid is like blood plasma in many respects but it also has some important differences, *i.e.*, it lacks blood cells and and its protein content is quite different. Since it also lacks the buffer systems that are present in blood, its pH is altered by changes in CO_2 concentration. This is probably an important factor in the controls of respiration (Sect.

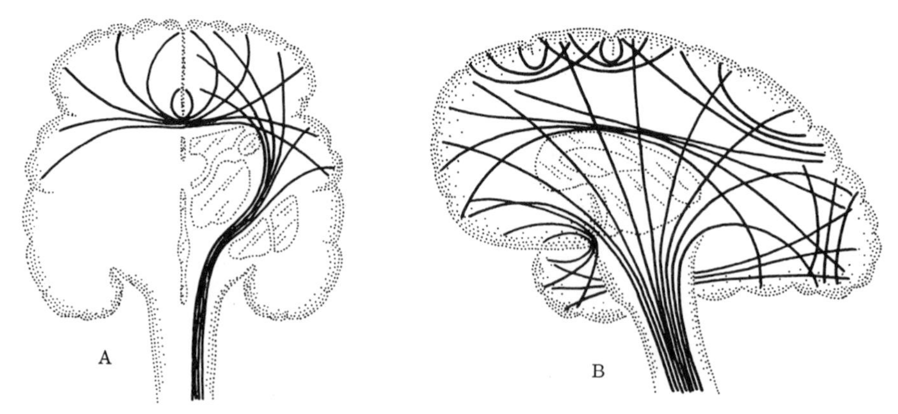

Figure 12–6 Nerve interconnections (association fibers) within the human central nervous system. *A*, frontal view; *B*, lateral view.

3.9), of blood pressure (Sect. 5.11), and perhaps in other ways as well. The special permeability characteristics of the choroid plexus have given it the designation of "blood-brain barrier." It is important, for example, to determine whether or not drugs of various kinds will cross the "barrier" and thus be effective in the treatment of certain central nervous system diseases.

Within the cerebrum are many nerve tracts leading to and from lower levels of the central nervous system and interconnecting various parts of the cerebrum itself (Fig. 12–6). Besides these tracts there are also several internal concentrations of nerve cell bodies (nuclei or ganglia) including the thalami, the caudate nuclei, the lentiform nuclei, the hypothalami, and others.

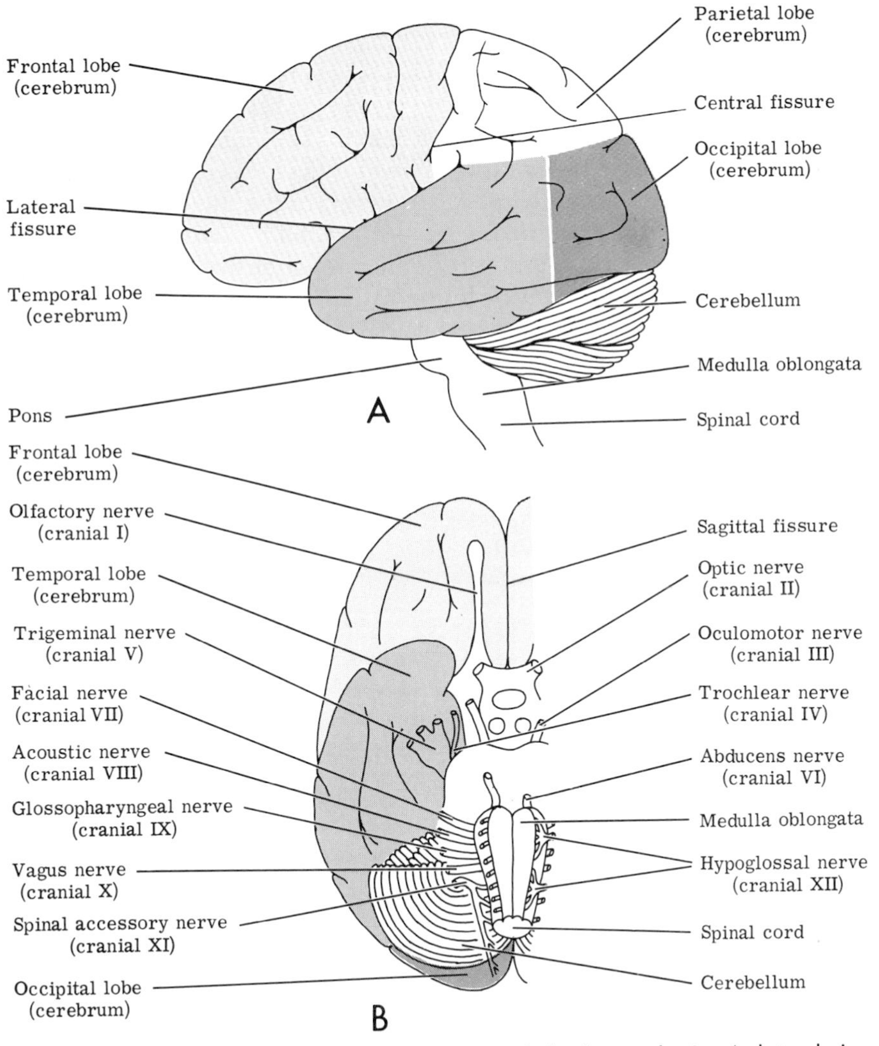

Figure 12–7 The general surface features of the human brain. *A*, lateral view; *B*, view from below. (From Cockrum and McCauley: *Zoology*. Philadelphia, W. B. Saunders Company, 1965.)

From the floor of the forebrain extends the stalklike infundibulum leading to the pituitary gland or hypophysis (Fig. 12–7). Also prominently visible on the floor are the olfactory bulbs and tracts, the optic chiasma (where the right and left optic nerves meet to exchange fibers), the small round mammillary bodies, and the other pairs of cranial nerves.

The cerebrum and other portions of the forebrain are connected through the midbrain, or mesencephalon, to the hindbrain. The mesencephalon contains, of course, many nerve tracts connecting higher and lower portions of the central nervous system but it also contains several important nuclei that have special functions. Dorsal swellings on the surface of the mesencephalon, the corpora quadrigemina, contain nuclei that function in the integration of reflexes of body balance and optical orientation. Other nuclei within the mesencephalon serve other fundamental integrative purposes.

The medulla oblongata, like the midbrain, contains nerve tracts leading to higher and lower portions of the central nervous system as well as concentrations of nuclei that serve very fundamental nervous reflexes. The centers of control for respiration (Sect. 3.9) and blood pressure (Sect. 5.11), for example, are found in the medulla oblongata. Other nerve tracts pass horizontally through and around the medulla, linking together the right and left hemispheres of the cerebellum and forming the prominent anterior bulge called the pons cerebelli.

The cerebellum is the other part of the hindbrain and extends as paired hemispheres from the dorsal surface of the medulla oblongata. It is marked by long, narrow surface gyri separated from each other by deep sulci or fissures. These sulci branch below the surface, giving the cerebellum a very large surface area (cortex). Nerve tracts linking the cerebellum with other portions of the central nervous system pass upward and downward in the roof (dorsal wall) of the fourth ventricle (Fig. 12–4). The cerebellum also contains concentrations of nerve cell bodies that function in connection with the fine integration of skeletal muscle activity.

12.3 THE HUMAN BRAIN: FUNCTION

Through experimentation and by the careful examination of the brains of persons with known functional losses, it has been possible to map the functional areas of the cerebral cortex (Fig. 12–8). Parts of the cerebral cortex function in ways that are difficult to quantify because they involve psychic factors that are not easily defined. Such characteristics as introspection, foresight, moral inhibition, ambition, conscious pleasure, and so on are relegated to these so-called psychic areas of the cerebral cortex. The cortex of part of the frontal lobes, most of the temporal lobes, and much of that within the sagittal fissure is psychic in function.

The occipital cortex is the region of visual reception and interpretation. A central portion of it receives, initially, nerve impulses from the

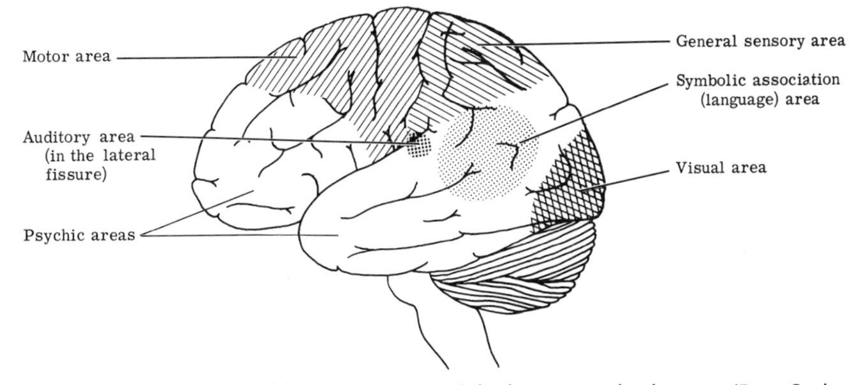

Figure 12–8 Principal functional areas of the human cerebral cortex. (From Cockrum and McCauley: *Zoology*. Philadelphia, W. B. Saunders Company, 1965.)

retinas of the eyes, and a surrounding portion interprets this pattern of impulses in terms of visual experience and learning. It is in this surrounding area, which we may call the visual association area, that conscious impressions of vision occur.

A similar primary auditory receiving area, surrounded by an auditory association area for conscious interpretation of sound, occurs on the superior portion of the temporal lobe. This area is not immediately visible in a lateral view of the brain, however, since it lies within the lateral fissure.

The primary cortex for the receipt of general cutaneous sensations, including touch, temperature, and so forth, occupies the postcentral gyrus – the gyrus which is immediately parallel to the central fissure and is a part of the parietal lobe of the cerebrum. Posterior to this primary area and covering much of the anterior and superior portions of the parietal lobe is the conscious association area for these same sensations.

The posterior and superior portions of the frontal lobe of the cerebrum contain the cortical areas concerned with the conscious production of muscular activity. These are arranged in much the same way as the sensory areas just described, although these are motor in function. It is the cortex of the precentral gyrus – the gyrus immediately anterior and parallel to the central fissure – that initiates the nerve impulses causing voluntary muscle contractions. The remaining motor association area of the frontal lobe cortex contains the learned patterns of activity, *i.e.*, the memory of how to carry out various muscle activities.

Probably the most outstanding human attribute not shared by other vertebrates is his ability to use symbolism. The words on this page are meaningful to the reader only because he has learned to interpret the alphabetical symbols of language as he sees them in the particular sequences that are symbolic of words. The vocal sounds which we use in speaking are similarly only auditory symbols, and they acquire meaning for us only after we learn to associate them with the things and the activities which they represent in the real world about us. Not only do we learn to interpret these symbols of printing and of sound, but we also learn to reproduce them through the physical activities of writing and speaking. In addition, the blind learn to interpret Braille symbols through the sense of touch, and deaf-mutes come to express the symbols of language through movements of the fingers. Thus, virtually all of the conscious senses, as well as the voluntary motor regions of the cerebral cortex of humans, are potentially involved in language symbolism. It is interesting, therefore, to note that a central area of the parietal lobe cortex is apparently involved in this special interpretive function. This symbolic language association area, because of its central position, lies immediately adjacent to the visual, auditory, and cutaneous sensory association areas (Fig. 12–8). Moreover, there is a special nerve tract linking it with the frontal cortex areas, which are concerned with movements of the vocal muscles and those of the hands, *i.e.*, the muscles used in producing language symbols for others to see or hear.

Each of the functional areas of the human cerebral cortex has a

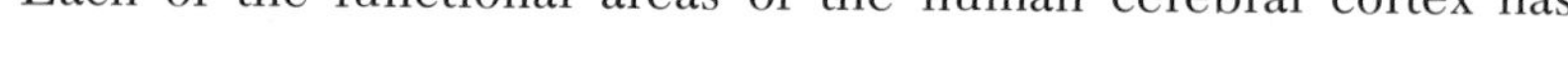

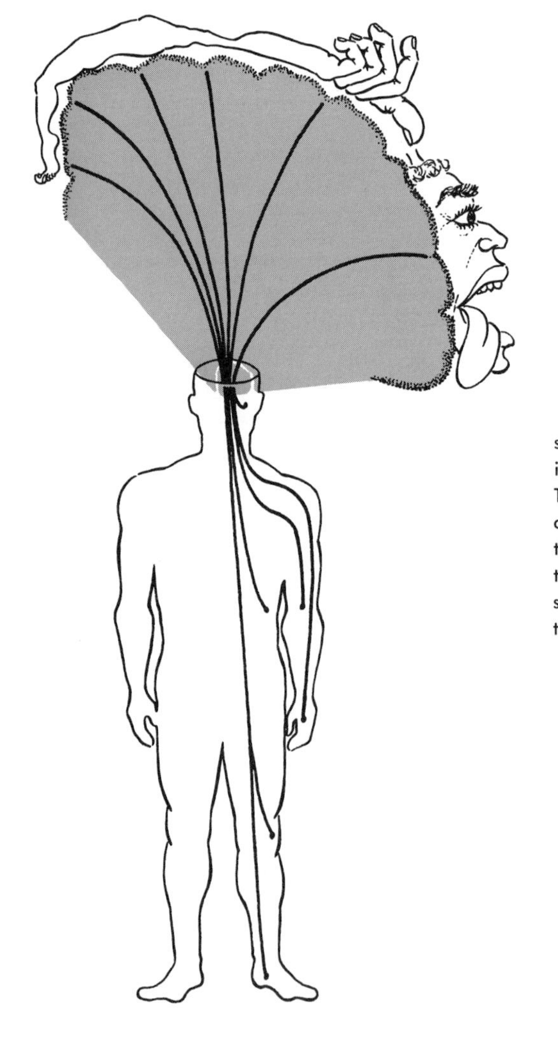

Figure 12–9 Lamination of sensory and motor nerve tracts leading to and from the cerebral cortex. The human figure sketched on the cortex is distorted in proportion to the relative cortical areas devoted to each body part, and this corresponds with the density of innervation of each body part.

more or less definite internal organization in which body parts are arranged in sequence. For example, both the primary motor area (precentral gyrus) and the primary cutaneous sensory area (postcentral gyrus) contain sequential regions connected with the various body parts in almost exactly the same order as these body parts occur (Fig. 12–9). This is not really surprising when one considers that nerve tracts remain in laminar organization as they ascend or descend through the central nervous system. It is interesting to note, moreover, that body portions which are heavily innervated with either cutaneous sensory or skeletal muscle motor nerves are represented by correspondingly larger cortical regions in these primary areas. In a similar way the various portions of the retinas of the eyes are represented by corresponding portions of the occipital cortex, and the various portions of the organ of hearing of the ears (Sect. 14.5) are represented by sequential portions of the auditory cortex.

The internal nuclei or ganglia of the cerebrum (thalamus, lentiform nucleus, and so on) are functional in the control of gross, stereotyped motor activities in man. In lower vertebrates, in which the cerebral cortex is less well developed, the thalami apparently have greater importance. In these animals the thalami are nearly the highest brain center and they control all motor activity; they probably also interpret most sensory input as well. In humans sensory input passes through the thalami on the way to the cortex and, similarly, cortical motor output is relayed through the thalami on its way to lower levels of the central nervous system. Much remains to be learned about the details of function of these internal accumulations of nerve cell bodies in the human cerebrum.

Beneath the cerebral ganglia and nuclei just discussed is another complex group of nuclei collectively called the hypothalamus. This contains centers that control the activities of the pituitary gland (Sect. 17.2) and a variety of other fundamental processes. It is apparent that sleep and wakefulness, thermoregulation (Chapter 18), hunger, and perhaps other states are controlled by these subconscious centers in the hypothalamus.

As was pointed out in the preceding section, the functions of the midbrain and medulla oblongata, other than the conduction of nerve impulses between higher and lower levels of the central nervous system, are associated with very fundamental life processes. These include the regulation of respiratory and cardiovascular activities, postural balance, and control of the visual reflexes involved in focusing the eyes (Chapter 15).

The role of the cerebellum is an extremely interesting one. It will be recalled that the precentral gyrus of the cerebral cortex initiates nerve impulses that produce voluntary movements of skeletal muscles. A pattern of impulses originates there and is conducted into the spinal cord and out through various spinal nerves to appropriate muscles of the body. This pattern of impulses arising in the cerebral cortex, however, merely produces a rather gross muscular activity, and it is necessary to modify the pattern as the activity progresses in order that it be smooth and effective. The cerebellum monitors the motor output of the cerebral cortex and also monitors its effects. It receives feedback from the various sensory receptors within the muscles, their tendons, and the joints of the skeleton (Sect. 13.11) so that it is constantly "informed" of the progress of the activity induced there. From this information it adjusts and modifies the motor output so that the activity is carried out smoothly and purposefully. For example, if one reaches to pick up a pencil from the desk, the gross movement of reaching and grasping arises from the cerebral cortex, but the fine coordination of the act is provided by the cerebellum. Following certain types of cerebellar damage, the individual tends first to overshoot and then to undershoot the "target" with the result that tremors of the hand and fingers make the fine movements impossible so that the pencil cannot be picked up.

Many more details of human brain function are known, of course, and the interested student is referred to any good textbook of neuroanatomy or neurology for further information on the subject.

12.4 THE HUMAN SPINAL CORD

The spinal cord is composed in part of tracts of nerve fibers carrying both ascending and descending nerve impulses. It also contains regions relating to the coordination of simple reflex responses (Sect. 11.2).

In the adult human the spinal cord passes through the neural canal of the vertebral column only as far as the last thoracic vertebra (Fig. 12–10). In the embryo it extends the full length of the vertebral column, but the bony vertebral column grows more rapidly in length than does the spinal cord, thus leading to the adult condition. Also in the embryo, the paired spinal nerves leave the cord laterally through intervertebral foramina and travel thence to structures of the body which they innervate. As the spinal cord falls behind in its lengthwise growth, however, the lower spinal nerves must travel some distance beyond the end of the cord and within the vertebral canal before reaching the intervertebral foramina through which they exit. Because of this the vertebral canal of the lumbar and sacral regions contains a bundle of spinal nerves rather than the spinal cord itself. This bundle of spinal nerves is called the cauda equina or "horse's tail." Since the meninges covering the brain extend over the spinal cord and for a short distance along all spinal nerves, the cauda equina is also contained within a meningeal sac. This sac is, of course, filled with cerebrospinal fluid.

In cross section the spinal cord varies somewhat in shape and size from level to level but it presents the same general features everywhere (Fig. 12–11). The outer portions are composed of "white matter" and contain the ascending and descending nerve tracts. This portion may be divided into dorsal, lateral, and ventral white columns. The more centrally located "grey matter' is made up of nerve cell bodies and their connective (synapsing) processes. This region is in the form of a butter-

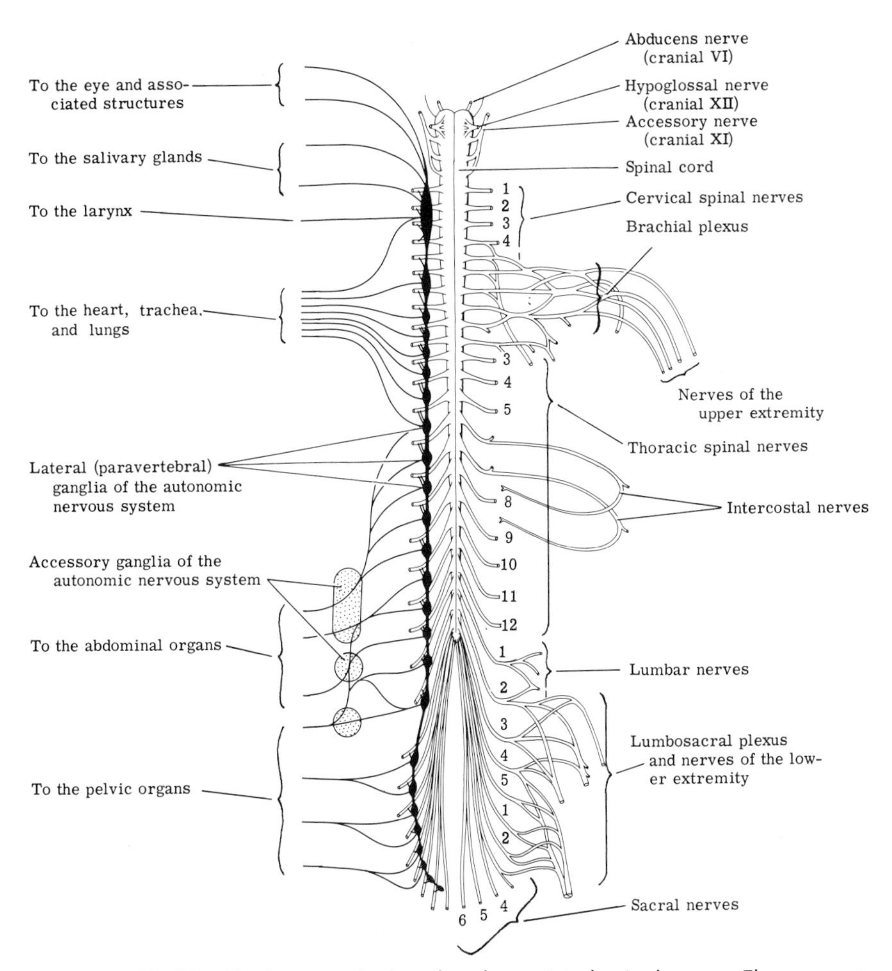

Figure 12–10 The human spinal cord and associated spinal nerves. The paravertebral ganglia and their connectives are shown in black on the left side, and the nerve plexuses associated with the body extremities are shown on the right side (diagrammatic). (From Cockrum and McCauley: *Zoology*. Philadelphia, W. B. Saunders Company, 1965.)

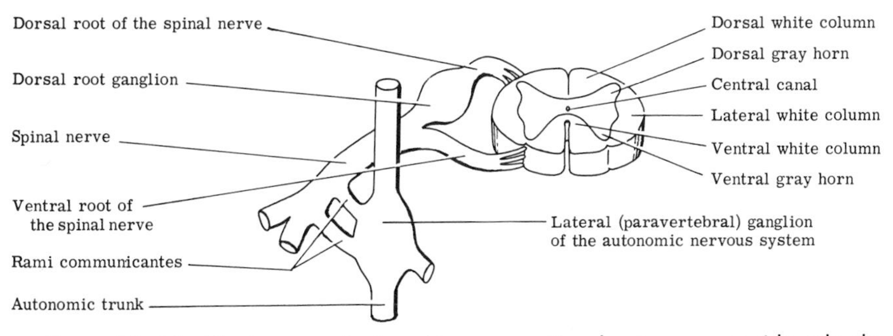

Figure 12–11 The human spinal cord in cross section showing, on one side only, the spinal nerve connectives. (From Cockrum and McCauley: *Zoology*. Philadelphia, W. B. Saunders Company, 1965.)

fly, or the letter H, and can be subdivided into dorsal, lateral, and ventral grey horns. It is penetrated at its center by the tiny central canal of the spinal cord. This, it will be recalled, is continuous above with ventricle IV of the brain.

Near the end of the dorsal grey horns are attached the dorsal roots of the spinal nerves on each side of the spinal cord. Similarly, near the ends of the ventral grey horns are attached the ventral roots of the spinal nerves. The dorsal and the ventral roots join each other on each side of and lateral to the spinal cord to form the spinal nerves themselves.

12.5 THE PERIPHERAL SOMATIC NERVES

The dorsal roots of the spinal nerves (Fig. 12–11) are each marked by a prominent swelling called the dorsal root ganglion. This ganglion contains the cell bodies of all sensory nerves. Only sensory neurons pass through the dorsal root, and all such neurons, therefore, enter the spinal cord near the dorsal grey horn. The ventral roots of the spinal nerves, on the other hand, contain only motor neuron tracts. The cell bodies of these neurons lie in either the lateral or the ventral grey horns of the spinal cord. Those associated with the somatic nerves (nerves carrying motor impulses to skeletal muscles) are located in the ventral grey horns, and those associated with the visceral nerves or autonomic system (nerves carrying impulses to smooth muscles) are located in the lateral grey horns. The spinal nerves, which are formed by the junction of the dorsal and ventral roots, then, contain mixtures of sensory and motor tracts of all kinds.

Laterally, each spinal nerve divides into dorsal and ventral primary rami, which lead, with many subdivisions of each, to the dorsal and ventral portions of the body respectively. Sensory nerves pass by these routes from receptors located in the skin (exteroceptors), which are sensitive to touch, temperature, pressure, and pain. Other sensory tracts arise, instead, from receptors located within skeletal muscles, joints, and tendons (proprioceptors) and carry impulses related to stretch, pressure, and the like. Somatic motor neurons also send processes through these same nerves to innervate the skeletal muscles of the body. In addition, some visceral sensory and motor nerves make use of the same routes since they lead to the smooth muscles of arteries and glands in those regions of the body.

12.6 THE PERIPHERAL VISCERAL NERVES

Associated with the vertebral column on each side is a connected series of lateral or paravertebral ganglia (Fig. 12–10). Each ganglion of these series is a collection of nerve cell bodies and synapses associated with the autonomic (visceral motor) system (Fig. 12–11). Each paravertebral ganglion connects with the corresponding spinal nerve through one or two (usually two) rami communicantes (Fig. 12–11). Each is also connected, by way of a peripheral nerve, to some visceral (or cardiac) muscle of the body.

Autonomic nerve fibers originating from nerve cell bodies in the lateral grey horns of the thoracic and lumbar portions of the spinal cord are known as sympathetic nerves. The preganglionic neuron, as the sympathetic cell body in the cord is called, sends its process out through the ventral (motor) root of the spinal nerve and through one of the rami communicantes to the correspondlng paravertebral ganglion. Here most preganglionic neurons end by synapsing with the cell body of a postganglionic neuron. The axon of the postganglionic neuron then passes through a peripheral nerve to innervate some smooth or cardiac muscle. There are minor anatomical variations which have no real physiological significance. For example, the pre- or postganglionic neuron may either ascend or descend for some distance through the connected series of paravertebral ganglia to synapse or to exit at a different level. Also, the axon of a postganglionic neuron may reenter the spinal nerve through the second of the rami communicantes if it is destined to innervate a smooth muscle which is best reached by way of a branch of the spinal nerve, such as the smooth muscle in the walls of an artery associated with skeletal muscle or a gland located in the skin.

Autonomic nerve fibers originating from cell bodies in the sacral portion of the spinal cord or in the cranial portion (the brain) are called parasympathetic nerves. Those of the sacral region follow pathways similar to those described for sympathetic nerves in the preceding paragraph except that no synapsing occurs in the paravertebral ganglia. Instead, the axon of the preganglionic neuron passes on through the paravertebral ganglion and its peripheral nerve and reaches the postganglionic nerve cell body only when it is in or very close to the smooth muscle to be innervated. There it contacts a very short postganglionic neuron. Parasympathetic neurons arising in the brain have similar connections but travel through cranial nerves rather than spinal nerves. The tenth cranial nerve, the vagus, carries a large number of these and descends through the neck and trunk to innervate most of the viscera of the thoracic and abdominal cavities. Acetylcholine is the transmitter substance operative at all pre- to postganglionic autonomic synapses (Sect. 11.4).

All of the smooth muscles of visceral organs located in the thoracic, abdominal, and pelvic cavities receive both sympathetic and parasympathetic innervation. The two divisions of the autonomic system are antagonistic in their effects upon any given visceral organ. Sympathetic stimulation dilates the pupil of the eye, inhibits salivation, and raises the blood pressure by speeding the heart and contracting the smooth muscles in the walls of arterioles. Parasympathetic stimulation, on the other hand, constricts the pupil of the eye, stimulates salivation, and lowers the blood pressure by slowing the heart. Probably it does not cause active dilation of arterioles. No parasympathetic innervation of arterioles is necessary since the distending blood pressure within will cause vascular dilation if only active, sympathetic constriction ceases (Sect. 5.11). Most likely in any given instance of dual autonomic innervation, the end result depends upon the ratio of nerve impulse frequencies reaching the muscle by way of the two routes. That is, the pupil diameter or the heart rate at any given moment is probably established by the "balance of power" between the two opposing innervations. When events make a speeding of the heart physiologically desirable, for example, it is probable that there is a simultaneous increase in sympathetic impulse frequency and a decrease in parasympathetic impulse frequency. Presumably this would lead to a different percentage of transmitter substance being released by each nerve ending at the neuromuscular junctions (Sect. 11.4). The transmitter substance at most sympathetic neuromuscular junctions is norepinephrine; the transmitter substance at most parasympathetic neuromuscular junctions is acetylcholine (Fig. 5–16). As we have already seen, the transmitter substance at all somatic motor neuromuscular junctions and at the pre-to-postganglionic synapses of the autonomic nervous system is also acetylcholine.

A given cardiac or smooth muscle responds differently to impulses reaching it by way of the sympathetic or the parasympathetic routes because the resulting transmitter substance is different. The muscle, then, is actually responding to the chemical transmitter, not to the nervous impulse *per se.* Transmitter substances acting at neuromuscular junctions in this way are often known as neurohumors.

Besides the exception to dual innervation noted for the smooth muscles of arteriole walls, another notable exception exists. The medulla of the adrenal gland (Sect. 17.9) receives only preganglionic sympathetic neurons. This gland is derived during embryonic development from the same tissues which give rise to sympathetic neurons, and it responds to stimulation by producing epinephrine, a compound closely related to norepinephrine. Thus, one might say that it is composed of modified postganglionic sympathetic cells—cells specialized for the production of large quantities of "transmitter substance" rather than for conduction of nerve impulses. The "transmitter substance" in this case, however, is not released into any neuromuscular junction but, rather, into the blood stream. Its effects everywhere in the body are to augment the action of the sympathetic nervous system.

As the student has perhaps already noted, the effects of sympathetic stimulation are, in every case, those associated with emergency stress situations. Dilation of the pupils of the eyes, inhibition of salivation, rising blood pressure accompanied by increased heart rate, increased ventilation rate and depth, and so forth are all responses preparing the individual to "fight or flee." In such emergency situations, the additional output of epinephrine from the adrenal medulla augments these responses.

12.7 COMPARATIVE NERVOUS SYSTEMS OF VERTEBRATES

The general trend throughout animal evolution has been toward a larger and more complex nervous system, and this trend is particularly marked in vertebrate evolution. With increasing complexity of the nervous system has come increased control of all physiological functions. Correlated with nervous system development has been the transfer of many functions from primitive lower centers to the more recently developed higher centers of the central nervous system. Another characteristic of nervous system evolution has been, of course, adaptation of development to the particular requirements of each vertebrate group (Fig. 12–12).

Olfaction is a more useful sense in water than is vision. The lighting, turbulence, and other factors of the aquatic environment limit optical range in most cases. Correlated with this situation is the relatively large development of olfactory lobes and the rather small optic lobes of typical fish brains. Amphibia retain a well-developed olfactory apparatus as do all higher vertebrate groups except birds. In addition, however, the characteristics of the terrestrial environment renders vision an extremely useful sense, and, consequently, greater development of the optic lobes of the brain is also seen in all terrestrial forms. The optic lobes of birds are particularly large since birds depend upon this sense more than do other vertebrates (Chapters 13 and 15).

Fishes have well-developed midbrain regions, and it is in this portion of the brain that integration of the acoustico-lateralis apparatus (Sect. 13.13) is achieved. The lateral line portion of this apparatus is an extremely important sensory system for the coordination of swimming activity in fishes. It is of less importance in amphibia and does not occur in higher vertebrates, however; thus, the midbrain region can be proportionately smaller.

Fishes also have well-developed cerebellar lobes since these too are involved in the coordination of swimming. The cerebellum, it will be recalled, functions in fine muscle coordination, an activity of importance to an animal that moves about in three dimensions. Fish can be thought of as flying in water, or, conversely, birds may be thought of as swimming through air. In either case the problems of muscle coordination are greater for these animals than for those which merely move about on the land surface. The cerebellar development is greatest in these animals and in representatives of other groups (reptiles, mammals) that also swim or fly. It is, of course, also greatly developed in humans since it is associated with the fine muscular coordination required for the precise use of the hands, vocal apparatus, and so forth.

The great developmental change appearing in mammals is the enlargement of the cerebral hemispheres and, particularly, in the development of the cerebral cortex. These developments are especially marked among the primates and in them increased cerebral size and cortex surface area reaches its peak. These developments are associated

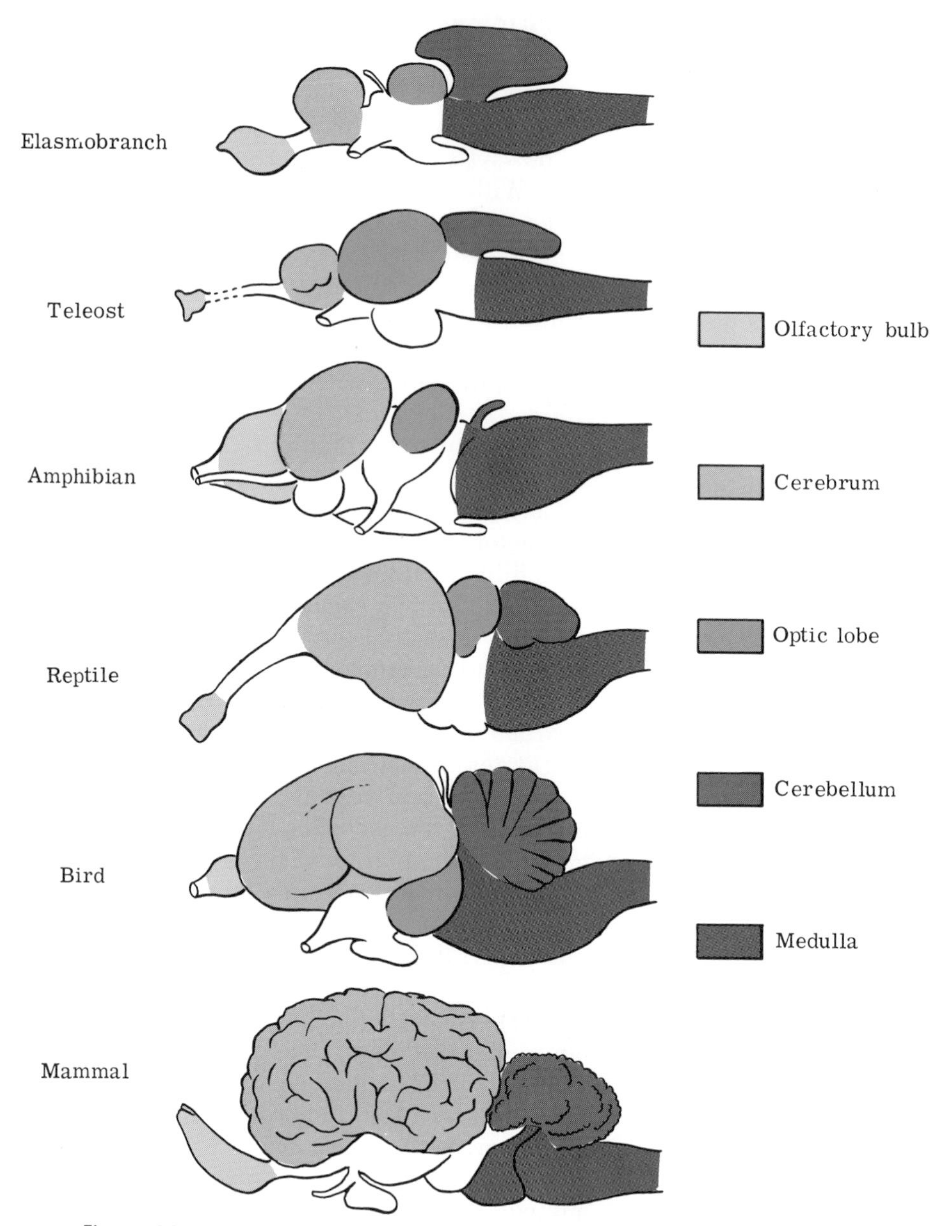

Figure 12–12 A diagrammatic comparison of the brains of various vertebrates.

with increased problem-solving and communication abilities. Interestingly, whales and dolphins also have especially large and convoluted cerebral hemispheres, and it appears that these animals may also have unusual abilities of the same kinds. Certainly they seem to be capable of complex communication and they give every evidence of intelligence. Man, perhaps, does not have to wait to contact some extraterrestrial life form to study an animal of the same order of intelligence as himself!

CHAPTER 13

GENERAL RECEPTORS

13.1 INTRODUCTION

We have seen that it is necessary for the central nervous systems of vertebrate animals to continuously monitor many different conditions of both the internal and the external environments. In either case, some specific type of receptor responds to a particular modality of environmental stimulus such as light, pressure, or temperature. The receptor response leads, in turn, to stimulation of an associated neuron, which conducts nerve impulses to the central nervous system. There the impulses are interpreted in terms of the receptor which originated them. For example, action potentials arriving in the optic nerve are interpreted as light, while those arriving in the auditory nerve are interpreted as sound, even though the nerve impulses themselves are identical in both cases.

Receptors respond to environmental stimuli because, like all living cells, they have the property of irritability. Except for the physical location of the receptor cells and the modifying effects of surrounding tissues, each receptor could respond in the same way to modalities of all kinds. That is, a receptor that ordinarily responds only to light retains the capacity to respond to pressure, temperature, and so forth (Sect. 11.1). It is located, however, where stimuli other than photic stimuli are unlikely to reach it. Moreover, the specialized surrounding tissues of the eye focus light specifically on retinal receptor cells.

The number and frequency of nerve impulses and the duration of activity of a sensory neuron may be quite different from the strength and duration of the environmental stimulus that gives rise to them. This is true because, generally, some information processing takes place at the receptor itself. Such receptor processing makes it possible to encode additional information in the pattern of nerve impulses conducted to the central nervous system. Moreover, the central nervous system, by sophisticated analysis of the input, is often able to extract more useful information than was apparent in the nerve transmission. That these things are true will become apparent from the specific examples given in this chapter.

The activity of individual receptor cells can often affect the sensitivity of other nearby receptors of the same type. To some degree the central nervous system can also adjust the sensitivity of receptors. Neurophysiological studies of individual receptors, then, yield only partial answers, and it is ultimately necessary to conduct behavioral studies to assess the actual sensory abilities of an animal.

Similar receptors often occur in unrelated animals, having appeared independently during evolution. A commonly cited example is the similarity between the eyes of cephalopod mollusks and those of vertebrates. The eyes of each group have remarkably similar structures, although they certainly do not have a common phylogenetic origin.

More than a single receptor of a given kind is generally present; in fact, usually many similar receptors occur in an individual animal. Also, more than a single kind of receptor often monitors the same or a closely related environmental stimulus. These facts greatly reduce the possibility that an unusual situation may cause false, misleading, or confusing information to be sent to the central nervous system. For example, vision is not dependent upon a single receptor or even upon a single "field" of receptors (retina). Moreover, the evidence of vision is corroborated by other sensory inputs to the central nervous system. We can verify our own upright position by means of our eyes, our organs of balance in the inner ear, our proprioceptive senses, and by the touch receptors in the soles of our feet.

Receptor organs are commonly classified in one of two different ways:

1. Exteroceptors are located on or near the surface of the body and they monitor conditions in the external environment. These include receptors for vision, olfaction (smell), hearing, taste, touch, and temperature. Proprioceptors are located within muscles, tendons, and joints and respond to stretching, pressure, and the like. They monitor conditions relating to body posture and motion—the kinesthetic senses. Interoceptors are located within organs and tissues other than skeletal muscle and they monitor various modalities of the internal environment. These include the pressure and chemoreceptors of the carotid sinus (Sects. 3.9 and 5.11), the stretch receptors of the lungs, pain receptors of the intestinal tract, and many others.

2. Receptors may also be classified on the basis of the type of stimulus modality that they are especially adapted to monitor. Mechanoreceptors respond to physical forces such as pressure, vibration, and inertia. These include receptors for touch, hearing, equilibration (balance), and many others. Chemoreceptors respond to chemical substances and include olfaction and taste as well as the response to carbon dioxide and oxygen concentrations in the blood. Photoreceptors may be classed separately or as a special case of chemoreception, since photo-

receptors actually respond to a photochemical reaction set in motion by light rather than to the light itself.

Pain receptors do not fit well into either of these classification schemes because they are located almost everywhere and because they will respond to any modality so long as it is sufficiently intense. If the stimulus causes any damage to tissues, even though slight, pain receptors may be stimulated. In all probability pain receptors are actually chemoreceptors that respond to chemical substances released by damaged cells.

13.2 GENERATOR POTENTIALS

Receptor cells show membrane potentials much like those of nerve cells. The so-called generator potential (or receptor potential) is separate and distinct from the action potential of the associated neuron, however. Among other major differences is the fact that the generator potential is a sustained potential that bears a definite relationship to the ongoing intensity of the stimulus modality to which it is sensitive. For example, the generator potential of a pressure receptor in the carotid sinus is maintained in a steady state which is proportional to the existing blood pressure. If the blood pressure rises or falls, the level of the generator potential rises or falls proportionately. Another difference between the generator potential and the action potential of a neuron lies in the fact that changing generator potentials undergo transient overshoots in response to any sudden changes in stimulus intensity but then quickly settle down to a new steady-state level (Fig. 13–1). If blood pressure suddenly rises, the generator potential of the pressure receptors of the carotid sinus rises momentarily by an amount that is more than proportional to the change in blood pressure. It then quickly falls again to the proper level that is proportionate with the new blood pressure level.

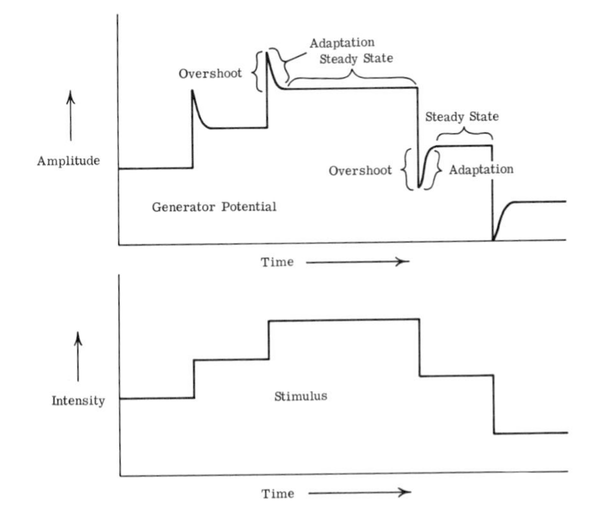

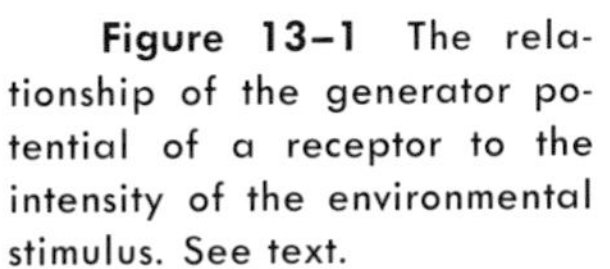
Figure 13–1 The relationship of the generator potential of a receptor to the intensity of the environmental stimulus. See text.

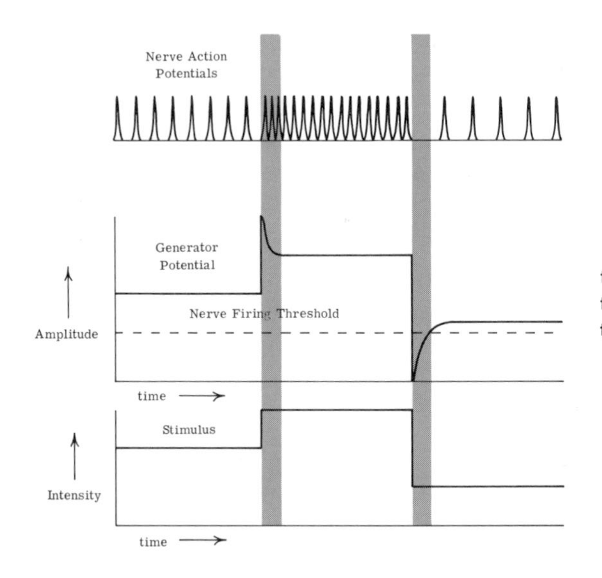

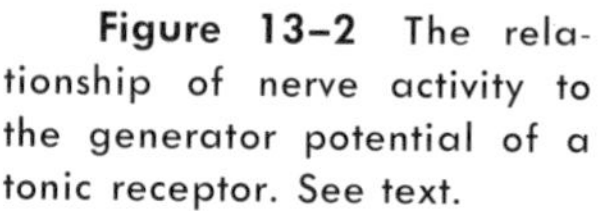
Figure 13–2 The relationship of nerve activity to the generator potential of a tonic receptor. See text.

The generator potential induces activity of the associated neuron but the character of this activity varies in different specific cases. In some receptors the generator potential is nearly always operating at a level which is above the firing threshold of the associated neuron. As rapidly as the resting membrane potential of the neuron is reestablished following each action potential, the continuing generator potential restimulates it so that it conducts another action potential. The frequency of these repeated action potentials is then proportionate to the steady-state level of the generator potential. Such tonic receptors, as they are called, constantly inform the central nervous system of the sustained level of the generator potential (Fig. 13–2), and since the level of the generator potential is proportional to the sustained intensity of the environmental stimulus, the central nervous system can constantly monitor that stimulus. Proprioceptive receptors in muscles,

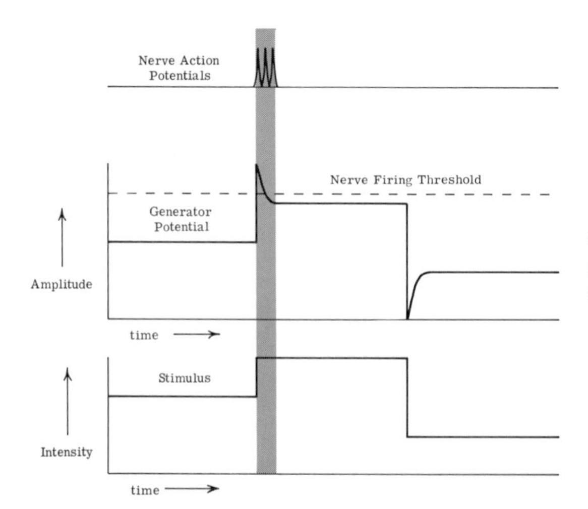

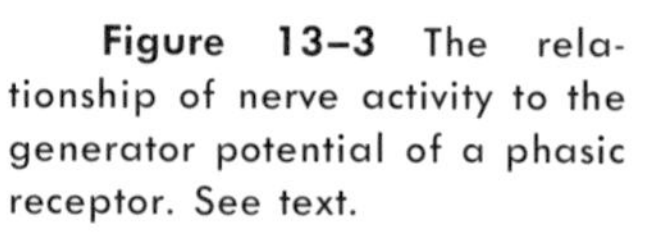
Figure 13–3 The relationship of nerve activity to the generator potential of a phasic receptor. See text.

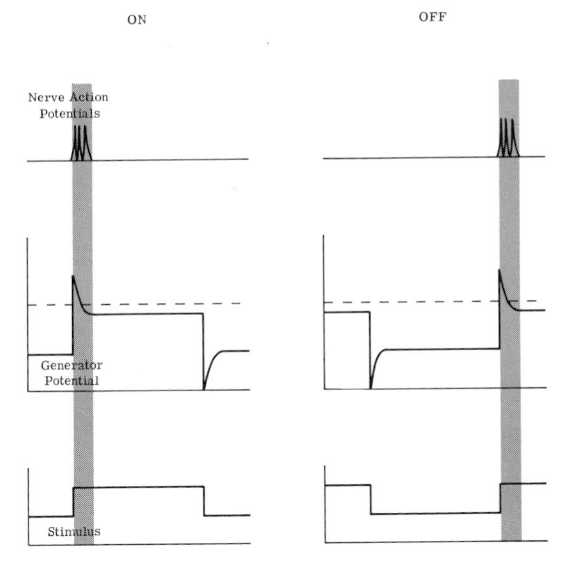

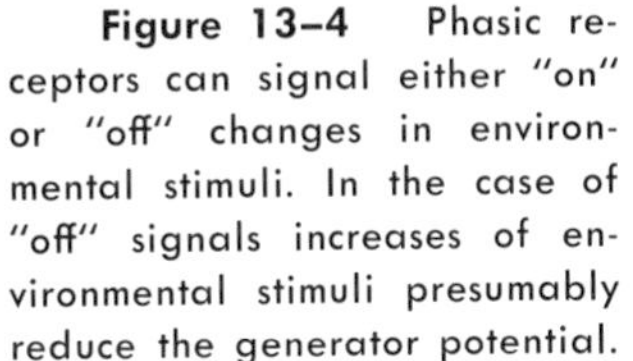

Figure 13–4 Phasic receptors can signal either "on" or "off" changes in environmental stimuli. In the case of "off" signals increases of environmental stimuli presumably reduce the generator potential.

tendons, and joints (Sect. 13.11) are of this type and constantly inform the central nervous system of the current body position. When body position changes, the intensity of the various stimuli (pressures, stretching, and so on) changes. This causes changes in the level of generator potentials, which in turn causes changes in the frequency of nerve impulses in the associated neurons. If the change in body position is abrupt, transient overshoots of generator potential cause temporary overshoots in nerve impulse frequencies so that the central nervous system is also informed of the abruptness of stimulus change.

In certain other receptors, the steady state of the generator potential is nearly always below the firing threshold of the associated neuron. In such phasic receptors, then, there is no ongoing frequency of nerve activity. Instead, action potentials occur as volleys only during the transient overshoots of generator potential (Fig. 13–3). They inform the central nervous system only of changes in the environmental stimulus intensity, not of its current state. A given phasic receptor may act either as an "on" signal, indicating an increase in stimulus intensity, or as an "off" signal, indicating a decrease in stimulus intensity (Fig. 13–4). During the brief volley of nerve action potentials resulting from such changes, the frequency of these impulses is proportional to the rate of change (abruptness) of the stimulus.

13.3 THE WEBER-FECHNER LAW

The proportionality of the generator potential with the stimulus intensity is exact only for relatively low levels (Fig. 13–5). Over most of the range it is more nearly proportional to the logarithm of the stimulus intensity. This means that a receptor can respond differentially to an extremely wide range of stimulus intensities and, in tonic receptors,

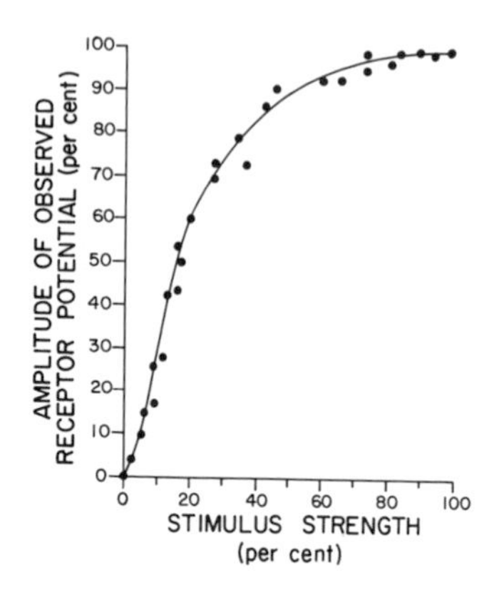

Figure 13-5 Relationship of the strength of a stimulus to the receptor potential produced in a pacinian corpuscle. (From Guyton: *Textbook of Medical Physiology,* 4th Edition. Philadelphia, W. B. Saunders Company, 1971.) Reprinted from Loewenstein: Ann. N. Y. Acad. Sci., *94*:510, 1961.)

encode these differences in a smaller range of nerve impulse frequencies. The softest sounds detectable by the human ear are only 1×10^{-10} as loud as the loudest sounds it can handle with accuracy. The possible range of nerve impulse frequencies is not nearly great enough to cover this extreme range of stimulus intensity on a linear basis but it is sufficient to cover the range on a logarithmic basis.

The power law is related to the Weber-Fechner law. It states that the conscious interpretation of nerve input to the central nervous system is proportional to the antilogarithm of impulse frequency. Therefore, conscious interpretation is directly proportional to the intensity of the stimulus itself. Thus, the stimulus intensity is first "translated" to the logarithm and then "retranslated" back to the original value during the course of sensory reception and transmission to the conscious levels of the brain.

13.4 PAIN RECEPTION

Pain receptors are free nerve endings located in the skin, in all internal organs except the central nervous system itself, in the covering membranes of bones (periosteum), in joints, and in the walls of arteries—in other words, virtually everywhere in the body. The sensation is consciously perceived as pain only when the stimulus is sufficiently intense to cause some amount of tissue damage. Probably the real stimulus in such cases is the release of substances from the damaged cells—substances to which these free nerve endings are sensitive. Lesser intensities of stimulus to the same receptors result in the conscious sensation of tickle or itch. Intense itching or tickling certainly borders on pain as a conscious sensation.

Visceral pain is not well localized consciously, perhaps because we are not as aware of our internal structure as we are of the external surface structures of our bodies. Consequently, in many cases internal pain tends to be generalized rather than localized. In other cases, internal pain may be referred to a more familiar surface area that is served by the same spinal or cranial nerve. The pain then seems to

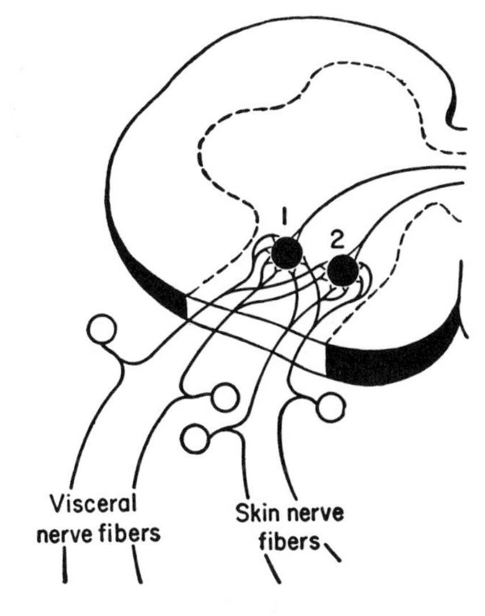

Figure 13–6 A diagram of the connections of sensory neurons in the spinal cord illustrating the mechanism of referred pain. (From Guyton: *Textbook of Medical Physiology,* 4th Edition. Philadelphia, W. B. Saunders Company, 1971.)

come from the surface area even though it really originates from within. A familiar example is the left-arm pain commonly resulting from coronary pain in the heart. The anatomical cross-connections of visceral and somatic neurons in the spinal cord explain how such referral of pain is possible (Fig. 13–6).

13.5 TEMPERATURE RECEPTION

The lateral line system of fishes (Sect. 13.13) is apparently sensitive to temperature changes as well as to certain other modalities of stimuli. Fishes react to environmental temperature changes of as little as 0.05° C. Elasmobranch fishes also have cold receptors on the head which are derived from portions of the lateral line system.

Rattlesnakes and other pit vipers are equipped with a small depression, called the pit organ, on either side of the head between the eye and the nostril. This pit contains cells that are sensitive to infrared radiation (heat). They are so very sensitive that a rattlesnake responds in less than one second to a rat-sized target having a temperature only 10° C warmer than the surrounding environment if that target is within 40 centimeters of the snake. This corresponds to a sensitivity to less than 0.002° C at the pit organ itself. Because the organs are bilateral, the snake also has accurate directional localization of the heat source and can strike at the target even in complete darkness. Boas also have infrared receptors located between the scales, but they are apparently much less sensitive than those of pit vipers.

The Australian bird *Leipoa* buries its eggs in warm sand to incubate. The male bird repeatedly inserts its bill into the sand and adjusts the cover over the eggs in order to maintain an even temperature around them. Apparently these birds have temperature receptors associated with the oral area so that the bill becomes a kind of thermometer.

All warm-blooded animals, and probably most cold-blooded animals, have internal receptors that monitor body temperature. In warm-blooded animals these receptors, operating through nerve reflex arcs, induce shivering (heat production), sweating (heat loss), and other thermoregulatory responses (Chapter 18). In cold-blooded animals similar thermoregulatory responses occur but they take the form of behavioral adjustments. The animal moves into the sunlight or the shade to avoid dangerous environmental temperature extremes. The location of internal temperature receptors is not known, but it is probable that they are within the central nervous system itself and respond to the temperature of the blood circulating there.

Humans and many other higher vertebrates have both warm and cold receptors present in the skin, especially in that of the tongue. These appear to be finely branched nerve endings which probably respond to changes in their own metabolic rates rather than to air temperature or infrared radiation directly. Gradations from cold through cool, warm, and hot involve differential responses of cold receptors, warm receptors, and pain receptors. Very cold and very hot are almost indistinguishable consciously because in both cases pain reception actually dominates.

CHEMORECEPTION

13.6 GENERAL CHEMORECEPTION

Chemoreception in its simplest form is a very primitive sense. It probably appeared at the very beginning of organic evolution and, in fact, may have predated life as such (Sect. 1.7). Actually, since both mechanical and chemical stimuli act, presumably, by altering the permeability of the receptor cell membrane, the distinction between chemoreceptors and mechanoreceptors is somewhat artificial. Lateral line receptors (Sect. 13.13), for example, are sensitive to both kinds of stimuli.

All living animal cells respond in some way to changes in the environmental concentration of oxygen and carbon dioxide. Similarly, all respond to various nutrients, toxic substances, and other materials such as hormones and secretions of organic origin. This is true even of single-celled animals. Lower vertebrates in aquatic environments retain this direct sensitivity in the surface cells of their skins, and this gives them a general but rather indiscriminate means of sampling the composition of the surrounding water. Simple chemoreception has been insufficiently studied but it appears to operate primarily to detect and avoid harmful substances. The receptors are not very sensitive; they detect only rather sharp changes in the immediately adjacent environment.

Some fishes have, in addition to the general chemoreceptivity of skin cells, specific taste receptors scattered in the skin. These are most numerous in the head region of the animal. They have greater sensitivity than the receptors just described and can probably discriminate between different chemical substances to some extent. Still more sensitive and more discriminating are the olfactory receptors which occur also among fishes. The development of more responsive chemoreceptors makes it possible for these animals to detect chemical substances in very low concentrations in the water, and, therefore, at some distance from their sources.

Perhaps paralleling the development of olfactory senses was the evolution in various species of fishes of skin glands that produce species- and sex-specific chemical substances. The odor of these materials permits recognition of friend, foe, or potential mate at some distance. An interesting example of such "chemical communication" occurs among certain cyprinid fishes whose skin contains special secretory cells called club cells. If the fish is injured, a chemical substance contained within these cells is released into the water. Other fishes of the same or closely related species respond to the odor of this material with a fright reaction and swim away from it. If the secretions of club cells are combined with the species odor of a predator fish, the attacked fishes are soon conditioned to associate the two odors. Thereafter, the odor of the predator alone is sufficient to elicit the fright reaction.

In non-aquatic vertebrates the taste cells remain as contact-sensitive cells with poor discrimination and are generally further restricted in locality. Wet-skinned amphibians retain some taste cells in the skin of the head, but higher vertebrates limit the distribution of these receptors to the tongue and palate regions (Sect. 13.7). Non-aquatic vertebrates also retain olfactory receptors of great sensitivity and discrimination and can, therefore, detect air-borne odors at some distance from the source (Sect. 13.8).

13.7 TASTE RECEPTION

The typical human taste bud contains about twenty modified epithelial cells grouped together (Fig. 13–7). It is estimated that adults have about 10,000 such taste buds, mostly located on the tongue. New cells constantly arise at the periphery of each taste bud and migrate toward the center where, as old cells, they die and are removed. From each taste cell project several microvilli (hairlike processes) which extend through a central taste pore in the overlying epithelium. The branches of a nerve ending lie between the cells of the taste bud.

Each taste bud contains cells which are particularly sensitive to one specific taste modality, although each can be stimulated by any modality if the stimulus is sufficiently intense. There is no detectable structural difference in the appearance of taste buds having different sensitivities, so the basis of modality discrimination is not known. A physiological difference in the buds themselves apparently exists, however, since the application of cocaine abolishes sensitivity to taste

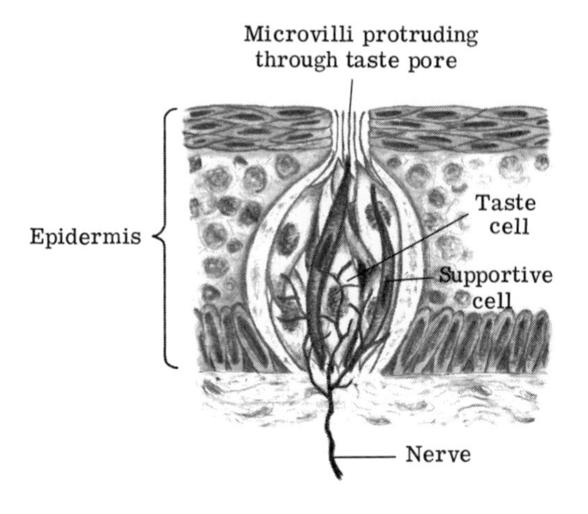

Figure 13–7 The structure of a taste bud.

modalities in a definite sequence. Sensitivity to bitter disappears first, then sweet, salt, and finally sour. These are the four modalities that human taste buds can differentiate. Some authorities believe that a fifth modality, metallic, is also separately discriminated by taste buds; others contend that the metallic flavor is a combination of the basic four modalities plus, perhaps, an ionic effect stimulating touch receptors.

Taste buds that are particularly sensitive to each of the basic four taste modalities are localized in different areas of the human tongue. Sweet-sensitive taste buds are located mostly near the tip of the tongue, while bitter-sensitive taste buds occur almost exclusively near its base. It is a matter of common experience that the sensation of bitter taste is a kind of aftertaste; the bitter substance has already been well chewed or, perhaps, swallowed before one becomes conscious of the bitterness. This is because the substance is not tasted until it has contacted the root of the tongue where these receptors are located. Sour-sensitive taste buds occur primarily along the lateral margins of the tongue; salt-sensitive taste buds are more uniformly scattered over the entire surface.

Sweet taste sensations are brought about by sugars, a number of different alcohols, aldehydes, ketones, esters, amino acids, and several other organic compounds as well as some inorganic salts. Many substances evoke stronger responses from the sweet receptors than does common table sugar. For example, nitrobenzene is about 5000 times sweeter than sucrose. The taste threshold for sucrose is about 0.01 molar; that is, a solution of sucrose need be only that concentrated to be detectable as a sweet taste to the average human.

Bitter tastes are also characteristic of a number of substances, most of which are long-chain organic compounds or alkaloids. Alkaloids are complex organic materials found in plants, many of which have physiological effects on animals and are used in medicine. Some substances that taste sweet at first also have a bitter aftertaste because they stimulate the sweet receptors at the tip of the tongue first and only later the bitter receptors at the base of the tongue. Saccharine is a familiar example. It has been used as a sweetener in low-calorie diets but is objectionable to many people because of its bitter aftertaste. The sensitivity of bitter receptors to the alkaloid quinine is 0.8×10^{-5} molar.

The sensation of sour is, of course, produced by acids of all kinds. The sensitivity of these receptors to hydrochloric acid is about 0.9×10^{-3} molar. Similarly, the sensation of saltiness is confined to a few inorganic compounds. The sensitivity of these receptors to sodium chloride is about 0.01 molar.

Little is known about which taste modalities can be differentiated by other vertebrate animals. Psychological (behavioral) studies indicate that some other mammals lack a specific sensitivity to sweet. Many, however, apparently have a specific response to the taste of water, a modality not separately distinguished by humans. As most people have learned at some time or other, pure distilled water is essentially tasteless. Natural waters vary in flavor only because of the dissolved substances contained in them.

13.8 OLFACTION

While taste is definitely a contact chemical sense, olfaction can be thought of as a distance sense. The statement is not quite accurate, of course, since both senses require that the substance be placed in actual contact with the receptor cells. Olfaction operates as a distance receptor only because it is some 10,000 times more sensitive than taste. Thus, the substance to be detected can be present in much lower concentrations, *i.e.*, in the concentrations present in the air (or water) at a distance from the substance.

The scientific study of olfaction is fraught with difficulties and much less is known about it than, for example, about vision or hearing. There are two major reasons for this.

1. Olfaction has its central nervous connections in the phylogenetically older, more primitive portions of the brain. The effects of olfactory stimulation often involve subconscious emotional attitudes and basic social behavior, which do not often lead to predictable modifications of activity on a conscious level, *i.e.*, activity that can be observed quantitatively. Scents are used by animals in their search for food, for species and sex recognition, for marking the boundaries of territories, and for other fundamental social purposes. Use of scent in this way reaches its peak of development among social insects. Individuals of an ant or bee colony band together as a sort of superorganism within which olfactory substances called phaeromones act to coordinate the activity of the colony as a whole. The situation is almost perfectly analogous with the role of hormones in coordinating the activity of the many cells banded together to form a vertebrate animal. For humans the odors of good perfume, new automobiles, libraries, or dental offices are tied to emotional attitudes derived through past experience. These odors evoke

subtle behavioral patterns rather than producing immediate, overt response patterns that can be easily studied.

2. Another reason for the difficulty encountered in the study of olfaction is the lack of any precise "language" or classification system for odors. Sound is nicely arranged in sequential frequencies ranging from 10 or 15 cycles per second (very low musical notes) to 15,000 or more cycles per second (very high notes) at the extremes of human hearing. A language based upon letters of the alphabet has been developed for talking about sounds; B-flat is a note that can be accurately reproduced and every trained person can recognize and identify it. Similarly, light occurs in sequential wave lengths which we know as colors of the spectrum. These range from violet through blue, green, yellow, orange, and red at the extremes of sensitivity of the human eye. We have no such language for odors, however. We can only say that a given substance smells "something like" some other more familiar substance or that it smells generally putrid, flowery, spicy, and so forth. None of these terms is sufficiently precise for good scientific application.

Humans are capable of identifying separately hundreds of different substances by their odors. It seems certain that there cannot be separate receptors for each of these. Rather, most of them must be combinations of a relatively few primary odors just as certain colors result from the mixture of primary colors.

Attempts have been made to identify the primary odors and some success has been achieved. It now seems likely that humans differentiate seven primary odors: camphoraceous, musky, floral, minty, ethereal, pungent, and putrid. Studies of frogs have yielded evidence that they distinguish five of the same modalities plus three additional primary odors not discriminated by humans.

Except in the cases of pungent and putrid odors, the molecular conformation of an odor-emitting substance probably enables it to "fit" a specific receptor surface and, thereby, stimulate that receptor. Thus, any substances with similar molecular shapes will be perceived as having the same primary odor (Fig. 13–8). Pungent and putrid odors, on the other hand, are apparently distinguished on the basis of the molecular charge. Combinations of more than a single odor are probably perceived as the algebraic sums of the primary odors present in the mixture. Commercial deodorizers presumably work only because they combine with and, therefore, mask the molecular shapes of the odorous materials present in the air.

It has been recognized for some time that certain kinds of chemical structures are associated with strong odors and that relatively minor chemical adjustments of the molecule could change the character of the odor without interfering with its strength. Ambergris, for example, is an extremely foul-smelling substance regurgitated by sperm whales. It has an extremely strong, penetrating odor. Minor chemical treatment can convert ambergris into a pleasant-smelling substance useful in the manufacture of perfume. The perfume retains the strength of the amber-

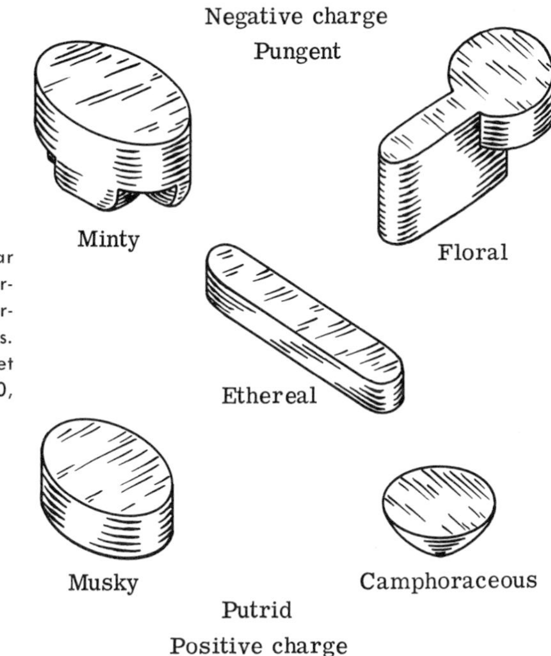

Figure 13–8 The molecular shapes of odorous molecules correlated with the primary odor perception they produce in humans. (After Amoore. *In* J. Proc. Toilet Goods Assoc., Sci. Suppl. 37:1–20, 1962.)

gris but is rendered pleasant in smell. This seems to agree with the observation that putrid and pungent are only different because of a difference in molecular charge, not molecular structure. It is also known that animal musks (sex-recognition odors) can be converted into pleasant floral odors by minor chemical treatment. Note that musk and floral molecules are somewhat similar in shape (Fig. 13–8). Animal musks have basically a steroid molecular shape (Fig. 17–6) since many of them are derived from sex hormones. There is some indication that even after conversion to a floral odor, they still evoke the fundamental sex-recognition, emotional responses, whereas other substances having nearly the same conscious perception fail to do so. Again, this demonstrates the subconscious role of odors and the difficulty encountered in the study of olfaction.

To be an effective olfactory stimulant, the substance in question must be volatile so that it can be transported through the air. It must also be water soluble and, to some extent, fat soluble. Presumably these solubilities are related to the manner in which it is bound to the receptor surfaces of the olfactory cells. It will be recalled that cell membranes are composed, in part, of fat (Sect. 2.2) and that tissues of all kinds contain large proportions of water (Chapter 8). Water solubility is also of importance, of course, in aquatic olfaction by fishes.

Once bound to the receptor surface, the substance may be immediately released again or it may be eliminated in some other way. In either case, it does not continue to induce nerve impulses for very long because olfactory cells are phasic receptors (Sect. 13.2).

13.9 OLFACTORY ORGANS

The human olfactory apparatus lies in the upper, back, and lateral portions of the nasal cavities. There are about 10 to 20 million nerve endings there which cover an area of about five square centimeters. The receptor cells are actually nerve cells which are modified by the possession of sensory hairs or cilia (Fig. 13–9). The hair surfaces are, presumably, the receptor sites for odorous substances; if this is so, there is an enormous receptor surface available.

Although human olfaction is apparently somewhat less sensitive than that of most mammals, the thresholds are surprisingly low. Methyl mercaptan can be detected in a concentration of only 2.5×10^{-10} milligrams per liter of air. For this reason the substance is regularly added to natural gas supplies so that even minute leaks become immediately evident. Since only a small amount of the material present in a liter of

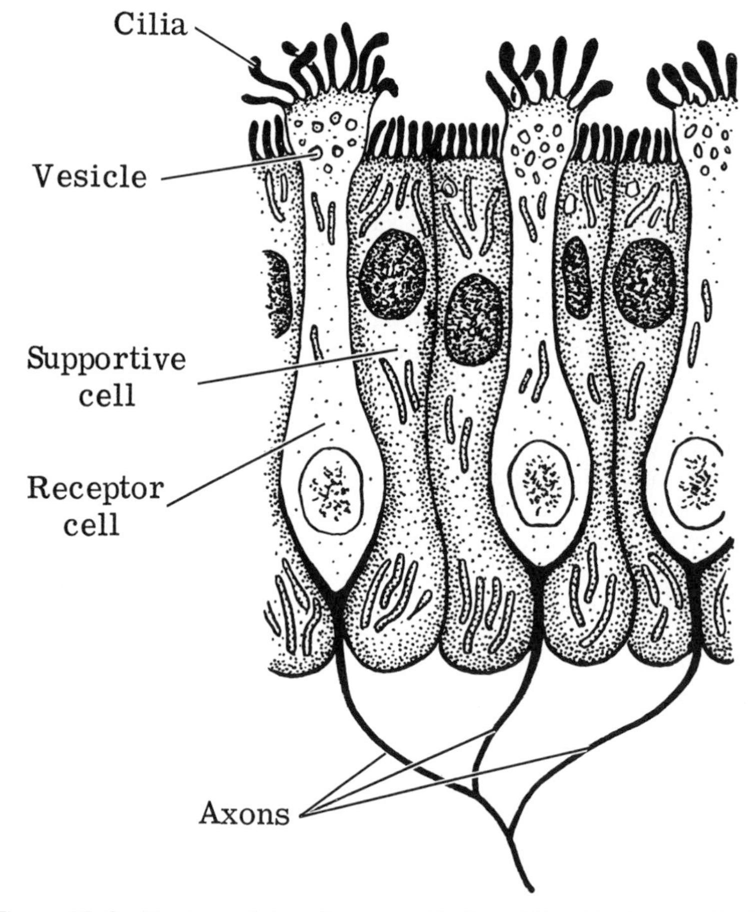

Figure 13–9 Structure of the olfactory epithelium. (After Gordon, *et al.*: *Animal Function: Principles and Adaptations.* New York, The MacMillan Company, 1968, and Hoar: *General and Comparative Physiology.* Englewood Cliffs, New Jersey, Prentice-Hall, Inc., 1966.)

air actually enters the nasal cavities and contacts the olfactory epithelium, probably only a few molecules of methyl mercaptan must contact the sensory hairs in order that it be detected. Since maximal stimulation is achieved with a concentration of 2.5×10^{-6} milligrams of the substance per liter of air, only a few additional molecules bound to receptor sites are required for maximal response.

The olfactory organs of elasmobranch fishes are housed in olfactory pits on the ventral surface of the snout. It should be recalled that many modern elasmobranchs (and, apparently, most ancestral elasmobranchs) are bottom feeders, so this position is appropriate for searching out food. Respiratory activity draws water through the olfactory pits and over the elaborately folded sensory epithelium within them.

The olfactory pits of teleost fishes are located dorsally on the snout and generally are not connected with the respiratory system. In some cases, however, the same muscular activity forces water through both the gills and the olfactory pits. Typically the olfactory pits have separate incurrent and excurrent openings, and water flows through them because of respiratory movements, ciliary activity within the pits, or forward swimming through the water. In the lungfishes the olfactory cavities open dorsally on the surface of the snout and ventrally into the oral-pharyngeal cavity, as they do in all higher vertebrates.

Fishes apparently make great use of olfaction and have very good sensitivity and discrimination. It has been shown conclusively, for example, that salmon return to the precise stream in which they hatched, presumably because of olfactory memory. The differences in dissolved materials in two adjacent tributaries of a stream must be exceedingly small, yet there is sufficient difference to allow the salmon to make the proper choice. Sharks are notoriously sensitive to the presence of even small amounts of blood in the water. They detect it at great distances and quickly gather at the site. Other examples of chemical communication between fishes have been discussed elsewhere (Sect. 13.5).

Amphibians also utilize olfaction to a considerable extent. Evidence suggests that they, as well as many reptiles, can locate bodies of water at fairly great distances and travel overland to reach them.

A different olfactory organ, called Jacobsen's organ, is found in many terrestrial reptiles. This organ consists of a pair of pits opening into the roof of the mouth. The typical forked tongue is protruded to sample the air and then withdrawn and inserted into the openings of Jacobsen's organ. Thus, chemical materials from the air are placed in direct contact with the olfactory sensory cells. Since reptiles generally require less respiratory ventilation than other vertebrates, this adaptation probably is an improvement over the dependence upon respiratory passage of air over the olfactory epithelium.

The sense of smell is very poorly developed in birds; olfaction is of little use in the sky. Vision is of much greater importance and the great development of visual acuity in birds (Chapter 15) has led to increasing dependence upon that sense to the virtual exclusion of olfactory development. Birds must keep the "payload" at a minimum and cannot, therefore, afford to carry organs that are of little use, especially when other organs must become more highly developed and more efficient. Hummingbirds select flowers on the basis of color alone and are attracted to a piece of red cloth as strongly as to a red flower. Most species of vultures will ignore a well-ripened carcass if it is covered and, therefore, not visible to them. Some vultures, which do presumably have a sense of smell, are attracted to leaks in cross-country natural gas lines. Gas line inspectors watch for accumulations of vultures along the lines as a means of locating leaks. Apparently, to the vulture, methyl mercaptan smells like rotting flesh. The kiwi is another exception to the rule. This flightless bird locates worms beneath the surface of the soil by olfactory means.

In the absence of effective olfactory organs, the bills of most flying birds have been adapted to sense air speed. In a manner analogous to the operation of a Pitot tube (which extends forward from the leading edge of an aircraft wing), these nasal structures respond to the air pressure generated by forward motion as the bird flies. When landing it is necessary for a bird to coordinate air speed with ground speed (which is estimated visually) just as an aircraft pilot must. Birds and aircraft must both reach stalling speed exactly at the moment of landing. This, of course, is particularly important for perching birds when landing on a twig.

Many mammals have extremely acute olfactory senses, and the olfactory abilities of dogs are particularly well known. In Europe dogs (and pigs) are trained to locate truffles (underground fungi of importance in gourmet cooking), and in this country dogs are trained to locate contraband drugs at border crossings. Dogs are also used extensively to track humans. The amount of residual odorous material remaining in footprints after days of elapsed time must be extremely small, yet it is sufficient to allow a trained dog to follow an individual without becoming confused by crossing trails left by other humans. Interestingly, dogs are confused when the trail of one individual crosses that of his identical twin if the dog has become acquainted with only the one individual. Given the opportunity to familiarize himself with both twins, however, the dog can then track either one or the other without confusion. Apparently, identical twins smell almost but not quite alike just as they closely resemble but are not exact duplicates of each other in appearance.

MECHANORECEPTION

13.10 CUTANEOUS MECHANORECEPTION

Any stimulus that stretches, wrinkles, or presses against the surface membrane is an effective stimulus to the simple mechanoreceptor cells of vertebrate skins. More complex mechanoreceptors of the skin are associated with elaborate accessory structures, which render them more specific in sensitivity. These are exemplified by the organs of hearing and equilibration since they were derived, during evolution, from structures that were originally on the surface (Sect. 13.13 and Chapter 14).

However, many simple mechanoreceptors, especially those of the skin, are not associated with any accessory cells to form receptors in the usual sense. Instead, they are merely free nerve endings. Still, in such cases the stimulus induces a receptor potential in the free nerve ending much like the generator potential of more complex receptors (Sect. 13.2). This is not conducted but evokes true nerve impulses arising at the first node of Ranvier (Sect. 2.4) when the receptor potential exceeds the firing threshold at that node.

Single-cell organisms commonly react to mechanical stimuli and, similarly, many non-receptor cells of higher organisms respond directly and individually to such stimuli. For example, isolated smooth muscle cells will contract in response to stretching stimuli. Mechanoreception in its simplest form, then, is a very primitive sense, one that is perhaps as old as life itself.

Human skin has a total of nearly three quarters of a million discrete receptors. They include specialized structures such as pacinian corpuscles, Meissner's corpuscles, and, in a few places, end bulbs of Krause (Fig. 13–10). They also include free nerve endings, some of which form rings about the bases of hair follicles so that mechanical disturbances of the hair cause stimulation of the associated nerve (Fig. 13–11).

Cutaneous receptors are unevenly distributed, being most concentrated in areas surrounding body orifices, the eyes, and the fingertips. The relative distribution can be tested by determining the two-point resolution in various skin areas of an individual. The test determines how far apart two stimuli must be in order to be distinguished as two, rather than as a single stimulus. The blindfolded subject is touched sometimes with one and sometimes with a pair of sharp points and asked to indicate whether he feels one or two separate stimuli. The distance between these points is slowly widened until he always gives the correct answer. Theoretically, in order to resolve them as two separate stimuli, each must produce a response in a different, single receptor. Furthermore, these two receptors must be separated from each other by at least one unstimulated receptor.

In many instances individual stimulated receptors may inhibit the sensitivity of surrounding receptors. This can produce greater contrast between strongly stimulated and less strongly or unstimulated areas, but for the reasons outlined in the preceding paragraph this contrast is achieved at the expense of some resolving power. Frequently stimulation of the skin represents a potential danger from which the animal should withdraw. Under such conditions, fine resolution is less important than sensitivity, and heightened contrast coming about in this way increases sensitivity. In other instances great resolution is also possible. One only has to recall how much detail his tongue can find in a chipped tooth or to observe a blind person reading Braille to demonstrate this fact.

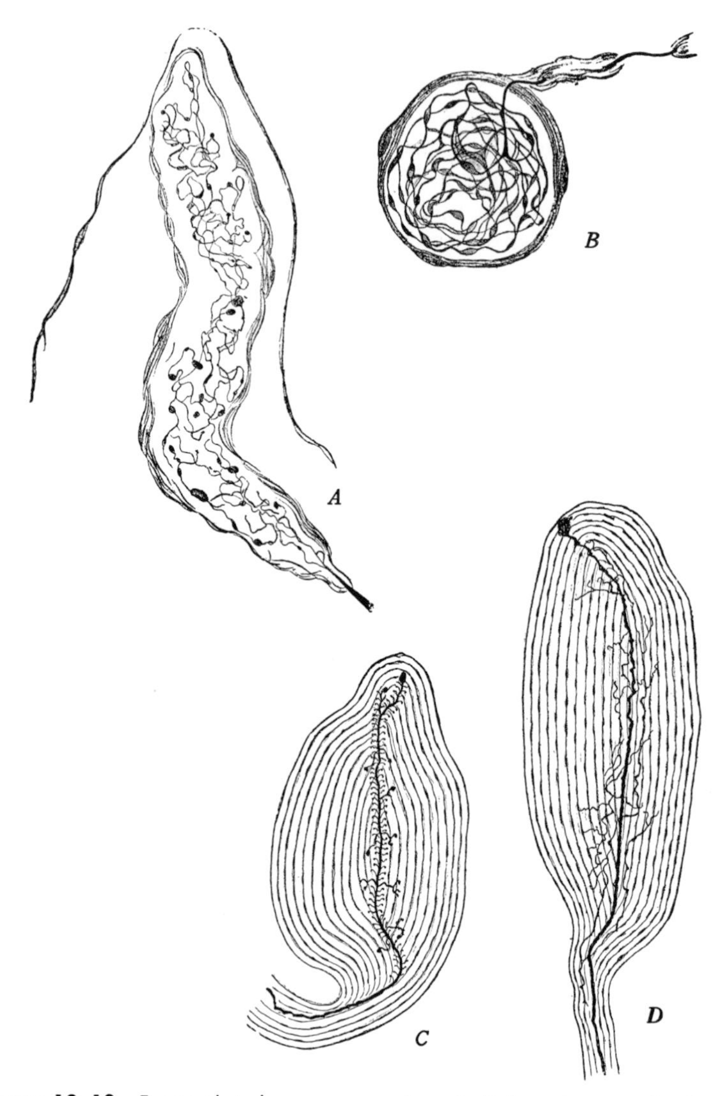

Figure 13–10 Encapsulated sensory receptors. *A,* Meissner's corpuscle; *B,* end bulb of Krause from the human conjunctiva; C and *D,* pacinian corpuscles. (From Ranson and Clark: *Anatomy of the Nervous System,* 9th Edition. Philadelphia, W. B. Saunders Company, 1953.)

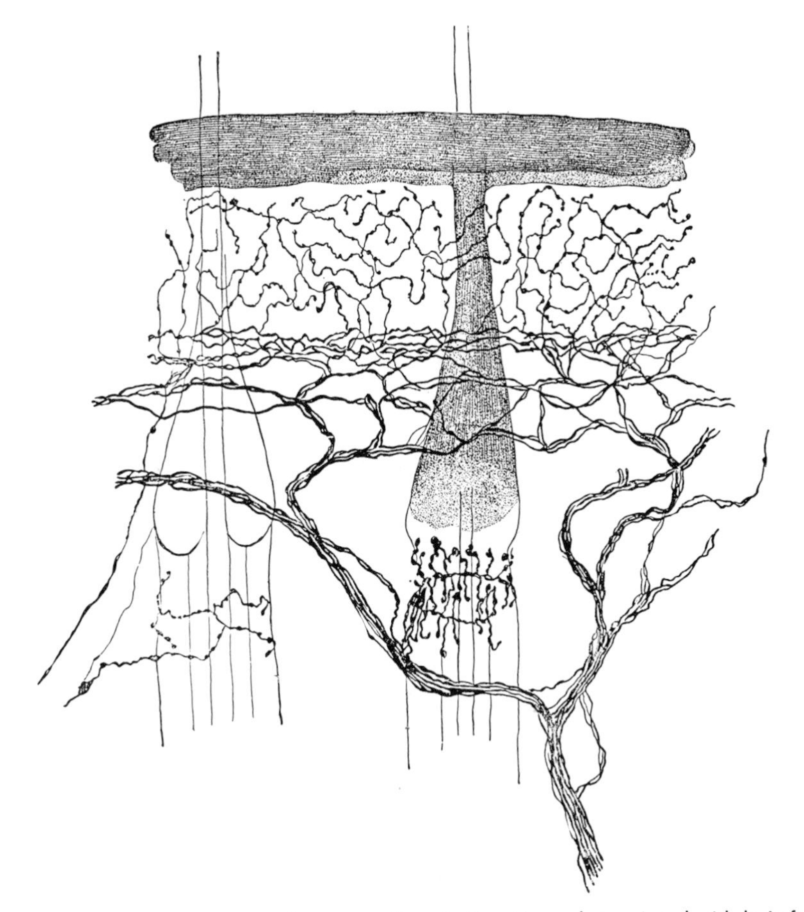

Figure 13–11 Nerves and nerve endings in the skin and associated with hair follicles. (From Ranson and Clark: *Anatomy of the Nervous System,* 9th Edition. Philadelphia, W. B. Saunders Company, 1953.)

13.11 PROPRIOCEPTION

The receptors located in skeletal muscles and in the associated tendons and joints are mechanoreceptors. They are stimulated, in most cases, by stretching of the tissues in which they lie. Three discrete receptor types are recognized (Fig. 13–8) and together they continuously inform the central nervous system of the current position and movements of body parts.

Joint receptors are located within the joint capsules of the skeleton. When the joint is moved the receptors are subjected to stretching, which is the effective stimulus to them. Probably each is stretched by movement of the joint in a given direction only but collectively they monitor all possible joint movements. They respond in a tonic manner (Sect. 13.2), indicating both the rate of movement and the final position of the joint.

Golgi tendon receptors (Fig. 13–12) are also tonic receptors sensitive to stretching of the tendon. An additional burst of action potentials in the associated nerve encodes the rate of change, and a subsequent steady state of discharge frequency indicates the final static degree of stretch. Two different kinds of activity may result in the stretching of a tendon, and the Golgi receptors cannot distinguish between the two: (1) contraction of the associated muscle stretches its tendon and (2) contraction of the antagonistic muscle also stretches it. For example, contraction of either the biceps brachii muscle or of the triceps brachii muscle will result in stretching of the tendons of both muscles. Together with the input from the joint receptors of the elbow, however, the central nervous system is given sufficient information to determine

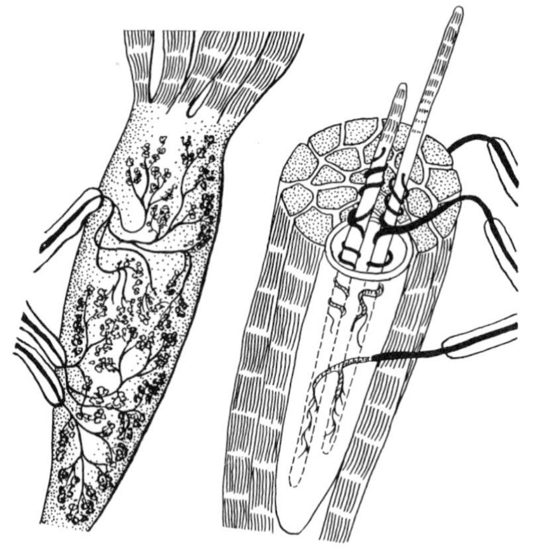

Figure 13–12 Proprioceptive receptors. (From Cockrum and McCauley: *Zoology*. Philadelphia, W. B. Saunders Company, 1965.)

which of the two muscles contracted and how rapidly that contraction occurred. It will be recalled too that after the biceps contracts, a reflex inhibition of the triceps soon occurs (Sect. 11.2). This, undoubtedly, reduces the stretching of the triceps tendon while that of the biceps continues.

Muscle spindle fibers are actually modified muscle fibers lying here and there within skeletal muscles (Fig. 13–12). Each spindle fiber has an independent motor nerve supply so that it can contract or relax independently of the surrounding muscle fibers. Only the ends of the spindle fiber are contractile and these are innervated by this motor nerve. The central portion of the spindle fiber is a sensory receptor and is innervated instead by a sensory nerve.

The central portion responds in a tonic fashion to the stretch imposed upon it by the surrounding muscle fibers. It is stretched, specifically, when the surrounding fibers relax, since relaxation of a muscle increases its length. When the surrounding muscles contract, the degree of stretch of the spindle fiber is lessened. Similar to a typical tonic receptor (Sect. 13.2), the spindle fiber responds to changes in the degree of stretch by inducing a burst of nerve impulses at a frequency that is proportional to the rate of change of stretch. It then settles down to a frequency level that is proportional to the final degree of stretch. Unlike many tonic receptors, however, the spindle fiber follows this reaction with an adjustment to a base frequency which is always the same. For example, if the surrounding muscle relaxes, the spindle fiber is stretched. An initial overshoot of generator potential encodes the rate of stretching in an initial burst of nerve impulses. Following the overshoot, the generator potential levels off, and nerve impulses proportional to the static degree of stretch reach the central nervous system. Finally, the contractile ends of the muscle spindle relax enough to exactly relieve the stretch, and the tonic frequency of impulses in the associated sensory nerve drops to a standard resting level. The reverse occurs when the surrounding muscle fibers contract. Thus, these receptors inform the central nervous system of the rate and amount of change in muscle contraction and then "reset" themselves so that they are ready to respond in a like manner to the next change in muscle contraction.

Whereas muscle spindles monitor only briefly the final degree of muscle contraction, the tendon receptors continue to monitor it, and, it will be recalled, the joint receptors indicate the direction and rate of resulting joint movement as well as the final position of the joint. Collectively, then, these three kinds of receptors inform the central nervous system of all aspects of physical movement with great accuracy. The various reflexes associated with locomotion and posture (Sect. 11.2) are all mediated, in part, by such proprioceptive receptor responses.

Another generally recognized type of receptor is the flower-spray receptor of skeletal muscle. These consist of branched free nerve endings within the muscle, and it is probable that they are responsible for the conscious sensation of muscle contraction.

13.12 INTERNAL MECHANORECEPTORS

Mechanical interoception is found in the hydrostatic pressure receptors of the carotid sinus and the arch of the aorta (Sect. 5.11). These are tonic receptors that constantly monitor the level of arterial blood pressure. The walls of swim bladders of fishes (Sect. 3.4) contain similar receptors. Another example of internal mechanoreception is found in the stretch receptors of mammalian lungs (Sect. 3.9). They are also tonic receptors which monitor the degree of inflation of the lungs in a continuous fashion. Related receptors occur in the pharyngeal walls of some fishes where they play an analogous role in the control of gill ventilation.

There are other internal mechanoreceptors, including pacinian corpuscles, that are apparently identical with those of the skin. A very large corpuscle occurs in the mesentery of the cat and it has been the object of much investigation because of its size and location.

Usually there is no conscious awareness of the central nervous input from internal mechanoreceptors; this input functions instead through reflex mechanisms at subconscious levels. Internal pressures evoke conscious responses only after special training. Examples are the learned control of urinary bladder voiding and of intestinal evacuation. Both of these processes naturally operate as cord-level reflexes, and their volitional control is a learned behavior pattern.

13.13 LATERAL LINE ORGANS

Aquatic vertebrates (fishes, amphibian larvae, and some amphibian adults) have rows of sensory receptors called neuromasts along their sides. These, collectively, comprise the lateral line organ (Fig. 13–13). The lateral line organ is generally marked by associated pigmentation so that its location is visible as a single stripe running lengthwise along either side of the animal.

Each neuromast is a cell having a hairlike extension that is embedded in a gelatinous mound or column called a cupola (Fig. 13–14). Sometimes the receptor cells lie in the bottom of a shallow groove that opens to the side of the animal, but more frequently the groove is covered over, thus forming a canal lying near the surface. These canals

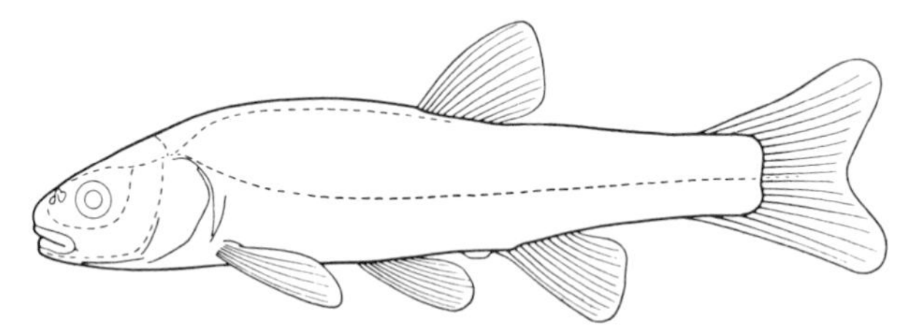

Figure 13–13 Diagram showing the position and extent of the lateral line system (dashed lines) of a fish. (From Florey: *An Introduction to General and Comparative Animal Physiology*. Philadelphia, W. B. Saunders Company, 1966.)

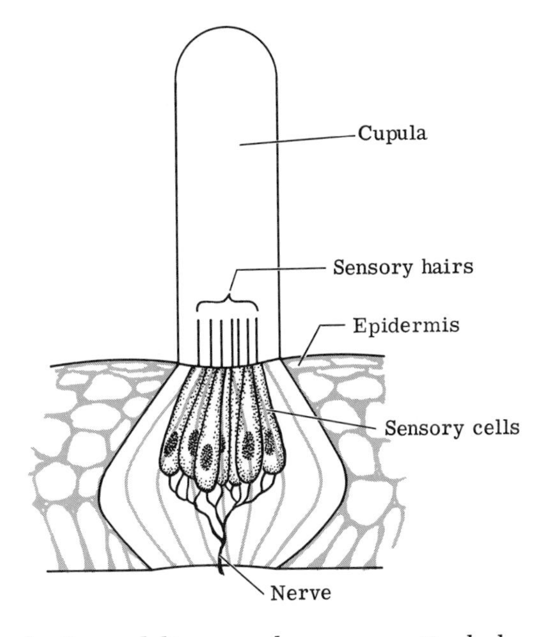

Figure 13–14 The structure of a neuromast (diagrammatic).

open to the surface at intervals. Lateral line canals occur particularly in more active forms, whereas the open groove system occurs in certain more sedentary forms.

The primitive function of neuromasts seems to have been as rheostatic receptors, *i.e.*, receptors sensitive to currents in the water. Water current, by displacing the cupola of the receptor cell, stimulates the embedded receptor hair. Lateral line systems seem to contain two types of receptors, and water currents moving from the head toward the tail of the animal stimulate some receptors but inhibit others. Thus, it is possible for the fish to detect the direction as well as the comparative velocity of the water current. This kind of sensory system enables a fish to swim upstream at a velocity exactly matching the velocity of the downstream movement of water so that the fish is able to maintain its position with respect to the bottom (determined visually, perhaps), or to actually move upstream or drift downstream at appropriate rates. Thus, a fish is able to detect its "water speed" by means of the lateral line system and its "ground speed" visually, a situation somewhat analogous to the simultaneous estimation of air speed and ground speed by birds (Sect. 13.9).

The same lateral line receptors are also able to detect disturbances in the water which produce deviant currents. A wriggling worm might produce such currents, and, moreover, they would be repeating currents (or pressure waves) of low frequency (Fig. 13–15). If the lateral line system were to attain greater sensitivity, it could presumably detect such disturbances at greater distances, and there is no reason why it might not attain the ability to detect higher frequencies (shorter wave lengths) resulting from such disturbances. Such sensitive reception would constitute "touch at a distance" or, ultimately, the detection of underwater sound of relatively low frequency. Thus, the lateral line system is fundamentally capable of developing as an organ of hearing.

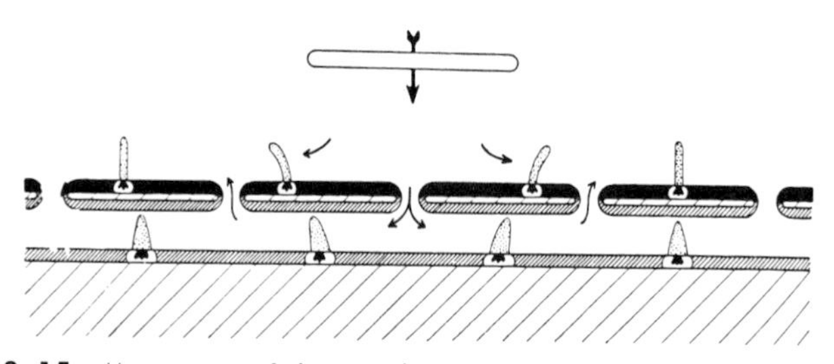

Figure 13–15 Movements of the cupulae of the lateral line organ caused by water currents and by pressure due to an approaching object. (From Florey: *An Introduction to General and Comparative Animal Physiology.* Philadelphia, W. B. Saunders Company, 1966.)

This has not actually occurred in the lateral line system itself, although some fishes can detect vibrations up to a frequency of about 100 vibrations per second through lateral line reception. Derivatives of the primitive lateral line system have developed in vertebrates as organs of hearing (Chapter 14), and with these humans can detect frequencies as low as about 10 vibrations per second. The distinction between the detection of vibration by the lateral line and the true reception of sound is, therefore, somewhat artificial.

Because movement through motionless water creates an apparent current and because this varies on both sides of the animal as it alters course, it is clear that the lateral line organ is fundamentally capable of monitoring linear (straight-line) and angular (turning) acceleration and deceleration. The structures of the vertebrate inner ear (Chapter 14) contain apparatus for monitoring linear and angular acceleration and deceleration, and these structures were also undoubtedly derived during evolution from portions of the lateral line system. Inner ear receptors are also capable of detecting the direction of the pull of gravity and therefore serve to orient the body in opposition to that pull (Sect. 14.3). This response to gravity is also implicit in the structure of neuromasts if the cupola were to sag a bit under the pull of gravity. To do this would require only that some material having a greater density than that of water be present in the cupola. The lateral line probably does not play this role, but the neuromastlike cells of parts of the inner ear are equipped with calcium carbonate grains and do respond in this fashion (Fig. 14–2).

In some fishes the neuromasts are sensitive to modalities other than those just discussed. Their sensitivity to electrical disturbances renders them useful to those fishes which depend upon low-output electrogenic organs (Sect. 16.9) for navigation. Likewise, in some fishes there are neuromasts in the head region which respond to changes in either the temperature or salinity of the surrounding water. It has even been suggested (without convincing evidence) that lateral line systems may sometimes be sufficiently sensitive to detect the magnetic field of the earth and, thus, be useful as a guidance system in long migrations. In any case, it is clear that the structure of the lateral line system and its neuromasts is potentially basic to the development of a number of more sensitive and more discriminating receptors. Some of these derivative receptors will be discussed in the following chapter.

CHAPTER 14

INNER EAR RECEPTORS

14.1 INTRODUCTION

During the embryonic development of vertebrates, an ectodermal sac called the otic vesicle forms on each side of the head. The lining of the otic vesicle contains certain tissues which in the lower aquatic vertebrates include parts of the lateral line organ (Sect. 13.13). In adult vertebrates the otic vesicle becomes the inner ear and contains sensory receptors that are derived from neuromasts like those of the lateral line organ. We have seen in the preceding chapter that such neuromast receptors are, potentially at least, sensitive to water currents, linear and angular acceleration or deceleration (inertia), low-frequency vibrations or sound waves, and gravity. If receptors of this general kind, then, could be modified so that their sensitivities were limited to a single modality from this list, they would be far more useful to the organism. Unrestricted response to each of a variety of stimulus modalities yields less precise information to the central nervous system. Limiting some neuromasts to one modality and others to a different modality has occurred in the vertebrate inner ear. The modifications are brought about by isolating them where extraneous stimuli are unlikely to reach them and by surrounding them with specialized accessory structures that serve to emphasize the effect of single modalities (Sect. 13.1).

In mammals the inner ear receptors lie within a cavity in the temporal bone of the skull, a cavity called the bony labyrinth (Fig. 14–1). This cavity contains a series of organs composed of soft tissues collectively spoken of as the membranous labyrinth. The membranous labyrinth does not completely fill the bony labyrinth and the remaining space is filled with a fluid called perilymph. The membranous labyrinth itself is filled with a similar fluid called endolymph. The endolymph is somewhat more viscous than the perilymph, and there is no point at which the two fluids are in direct contact with each other; they are separated everywhere by membranes of the membranous labyrinth.

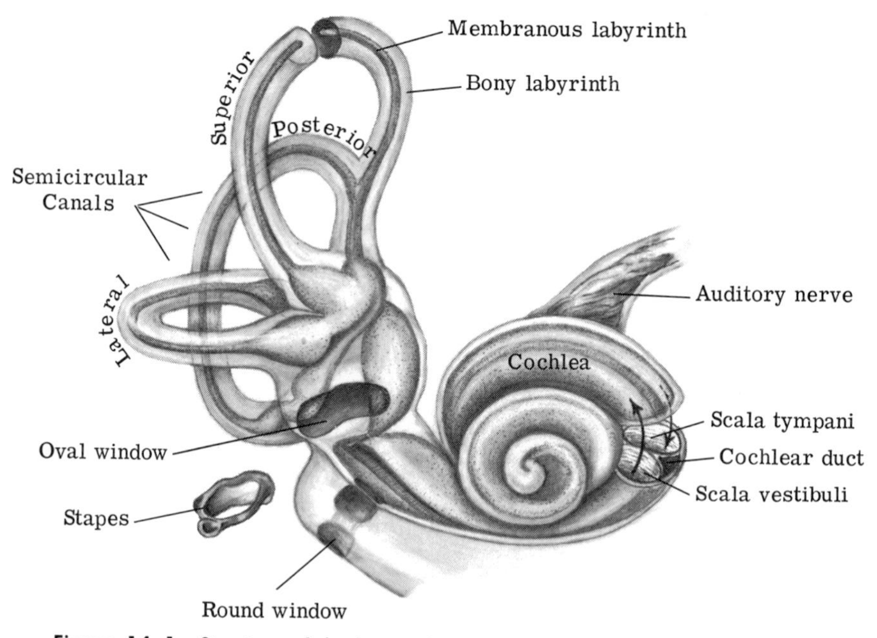

Figure 14–1 Structure of the human inner ear. (After Grollman: *The Human Body.* New York, The MacMillan Company, 1969.)

The coiled cochlear portion is the organ of hearing in mammals, which does not occur in lower vertebrates. The remainder of the mammalian inner ear is called the vestibular portion and contains receptors responding to linear and angular acceleration-deceleration, gravity, and perhaps low-frequency vibration. For the most part these receptors of the vestibular portion are concerned with the orientation of the animal in space, a sense commonly spoken of as equilibration.

VESTIBULAR APPARATUS

14.2 SEMICIRCULAR CANALS

Each semicircular canal resembles, as the name implies, a half circle. It is composed of a tubular extension of the membranous labyrinth lying within a larger tubular extension of the bony labyrinth. Each membranous tube is attached at both ends to a central membranous sac called the utricle (Fig. 14–1).

Hagfishes have only a single semicircular canal in each inner ear; cyclostomes have two such canals. All higher vertebrates, however, have three semicircular canals on each side. They are oriented in planes that are all at right angles to each other, so there are two vertical canals and one horizontal one. The horizontal canal is the one missing in the cyclostome ear.

At one of the ends of each membranous semicircular canal is an enlargement called the ampulla. This ampulla is the site of the receptor organ, a structure called the crista. The crista closely resembles lateral line neuromasts (Sect. 13.13), being composed of cells that have hair-like extensions embedded in a gelatinous cupola (Fig. 14–2). The cupola extends completely across the cavity of the ampulla so that any movement of endolymph in either direction moves the cupola backward or forward like a swinging door. Such movement results in either raising or lowering the generator potentials of the hair cells, depending upon the direction of movement. Since they are tonic receptors (Sect. 13.2), this results in an increase or decrease in the frequency of nerve action potentials to the central nervous system.

The function of the semicircular canals can, perhaps, be best understood by means of an analogous model. If a bucket of water is suddenly rotated about its own axis through a part of a rotation, the water within the bucket tends to remain at rest because of its own inertia. The bucket, in effect, rotates about the water while the water remains still. If a structure like the cupola were attached to the inside surface of the bucket so that it extended into the water, such a move-

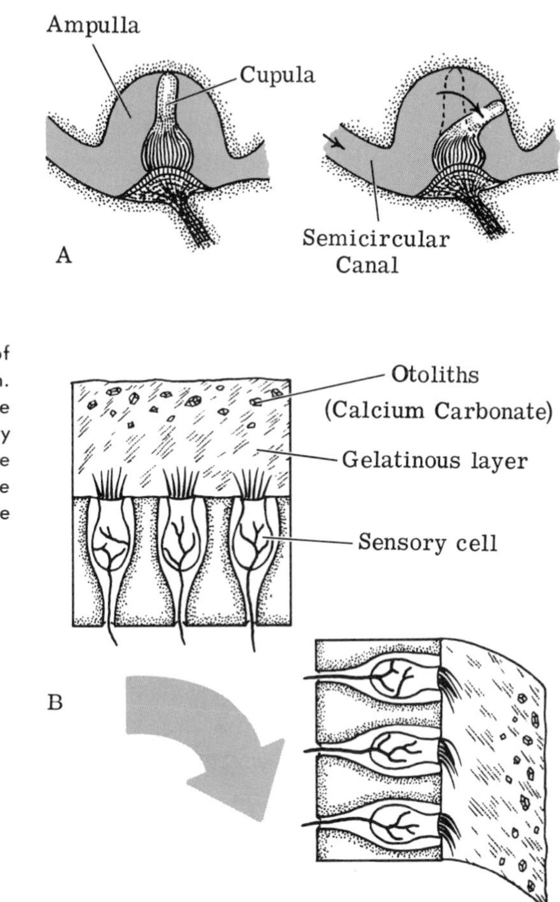

Figure 14–2 Function of the inner ear in equilibration. *A*, ampulla and crista of the semicircular canals; *B*, sensory organs of the saccule and utricle (see text). (After Grollman: *The Human Body*. New York, The MacMillan Company, 1969.)

ment of the bucket would cause deflection of the cupola in the opposite direction. Actually, the cupola would be moving through the water rather than the water flowing past the cupola. Nevertheless, the cupola would respond to this apparent current just as it would to a real current in the water.

Rotation of the human head about its own axis produces apparent currents of the same kind in the horizontal semicircular canal of each inner ear. There is not much actual fluid motion because endolymph is a fairly viscous fluid, but it is sufficient to cause deflection of the cupola and, consequently, a change in the frequency of action potentials in the associated neurons. Since semicircular canals lie in the other two dimensional planes as well, rotation of the head about any other axis will also produce apparent currents in at least one of the three semicircular canals. Rotation in most directions will, in fact, produce such apparent currents in two of the canals on each side. Such an event presents the central nervous system with input from both of the canals involved and, therefore, with a problem in vector analysis. The comparative amounts of cupola deflection in the two canals encodes all of the information necessary for the central nervous system to determine the true direction and speed of head rotation. The code, of course, is the frequency of nerve action potentials coming from the two receptors, since these frequencies are proportional to the generator potentials established in those receptors (Sect. 13.2). Such rotational movement of the head constitutes angular acceleration (starting the movement) and deceleration (stopping the movement).

Of course, the central nervous system also receives data regarding such angular or rotational movement from other receptors as well. The skeletal muscles producing the movement are equipped with proprioceptors (Sect. 13.11); the effects of head movements are also monitored by the eyes. As we have seen, such redundant sensory inputs reduce the possibility of false or misleading information reaching the central nervous system (Sect. 13.1). In this particular case, however, such misleading and conflicting inputs are quite common and familiar. Following a period of prolonged rotation, one feels dizzy for a time as though one were still rotating. This is because the semicircular canal organs are still being stimulated. Return briefly to the analogy of the bucket of water mentioned earlier in this section. If we rotate the bucket of water continuously through several rotations and then stop it, we will find that the inertia of the water at rest has been overcome and it has begun to rotate with the bucket. When the rotation of the bucket ceases, inertia again causes the water to continue in motion. The sensory organ that we visualized as being attached to the inner surface of the bucket is then stimulated by an actual water current for a time until the water again comes to rest. This requires some time for water but much less time for the more viscous endolymph.

Thus, following prolonged rotation, the organs of the semicircular canal continue to send information to the central nervous system indicating that rotation is continuing when, in fact, it has stopped. This causes the sensation of vertigo or dizziness. To a certain extent, the central nervous system acts on this false information. The eyeballs, for example, reflexly move back and forth, rapidly in one direction and more slowly in the other, in a characteristic pattern known as nystagmus. This is precisely the way in which the eyes would move if actual rotation of the head were continuing. While rotating, a person looks fixedly at some object, following it with his eyes until the rotation causes it to drift out of the visual field. Then he quickly moves his eyes in the opposite direction to fix on some new object which is just coming into view. During vertigo (dizziness), this reflex nystagmus makes it seem as though objects are really moving past, although, of course, they are not. Nevertheless, the "world seems to spin" because the eyes are moving in this fashion. Watching a passing freight train involves the same fast-slow, back-and-forth movement of the eyes. Each box car is followed until it moves out of the visual range; then the eyes quickly snap back to pick up another box car that is just coming into view. Curiously, this voluntary movement of the eyes arouses in many people the feeling of dizziness. For most people dizziness, whether produced by rotation or by watching a passing freight train, also causes a feeling of nausea if the dizziness persists. Sea sickness also causes a feeling of dizziness, but the relationship between illness and vertigo is not at all clear physiologically.

While it is a less common experience, it is also possible to induce vertigo in a vertical (or any other) plane by prolonged rotation in that plane. In vertical dizziness the world seems to be rotating in an upward or downward direction rather than horizontally. The movements of the eyes (nystagmus) also occur, but in this case movement is also in a vertical direction. Here still other muscular reflex responses also come into play. To see the world apparently passing upward gives one the impression that he must be falling forward. He responds to this by attempting to throw himself backward to regain his normal equilibrium.

The cyclostome *Petromyzon* compensates for the lack of a horizontal semicircular canal in an interesting way. Ciliated epithelium lining the large endolymphatic cavities of the animal continuously maintains a whirlpool-like motion of the enclosed fluid. Movements of the animal in the horizontal plane, because of a gyroscopic effect on this whirlpool, produce pressure changes in the two vertical semicircular canals. However, nothing is yet known about the functional limitations of the single hagfish semicircular canal.

14.3 THE UTRICLE, SACCULE, AND LAGENA

In all vertebrates but the very primitive fishes, the membranous labyrinths are differentiated as semicircular canals plus two saclike portions, the utricle and the saccule (Fig. 14–1). The semicircular canals communicate, as we have seen, with the utricle, which, in turn, opens through an endolymphatic duct to the saccule. In non-mammalian vertebrates the saccule is extended as a small flap of tissue called the lagena. In addition, in reptiles, birds, and mammals there is a tubular element extending from the saccule. This is called the cochlear duct, or scala media, and is associated with the organ of hearing.

The utricle, saccule, and lagena all contain sensory organs that resemble each other. Each consists of a patch of sensory hair cells much like lateral line neuromasts (Fig. 13–2). The sensory hairs are embedded in a gelatinous mass, which also contains grains of calcium carbonate. These particles, called otoliths, have a density that is about 2.5 times as great as that of the surrounding tissues and fluids. Therefore, they react differentially to the force of gravity. If the head is tipped to one side, the weight of the otoliths causes bending of the sensory hairs and therefore initiates a change in the generator potential of the receptor (Sect. 13.2). This, in turn, causes a change in the frequency of action potentials in the associated neurons. The central nervous system is thus informed of the rate and degree of head tipping as well as of the final resting position of the head after such movement takes place. Again, information from the proprioceptors of neck muscles as well as from vision provides confirmation to the central nervous system.

It is probable that the receptors of the utricle, saccule, and lagena also respond to linear acceleration and deceleration since such movements in any given straight line would also cause bending of the hair cells. In this case the greater inertia of the otoliths would be responsible. Some authorities also feel that the same inertial response would be adequate to make these receptors respond to low-frequency vibrations as well. This is probably particularly true of those receptors in the lagena and may constitute, therefore, primitive low-frequency hearing.

In fishes all three chambers show sensitivities which are broadly overlapping. Probably the utricle and the nearby portions of the saccule are mainly concerned with orientation sensing (acceleration, deceleration, and gravity), whereas the remainder of the saccule and the adjacent lagena function in vibration sensing. Amphibians and reptiles show increasing development of sensory receptors in the lagena and, in crocodiles, the addition of a cochlear duct as well. As these structures evolved and extended the frequency range for hearing, the saccule apparently became less functional.

AUDITORY APPARATUS

14.4 MAMMALIAN EXTERNAL AND MIDDLE EARS

The external ear (auricle or pinna) is a sound-gathering device for most mammals. Mobile ears—ears that can be turned from side to side—also serve to localize the exact source of the sound. The human ear has been so reduced in size and mobility, however, that it is virtually useless in this regard. Its function is more decorative (and to be decorated) than to gather or localize sounds.

Sound waves are alternate condensations and rarefactions of the air and they move in all directions from any vibrating source. Since they are waves in the air, they have variable frequencies (cycles per second), and these are inversely related to their wave lengths. Sounds with short wave lengths have high frequencies and are heard as treble notes. Sounds with long wave lengths have low frequencies and are heard as bass notes. Sound waves also differ in amplitude (loudness) and in timbre (the simultaneous presence of waves of different but related frequencies which produces complex patterns of harmonic sound). Different musical instruments sound different even when they are playing the same note because of differences in timbre. All sounds are, of course, complex patterns of wave lengths, and the resulting pattern is considered to be musical only when the various frequencies bear some interesting and pleasant relationship to each other. When these relationships are lacking, the pattern becomes discordant and unpleasant or, if the wave lengths present are totally random, it becomes pure "noise."

The black paper cone in the speaker of a phonograph or radio is caused to vibrate by electromagnetic forces; this vibration produces waves of corresponding frequencies, which spread through the air (or water) in all directions until they strike other objects. There they are either absorbed (damped) or reflected (echoed).

When sound waves enter the external auditory meatus of the human ear, they impinge upon the tympanic membrane (ear drum), which closes the innermost end of the auditory canal (Fig. 14–3). This membrane is a thin, cone-shaped structure somewhat like the paper cone in a radio speaker. Sound waves striking it cause it to vibrate in resonance with the frequencies of sound present. Thus, all wave lengths of sound striking the tympanic membrane at any given time produce patterns of vibration in it.

The point of the tympanic membrane cone is directed inward (medially) and is in contact with the handle of the first ossicle (malleus) of the middle ear. Another point of the malleus is attached to the wall of the middle ear cavity just beside the edge of the tympanic membrane.

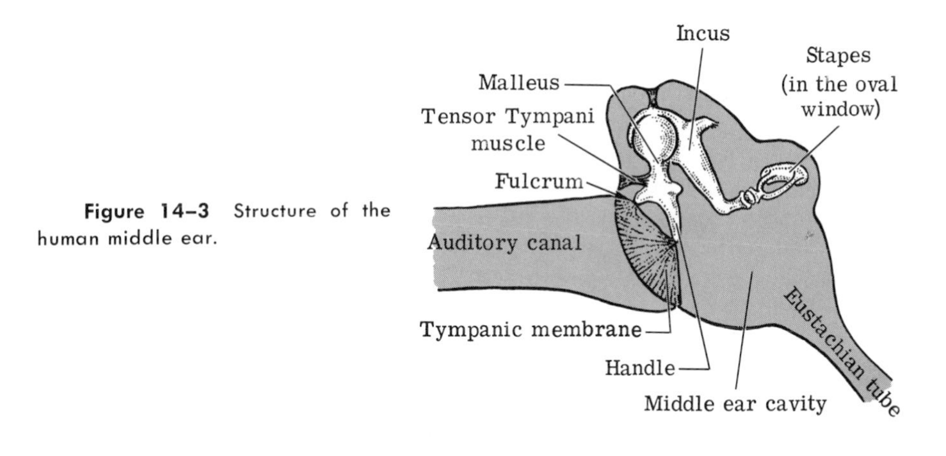

Figure 14–3 Structure of the human middle ear.

This point serves as a fulcrum on which the bone rocks in response to the vibrations of the tympanic membrane. The malleus is immovably attached to the second ossicle, the incus, so the two bones move together as though they were one.

The point of the incus articulates with the third ossicle, the stapes, by a movable joint. The stapes sits, like a piston in a cylinder, in the opening of the oval window, which is an opening through the bone of the skull leading from the middle ear cavity into the perilymphatic space of the inner ear (Sect. 14.1). Movements of the incus move the stapes in and out within the oval window—again, like a piston in a cylinder. Thus, air-borne sound patterns are finally translated into patterns of inward and outward movements of the stapes.

Since the footplate of the stapes is only about a twentieth as large as the tympanic membrane, there is a total concentration of the power of motion resulting from this linkage of ossicles. Thus, the strength of the thrust of the stapes against the perilymph of the middle ear is greatly increased.

The handle of the malleus is constantly pulled inward (medially) by contraction of the tensor tympani muscle. This action maintains the tympanic membrane under some tension. Similarly, the stapes is constantly pulled outward (laterally) by the stapedius muscle. The tonus of both muscles is under reflex nervous control and can be tightened to damp very loud sounds or loosened to increase the sensitivity to very soft sounds. Thus, the central nervous system is able to "tune" the sensitivity of the organ of hearing by controlling the movability of the tympanic membrane and ossicle linkage of the middle ear.

The eustachian tube connects the cavity of the middle ear with the pharynx. Air can pass through the eustachian tube in either direction to equalize the pressure on the two sides of the tympanic membrane. Everyone is familiar with the damping of sound that occurs during a fast ascent or descent in an elevator or unpressurized aircraft. This damping disappears when one swallows or yawns, activities which cause a momentary opening of the eustachian tube and, therefore, pressure equalization. A slight click is often heard as the tube opens.

14.5 MAMMALIAN COCHLEAR STRUCTURE

The cochlear portion of the bony labyrinth is shaped like a snail's shell; in man the spiral has a little more than two and two-thirds complete turns and is partially divided by a bony shelf extending outward from the axis of the spiral (Fig. 14–1). The contribution of the membranous labyrinth to the cochlea is a single, elongate tube, called the cochlear duct or scala media. This is roughly triangular in cross section and it follows the outermost wall of the bony cochlea (Fig. 14–4). The cochlear duct, together with the bony shelf mentioned previously, divides the bony labyrinth into two tubular passages, the scala vestibuli above and the scala tympani below. The cochlear duct itself is thus separated from the scala vestibuli by one of its walls, the vestibu-

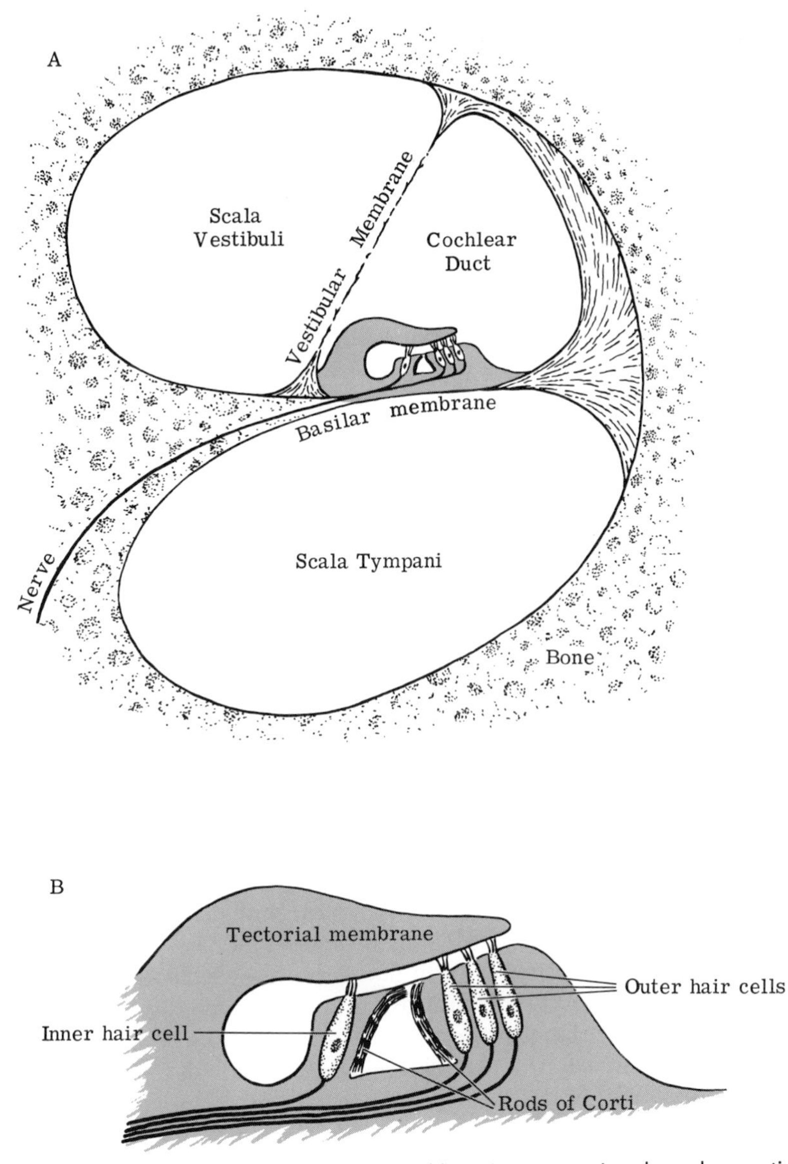

Figure 14–4 Structures of the human cochlea. *A*, cross section through a portion of the cochlea; *B*, the organ of Corti.

lar membrane, and from the scala tympani by a more complex wall, the basilar membrane. The basilar membrane is attached to and continuous with the bony shelf of the bony labyrinth. Thus, the entire cochlear passage is subdivided into three tubular passages, of which two (the scala vestibuli and scala tympani) are filled with perilymph; the third tubular passage (the scala media or cochlear duct) is filled with endolymph. The cochlear duct extends from the saccule and ends blindly at the apex of the cochlear spiral. The scala vestibuli and the scala tympani are extensions of the perilymphatic spaces surrounding the membranous labyrinth. At the apex of the cochlear spiral the scala vestibuli is continuous with the scala tympani; that is, at this point, which is called the helicotrema, there is no bony shelf separating these two passages from each other.

The oval window, containing the footplate of the stapes (Sect. 14.4), opens into the large perilymphatic space between the utricle and saccule. This space is continuous with the perilymph of the scala vestibuli and, by way of the helicotrema, with that of the scala tympani. At the basal end of the scala tympani there is another opening to the cavity of the middle ear. This opening, the round window, is covered by a membrane. Movements of the stapes are translated into sound waves in the perilymph which then pass into the cochlea through the scala vestibuli and thence back out again through the scala tympani. Having reached the basal end of this last passage, they are damped out by the membrane stretched over the round window. If a bony wall were present at this point, these waves would be reflected just as waves in a pond are reflected from the shore. Reflected sound waves passing through the cochlea in the reverse direction would create confusion, or echos, in sound reception. Another way to visualize the operation of this system is to consider that each time the stapes moves inward in the oval window it displaces a certain amount of perilymph and this, in turn, causes outward bulging of the membrane over the round window.

The receptors for hearing are located on top of the basilar membrane and therefore within the cochlear duct. Collectively the receptor cells and associated structures form what is known as the organ of Corti (Fig. 14–4). On top of the basilar membrane is a number of hair cells which are not unlike lateral line neuromasts except that they lack a gelatinous cupola (Sect. 13.13). These hair cells are supported in a framework of other cells. The sensory hairs are directed upward and their tips are actually embedded in the substance of the overhanging shelflike tectorial membrane. In humans the hair cells occur as a single inner row of about 3500 cells and three to four outer rows of more than 5000 cells each. Between the inner and outer rows is an opening bordered by pillarlike structures known as the rods of Corti. From each hair cell there originates a nerve fiber which passes, in the basilar membrane, toward the axis of the cochlear spiral. In this axis it connects, by way of a spiral ganglion, with fibers of the auditory nerve (eighth cranial nerve).

14.6 MAMMALIAN COCHLEAR FUNCTION

When the stapes is pushed into the oval window, the displaced perilymph moves through the scala vestibuli, the helicotrema, and the scala tympani; this causes the membrane over the round window to bulge outward into the cavity of the middle ear. The frequencies of sound waves make this kind of movement of the perilymph fluid impossible; the inertia of the fluid prevents actual flow. Instead, compression in the scala vestibuli causes the basilar membrane to bulge downward

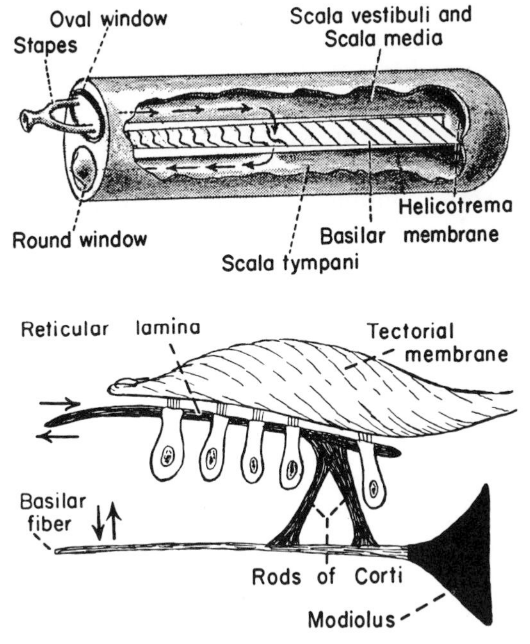

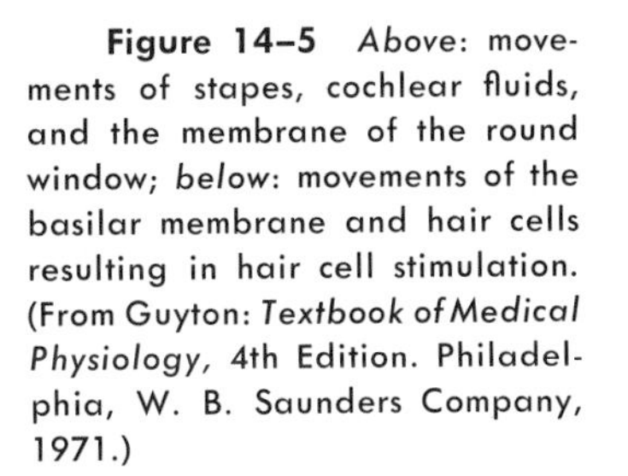
Figure 14–5 *Above*: movements of stapes, cochlear fluids, and the membrane of the round window; *below*: movements of the basilar membrane and hair cells resulting in hair cell stimulation. (From Guyton: *Textbook of Medical Physiology*, 4th Edition. Philadelphia, W. B. Saunders Company, 1971.)

toward the scala tympani. In other words, the compression wave takes a shortcut through the basilar membrane instead of passing through the helicotrema. The vestibular membrane and the endolymph of the cochlear duct move as a unit with the basilar membrane (Fig. 14–5).

Movements of the basilar membrane in response to sound waves in the perilymph cause inward and outward movements of the hair cells because of the way in which they are supported in the organ of Corti. Since the hairs are actually embedded in the substance of the tectorial membrane, this movement causes shearing forces on the sensory hairs, to which they respond with changes in generator potential.

In a manner that is not yet completely understood, the changes in generator potential of the hair cells give rise to volleys of action potentials in the associated neurons. The frequencies of action potentials for all the fibers of the auditory nerve collectively duplicate the frequencies of sound waves being received; but this is, apparently, not true for any single nerve fiber. Instead, each one (especially at high frequencies of sound) produces an action potential only for every second or third sound wave. The activity of nerve fibers, thus, must alternate in order to produce an accurate summation of the sound frequency. The mechanisms controlling this cooperative action between individual receptors remain to be thoroughly understood.

We have seen that sound waves, in effect, pass through the basilar membrane instead of going around by way of the helicotrema. This is particularly true for high-frequency sound waves, which tend to pass through the basilar membrane close to the point where they are generated by movements of the stapes. That is, high-frequency sound waves (treble notes) pass through the basilar membrane near the base of the cochlear spiral (Fig. 14–6). Sound waves of lower frequencies travel further into the cochlea and, thus, stimulate portions of the organ of Corti, which are further from the stapes and closer to the helicotrema. Thus, while these bass notes may stimulate to some extent the entire length of the organ of Corti, it is near the apex of the cochlear spiral that the maximum stimulation occurs. For every frequency of sound there is a corresponding region of the organ of Corti that is maximally stimulated, and apparently only the input from this region to the central nervous system is interpreted by the brain as significant input. The less intense stimulation to other portions of the organ of Corti is in some way ignored or damped out in the central nervous system. The actual physiology of tone discrimination is more complex than has been presented here and it is not yet completely understood.

Sound amplitude (loudness) is discriminated on the basis of greater amplitude of movement of the stapes, and therefore of the basilar membrane. The resulting shear forces on the hair cells presumably cross the sensitivity thresholds of more neurons of the auditory nerve. The Weber-Fechner law (Sect. 13.3) holds for amplitude discrimination and gives the ear an extremely wide range despite the narrower limitations in range of action potential frequencies in the auditory nerve. As we have seen, the sensitivity of the auditory apparatus to sound amplitude can also be adjusted by reflex contraction or relaxation of the tensor tympani and stapedius muscles of the middle ear (Sect. 14.4).

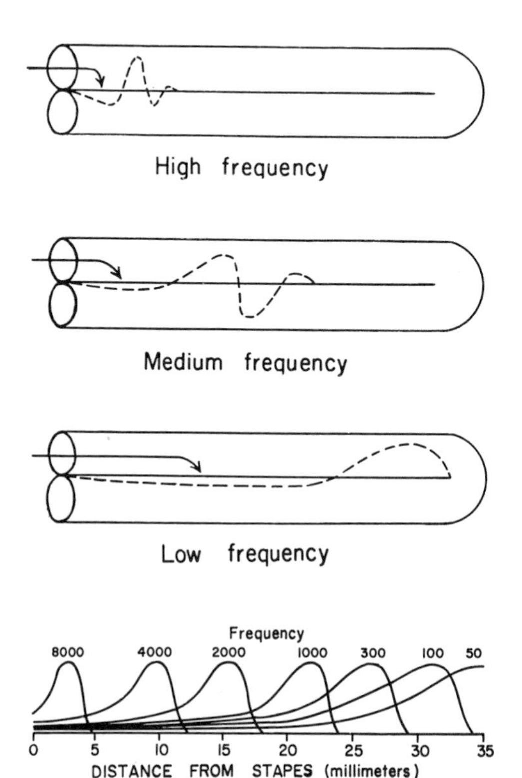

Figure 14–6 Amplitude patterns of "traveling waves" vibrating the basilar membrane of the cochlea. (From Guyton: *Textbook of Medical Physiology*, 4th Edition. Philadelphia, W. B. Saunders Company, 1971.)

14.7 NON-MAMMALIAN AUDITORY APPARATUS

Fishes have no middle ear cavity and no ossicles, except in the special and unrelated case of Weberian ossicles described later. Since fishes are immersed in water and are themselves composed of water to the extent of about 80 per cent, they are relatively "transparent" to water-borne sound waves, and there is no real necessity for any device to carry sound waves inward to the receptors. The tissues of the fish are so nearly like the surrounding aquatic medium that the sound waves continue through them as through the water.

We have already seen that fishes receive low-frequency sound waves in the lagena and, perhaps, in a part of the saccule organ (Sect. 14.3). Usually reception is limited to tones with frequencies of less than 1000 cycles per second, but in a few instances frequencies up to 4000 cycles per second are apparently distinguished. Such cases are attributed to the presence of a unique system of ossicles derived from the transverse processes of vertebrae. These are called Weberian ossicles (Fig. 14–7) and they are not homologous with the ossicles of the mammalian middle ear. The swim bladder (Sect. 3.4) acts as a resonating sound receiver which transmits its vibrations by way of the Weberian ossicles directly to the lagena. Because the swim bladder is filled with gas rather than water, it is not "transparent" to sound as are the other tissues of the fish.

Fishes produce sounds by rubbing fins or gills or by vibrating the swim bladder and they apparently use these sounds as a form of primitive communication. Alarm signals, sex and species recognition signals, and schooling behavior depend, at least in part, on such sounds.

Amphibians and reptiles possess what appears to be a rudimentary organ of Corti, even though it consists of only a few additional hair cells in the lagena. Like fishes, they are generally limited to the reception of low-frequency sounds. Animals of these groups, with a few exceptions such as male frogs, do not produce sounds as a means of communication. The crocodilian ear is more advanced than that of other reptiles because the saccule has a tubular extension (cochlear duct) with an organ of Corti running along its entire length.

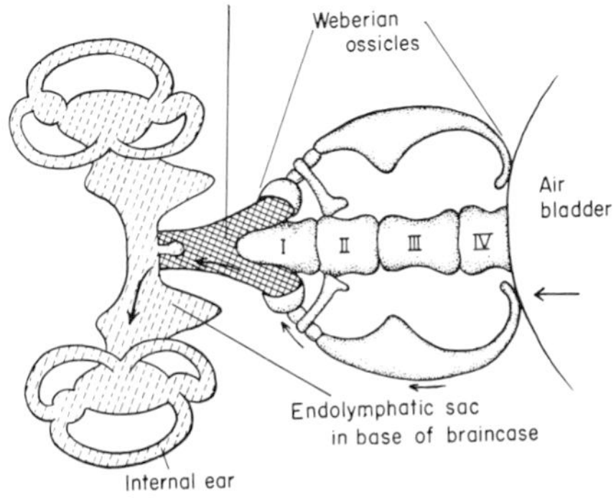

Figure 14–7 Diagrammatic horizontal section of a teleost fish showing the weberian ossicles. (From Romer: *The Vertebrate Body*, 4th Edition. Philadelphia, W. B. Saunders Company, 1970.)

The cochlear duct in birds is longer than in reptiles but it is not coiled as in mammals. An organ of Corti is present but it is not of great length. Birds, of course, make great use of sound for communication, and all evidence suggests that their hearing is generally as good as that of mammals. The relatively poor development of the receptor apparatus makes this high acuity difficult to understand, and further studies of auditory reception in birds are greatly needed. Another difference in the cochlear structure of birds is the absence of a helicotrema. The organ of Corti is much less delicate in structure and there are no rods of Corti as in the mammalian cochlea. Thus, the mechanism suggested for stimulation of hair cells in mammals cannot be operative in birds; some different and unknown mechanism must be involved. The organ of Corti in birds, on the other hand, has about ten times as many rows of hair cells as does the mammalian organ.

Amphibians, reptiles, and birds are alike in that they possess a single ossicle, called the columella, which transmits sound frequencies from the tympanic membrane to the perilymph of the inner ear. This single, rod-shaped bone performs the same task as the mammalian ossicle system by concentrating the power of the sound waves (Sect. 14.4), but muscular damping of sound amplitude cannot occur as it does in mammals.

Unlike fishes, terrestrial vertebrates communicate by means of sounds produced by the respiratory system. Undoubtedly, the hiss of a snake is the basic sound form which has been variously modified by the addition of accessory vocal cords, resonating chambers, and so on. Communication sounds are all of low frequency in cold-blooded vertebrates, but birds and mammals make use of wide frequency ranges. In general, smaller animals have shorter frequency ranges and these tend to be fairly high frequencies. This may well be due to the lack of adequate space in the head for a longer cochlea and for larger vocal organs. It will be recalled that lower frequencies are detected by those portions of the cochlea that are further removed from the stapes and longer cochleas are therefore required to receive these lower tones.

14.8 ECHOLOCATION

A few animals make use of sound as a navigational aid or in the capture of prey. In principle, this is much like the use of radar in air and identically like the use of sonar in water. Sound waves produced by the animal travel out through the air or water and reflect back from any objects which they encounter. The echos are received by the animal that produced the sounds and are interpreted as follows: (1) The time elapsed between the production of the sound and the reception of the echo is the basis for estimating the distance to the reflecting object. (2) Differences in the time of arrival of the echo to the two ears of the animal (plus certain other factors) provide a means for estimating the direction of the reflecting object. (3) Changes in the timbre of the sound result from the absorption of certain wave lengths and the reflection of others, thus giving clues to the texture of the reflecting object. (4) Very small objects will reflect only the short-wave sounds (high frequencies), while larger objects will reflect longer wave lengths as well. Thus, changes in the range of frequencies sent and received form the basis for estimating the size of the reflecting object.

Most common species of bats utilize frequencies of 100 to 200 kilocycles (100,000 to 200,000 cycles per second), since very short wave lengths are required to locate the small insects that they prey upon or to locate with precision the roosting places of the bats. Frequently the echoranging sounds are emitted in modulated bursts, *i.e.*, sounds that begin on very high frequencies and rapidly descend to lower frequencies in each "beep." In other cases, several simultaneous frequencies are produced. The difference between these emitted frequencies and those received as echos is important in echolocation as outlined in the preceding paragraph.

The necessity of rapid and accurate flight in complete darkness makes echolocation a vital adjunct to vision for bats. Just as a bird or an aircraft must reach its stalling speed at exactly the moment of landing (Sect. 13.9), so must a bat. Moreover, this must be done in complete darkness and at the top of a half-loop executed in flight! Thus, extreme accuracy of echolocation is a necessity.

Echolocation has also been developed in other animals inhabiting very dark environments. The oil bird *Steatornis* of Venezuela and the Oriental swift *Collocalia* both live in dark caves and both make use of echolocation in the same ways. Similarly, marine mammals such as whales and porpoises use echolocation for navigation and prey location. These mammals also use high-frequency sounds for engaging in relatively complex communication. Indeed, accumulating evidence now suggests that their communication may be more advanced than that of any vertebrate other than man.

Recent experiments have shown that blind humans also learn to utilize sounds for echolocation. In some cases they produce tiny tongue clicks; or they may rely upon the tapping sound from a cane or upon general background noises. Generally they are unaware of their echoranging ability but practice it subconsciously. It has been shown that people with normal vision can learn to echolocate with some precision after spending a few days blindfolded. Thus, it appears that the ability to echolocate is not uncommon among higher vertebrates but has developed to a great extent only in those species with great need for it.

CHAPTER 15

PHOTORECEPTION

15.1 INTRODUCTION

Light is a form of radiant energy which behaves physically both as a wave, like sound, and as a particle. The wave length varies from 2000 to 100,000 Angstrom units (Å) or from 200 to 10,000 millimicrons (mμ). The human eye is sensitive to only a part of this range—400 to 750 mμ (deep violet to far red). Between these limits lie intermediate wave lengths which we perceive as various colors in the sequence violet, indigo, blue, green, yellow, orange, and red.

The energy of light behaves as though it were in packets or quanta the sizes of which are inversely proportional to the wave length. The quanta of short wave lengths (ultra violet) contain sufficient energy to break chemical bonds and are, therefore, biologically destructive. Longer wave lengths have smaller quanta but they are still large enough to have chemical effects. By absorbing a quantum of light energy a compound may be raised to a higher level of chemical energy where it may be able to undergo reactions that would otherwise be unlikely if not impossible. Also, absorbed quanta of energy may be secondarily passed on from compound to compound, raising the energy level of each in sequence. A mechanism of this kind is involved in the process of photosynthesis in green plants, a mechanism not unlike the reverse of the hydrogen transfer series in animal metabolism (Sect. 10.7). The absorption of quanta of light energy is also responsible for the excitation of visual receptor cells in the vertebrate eye.

Protoplasm in general is affected by the absorption of light energy. Consequently, a generalized response to light is present even among the Protozoa in which no specific photoreceptor structures are present. Protozoans generally avoid bright light but are incapable of detecting, of course, any actual images; that is, they cannot "see." Many higher animals, including the mollusks, arthropods, vertebrates, have receptor organs that can discriminate varying light intensities and, in some cases, actually form images. In addition, there is evidence of non-visual light reception by these animals. The 24-hour "biological clock" which regulates the diurnal activity of many animals is generally "set" daily by dawn or by dusk so that it keeps pace with seasonal changes in day length. The annual cycles of reproduction, migration, hibernation, pelage change, and so forth are also "set" by changes in the relative lengths of darkness and light as the seasons change.

Individual pigment cells in the skins of amphibians respond to light changes by dispersing or by concentrating the pigment within them. Consequently, the relative lightness or darkness of the animal can be changed (Fig. 15–1), and in animals in which more than a single

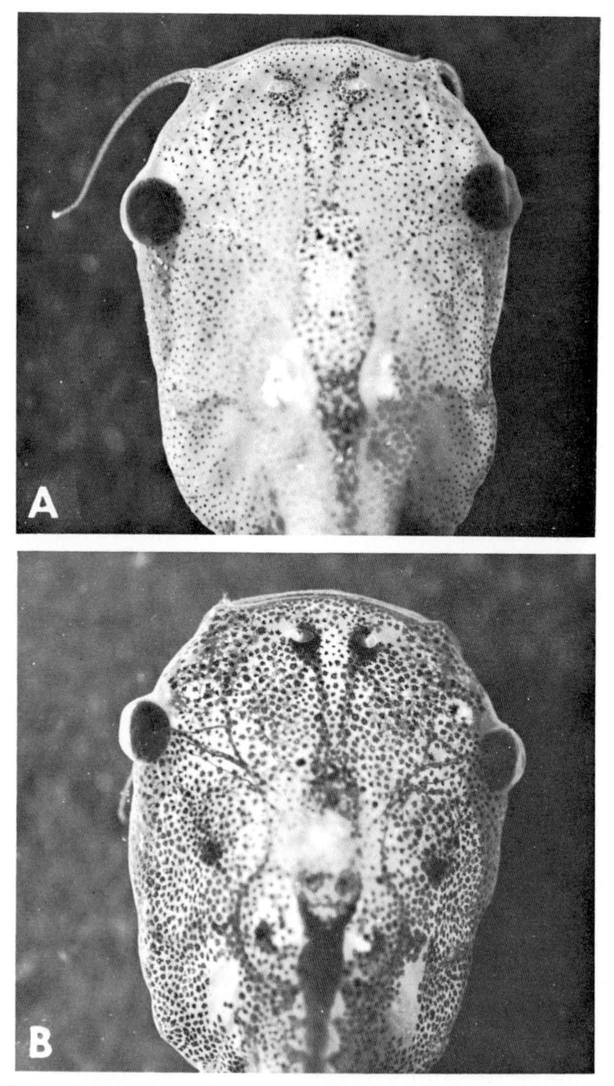

Figure 15–1 Melanophores in the skin of a larval African clawed toad, *Xenopus*. *A*, melanophores contracted; *B*, melanophores expanded. (Photo by Joseph T. Bagnara.) (From Cockrum and McCauley: *Zoology*. Philadelphia, W. B. Saunders Company, 1965.)

pigment is present, striking changes in skin color are also possible. These responses occur as a result of hormonal influences (Sect. 17.5) in the intact animal, but in addition there is a direct effect of light upon the individual pigment cells. The severed and isolated tails of *Xenopus* larvae have no source of endocrine control, yet the melanophores (melanin-containing pigment cells) disperse their pigment in the dark and concentrate it in the light.

Visual receptors of vertebrates are variously adapted to meet the requirements of different species in a variety of ways. Frequently such special adaptations mean the striking of a compromise between two opposing situations. For example, sensitivity to low levels of light requires the additive effects of several receptors, a form of spatial summation (Sect. 11.4). This, in turn, means a reduction in two-point discrimination (Sect. 13.10) and, therefore, only rather crude image formation. The formation of a fine-grained, detailed image requires good lighting and the differential response of many individual receptors. Thus, nocturnal and diurnal animals have different and opposite requirements in this regard. Those animals which are active both day and night (*e.g.*, humans) must strike some compromise that is adequate in both situations.

It is pointless for a fish to have great distance vision since the light-transmitting qualities and the turbidity of most natural waters are limiting factors. For a high-flying predacious bird such as an eagle, however, such distance vision is essential. Moreover, the optical qualities of water and of air are quite different, so the structure of the eyes of fishes and eagles must be quite different in still other ways.

Depth perception depends upon stereoptic or binocular vision in most cases. This type of vision is advantageous to a predator that must accurately judge his distance from the prey before attacking. On the other hand, a very wide visual field is ordinarily possible only with non-stereoptic or monocular vision, a type which is advantageous to an animal that is preyed upon. Such an animal merely needs to detect the presence of the predator, in whatever direction it may be, and move away from it. Thus, the eyes of a cat are placed anteriorly where the visual fields can broadly overlap to give stereoptic vision straight ahead, whereas those of a rabbit are placed laterally where they can together survey the entire horizon. Animals, such as predacious birds, that need both generally searching and localized attacking vision must compromise in some way between these extremes.

Similarly, the ability to perceive motion, color, and contrast is variously present or absent in different vertebrates. The presence of such adaptations corresponds more closely with the ecology of the animal in question than with its phylogeny, a fact which demonstrates that similar adaptations have arisen independently many times in vertebrate evolution.

Probably most important of all is the relative excellence of the central nervous system and its consequent ability to utilize the visual image with greatest efficiency. The actual visual input from the human eye, for example, is poor compared to the conscious perception that it evokes. The difference is due to the ability of the human brain to utilize other sensory inputs plus visual memory to develop and improve upon the actual input from the eyes.

OPTICAL PRINCIPLES

15.2 REFRACTION

Light travels through space at the rate of 186,000 miles per second. It is slowed slightly when passing through air and still more when passing through glass, water, or other transparent substances. The amount by which the speed of light is reduced is spoken of as the optical density of the substance in question.

When light passes from a medium of one optical density into a medium of a different optical density, it is either slowed or accelerated at the interface between the two media. If it strikes this interface at an angle of less than 90°, the change in speed is accompanied by a change in the direction of light as well. This change in direction is known as refraction (Fig. 15–2). Since light travels through glass at only about two-thirds its speed through air, glass is said to have a refractive index of 3/2 or 1.5 as compared with the refractive index of air, which is the standard at 1.0.

The phenomenon of refraction is a common one, applying to many things other than light. For example, if an automobile is driven over a paved surface and into loose sand at an angle of 90°, the sand grips the tires and slows the forward motion of the automobile. If the automobile enters the sand at an angle of less than 90°, one front wheel is gripped and slowed before the other (Fig. 15–3) and, as a result, the automobile is turned (refracted) onto a new course. Again, when the vehicle leaves the sand to return to the pavement, one front wheel is released from the sand first and resumes normal speed before the other front wheel is released. Consequently, the automobile is "refracted" again. If the two pavement-sand interfaces are parallel, the two refractions will be equal

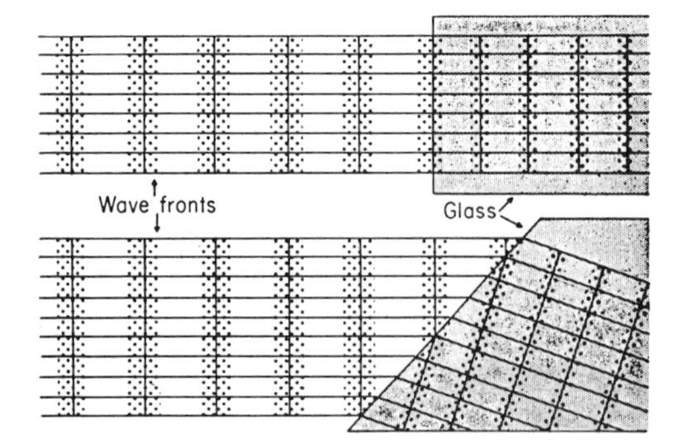

Figure 15–2 Wave fronts entering glass at 90° (*top*) and at an angle less than 90° (*bottom*). Refraction in the latter case results in a change of direction and of wave length. (From Guyton: *Textbook of Medical Physiology*, 4th Edition. Philadelphia, W. B. Saunders Company, 1971.)

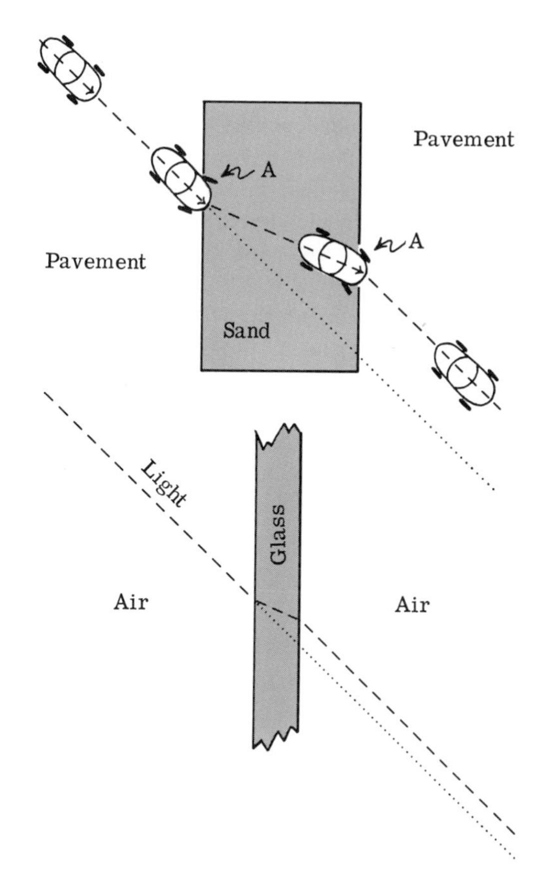

Figure 15-3 The refraction of light by the two surfaces of a pane of glass (*below*) illustrated by the analogous "refraction" of an automobile by the two sides of a sand pile (*above*). See text.

and opposite so that the final course of the car is parallel with the original course.

Another example may be helpful in visualizing the phenomenon of refraction. Imagine a column of soldiers four abreast marching across a smooth field and entering, at an angle of 90°, a strip of deep sticky mud. As each rank enters the mud its forward progress is slowed and as each rank leaves the mud on the opposite side, it resumes its former speed. Imagine, now, this same column entering the strip of mud at an angle of less than 90°. The man on one end of the first rank enters the mud first and slows down. Then the same thing happens to the man next to him and so on until the entire rank has entered. This will have brought about a wheeling maneuver so that any further march through the mud will be at an angle to the original course of the column. The same thing happens to each succeeding rank so that the entire column is "refracted." Again, as each rank leaves the mud on the opposite side, similar refraction turns it again. If the two dry ground–mud interfaces are parallel, the refractions will be opposite and equal and the new course of the column will be parallel with its original course.

Incoming waves on the surface of the sea are refracted when the bottom becomes shallow enough to interfere with the forward progress of the wave. If the line along which this occurs is at an angle to the course of the wave, it will be refracted in a totally similar manner. Many other examples of refraction of waves can be seen in blowing sand, washboard roads, and elsewhere in nature. The same thing is true of light passing through a pane of glass, *i.e.*, it is refracted at both air-glass interfaces unless it strikes these interfaces at an angle of 90°.

The angle at which light strikes such an interface is known as the angle of incidence and the degree by which its course is changed is called the angle of refraction. These two angles are inversely related; as the angle of incidence drops below 90°, the angle of refraction progressively increases. Note that the angle of refraction is zero when the angle of incidence is exactly 90° (Fig. 15–2). Of course, light is also reflected from such an air-glass interface. Reflection, as such, plays no role in the optics of vision, however, so the physical laws applying to it need not be reviewed at this time.

15.3 LENSES

If the two surfaces of a block of glass are not parallel, the refraction at the two air-glass interfaces need not be equal or opposite. For example, light rays passing through a prism (Fig. 15–4) are refracted in

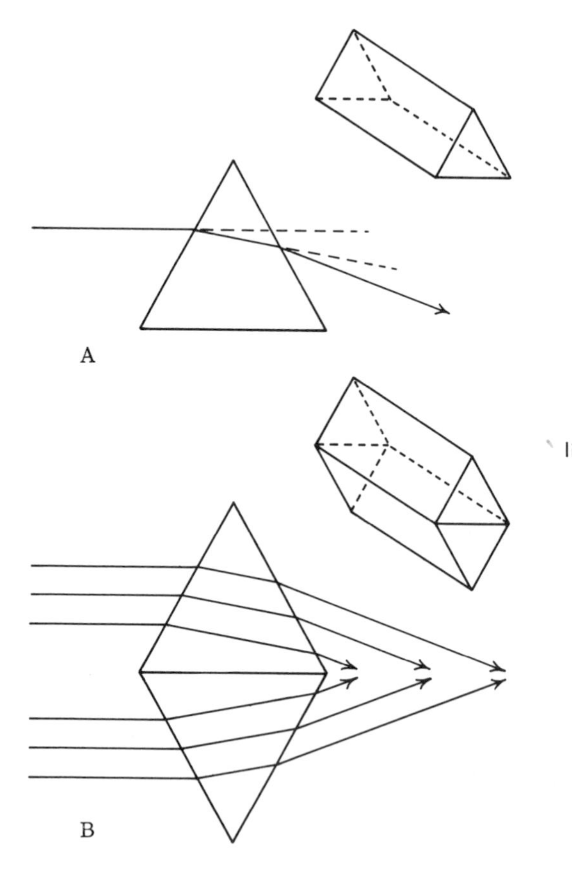

Figure 15–4 Refraction of light by a prism. See text.

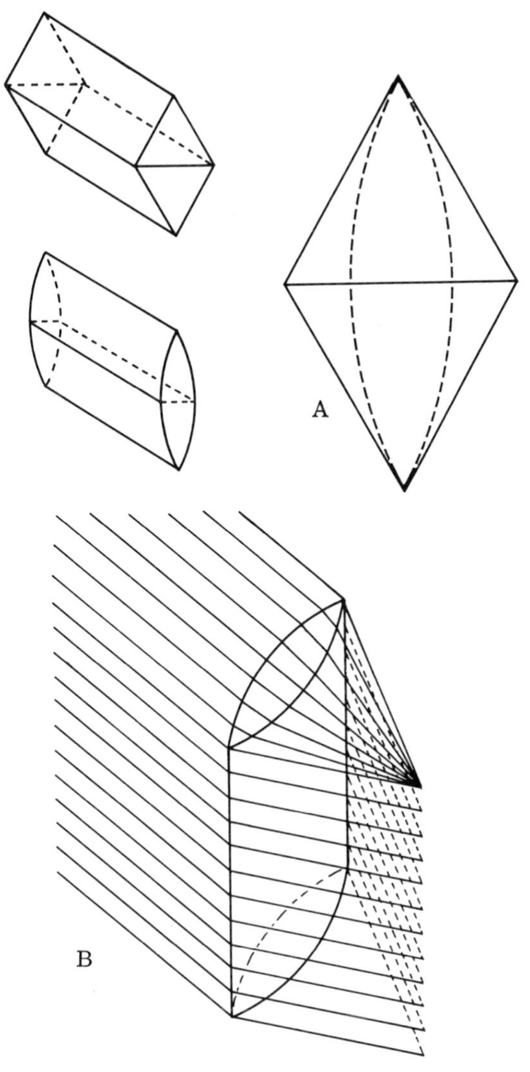

Figure 15–5 If two prisms are cemented together and then ground to a rounded shape (*A*), a cylindrical biconvex lens is formed that will refract light to a line in space (*B*) (see text).

the same direction at each interface. Thus, the final course of the light is not parallel with its original course. If two such prisms were glued base to base, then, parallel light rays entering each would be refracted toward each other. If the flat surfaces of the prisms were then ground to resemble arcs of circles, the angles of incidence of parallel light rays would progressively decrease from 90° at the center to somewhat less at the edges (Fig. 15–5). Light rays striking the prisms near the edges, then, would be refracted more than light rays striking near the center. This would result in all of the light rays being refracted toward a common line at some distance from the glass. Now, if the glass was ground so that both interfaces resembled segments of spheres, the result would be to bring all the light together at a point in space. Such a block of glass is called a biconvex lens, and its shape resembles an

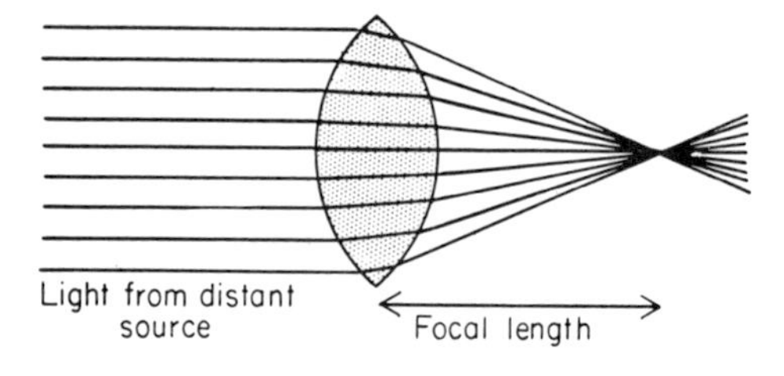

Figure 15–6 The refraction of light by a biconvex lens. (From Guyton: *Textbook of Medical Physiology*, 4th Edition. Philadelphia, W. B. Saunders Company, 1971.)

ordinary reading glass or magnifying glass. It is a matter of common experience that this lens can be used to focus all the parallel rays of sunlight to a point. If a piece of paper is placed at this point, the concentrated energy of sunlight is sufficient to burn the paper. The distance from the central point (node) of the lens to the point at which parallel rays of light are focused is known as the focal length of the lens (Fig. 15–6).

If the curvature of the surfaces of a biconvex lens is increased—that is, if they are segments of a smaller sphere—the angle of incidence at any point other than the center is decreased (Fig. 15–7). This means that the angle of refraction at each such point is increased and, therefore, the focal length of the lens is reduced.

Opposite to the biconvex lens in nearly every aspect is the biconcave lens (Fig. 15–8). Light rays entering such a lens on parallel courses are refracted away from each other in all directions. Therefore, instead of concentrating light to a point, such a lens disperses it in all directions. No such concave interfaces occur in vertebrate eyes but lenses of this kind are used to correct certain refractive defects in human eyes (Sect. 15.8).

A combination of convex and concave interfaces may occur in the same glass lens (Fig. 15–9). Such a lens is called a convexo-concave lens, or a concavo-convex lens. If both surfaces of such a lens have identical curvatures, they cancel each other in effect and no net refraction of light occurs. If either surface has a greater curvature (*i.e.*, is a segment of a smaller sphere) than the other, however, the refractive effect of that surface will more than counteract the refractive effect of the other surface. For example, if the convex side has a greater curvature than the concave side, the total effect will be convergence of the

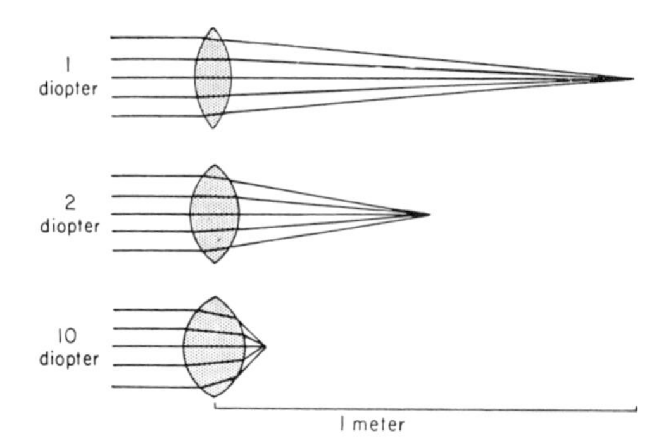

Figure 15–7 The effect of curvature on the focal length of a biconvex lens. (From Guyton: *Textbook of Medical Physiology*, 4th Edition. Philadelphia, W. B. Saunders Company, 1971.)

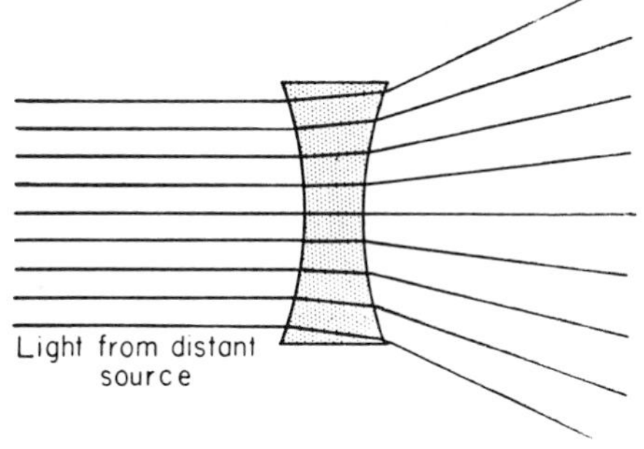

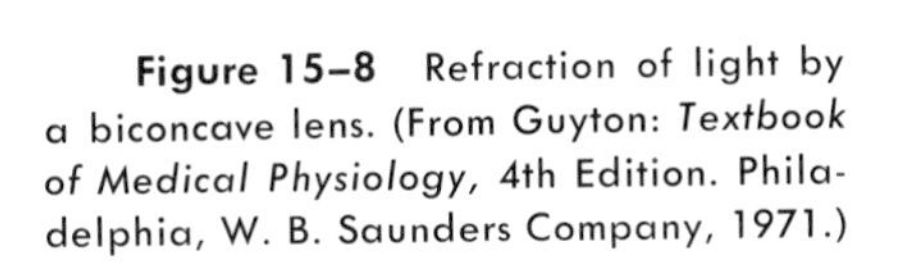

Figure 15–8 Refraction of light by a biconcave lens. (From Guyton: *Textbook of Medical Physiology*, 4th Edition. Philadelphia, W. B. Saunders Company, 1971.)

light. The net refraction is the algebraic sum of the effects of the two interfaces. Eyeglass lenses are commonly of this type since they present a more attractive appearance and interfere less with the eyelashes.

One additional lens type, the astigmatic lens, needs to be described here. In its simplest form this lens has different curvatures along different meridians (Fig. 15–10). If the lens is cut in half along line B–C, it will have a cut surface like that pictured to the right in Figure 15–10. If, however, it is cut along line D–E, it will have a cut surface like that shown at the bottom of the same figure. Since point A is common to both sections, both lens sections have the same maximum thickness. The curvature of the lower section, however, is greater than that of the right-hand section. Sections made at lines intermediate between B–C and D–E will have intermediate curvatures. We have seen that the curvature of the lens affects the angles of incidence and refraction and, therefore, the focal length of the lens. The focal length of an astigmatic lens, then, varies depending upon which meridian (section) of the lens one is considering.

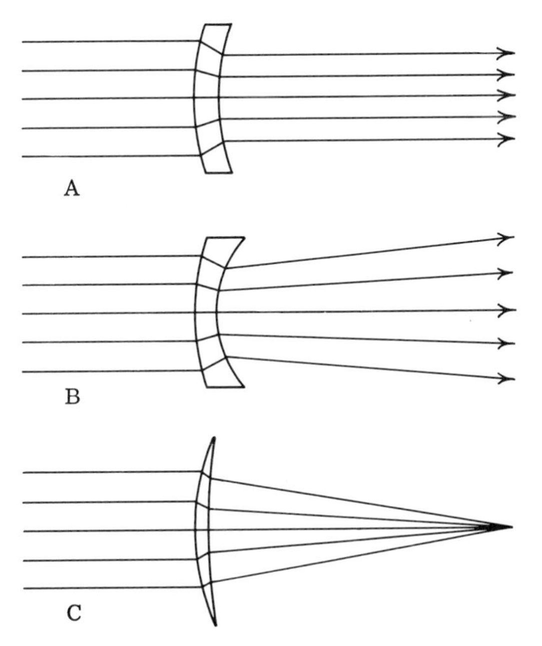

Figure 15–9 Refraction of light by combination lenses. The actual angles of refraction are exaggerated somewhat incorrectly to emphasize the total effect. *A*, a convexo-concave lens with equal curvature on both faces; *B*, a convexo-concave lens with a greater concave curvature; *C*, a convexo-concave lens with a greater convex curvature.

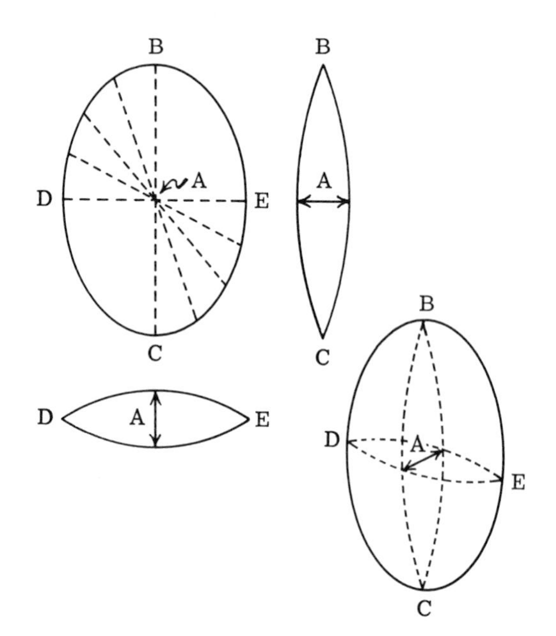

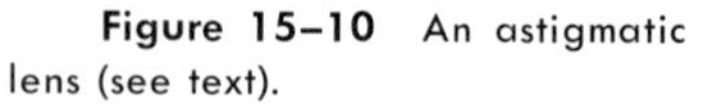
Figure 15–10 An astigmatic lens (see text).

Thus, an astigmatic lens is one which is oblong in shape—somewhat like a flattened egg or flattened football. Since the refractive surfaces of the human eye sometimes present similar uneven curvatures, it is important to understand the ways in which they refract parallel rays of light. To correct such an abnormal refractive surface and restore normal vision requires that the lens placed in the eyeglasses also be astigmatic so that the additive effect exactly cancels the abnormality.

15.4 IMAGE FORMATION

When light strikes a non-transparent object, most of it is reflected in all directions; it is this reflected light subsequently entering the eye that makes it possible for us to see the object. Any object can be thought of as being composed of an infinite number of points, each reflecting light in this way. Let us consider, for example, what happens when light reflects from a person and passes through a biconvex lens (Fig. 15–11). It will be sufficient for us to consider just two reflecting points on the person, one at each end. Light reflecting from all the other points between these extremes will follow intermediate pathways. For each of these two reflecting points, moreover, we need to follow the courses of only two light rays. One of these enters the lens near its edge and the other passes through the center (node) of the lens. Again, light rays intermediate between these will follow intermediate pathways.

Light rays that pass through the node of the lens show no net refraction because the angles of incidence at the two interfaces are

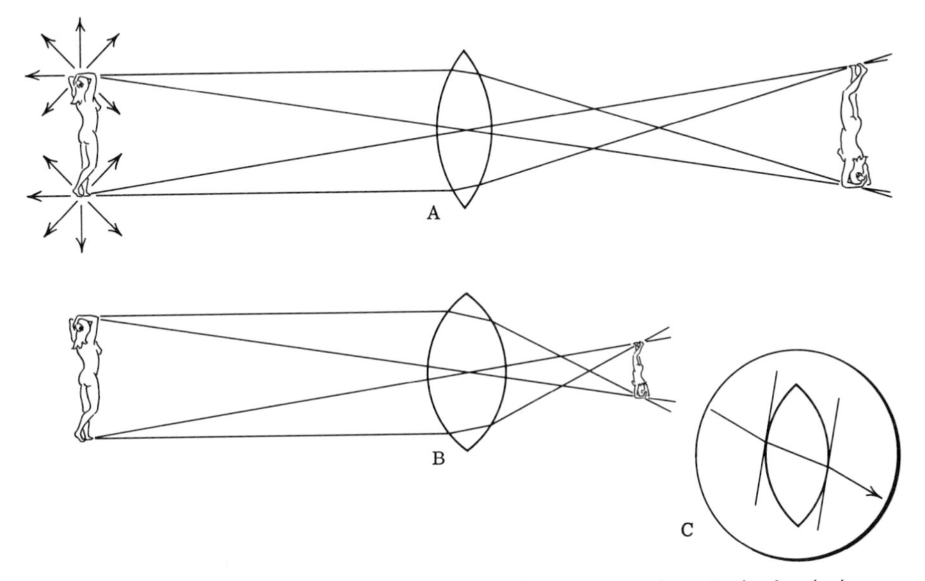

Figure 15–11 *A*, the formation of an image by a biconvex lens; *B*, the focal plane draws closer to the lens as the curvature (refractive power) of the lens increases; *C*, light rays passing through the center of the lens may be considered as unrefracted (see text).

equal and opposite and therefore cancel each other. Even though this ray passes through the node at an angle, the situation is as though the ray were passing at an angle through a pane of glass with parallel surfaces (Sect. 15.2). This fact can be demonstrated by constructing a line tangent to the interface at the point of passage of the light ray on each surface. It will be found that these tangents are parallel with each other.

It is apparent from the diagram (Fig. 15–11) that all light which reflects from the person's head and passes through the lens is brought together again at a point beyond the lens and near its bottom edge. Similarly, all light reflecting from the person's foot and passing through the lens is brought together again beyond the lens and near its top edge. Light reflected from intermediate points along the body of the person is brought together at intermediate points beyond the lens. Thus, the infinite number of points representing the total body surface reflect light through the lens, and each of these points is reestablished along a plane located beyond the lens. If a projection screen is placed in this plane, an upside-down image of the person will be cast upon it.

The plane in which the image is formed is called the focal plane of the lens in question. The distance from the lens to this focal plane is related to the curvature of the lens, as is the focal length of the lens (Fig. 15–7). The focal plane is closer to a lens with greater curvature (a segment of a smaller sphere) than to one with lesser curvature. Differences in the size of the projected image are similarly effected by lens curvature.

If an object is sufficiently far away from a lens, all of the light

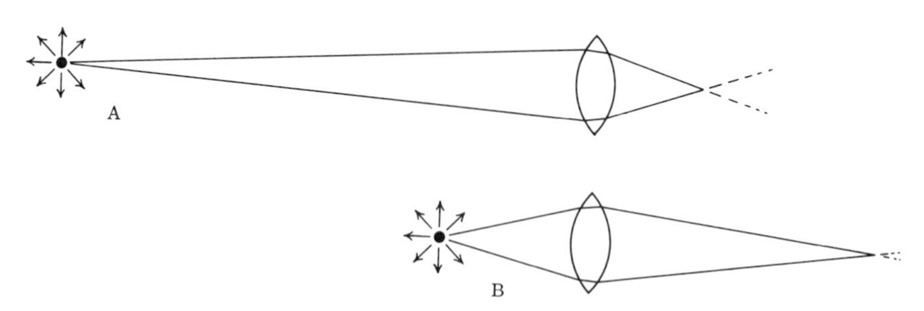

Figure 15–12 Diagram illustrating the effect of the distance of the object viewed upon the focal distance. See text.

reflected from it reaches the lens on essentially parallel courses (Fig. 15–12). If, however, the object is close to the lens, light reflected from it enters the lens on definitely divergent pathways. A lens of given curvature is capable of refracting light by a definite amount regardless of whether that light enters the lens on parallel or divergent pathways. Thus, the point at which the lens forms the image will be further from it when divergent light is passing through the lens and closer to it when parallel light rays are passing through the lens.

All of the principles outlined thus far in this chapter will be of importance in understanding the optical function of the vertebrate eye. Other optical principles that have no immediate application to this subject have not been discussed but will be of interest to the serious student. These are, of course, thoroughly discussed in any college textbook of optics or physics.

THE EYE AS A CAMERA

15.5 STRUCTURE OF THE HUMAN EYE

The human eye is a nearly spherical, fluid-filled, hollow organ. Its walls are composed of three general layers variously modified in different regions of the eyeball to perform different functions (Fig. 15–13). The outermost layer is composed of tough connective tissue that gives rigidity to the spherical body of the eye. This portion, called the sclera, is modified anteriorly to form the optically clear cornea. The corneal bulge has a greater curvature than that of the sclera and is closely covered by a thin conjunctiva. The latter is a transparent continuation of the skin of the face which also lines the spaces behind the eyelids. Together, the cornea and the conjunctiva form the only air-tissue interface in the vertebrate optical system and, as such, the chief refracting surface of the eye.

The innermost layer of the eyeball is the retina; it is present only on the sides and the posterior portion of the eyeball. The retina is, of course, the layer containing the photoreceptor cells about which we shall have more to say presently.

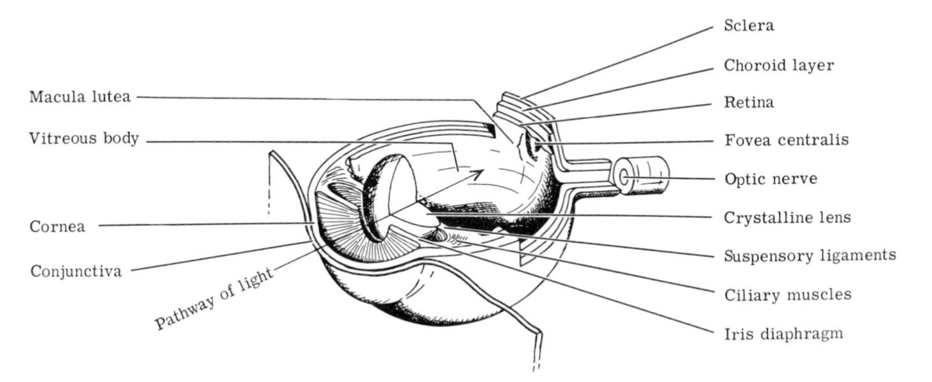

Figure 15–13 The general features of the human eye (diagrammatic). (From McCauley: *Human Anatomy: A Lecture and Laboratory Manual.* Minneapolis, Burgess Publishing Company, 1962. Reprinted in Cockrum and McCauley: *Zoology.* Philadelphia, W. B. Saunders Company, 1965.)

Between the sclera and the retina is an intermediate vascular layer called the choroid. Besides carrying the major blood supply to structures of the eye, this layer contains on its inner surface a heavy deposition of pigment. Continuous with the choroid anteriorly are two sets of muscles. One of these forms the iris diaphragm, the colored portion visible in the eyes of others. It is composed of concentrically and radially arranged muscles which control the size of the pupil of the eye. This, in turn, controls the amount of light entering the eye.

Just behind the pupil is the crystalline lens of the eye. This is held in position by suspensory ligaments which are anchored into the connective tissue framework (sclera) of the eyeball. Also attached to the suspensory ligaments are fibers of the ciliary muscle. This, the second muscle group (besides the iris mentioned above), is more or less continuous with the choroid layer of the eyeball. These muscles function in the control of the curvature of the crystalline lens (Sect. 15.7).

The anterior and posterior chambers of the eye lie on either side (anterior and posterior) of the iris diaphragm and are both filled with a fluid called the aqueous humor. The larger cavity, which is posterior to the crystalline lens, is filled with a semisolid transparent mass called the vitreous humor.

The retina is thinnest at its edges and becomes progressively thicker toward the posterior and medial point, which is known as the optic disc. At the optic disc the retina becomes continuous with the optic nerve. The optic nerve is covered by a connective tissue sheath, which is a continuation of the sclera, and passes posteriorly along the base of the brain. At the posterior pole of the retina there is a slight depression, or thin spot, known as the fovea centralis. This is the area of most acute vision. It is surrounded by a raised, thickened region called the macula lutea.

Attached to the sclera externally are six skeletal muscles which rotate the eyeball in various directions. Another extrinsic muscle, the levator palpebrae, lifts the eyelid to open the eye. The eyelid is closed by contraction of the facial muscles which surround the opening.

15.6 PUPILLARY APERTURE CONTROL

The iris diaphragm is the pigmented ring, which may be black, brown, blue, or, less commonly, another color, surrounding the pupil of the eye and situated just anterior to the crystalline lens. It is composed, as we have seen, of two sets of muscle fibers. One set is arranged concentrically and by contracting decreases the diameter of the pupil. The other set of muscles radiates in all directions from the edge of the pupil to the edge of the iris. Contraction of these muscles increases the diameter of the pupil. Thus, the relative degree of contraction of the concentric muscles and the radiating muscles determines pupillary aperture at any given time. These two sets of muscles are controlled by the autonomic nervous system (Sect. 12.6). Sympathetic stimulation enlarges the pupil while parasympathetic stimulation constricts it.

Pupillary size plays two important roles in vision. Most obvious is the limitation of the total amount of light entering the eye. Light glare on the retina acts through reflex nervous pathways to cause pupillary constriction. Dim light acts through similar pathways to cause pupillary expansion. In addition, the pupillary aperture is important in determining the depth of focus of the eye and must be adjusted when the eye views objects that are close to it (Sect. 15.7).

The human pupil is adjustable between diameters of 1.5 and 8.0 millimeters. The amount of light entering the eye is proportional to the square of the diameter, so this range permits a 30-fold difference in the light intensity reaching the retina.

15.7 ACCOMMODATION TO NEAR VISION

We have seen that when objects are sufficiently far away from the eye, light rays reflecting from them enter the eye on essentially parallel courses (Fig. 15-12). For all practical purposes this is true for any object which is slightly beyond arm's reach. The refractive power of the air-tissue interface (corneal and conjunctival surfaces) plus that of the crystalline lens in its natural resting shape is such that the images of distant objects fall exactly upon the retina. When an object is moved closer to the eye, however, light rays reflecting from it enter the eye on divergent courses. If no adjustments were made to correct for this, the image would fall further behind the lens—that is, actually behind the retina (Sect. 15.4). If the refractive power of the eye could be increased in some way, the plane in which the image is formed could be moved forward so that it fell exactly upon the retina as it should. This can be accomplished by increasing the curvature of the crystalline lens (Figs. 15-7 and 15-11). Note that the same end is achieved in a camera by pulling out the bellows or lens tube to increase the lens-to-film distance. This kind of adjustment is impossible in the human eye, but a similar kind of adjustment is actually made in some vertebrate eyes (Sect. 15.11).

The mechanism by which the shape of the human crystalline lens is changed to accommodate for near vision is perhaps best understood by means of an analogy. Visualize for a moment a spherical rubber balloon upon which we have marked an equator. Strings are attached at close intervals along this equator. Now, we place the balloon in the center of a ring of people and each person holds one of the strings (Fig. 15-14). If all the people back up sufficiently to create some tension on the strings, the spherical balloon will assume the shape of a biconvex lens. Like the balloon, the crystalline lens of the eye is also an elastic structure, and the suspensory ligaments holding it in position exert pull on it in the same manner as the people pulling on the strings. If the suspensory ligaments were all severed, the crystalline lens would become spherical, as would the balloon if all of the people dropped their strings. Let us assume, however, that our balloon has been stretched to form a biconvex lens roughly the shape of the crystalline lens when an object four feet from it is being viewed. If all of the people in the ring take a short step forward, the balloon will acquire a greater curvature; that is, it will become a thicker biconvex lens. If the suspensory ligaments could be made to relax some of the tension on the crystalline lens, it too would become a thicker, more refractive lens. Then it would be able to focus an image of an object perhaps three feet away exactly

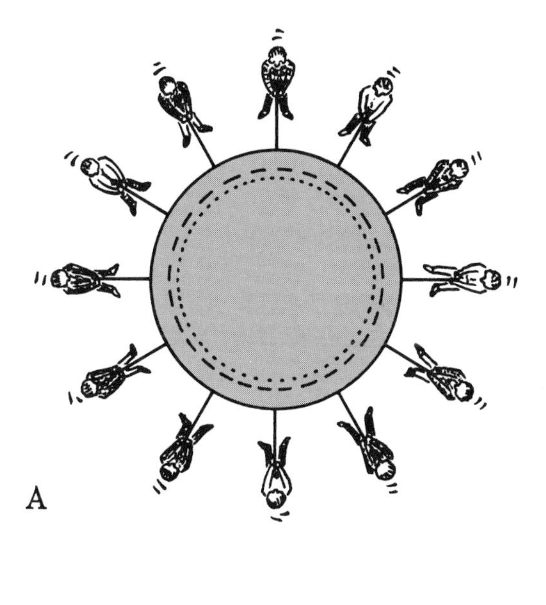

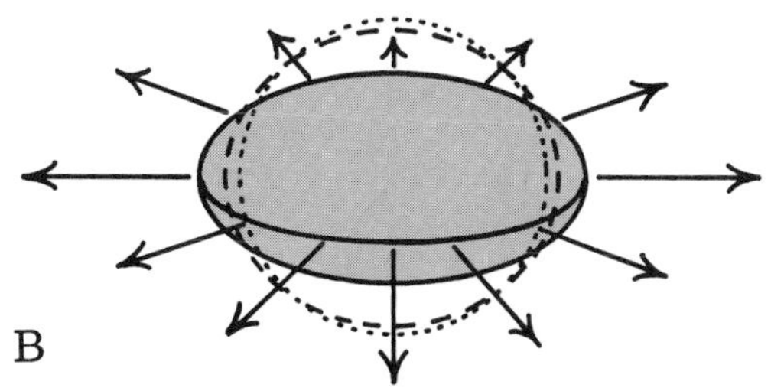

Figure 15–14 An analogy illustrating how the lens of the eye can change in shape and, consequently, in refractive power (see text).

on the retina. Further relaxation of tension of the suspensory ligaments would, similarly, accommodate the eye to viewing objects at still smaller distances.

Relaxation of the suspensory ligaments is accomplished by contraction of the ciliary muscles of the eye (Sect. 15.5). The fibers of these muscles are arranged in such a way that contraction pulls the scleral attachments of the suspensory ligaments forward and centrally toward the crystalline lens (Fig. 15-15). This lessens the pull of the suspensory ligaments on the crystalline lens and allows the lens to become thicker and more refractive. Thus, while the air-tissue interface at the surface of the cornea and the conjunctiva is the greatest refractive surface of the eye, changes in shape of the crystalline lens are sufficient to make the small adjustments in refraction that are necessary to accommodate the eye to near vision.

Because close work requires constant contraction of ciliary muscles, such work is tiring to the eyes. Distant vision requires no such muscular activity and can be carried on for prolonged periods without noticeable fatigue.

With increasing age the crystalline lens tends to lose much of its elasticity. As a result, contraction of the ciliary muscles has little or no effect on the shape of the lens. It becomes increasingly difficult for persons older than about 45 years to focus accurately upon objects held close to the eyes, and they commonly tend to hold reading matter as far away as they can while examining it. This condition, known as presbyopia, frequently requires the use of reading glasses or, if glasses are already worn for the correction of other defects (Sect. 15.8), the addition of a bifocal lens to the eyeglasses. Bifocal lenses have an upper region for distance viewing and a lower region, which is more refractive, for reading and other close vision.

When an object is held close to the eye, a greater proportion of the total amount of light reflected from that object is able to enter the eye (Fig. 15-12). This means that more total light enters the eye and pupillary constriction must occur to reduce it to appropriate levels. Pupillary

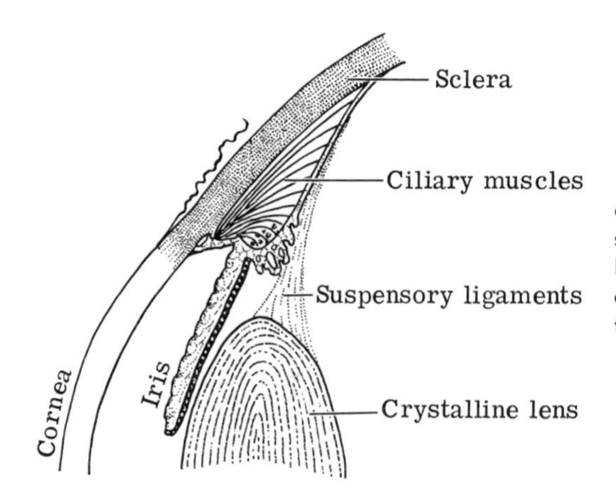

Figure 15–15 Contraction of the ciliary muscles pulls the suspensory ligaments toward the lens. This allows the lens to become more nearly spherical. See text.

constriction also increases the depth of field, making it easier to focus the image clearly on the retina. Thus, pupillary constriction is also a part of the process of accommodation to near vision.

15.8 COMMON VISUAL DEFECTS AND CORRECTIONS

We have already seen that astigmatism results from uneven curvatures of the major refractive surfaces of the eye, *i.e.*, the air-tissue interface of the cornea and conjunctiva (Sect. 15.3). Such uneven curvatures can be corrected by means of lenses placed in the eyeglasses that have opposite, and therefore cancelling, astigmatic curvatures. The use of contact lenses automatically corrects for astigmatism since the air-glass interface of the lens actually replaces the faulty air-tissue interface of the eye.

We have already considered presbyopia, another common visual defect, and its correction in the preceding section.

Two additional common visual defects are myopia and hyperopia. These usually result from an eyeball which is out-of-round, that is, in which the lens-to-retina distance is incorrect (Fig. 15-16). In cases of myopia, this distance is too great, which results in an image plane that lies in front of (anterior to) the retina. Myopia can be corrected by decreasing the refractive power of the eye so that it forms its image further behind the lens. This, in effect, means that we need some initial divergence of the light rays, that is, a biconcave lens in the eyeglasses.

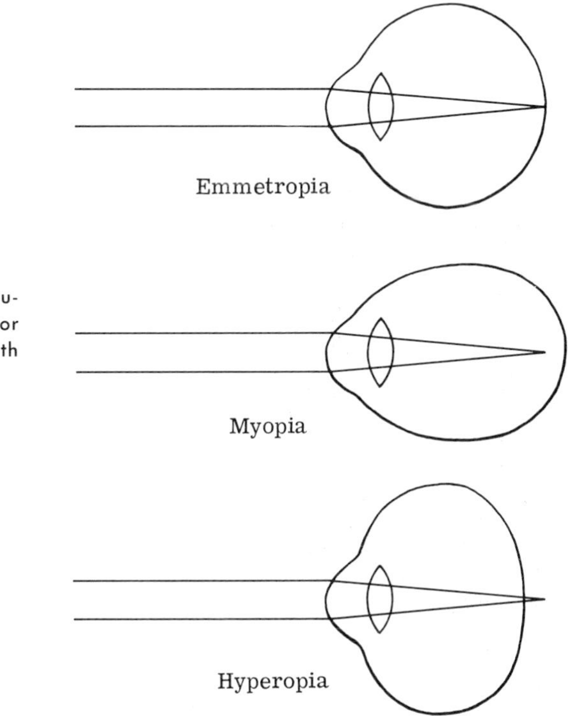

Figure 15–16 Common human visual defects; the normal or emmetropic eye compared with myopic and hyperopic eyes.

Hyperopia is due to the opposite condition: when the lens-to-retina distance is too short, the image needs to be brought forward. This can be done, of course, by increasing the refractive power of the system, and a biconvex lens in the eyeglasses accomplishes this. Actually, biconcave and biconvex lenses are not commonly used in eyeglasses; rather, convexo-concave lenses of appropriate design (Fig. 15-9) are used to accomplish the same purposes.

The common terms near-sightedness and far-sightedness are not very descriptive and, therefore, are not used scientifically. Both hyperopia and presbyopia are types of far-sightedness, but these are very different conditions and are corrected in very different ways. The student will be wise to forget the common terms near-sightedness and far-sightedness and concentrate upon learning the proper names for visual defects; in the long run they are much less confusing.

THE RETINA

15.9 RETINAL STRUCTURE

The human retina contains receptor cells of two types, rods and cones (Fig. 15-17), as well as associated neural cells and their nerve processes. Each human retina contains about 1.2×10^8 rods and about 5.5×10^6 cones. The nerve cells and processes are innermost so light must penetrate through these layers before reaching the sensory cells themselves. This seemingly inefficient arrangement is a consequence of the embryonic formation of the retina as an extension of the central nervous system. The student should consult any good textbook of ver-

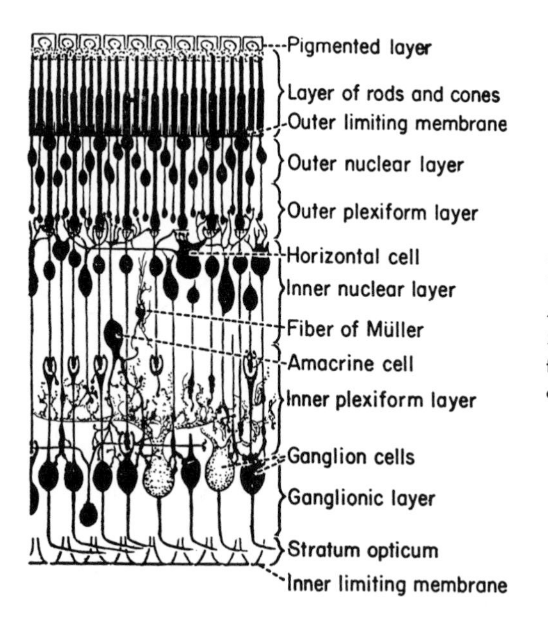

Figure 15–17 Organization of the human retina. (From Guyton: *Textbook of Medical Physiology*, 4th Edition. Philadelphia, W. B. Saunders Company, 1971. Reprinted from Polyak: *The Retina*, University of Chicago Press.)

tebrate embryology for the details of retinal development. It is interesting to note in passing that the retinas of the cephalopod mollusks, which arose independently in evolution, are very similar to those of vertebrates except that the layers of the cephalopod retina are arranged in the opposite order. In these animals the receptor cells are exposed on the innermost layer of the retina so that light entering the eye strikes them without passing through any neural tissues en route.

The area within the fovea centralis of the retina (Fig. 15-13) contains closely packed cones but no rods. The retina is thinner at this point and thicker in the immediately surrounding area because the neural elements are diverted to one side. In this region, then, light penetration to the receptor cells is greatly improved. From the fovea centralis outward in all directions there are progressively fewer cones and more rods until at the periphery of the retina only rods occur.

Both rods and cones contain photosensitive chemical substances which undergo reactions when they absorb photic energy. Much less is known of the chemistry of cone reception, but it is undoubtedly similar in principle to the chemistry of the rods, about which a great deal is known (Sect. 15.10).

In the region overlying the optic nerve there are no receptor cells at all. Consequently, this part of the retina, called the optic disc, is blind. We are not consciously aware of this blind spot in the visual field, but its presence can be demonstrated by means of an experiment commonly seen in children's books of games and tricks (Fig. 15-18).

Cones require good light intensity and are, therefore, only stimulated in daylight (photopic vision). Each cone is associated with a separate neuron leading to the visual cortex on the occipital lobe of the cerebrum (Fig. 12-8); this guarantees good resolution, or two-point discrimination. As in the case of tactile discrimination (Sect. 13.10), resolution in vision requires that one unstimulated cell be present between two stimulated cells if two separate points are to be resolved. Because of these characteristics and of the distribution of cones on the retina, we see in detail only that portion of the visual field which falls on the fovea centralis. If one looks fixedly at some object and then shifts his attention (but not his eyes) to the side, it becomes immediately apparent that detailed vision (good resolution) is limited to the center of the field.

Cones are capable of distinguishing color, whereas rods are not. Again, if one looks fixedly at an object and then shifts his attention to the side, he will find that colors are seen only in or near the center of

+ O

Figure 15–18 A demonstration of the blind spot on the human retina. Holding the book at arm's length, cover the left eye and look fixedly at the + sign with the right eye. Now move the book slowly toward the eye. Note that at a particular distance from the eye the 0 sign is not visible. At that distance the image of the 0 sign falls on the blind spot. The 0 sign is visible when the book is held either closer to the eye or further from it since in both cases the image falls on sensitive retinal areas.

the visual field. This is true because the peripheral portions of the visual field fall upon peripheral portions of the retina where only rods are present.

Rods, on the other hand, are relatively sensitive and require only small amounts of light to be stimulated. This sensitivity permits scotopic or night vision, and it is a matter of common experience that in darkness one sees the periphery of the visual field better than the central portions. The sensitivity of the rods is probably related to the fact that several rods typically converge upon a single neuron. Such an arrangement makes spatial summation possible (Sect. 13.2) so that low intensities of stimulus are sufficient to produce action potentials in the common neuron. At the same time, however, resolution cannot be as good because several receptors effectively serve as a single receptor. Night vision, therefore, lacks the precise two-point discrimination of day vision. Also, of course, night vision lacks color discrimination since the rods are incapable of it.

The structure of rods and cones is highly specialized. Rods have a terminal portion that is apparently derived from a cilium (Fig. 15-19). The nine typical peripheral fibers of a cilium (Fig. 2-7) are present but, in addition, there are many disclike plates having what appears to be a unit membrane structure (Sect. 2.2). These plates also contain a specific

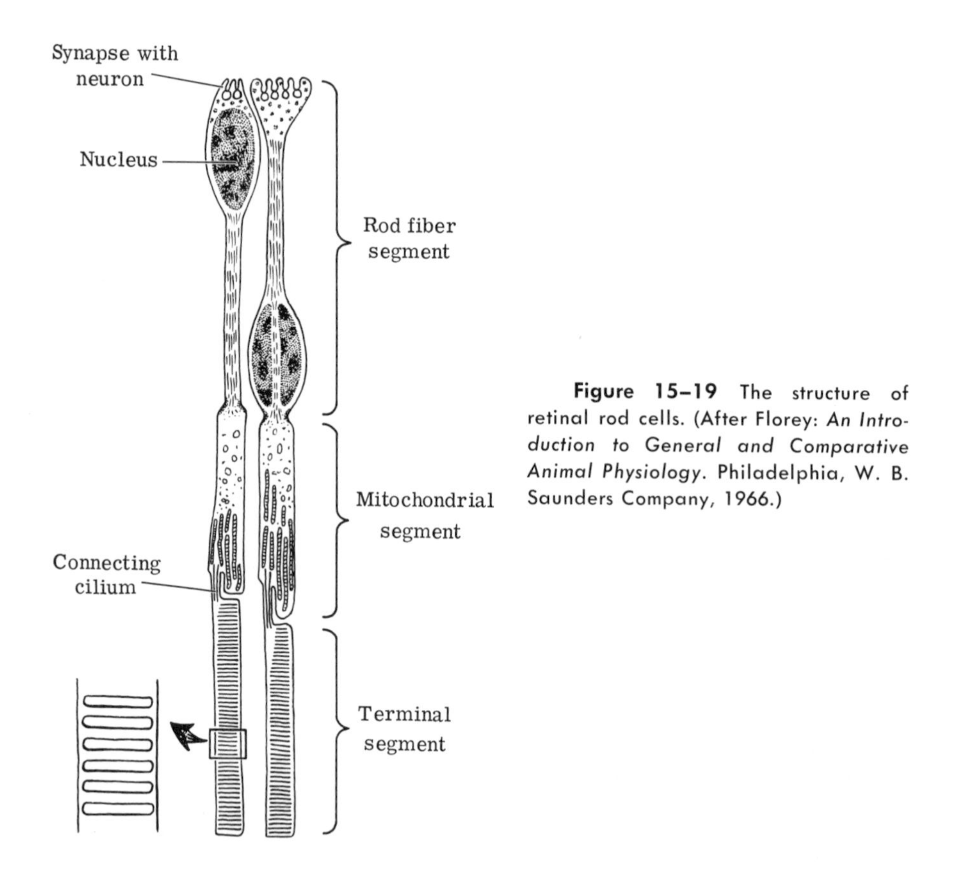

Figure 15–19 The structure of retinal rod cells. (After Florey: *An Introduction to General and Comparative Animal Physiology*. Philadelphia, W. B. Saunders Company, 1966.)

photoreactive substance called rhodopsin. The adjacent segment of the rod cell contains numerous large mitochondria. It is interesting to note in passing that the light-receptive cells concerned with photosynthesis in green plants also have layered, platelike organelles in which the light-absorbing chlorophyll is found.

15.10 RETINAL FUNCTION

The terminal segment of each rod contains the photosensitive pigment rhodopsin, or visual purple as it is sometimes called. When this substance absorbs a sufficient amount of light energy, it splits into its constituent parts, opsin (a protein) and trans-retinene (Fig. 15-20). This reaction alters the generator potential of the rod, which, in turn, produces activity of the associated neuron. Rhodopsin can be resynthesized by recombining the retinene with the opsin but this recombination requires the prior isomerization of trans-retinene to cis-retinene. The enzyme retinene isomerase catalyzes the reaction. Both trans- and cis-retinene can be derived from the corresponding isomers of vitamin A (Fig. 15-21). This relationship explains the important role of vitamin A in night vision (Sect. 9.5).

Presumably, similar photochemical processes take place in cones, and one photosensitive substance, iodopsin, has been identified there. In order for color discrimination to take place it is postulated that three different types of cones must be present. Each type, theoretically, contains a different photosensitive compound which absorbs light within a different range of wave lengths. One of these absorbs red light, one absorbs green light, and one absorbs in the blue-violet portion of the spectrum (Fig. 15-22). The ranges of wave lengths absorbed overlap broadly, but the maximum absorption occurs in a fairly narrow band of the color spectrum. Just as an artist can mix these primary colors and obtain any other color of the spectrum, so can the mixed stimulation of

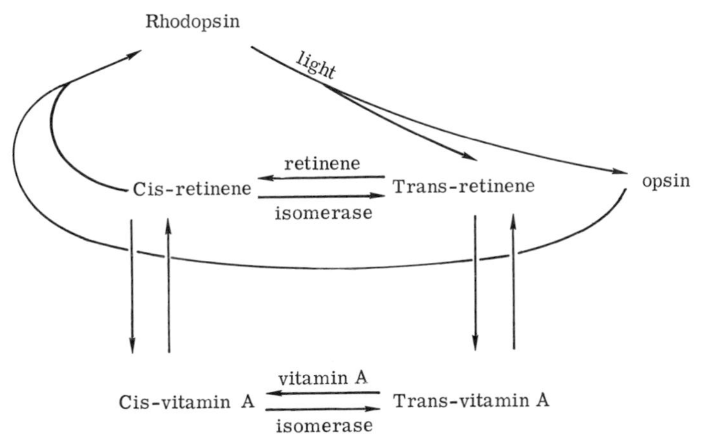

Figure 15–20 The chemical reactions involved in rod vision (diagrammatic).

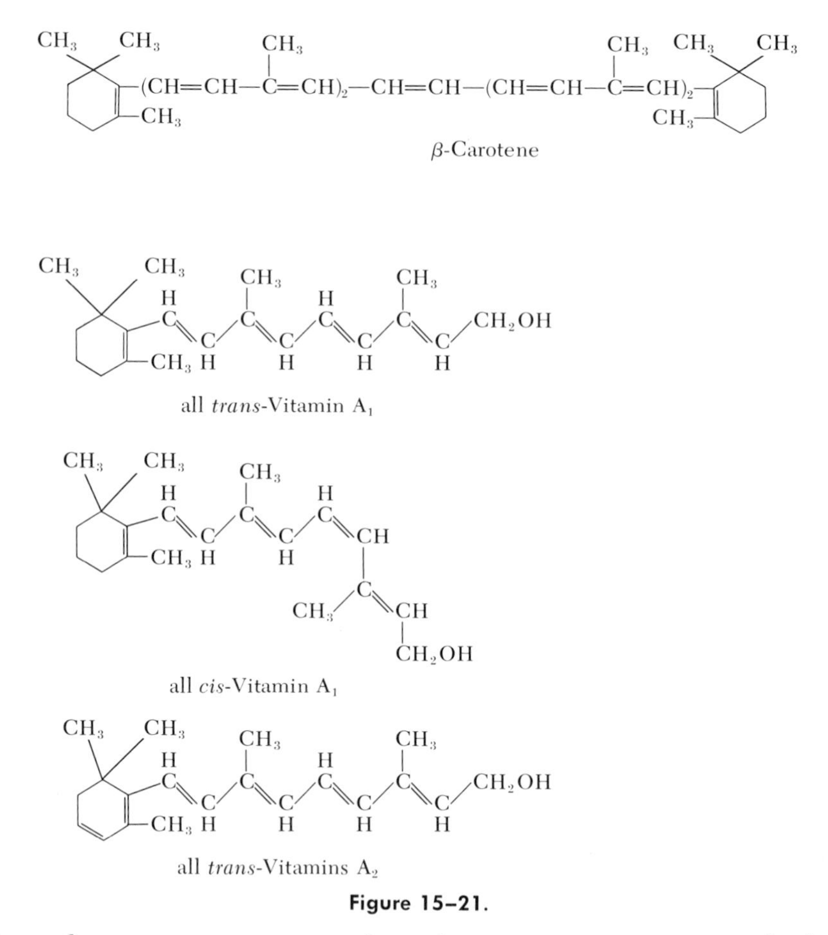

Figure 15–21.

these three receptor types produce the conscious perception of other colors in the visual cortex of the cerebrum. There is no visible difference in the structure of the three cone types, and the differences in photosensitive substances have not yet been determined experimentally.

After exposure to bright daylight most of the rhodopsin in the rods has been broken down to form opsin and trans-retinene. Consequently, when one first enters a darkened room he is for a time nearly blind. Light intensity is below the threshold for cones and the resynthesis of rhodopsin is a time-consuming process. The delay is due, in part, to the fact that several different isomers of retinene are formed under the influence of retinene isomerase. The 11-cis form is produced in the smallest quantity, yet this is the only one which can be used in the synthesis of rhodopsin. More than half an hour is ordinarily required to attain maximum rod sensitivity, although more than half of the recovery is attained within about ten minutes. Following such dark adaptation of the retina, a return to bright daylight produces a few seconds of virtual blindness due to dazzle. During these few seconds all of the rods are stimulated maximally, and, moreover, the light intensity is sufficient to stimulate cones as well. The result is a confusion of nerve impulses reaching the brain along almost all of the neurons of the optic nerve.

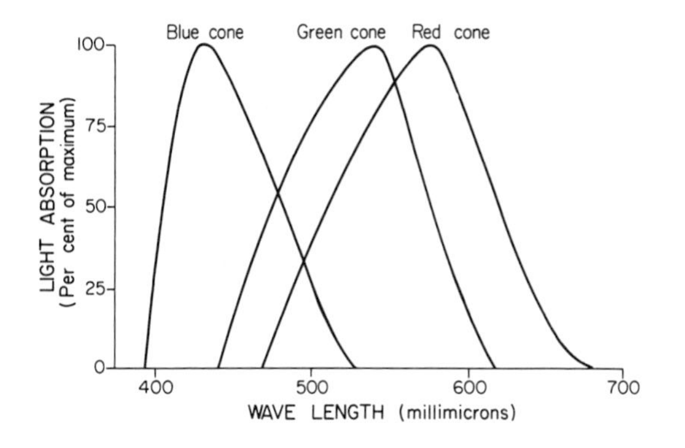

Figure 15–22 Light absorption by the respective pigments of the three cone types of the human retina. (From Guyton: *Textbook of Medical Physiology*, 4th Edition. Philadelphia, W. B. Saunders Company, 1971.)

Figure 15–23 A demonstration of negative after-image. Stare fixedly at the white butterfly for a few minutes and then look at a blank white wall. The after-image of the butterfly will appear to be black.

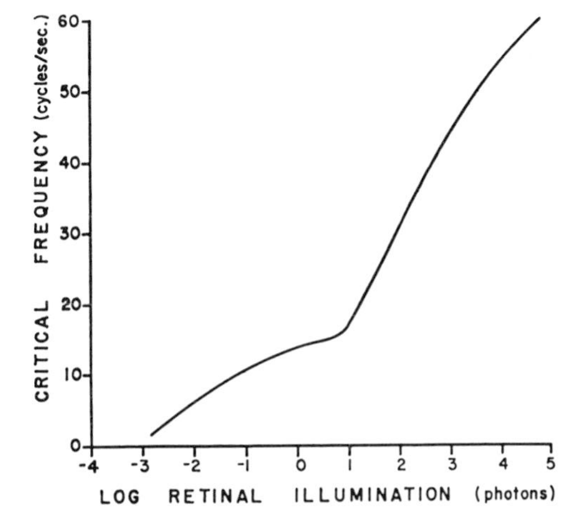

Figure 15–24 Relationship of the intensity of illumination to the critical frequency for fusion. (From Guyton: *Textbook of Medical Physiology*, 4th Edition. Philadelphia, W. B. Saunders Company, 1971.)

Within a few seconds, however, all of the rhodopsin is again broken down into opsin and trans-retinene and the rods become less functional. Thus, while dark adaptation is a slow process, light adaptation occurs very quickly. When driving at night the lights of oncoming vehicles, especially of those who forget to dim their lights, blind one for a few seconds because of the dazzle effect. More important, however, is the fact that by breaking down large amounts of rhodopsin in the rods, this flash of bright light reduces rod activity for more than a half-hour. This, of course, means that night vision is impaired. Since the edges of the highway are seen by the peripheral (rod) portions of the retina, this reduction of night vision ability adds substantially to the dangers of night driving.

If one looks fixedly at a black and white picture that has good contrast (Fig. 15-23), the black portions cause dark adaptation and the white portions cause light adaptation of corresponding parts of the retina. If one then looks at a blank white surface, the dark adapted regions are stimulated more than the light adapted regions. Consequently, one seems to see a negative image: the black regions of the original picture appear white and the white regions appear, now, to be black. Similarly, by looking fixedly at a picture or design consisting of blocks of primary colors (red, green, and blue), one can obtain an after-image on a blank white surface that appears to be in complementary colors; that is, red areas of the picture will appear to be green, and so forth. This effect is due to fatigue of specific cones which are sensitive to the respective primary colors.

Since the excitation of a photoreceptor cell outlasts the duration of the stimulus, successive light flashes may be perceived as being uninterrupted. Motion pictures consist of individual, non-moving pictures which are presented successively at the rate of 24 per second. Between the presentation of pictures there is actually a moment of darkness when there is no picture on the screen. The excitation of the receptor cells from the preceding picture continues through this period of darkness, however, so that we are not aware of the darkness at all. Instead, the preceding picture is changed slightly by the presentation of the next, slightly different picture, and the result is that the picture seems to move. This phenomenon of successive light flashes fusing into a continuous perception is known as flicker fusion. The frequency at which flashes must be presented to be fused in this way depends upon the amount of illumination of the human retina (Fig. 15-24). The frequency of flicker fusion differs in different animals.

VERTEBRATE ADAPTATIONS

15.11 REFRACTIVE ADAPTATIONS

Since the difference between optical densities of water and tissue is not great, the corneal surface in aquatic vertebrates is not a very refractive interface. Consequently in fishes the corneal surface is more flattened and the lens is more spherical in shape. The lens in these animals plays a more important role in refraction than it does in terrestrial species. Often the corneal surface is uneven (astigmatic or somewhat roughened), but this makes little difference in aquatic vision.

The opposite is true, of course, in terrestrial animals. The air-cornea interface is extremely refractive because of the great difference in the optical densities of air and tissue. In these animals, as we have seen, the lens is less important as a refractive structure and only operates to make small adjustments in focusing on objects that are close to the eye (Sect. 15.7). *Anableps anableps*, the "four-eyed" minnow of Central and South America, preys upon flying insects. It swims with the eyes half in and half out of the water and has, in effect, two separate optical systems. The curvatures of both the cornea and the lens are pear-shaped so that half serves to form an image of that portion of the visual field which is under water and the other of that portion which is above the water. Each eye has two separate retinal areas as well.

In general, nocturnal animals tend to have larger corneal surfaces than do diurnal forms. This results in more efficient light-gathering properties and, therefore, more acute night vision.

The eyes of cyclostomes and teleosts are generally focused for near vision at rest and must be actively accommodated for distance vision. It

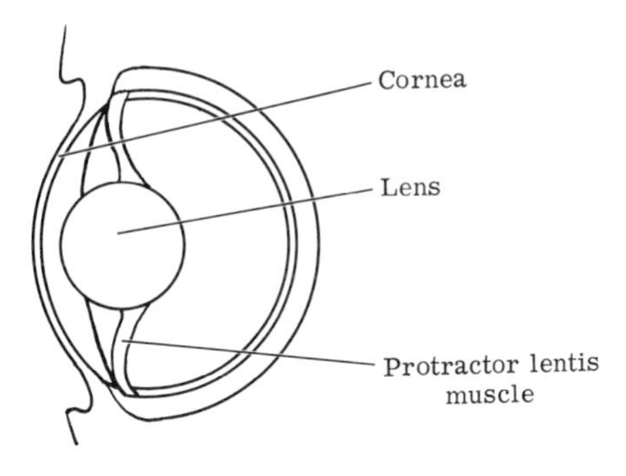

Figure 15–25 Comparison of fish eyes to show the differences in accommodation mechanisms. (After Cockrum and McCauley: *Zoology*. Philadelphia, W. B. Saunders Company, 1965.)

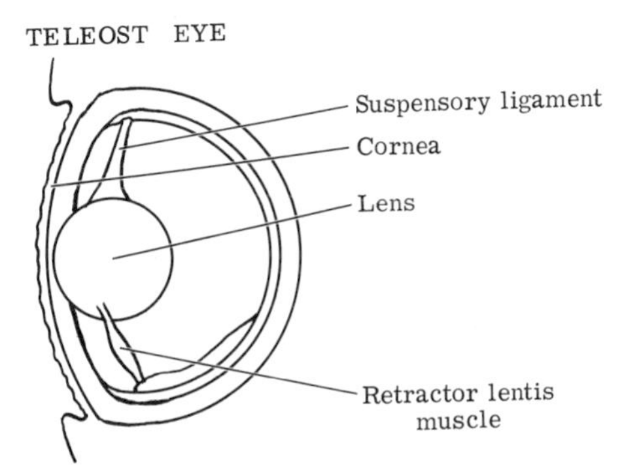

will be recalled that this is the opposite of the situation in humans (Sect. 15.7). Accommodation for distance vision in these fishes is accomplished not by a change in the lens shape as in humans, but by a change in its distance from the retina. A special muscle (retractor lentis) pulls the lens backward to shorten the lens-to-retina distance (Fig. 15-25). Conversely, the eyes of elasmobranchs, amphibians, and snakes are focused at rest for distance vision and must be actively accommodated for near vision. In these cases too the position of the lens, rather than its shape, is changed. A special muscle (protractor lentis) moves the lens forward to increase the lens-to-retina distance. In all other reptiles, and in most birds and mammals, accommodation to near vision occurs as it does in humans.

In a few animals, some parts of the retina are further from the lens than other parts. In such eyes both distance and close vision are simultaneously possible. The horse retina, for example, is a slanting structure, having one end further from the lens than the other. Also, the retinas of some bats are corrugated so that alternate regions are further from and closer to the lens. In all such cases, of course, resolution is decreased because a smaller retinal area is available for that portion of the image on which the animal has its attention. At the sacrifice of some resolution, however, the animal gains greater depth of focus.

15.12 ILLUMINATION CONTROL ADAPTATIONS

More light is required to adequately illuminate a large screen than a small one. Consequently, nocturnal and deep-sea forms often have very large light-gathering eyes but rather short lens-to-retina distances and small retinas. Again, the increased sensitivity gained is at the expense of good resolution, since resolution is better on a large screen than on a small one.

Many nocturnal forms, including members of all vertebrate classes, have increased the total illumination of the retina by the presence of light reflecting substances in the pigment layer of the choroid. Usually the tapetum lucidum, as the reflecting layer is called, is composed of crystalline guanine, but in some instances other different reflective materials occur. In any case, while reflection increases the total illumination and, therefore, the sensitivity of the retina, it does so at the cost of good resolution. The tapetum lucidum is responsible for the "nightshine" seen at night in the eyes of many animals, including cats and dogs, when their eyes are illuminated by the headlights of an automobile or by a flashlight.

In elasmobranchs black pigment cells mask the tapetum lucidum in daylight but concentrate their pigment to expose it at night. Thus, the eye is capable of good resolution in daylight and heightened sensitivity in the dark. Such pigment migration also occurs in a number of other submammalian vertebrates. In other cases, the pigment is unmoving but the receptor cells themselves move into or out from the pigment layer, thus accomplishing the same ends. In general, such changes occur rapidly in lower animals, more slowly in higher forms, and not at all in mammals.

During vertebrate evolution there has been a progressive development of pupillary responses as an alternative and better mechanism for controlling retinal illumination. Only a few fishes have variable pupil diameters and in these the response is slow. In these fishes, apparently, the iris responds directly to light intensity, whereas in higher forms the response, which is fast and effective, is mediated by way of nervous reflex arcs. The slit-shaped pupils characteristic of many reptiles and some mammals (*e.g.*, cats) can be closed more completely than those which are round, as is the human pupil. This is an advantage in an eye that is highly adapted to night vision but capable of some day vision with subdued retinal illumination.

Some animals, particularly birds, utilize colored filters in various ways. Tinted lenses or colored oil droplets overlaying the retinal cells can act as ultraviolet filters or haze-piercing "sun glasses." These are of great advantage especially to high-flying birds. In some cases retinal oil droplets of red, orange, yellow, or even green occur which apparently function in food gathering. For example, if a red insect or berry is surrounded by green foliage, the image falls first on a red-filtered receptor, then on some other receptor, then on another red-filtered receptor, and so forth, as the bird flies by. The red-filtered receptors do not see the red object but the others do; therefore, to the bird it must appear that the object is flashing off and on like a neon sign.

15.13 RETINAL ADAPTATIONS

Nocturnaal forms have increased numbers of rods while diurnal animals have increased numbers of cones. In a few cases retinas having 100 per cent rods or cones exist, but most vertebrates have mixtures of the two cell types. Generally, in such mixed retinas the cones tend to be concentrated in one area, the fovea centralis, as they are in humans. In lower vertebrates the fovea is merely a thickening due to increased numbers of tightly packed cones, but in higher forms it becomes a depression because overlying neural elements are diverted to one side. This, as we have seen, improves the sensitivity of the retina at that point (Sect. 15.9). Cone density in the fovea centralis reaches its greatest development in birds, some of which probably have up to eight times the visual resolving power of humans. Birds have as many as 1.0×10^6 cones per square millimeter as compared with about 1.4×10^5 cones per square millimeter in the human fovea. Such packing of cones leaves little room for blood vessels, so a special structure is necessary to provide nutrients and oxygen to the retina. The pecten, as this structure is called, is a comblike, folded vascular structure arising from the optic disc and projecting into the vitreous humor. It diffuses nutrients and oxygen into the vitreous humor and absorbs carbon dioxide and wastes from it.

Foveas in birds are comparatively large, a feature allowing for greater image size and, consequently, better resolution. In order to accomplish this without making the eyeballs unnecessarily large, they are often modified as tubular (rather than spherical) eyes. The eyes of birds are enormous in relation to the size of the head in any case, and greater bulk would impose additional payload to flight as well as further restricting the space remaining for the brain and other structures.

Some birds (*e.g.*, hawks, hummingbirds) must be able to search in all directions by the independent use of both eyes (monocular vision) and also to be able to "attack" using the depth perception that comes from the simultaneous and coordinated use of the eyes (binocular vision). In such cases, each eye contains two separate foveal areas, one central fovea for monocular vision and one temporal fovea for binocular vision.

Birds which float on the water and other vertebrates which live on open plains often have foveas that are extended, horizontal strips, corresponding with the image of the horizon.

In some birds the retina has greater cone density in one half of its area than in the other half. For example, *Acipiter acipiter,* the goshawk, is a high flying bird which normally hunts ground-dwelling prey. The top half of the retina receives the image from the ground below in great detail because of the concentration of cones in that half. The lower half of the retina has such poor resolution that, when the bird is perched, it must turn its head upside down in order to inspect the sky above.

We have seen that the convergence of several rods upon a single optic nerve fiber results in increased sensitivity to low illumination at the expense of good resolution. This condition reaches its highest degree in certain bats in which the convergence is estimated to be as much as 1000 rods per single neuron.

15.14 PHOTOCHEMICAL ADAPTATIONS

Rhodopsin is the characteristic photosensitive pigment in the eyes of both terrestrial and marine vertebrates. Rhodopsin is composed of opsin plus the 11-cis isomer of retinene$_1$, a derivative of vitamin A_1. In fresh-water fishes, however, the visual pigment of the rods is called porphyropsin and contains the 11-cis isomer of retinene$_2$, a derivative of vitamin A_2 (Fig. 15-20). Among the amphibians which are primarily terrestrial as adults, the pigment changes from porphyropsin to rhodopsin at the time of metamorphosis, when the animal leaves the fresh-water habitat. Fishes that migrate between fresh and marine waters have mixtures of the two pigments but the ratio shifts in the appropriate direction during the migration—more rhodopsin when in the sea and more porphyropsin when in fresh waters. The two pigments have different maximum absorption wave lengths, rhodopsin being better adapted to receive the shorter wave lengths of light.

The opsin portion of either rhodopsin or porphyropsin also has a great influence on the wave length of maximum light absorption. In marine fishes, for example, greater habitat depths are generally associated with rhodopsins, which absorb shorter wave lengths. These short wave lengths, in the blue-violet end of the spectrum, penetrate deepest into the sea. Many organisms which live on the sea bottom are bright red or orange in color, and one would think that this would make them apparent to predators. Since the longer wave lengths (red-orange) of light do not penetrate deeply into the sea, however, there is none to reflect from these brightly colored animals. Consequently, they are virtually invisible to predators. Blues and violets show up much more clearly for the opposite reason.

It is interesting to note in passing that no animal is known that makes any use of retinene isomers other than the 11-cis form. Though photochemistry evolved independently in the mollusks, arthropods, and vertebrates, all three groups utilize this isomer exclusively.

Color vision involves, theoretically, the presence of at least three different photosensitive pigments. In all probability these also are retinene-opsin combinations in which different opsins limit light absorption to specified bands of the color spectrum. Color vision is known to be present in teleost fishes, frogs, turtles, lizards, and birds, but it is apparently rare among mammals. The primates, of course, and possibly cats and certain squirrels have color vision but it has not been demonstrated for other mammals. Certainly, the fighting bull is not attracted to the red cape by its color as is popularly supposed. He is attracted solely by the motion of the cape, and it is by controlling that motion that the matador is able to guide the bull past him. There have been no good ideas advanced to account for the lack of color vision in diurnal mammals since they are equipped with cones. Color perception can only be demonstrated by carefully controlled behavioral (psychological) testing, and it may well be that further studies will show that color vision is more common than we think.

CHAPTER 16

MUSCLES AND LOCOMOTION

16.1 INTRODUCTION

In the last few chapters we have considered the ways in which environmental conditions, both outside and within the body, are sensed and encoded in the form of nervous impulses sent to the central nervous system. We have also noted some of the ways in which this input is integrated within the central nervous system and how this integration results in purposeful patterns of motor nerve output. Most of the motor output is carried by nerves to various muscles of the body where it causes appropriate adjustments of muscle activity—activity which is appropriate to the initial environmental changes that activated the entire process in the first place. In this chapter we shall concentrate on the muscles and their responses to these controlling motor impulses.

There are three distinct kinds of muscle present in the vertebrate body, namely, skeletal, cardiac, and smooth. We shall consider the skeletal muscles in greatest detail because there are more of them, because they have been more thoroughly studied, and because the other kinds differ from them in function in only a few ways.

All three kinds of muscle exhibit contractility, a property exhibited by almost all cells but developed in muscle to a much greater degree. Muscle cells are all elongate cells which shorten in length when they contract. It is important to remember that the muscle cell accomplishes physical work only when it is contracting. The process of relaxation is not an active physical process. Therefore, a muscle can only pull two points (its end attachments) closer together; it cannot push them further apart.

Skeletal muscles are attached at each end, sometimes by way of an elongate tendon, in most cases to bones of the skeleton. In a few cases they may be attached to skin rather than to bone. Usually, then, the contraction of skeletal muscle pulls the attached parts of the skeleton closer together. Generally one attached end of the muscle, called its origin, remains relatively fixed in position, while the other end, the insertion, moves when the muscle contracts. For example, the biceps brachii muscle of the arm originates from the scapula bone in the shoulder and inserts on bones of the forearm just beyond the elbow. When this muscle contracts the origin (shoulder) remains fixed but the insertion (forearm) is moved upward toward the shoulder. As a result, the elbow joint is bent. In this example the distal end of the humerus serves as a fulcrum, about which the radius and ulna move as a lever (Fig. 16-1).

Since levers move in both directions, different skeletal muscles oppose each other. Such muscle pairs are called antagonists; one moves the lever in one direction and the other moves it in the opposite direction. The biceps brachii muscle shown in Figure 16-1 is opposed by the triceps brachii muscle located on the opposite side of the humerus. The biceps brachii flexes the elbow and the triceps brachii extends it.

Sometimes more than a single muscle is involved in a given movement of a skeletal lever. Such cooperating muscles are called synergists. Often in groups of muscles acting generally as synergists, individual muscles actually produce minor variations of movement, depending upon which members of the group are contracting and to what degree. A series of muscles located in the forearm cooperate in various combinations to bend the wrist in any direction desired.

Cardiac and smooth muscles are generally found in the walls of hollow organs. Contraction of these muscles causes constriction of the organ so that its lumen becomes smaller. Rhythmic, repeated contractions result in expelling the blood contained in the heart. In blood vessels the contraction of smooth muscle constricts the diameter of the vessel and thus raises the pressure and velocity of the blood (Sect. 5.6). In the intestinal tract smooth muscle contractions result in peristalsis and related kinds of movement that mix and drive the contents of the gut forward (Sect. 9.16). For these kinds of muscle there are often

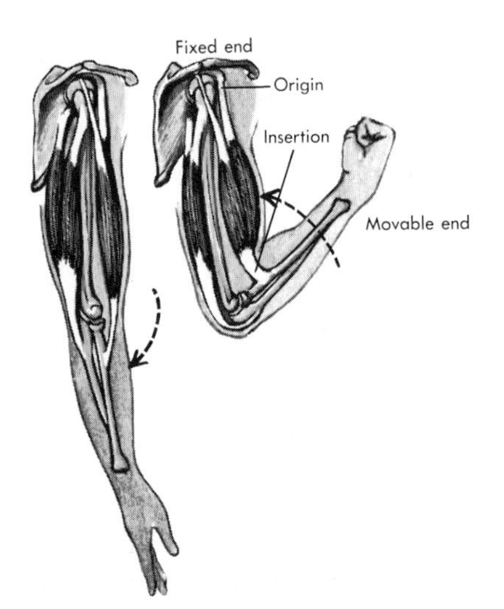

Figure 16–1 Pairs of antagonistic muscles move bones as levers in opposite directions using the joint between bones as a fulcrum. The biceps brachii muscle flexes the elbow while the triceps brachii muscle extends it. (From Gardner & Osburn: *Structure of the Human Body*. Philadelphia, W. B. Saunders Company, 1967.)

no antagonists since the distending force of materials inside the heart, vessels, or gut has an opposing force. Dilation occurs without the necessity of any muscular activity. There are exceptions to this, of course.

SKELETAL MUSCLE

16.2 STRUCTURE OF SKELETAL MUSCLE

A typical skeletal muscle is an elongate structure having each end attached to bones. It tapers toward the ends and has an enlarged belly between these attachments. There are variations in shape in different muscles, but for our purposes this is an accurate picture. Within the belly of the muscle are definite bundles of tissue separated from each other by connective tissue sheaths, and within each muscle bundle are smaller bundles known as fascicles (Fig. 16-2). Fascicles are similarly separated from each other by delicate connective tissue sheaths. Each fascicle is composed of many muscle cells, each of which is referred to as a single muscle fiber. Individual muscle fibers range from 10 to 100 microns in diameter and up to several centimeters in length. Some may even be as long as the muscle in which they occur. Each muscle fiber is innervated by a branch of a motor neuron by way of a neuromuscular junction near its midpoint (Sect. 11.4).

Each muscle fiber contains several hundred longitudinally arranged myofibrils, which are the contractile units of the muscle. Under the optical microscope each myofibril appears to be marked by prominent cross-striations (Figs. 2-9 and 16-3). These cross-striations occur in a definite repeating sequence and have been designated by letters of the alphabet. The dark A band, with a lighter H band through its center, alternates with a light I band, which has a darker Z band through its center. The portions between successive Z bands are considered to be units of the myofibril and are called sarcomeres. It is this appearance of the cross-striations, of course, which has given skeletal muscle its other common name: striated muscle.

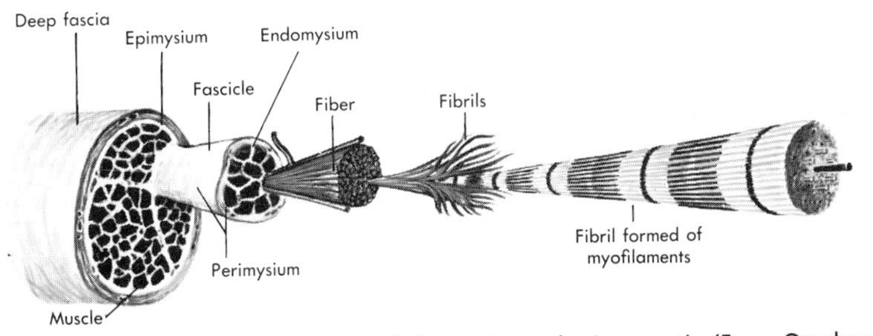

Figure 16–2 The gross structure of skeletal muscle (see text). (From Gardner and Osburn: *Structure of the Human Body*. Philadelphia, W. B. Saunders Company, 1967.)

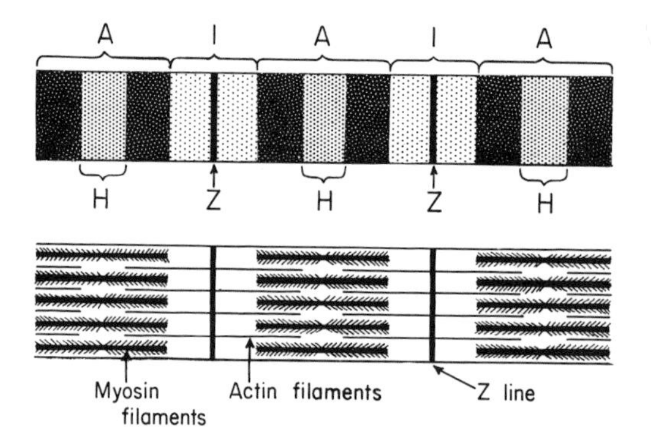

Figure 16-3 Structure of skeletal muscle as seen with the optical and electron microscopes (diagrammatic; see text). (From Guyton: *Textbook of Medical Physiology,* 4th Edition. Philadelphia, W. B. Saunders Company, 1971.)

The reason for the apparent cross-striations is made clear when one looks at a myofibril by means of the electron microscope. The greater resolution of this instrument permits one to see that each myofibril contains about 2500 longitudinally arranged filaments of two different kinds. There are thin filaments composed of a protein called actin and thicker filaments composed of a different protein called myosin. The cross-striations are apparent because these two kinds of filaments interdigitate and partially overlap with each other (Fig. 16-3) and because of the presence of interrupting walls that correspond to the Z bands. These walls, because they completely separate one sarcomere from another, are called Z discs or Z membranes. The electron microscope also reveals that the thick myosin filaments have tiny spurlike processes extending from them toward the thin actin filaments.

When the myofibril contracts, the myosin and actin filaments slide together, increasing the amount of interdigitation or overlap (Fig. 16-4). This is believed to be due to electrostatic forces between the actin filaments and the "spurs" of the myosin filaments (Sect. 16.3). Relaxation, then, results from a disappearance of these electrostatic forces and a consequent sliding apart of the filaments in response to elastic forces in the muscle as a whole.

The endoplasmic reticulum (Fig. 2-4) of skeletal muscle is a specialized structure known as the sarcoplasmic reticulum (Fig. 16-5). It consists of two sets of tubules: (1) Fine tubules extend lengthwise

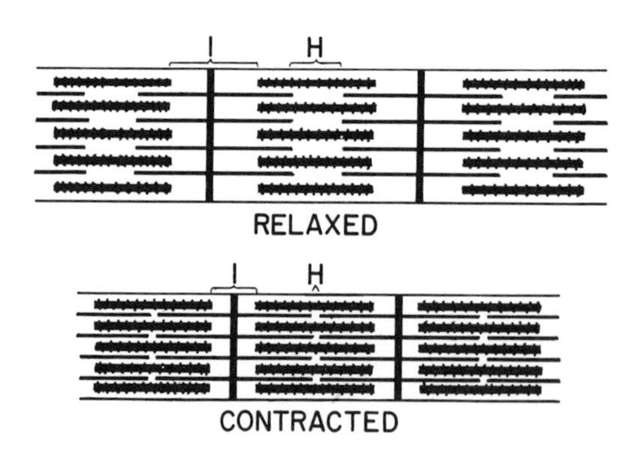

Figure 16-4 Changes occurring in skeletal muscle with contraction and relaxation (diagrammatic; see text). (From Guyton: *Textbook of Medical Physiology,* 4th Edition. Philadelphia, W. B. Saunders Company, 1971.)

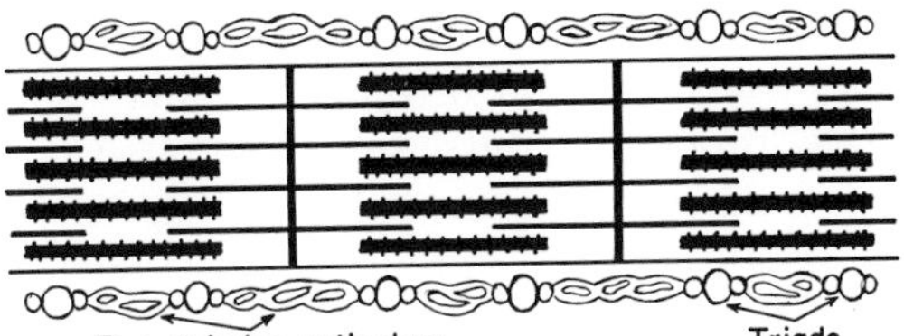

Figure 16–5 The sarcoplasmic reticulum surrounding a skeletal muscle fiber (diagrammatic). (From Guyton: *Textbook of Medical Physiology,* 3rd Edition. Philadelphia, W. B. Saunders Company, 1966.)

Fine tubular reticulum Triads

parallel with the myofibrils; (2) these are connected by larger tubules that are oriented at right angles to them. The larger tubules have expansions called triads in the regions next to the interdigitation or overlap of the actin and myosin filaments. The sarcoplasmic reticulum is believed to function in the triggering of muscle contraction following the delivery of a motor nerve impulse to the muscle.

Groups of 10 to 800 muscle fibers (cells) are innervated by terminal branches of a single motor neuron to form, collectively, what is known as a motor unit. Nerve impulses in that neuron, then, result in simultaneous contraction of all of the muscle fibers in the motor unit. Typically, fast-acting muscles, like those which move the eyeball, contain few muscle cells per motor unit, while slow-acting muscles contain greater numbers. The average number of muscle fibers per motor unit is probably about 150.

16.3 FIBRIL CONTRACTION

As we have seen, nerve impulses result in the release of transmitter substances at synapses and neuromuscular junctions (Sect. 11.4). The transmitter substance at skeletal neuromuscular junctions is acetylcholine (Fig. 5-16). The transmitter substance, in turn, produces an action potential in the membrane surrounding the muscle fiber; this action potential is essentially like the action potential of a nerve membrane (Sect. 11.3) but lasts about 15 times as long. It is carried throughout the muscle fiber probably by way of the sarcoplasmic reticulum.

Theories explaining the subsequent events in the induction of muscle contraction vary in detail but the following pattern is generally accepted as correct: (1) The arrival of the action potential at the triads of the sarcoplasmic reticulum causes the release of quantities of calcium ion (Ca^{++}), which at other times is bound to the walls of the sarcoplasmic reticulum. (2) The released Ca^{++} diffuses into the muscle cell and alters the chemical properties of the myosin filaments. In the presence of Ca^{++}, myosin becomes an ATPase; that is, it becomes an enzyme which catalyzes the release of a phosphate group from ATP (Fig. 10–2). This converts ATP to ADP and makes available a quantity of energy to be used for contraction. (3) The ATP in the preceding step was formerly attached to the actin filaments. The conversion to ADP releases it from that protein and, as a result, also alters the chemical properties of actin. (4) The altered properties of myosin and actin produce electrostatic attraction between the two proteins. (5) Finally, because of the electrostatic forces created, the myosin and actin filaments interdigitate further with each other, pulling the adjacent Z discs closer together. Since the same process occurs simultaneously in each sarcomere of the cell, the entire cell is shortened.

Following contraction, the released Ca^{++} is again bound to the membranes of the sarcoplasmic reticulum and thus is removed from the cell. Myosin loses its ATPase character and new ATP produced by the mitochondria of the cell becomes attached to the actin filaments. These events destroy the electrostatic attraction between actin and myosin, and the filaments are drawn apart again by elastic forces in the muscle as a whole. The cell thus returns to its resting length and relaxation is accomplished.

There is some evidence to suggest that magnesium ions (Mg^{++}) are normally present in some concentration in skeletal muscle cells and that this ion inhibits the ATPase activity of myosin. The release of Ca^{++}, then, may simply overcome the inhibiting effects of Mg^{++}. There are many other instances of such interactions between different ions in animal metabolism (Sect. 9.6).

Since individual action potentials in skeletal muscle last for about 5 to 10 milliseconds, nerve impulses arriving more frequently than this result in sustained contractions of the muscle; the muscle has no opportunity to relax before it is restimulated by another nerve impulse. Motor nerve action potentials last for only about 0.5 millisecond, so it is quite possible for several of them to arrive during the duration of a single muscle action potential. Conduction of action potentials through skeletal muscle has a velocity of only about 5 meters per second, while those of motor nerves reach speeds as great as 100 meters per second.

During contraction a great deal of energy is liberated from ATP, but most of this appears in the form of heat. The efficiency of the muscle in converting ATP energy into actual muscular work is only about 20 to 25 per cent, depending upon the velocity of the contraction. The loss of energy is largely due to frictional resistance within the muscle cells. Interdigitation of filaments necessarily displaces the fluids of the cell matrix or sarcoplasm; this material has a fairly high viscosity. Other mechanical factors also contribute to the internal frictional resistance of muscle.

Each individual muscle fiber has a stimulus-response threshold. That is, an artificial stimulus delivered to a muscle fiber must have some minimal intensity or magnitude in order to trigger a contraction of that muscle. Stimuli of lesser magnitudes are without visible effects. If the stimulus is of sufficient magnitude to produce a contraction, that contraction occurs to a degree characteristic of the muscle fiber in question. Stimuli of greater magnitudes do not produce greater contractions. This is known as the all-or-nothing effect; the muscle fiber contracts completely or not at all, depending upon whether its stimulus-response threshold is or is not exceeded. The magnitude of contraction of a muscle fiber may change, however, with changing conditions of the experiment, as we shall see presently.

If a series of artificial stimuli, each barely subthreshold in magnitude, is delivered in rapid succession, the muscle fiber will finally contract. One stimulus is insufficient to exceed the muscle threshold but a number of such stimuli summate their effects and contraction occurs. This indicates that even a subminimal stimulus creates some degree of "excitation" in the muscle membranes, and this "excitation" can be repeatedly boosted by successive stimuli until it exceeds the muscle threshold. Excitation, in this case, undoubtedly involves altering the permeability of the muscle membrane to sodium and potassium ions in much the same way as occurs in nerves (Sect. 11.3).

16.4 WHOLE MUSCLE CONTRACTION

The contraction of a whole muscle is, of course, the sum of the contractions of its component fibers. Generally in normal muscle contraction, only a few of the contained motor units are stimulated, and therefore only a few respond. The degree of force of the total muscle contraction depends upon the number of motor units simultaneously involved.

If an isolated skeletal muscle is arranged in such a way that its responses can be recorded, a number of interesting characteristics can be demonstrated. One such arrangement involves the use of the smoked kymograph (Fig. 16-6), although today modern electronic recording equipment is more commonly used.

If a muscle in such an apparatus is stimulated artificially with a

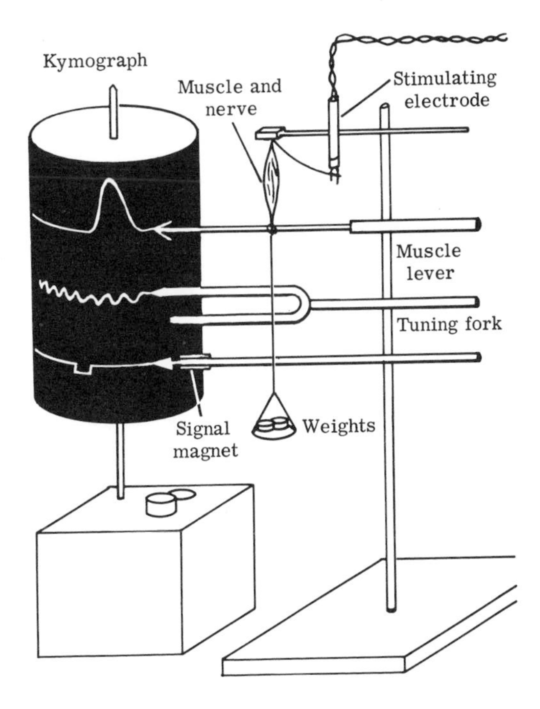

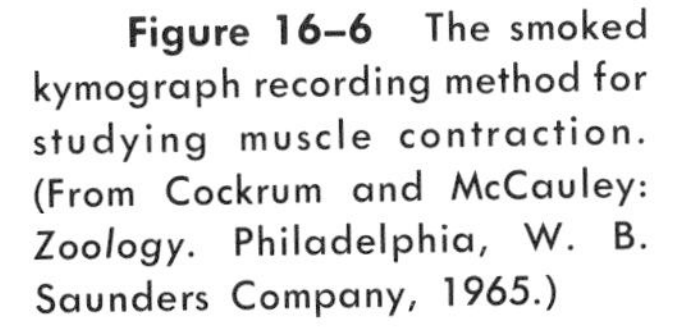
Figure 16–6 The smoked kymograph recording method for studying muscle contraction. (From Cockrum and McCauley: *Zoology*. Philadelphia, W. B. Saunders Company, 1965.)

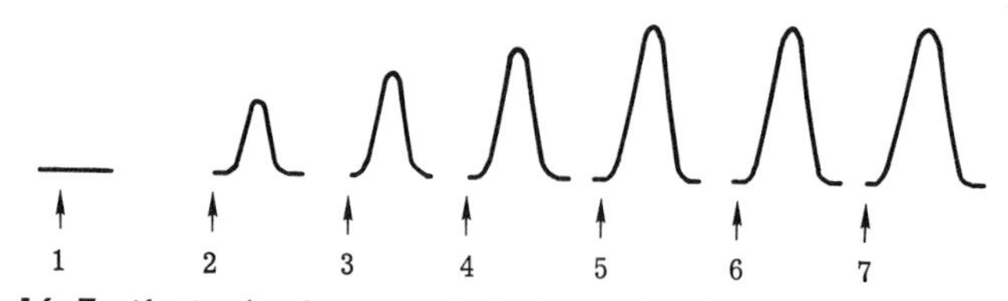

Figure 16–7 If stimuli of progressively increasing intensity are administered to a whole skeletal muscle, more and more motor units are recruited, resulting in progressively greater responses, until all of the motor units are responding. Further increases in stimulus intensity, then, have no further effects upon contraction amplitude.

short stimulus of an intensity below the threshold of any of its fibers, no visible response occurs. If it is then stimulated again with a stimulus of slightly greater intensity, a contraction may occur. Such a minimal intensity exceeds the thresholds of at least some of the muscle fibers within the muscle. If, now, the muscle is again stimulated with a stimulus of still greater intensity, the magnitude of the response may be greater since the thresholds of a greater number of muscle fibers have been exceeded. Successive increases in the intensity of the stimulus may produce successive increases in the magnitude of response until a stimulus intensity which exceeds the thresholds of all of the muscle fibers is reached (Fig. 16-7). This is called the maximal stimulus intensity. Further increases in stimulus intensity (supramaximal) do not, then, produce any greater magnitude of muscle response since every fiber is contracting in accordance with the all-or-nothing principle.

If the kymograph drum is revolving when a muscle is stimulated, its contraction can be recorded as a curve in which the magnitude (amplitude) of response is represented vertically and the duration of response is represented horizontally (Fig. 16-8). A single, short stimulus of at least maximal intensity will produce a single contraction of all

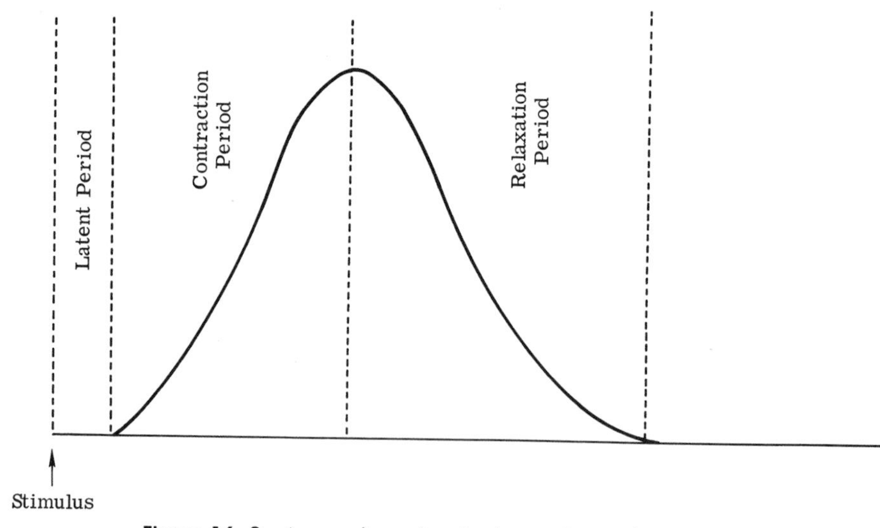

Figure 16–8 A recording of a simple muscle twitch. See text.

of the contained muscle fibers simultaneously. Such a contraction is known as a simple muscle twitch. By measuring the time intervals of various portions of the recorded curve we can learn something of the characteristics of a single contraction. For mammalian skeletal muscle a latent period of about 0.001 second elapses between the delivery of the stimulus and the beginning of the response. This is accounted for by the time required for the action potential to spread through the sarcoplasmic reticulum and for the diffusion of the released Ca^{++} into the muscle fibers (Sect. 16.3). Following the latent period, the time required for contraction is about 0.004 second, and the time required for relaxation is another 0.005 second. Thus, completion of the entire process requires about 0.01 second. Often in experimental laboratories the gastrocnemius muscle of a frog is used instead of a mammalian skeletal muscle. In this muscle all of the time periods mentioned above are about 10 times as long. The simple muscle twitch of a frog gastrocnemius therefore has a total duration of about 0.1 second, and this slower response makes the muscle easier to study than mammalian muscle. Probably simple muscle twitches do not occur naturally in the living animal; a possible exception is a single blink of an eyelid. Instead, most skeletal muscle responses are sustained contractions resulting from rapid series of nerve action potentials. Also, it is probable that natural contractions seldom, if ever, involve all the muscle fibers within any given muscle.

Suppose, now, that an isolated skeletal muscle, placed in the same apparatus as used to demonstrate the simple muscle twitch, is stimulated by two successive stimuli, of which each is at least of maximal intensity. If these two stimuli are sufficiently far apart in time, two separate simple muscle twitches will be recorded (Fig. 16-9). If, however, the time interval between the two stimuli is shortened, the second contraction may begin before the first twitch has been completed. The effects of the two contractions are then summated, and the total magnitude (amplitude) of the response is greater than that of a simple muscle twitch. At first this may seem to violate the all-or-nothing principle, since all muscle fibers were contracting under both sets of conditions. The difference in the magnitude of the responses is explained, however, by the fact that a muscle fiber which is already under some tension is capable of greater response than one which is completely relaxed. It

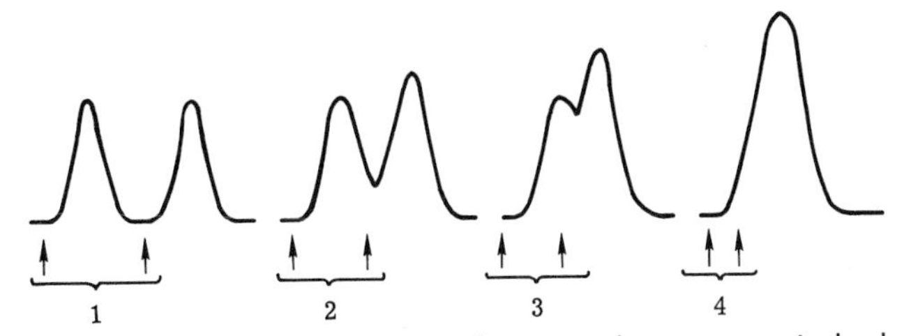

Figure 16-9 If two stimuli are delivered to a muscle at progressively closer intervals of time, the muscle responds with progressively greater amplitudes of contraction. See text.

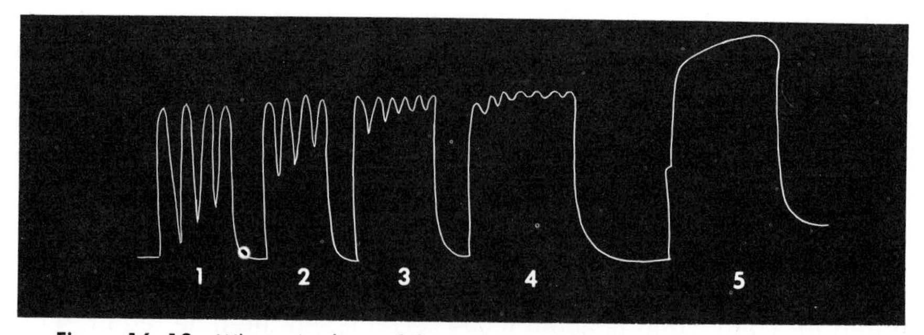

Figure 16–10 When stimuli are delivered to a muscle at shorter and shorter intervals, the muscle approaches a state of sustained or tetanic contraction. (From Cockrum and McCauley: *Zoology*. Philadelphia, W. B. Saunders Company, 1965.)

will be recalled that the response of a single muscle fiber can be altered by changing the conditions of the experiment, and this is a case in point. A study of Figure 16-8 will show that within limits the shorter the time interval between the two stimuli, the greater is the summated effect on contraction.

If, instead of just two stimuli, a series of supramaximal stimuli are delivered at short time intervals, the resulting series of muscle twitches fuse to produce a sustained response. This is called a tetanic contraction and is of greater magnitude (amplitude) than a single muscle twitch (Fig. 16-10). Probably most natural contractions in the living animal are of this type.

The form of the simple muscle twitch is also affected by fatigue of the muscle. If repeated twitches are evoked in an isolated frog muscle and the resulting recorded curves are superimposed on each other (Fig. 16-11), a series of changes in the form of successive curves is seen. Assuming that the muscle has not been previously stimulated before the experiment began, the first three or four twitches will have progressively greater amplitudes. Again, this is not a violation of the all-or-nothing principle since it can be explained on the basis of changing conditions within the muscle. Since each contraction produces some heat, the temperature of the muscle rises slightly, and the chemical reactions associated with energy production (Sects. 10.5 through 10.7) all proceed more rapidly (it will be recalled that all chemical reactions

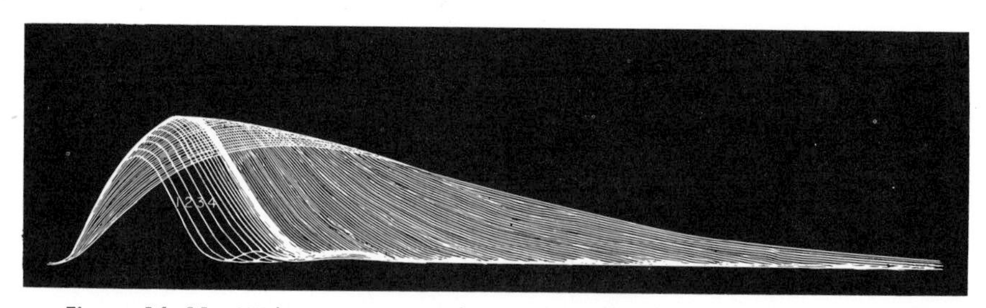

Figure 16–11 With successive stimulation of a skeletal muscle the form of the simple muscle twitch changes progressively (see text). (From Cockrum and McCauley: *Zoology*. Philadelphia, W. B. Saunders Company, 1965.)

are speeded by increases in temperature). Also, changes in the pH of the muscle occur as a result of the accumulation of pyruvic and lactic acids (Sect. 10.5), and this also affects in a favorable way the enzyme systems involved in energy production. All enzymes, being proteins, have optimum pH ranges within which they are most active. Because resting muscle has a pH near the alkaline end of this optimum range, the accumulation of some acid increases the effectiveness of these enzymes.

Following the initial increases in contraction amplitude, the muscle reaches a plateau of activity and for some time thereafter successive twitches are almost identical. If the experiment is continued, however, the curves eventually begin to change in form again. Each succeeding curve has a lower amplitude and a greater duration, particularly the duration of the relaxation period. These changes are easily explained on the basis of the exhaustion of available chemical energy stores within the muscle cells. Recall that new ATP must be produced and attach itself to the actin filaments before relaxation can take place (Sect. 16.3).

16.5 MUSCLE ACTION IN THE BODY

As we have seen (Sect. 16.1), skeletal muscles typically move one bone with respect to another, and frequently these bones are hinged at a joint so that the moving bone serves as a lever (Fig. 16-1). For the biceps brachii muscle the shoulder attachment is the non-moving origin and the forearm attachment is the insertion. The forearm, then, is moved as a lever, the fulcrum of which is at the elbow joint. Since the forearm attachment is about two inches from the fulcrum and the total forearm length is about 14 inches in an average human, we can calculate the forces developed. A large biceps muscle is capable of some 250 pounds of force but, because of the closeness of the insertion to the fulcrum, this results in only about 36 pounds of force at the hand. That is, with the forearm held at right angles to the upper arm, 250 pounds of biceps force will lift only 36 pounds held in the hand. If the arm is straightened out (fully extended) the amount that can be lifted is even less since the insertion is then even closer to the fulcrum.

Two general kinds of muscle contraction are recognized: isometric and isotonic. When a muscle contracts against a load that is too great for it to move, it cannot shorten. It nevertheless develops internal tension in the effort to contract. This is known as isometric contraction. The length of the muscle remains constant while the tension within it increases. This kind of contraction is popularly used in exercises designed to increase the strength of muscles. When, on the the other hand, a muscle actually lifts a load, it shortens in length but the internal tension remains constant. This is an isotonic contraction, which is the kind more usually involved in body movements.

Regular exercise, especially of the isometric type, of a muscle results in increased strength of that muscle. The increase in strength is due to hypertrophy of the muscle tissue. Individual muscle fibers become larger as they accumulate greater quantities of actin, myosin, ATP, mitochondria, and glucose, but there is no increase in the total numbers of fibers within the muscle. Disuse of muscles, on the other hand, results in progressive atrophy, the opposite of hypertrophy. Atrophy is particularly noticeable in the muscles of a limb enclosed in a cast during recovery from a bone fracture. Even a few weeks of such inactivity result in conspicuous reductions in total muscle size and strength. Permanent disuse due to denervation results in extreme atrophy.

A normal muscle within the body is almost never completely relaxed; a few of its motor units are nearly always contracting. This constant partial contraction of the muscle at rest is called resting tonus. The amount of resting tonus is also related to the amount of exercise that muscle receives. The muscles of a trained athlete feel hard to the touch because the resting tonus is greater than in those of an untrained person. Discontinuation of exercise decreases the tonus and the muscle becomes softer and flabby. When a muscle is stimulated to contract, the number of responding motor units increases; as a result, more tonus develops and contraction results. At the same time, the resting tonus of antagonistic muscles is decreased and consequently they do not oppose the action (Sect. 11.2). This comes about through reciprocal innervation of antagonistic pairs of muscles and reflex adjustments of tonus mediated through the spinal cord. For example, at the same time that the biceps brachii is stimulated to contract, the resting motor nerve activity which maintains resting tonus in the triceps brachii is decreased. Similarly, when the triceps is stimulated to contract, the resting tonus of the biceps is decreased. This makes it easier for the contracting muscle to stretch the antagonistic muscle and, thus, move the forearm lever without undue opposition.

CARDIAC AND SMOOTH MUSCLE

16.6 CARDIAC MUSCLE

Cardiac muscle is composed of striated cells that are much like those of skeletal muscle except for their interrelationships. Single cells do not extend for great distances but are, rather, relatively short and frequently branched (Fig. 16–12). Adjacent cells are separated from each other by intervening membranes known as intercalating discs. The branching cells form a latticework of fibers often referred to as a functional syncytium. A true syncytium is a tissue in which there are no cell separations of any kind. Heart muscle is not a true syncytium but in many ways it behaves like one. Action potentials, for example, spread easily from cell to cell so that individual innervation of each fiber is not

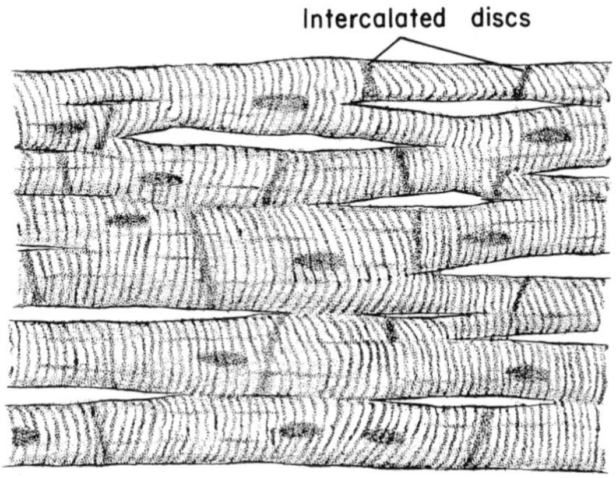

Intercalated discs

Figure 16–12 The structure of cardiac muscle. (From Guyton: *Textbook of Medical Physiology,* 4th Edition. Philadelphia, W. B. Saunders Company, 1971.)

necessary. In skeletal muscle, it will be recalled, no such spread of excitation occurs. Instead, each muscle fiber is innervated by a terminal branch of a motor nerve.

Since the action potentials of cardiac muscle arise spontaneously and rhythmically, contractions also occur at regular intervals. The normal origin of the successive action potentials is the sinoatrial node of mammals, as we have already seen (Sect. 5.10). While the action potential lasts for less than 0.01 second in skeletal muscle, it has a duration of 0.15 second in the muscle of the atria (atrial myocardium) and more than 0.3 second in the muscle of the ventricles (ventricular myocardium). Thus, all cardiac contractions (systoles) are somewhat sustained, and those of the ventricles are sustained longer than atrial contractions (Fig. 5-7).

When a skeletal muscle is in the process of contracting or relaxing, it will still respond to stimuli. Consequently, sustained tetanic contraction is possible (Fig. 16-10). Cardiac muscle, however, cannot be restimulated until complete relaxation has occurred. That is, during the process of contraction and relaxation, the muscle is said to be refractory, or non-irritable. Therefore, it is impossible for the heart to show tetanic contraction, which would not be compatible, of course, with the rhythmic, repeated beating of the heart. During contraction the refractoriness of the heart is absolute; it cannot be stimulated at all. During the relaxation phase, however, the refractoriness is relative. That is, there is

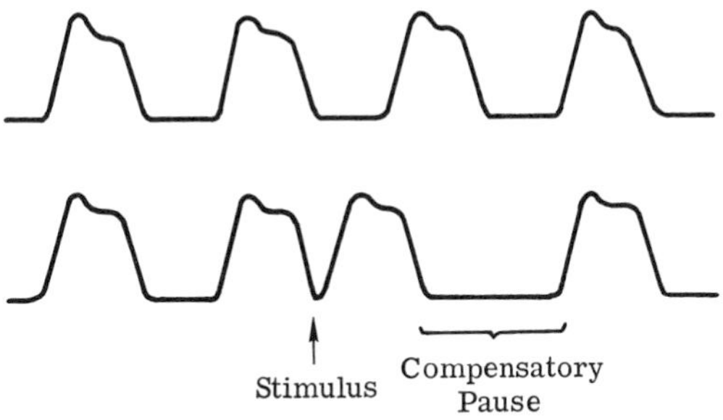

Stimulus
Compensatory Pause

Figure 16–13 A recording of premature systole and compensatory pause of cardiac muscle. See text.

a progressive return toward the completely irritable state, and late in relaxation stimuli of much greater intensity can evoke a response. Skeletal muscle also shows refractoriness, but it lasts through only a portion of the latent period following stimulation (Fig. 16–8).

It is possible to stimulate cardiac muscle to contract somewhat sooner than its own rhythmicity would dictate because relaxation becomes complete and the heart remains totally inactive for a time before contracting again. Thus, it is no longer refractory and an artificial stimulus can produce a "premature systole" (Fig. 16–13). Following a premature systole, there is an extended compensatory pause that allows the subsequent systole to occur at the proper time. Sometimes books refer to such displaced systoles as "extra systoles," but this is really an incorrect term. The total number of systoles has not been increased, although one has been made to occur prematurely. Sometimes series of several rapid premature systoles occur and are followed by a very long compensatory pause. Certain drugs, including nicotine, may bring about such episodes.

16.7 SMOOTH MUSCLE

Two general types of smooth muscle are recognized: multiunit and visceral. The multiunit type is composed of fusiform cells about 10 microns in diameter and up to 200 microns in length (Fig. 16-14). These are arranged in a loose network and each cell is separately innervated. Contraction occurs only when nerve action potentials reach the muscle cells. This kind of smooth muscle occurs in the walls of blood vessels and in the iris of the eye. Visceral muscle, on the other hand, has similar fibers which are closely packed in a parallel arrangement. Not every cell is innervated but action potentials spread easily from cell to cell. This is the type encountered in the walls of the digestive tract, the urogenital tract, and in other organs of the body cavities.

Both kinds of smooth muscle contain actin and myosin filaments, similar to those of skeletal muscle, but they are not arranged in a precise parallel order. Apparently the chemistry of contraction and relaxation is similar to that occurring in skeletal muscle as well, but

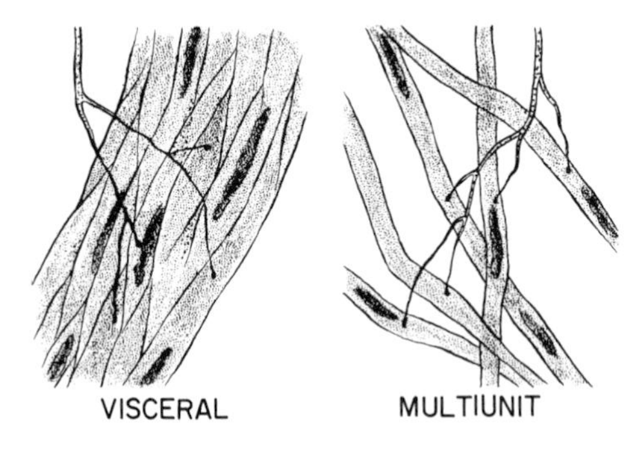

Figure 16–14 The structure of the two types of smooth muscle (see text). (From Guyton: *Textbook of Medical Physiology*, 4th Edition. Philadelphia, W. B. Saunders Company, 1971.)

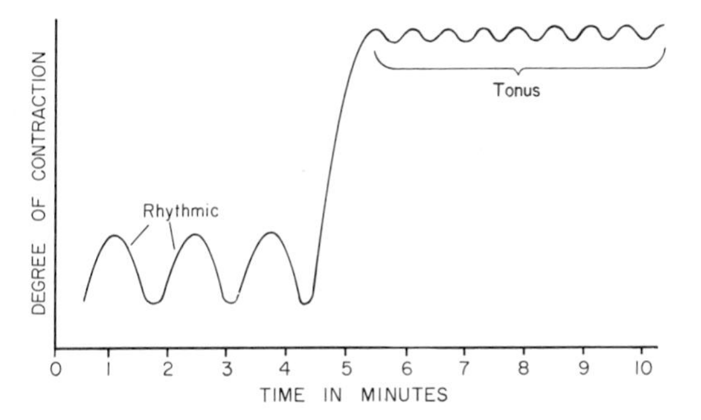

Figure 16–15 Smooth muscle shows spontaneous rhythmicity and can maintain tonus at various levels (see text). (From Guyton: *Textbook of Medical Physiology*, 4th Edition. Philadelphia, W. B. Saunders Company, 1971.)

because of the more random arrangement of filaments, contraction in smooth muscle is much slower. The same factor introduces additional resistance within the cell because of the dragging of filaments over each other during contraction. Probably this too is an important factor in the decreased rate of contraction.

Smooth muscle is capable of sustained contraction or tonus. In fact, typical smooth muscle can sustain tonus at various levels, depending upon the amount of nervous stimulation it is receiving (Fig. 16-15). Smooth muscle also has intrinsic rhythmicity (particularly visceral muscle) as does cardiac muscle, but the frequency of contraction is not as great. Successive contractions of smooth muscle occur at intervals of a few seconds.

VERTEBRATE LOCOMOTION

16.8 MUSCLES AND LOCOMOTION IN FISHES

The body musculature of fishes is typically arranged as a series of overlapping blocks, or myotomes, each composed of longitudinally oriented fibers. During swimming contraction waves progress from anterior to posterior and thus involve successive myotomes. Contraction waves alternate on the two sides of the body so that the body is bent first in one direction and then in the other (Fig. 16-16). Successive waves occur at the rate of 50 per minute in the dogfish shark and at the rate of 170 per minute in mackerel. They travel posteriorly at the rate of about 35 centimeters per second. These waves tend to force the adjacent water backward, or in other words to push the fish forward through the water. The fins play no part in this propulsion; they merely act as stabilizing surfaces to prevent rolling, pitching, and yawing movements. In this they act much like the wings and tail assembly of aircraft (Fig. 16-17).

The first fishlike forms to appear in evolution were undoubtedly bottom dwellers and slow swimmers. The large upward directed heterocercal tails (Fig. 16–18) tended to turn the nose downward in swimming, probably an advantage to such a bottom-dwelling organism. As

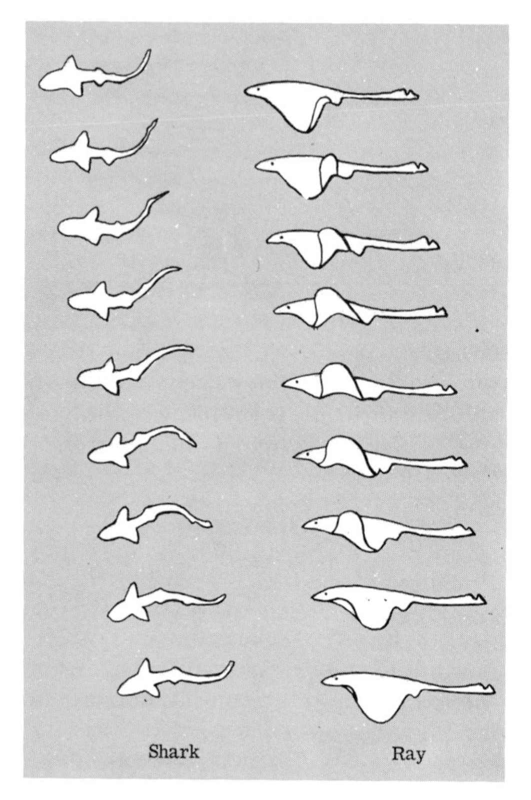

Figure 16–16 Characteristic locomotion of fishes. Teleosts and sharks swim with alternate lateral undulations of the body. Rays are so modified that they must use dorsoventral undulations of the lateral body margins. The sequence of pictures should be read from bottom to top. (From Cockrum and McCauley: *Zoology*. Philadelphia, W. B. Saunders Company, 1965.)

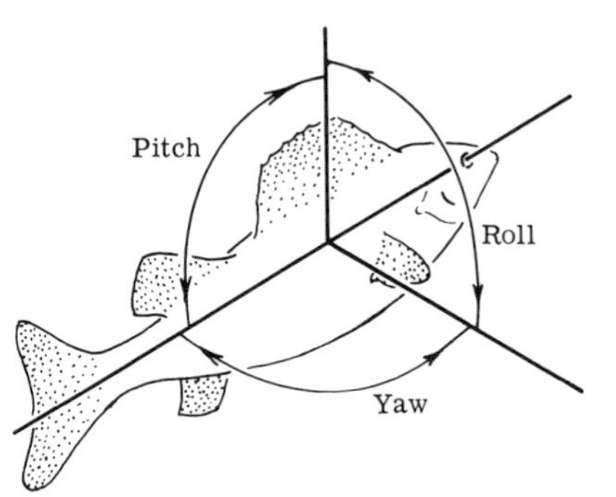

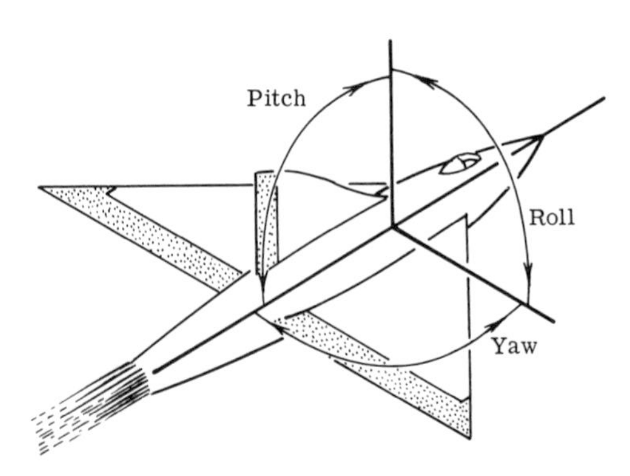

Figure 16–17 The stabilizing surfaces of aircraft and fishes prevent roll, pitch, and yaw. (From Cockrum and McCauley: *Zoology*. Philadelphia, W. B. Saunders Company, 1965.)

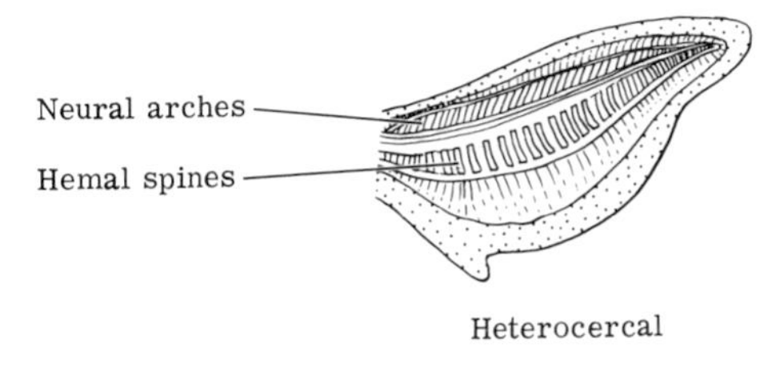

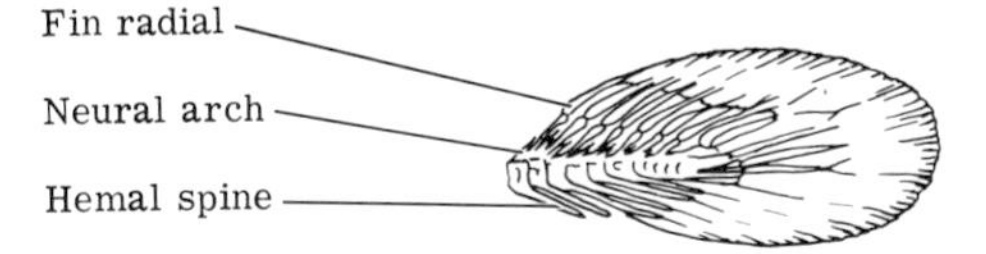

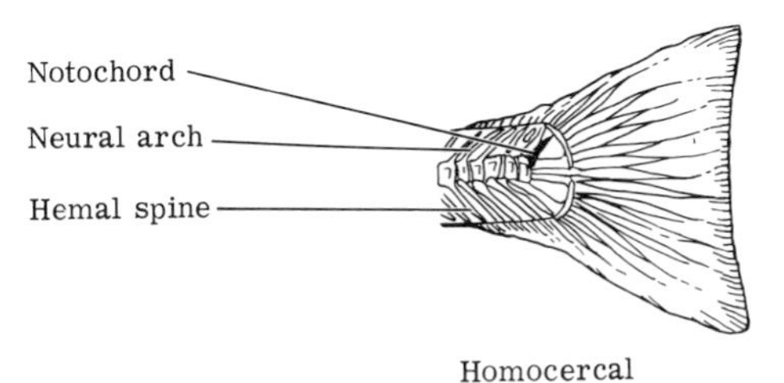

Figure 16–18 Types of fish tails. (From Cockrum and McCauley: *Zoology.* Philadelphia, W. B. Saunders Company, 1965.)

the means for more rapid swimming developed, more stability was required and more evenly divided diphycercal and homocercal tails appeared. Many sharks that are not bottom dwellers retain the heterocercal tail, but in these forms the dorsoventral flattening of the head and the addition of lateral fins offsets the uneven effect of the tail.

The lateral fins of sharks and other fishes also act as hydrofoils, used to execute banking and turning maneuvers or as brakes to achieve sudden stops. The development of such independent movement of lateral fins required, of course, the specialization of certain myotomes in the region of the fins.

The streamlining of fishes is, of course, an important factor in efficient and rapid swimming. Variations in body shape occur frequently but they are nearly always associated with a reduction in maximum swimming speed. Often protective gains are achieved by such atypical shapes, however, and these have survival value as well. The bilateral flattening of fishes, for example, is quite common. Angel fish are bilaterally flattened to such an extent that they resemble the plants growing in their native South American rivers. In other cases extreme flattening makes the animal more difficult for predators to detect when seen edge on. Typically, such fish are bottom feeders and are not called upon to swim rapidly. Rays and skates are dorsoventrally flattened and are also bottom feeders. Rays are so highly modified that typical swimming movements are no longer possible; instead, successive waves of the lateral body margins propel the fish forward. This movement still involves the same successive myotomes but the motion is up and down rather than side-to-side as in typical fishes. The body shape of the sea horse has been so completely modified that myotomes are no longer used for swimming at all. Instead, these very poor swimmers stand vertically in the water and propel themselves by wavelike motions of the dorsal fin. The elongate bodies of eels reduce swimming efficiency but have the advantage of allowing snakelike locomotion on land.

Some fishes can swim so rapidly that they are able to leap for considerable heights out of the water. Salmon and tarpon, in particular, are capable of such leaping. Flying fishes prolong the leap by using greatly enlarged lateral fins as gliding wings, and the flying gurnards actually flap the fins while in flight.

The lateral fins of certain bottom dwellers are modified as spine-like levers and are moved as legs. Such fins are of little use as stabilizers or steering mechanisms during swimming but they permit easy locomotion across the bottom. Of course, the use of such fins has necessitated the development of stronger muscles to move them, and in some cases such fishes are also able to walk across land. The mudskipper is a well-known example. It is interesting and important to note that the walking fins of these fishes are coordinated with typical swimming movements of the body. That is, each is moved backward as the myotome contraction wave passes it. The fins of the opposite sides, therefore, alternate to produce a walking pattern of fin movement. This kind of locomotion reaches greater degrees of development in the lungfishes.

16.9 ELECTROGENESIS

Several different fishes, both teleosts and elasmobranchs, have electrogenic organs that generate relatively large electric potentials. These are used by the animal in predation and defense. Although certain individual differences exist, the structure of electrogenic organs is generally quite similar in all fishes in which they occur (Fig. 16-19). Each consists of a series of thin modified muscle cells called electroplaxes arranged in columns like stacks of coins. Each cell is innervated on one side only.

At rest, each electroplax carries a membrane change much like that of any other cell—negative on the inside with respect to the outside (Sect. 11.3). Each column of electroplaxes, then, consists of alternating membranes of opposite polarities. The first membrane is charged positively above and negatively below; the next membrane (actually the other surface of the same electroplax) is charged negatively above and positively below, and so forth. When stimulated by way of the nerve on one side of the electroplax, the polarity of that side is momentarily reversed, as is that of a nerve membrane. At that moment, then, the column of electroplaxes consists of a series of membranes which are all charged in the same direction. Every membrane is then negative above

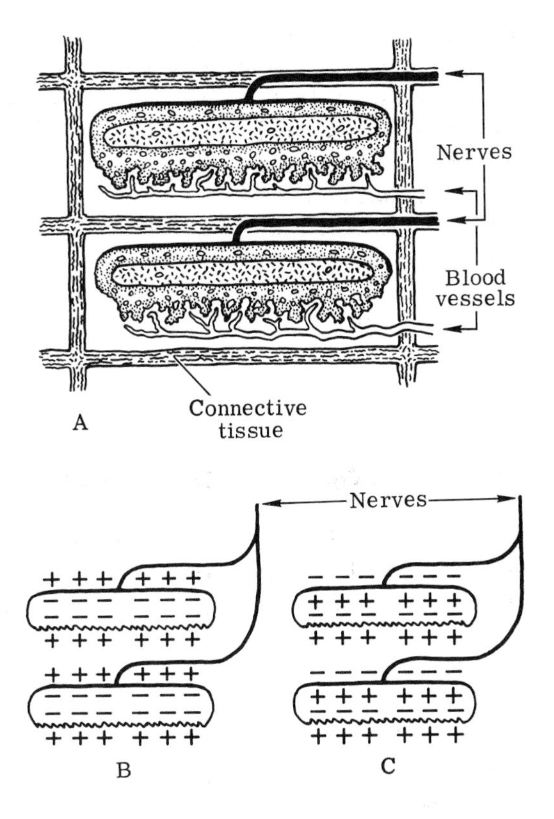

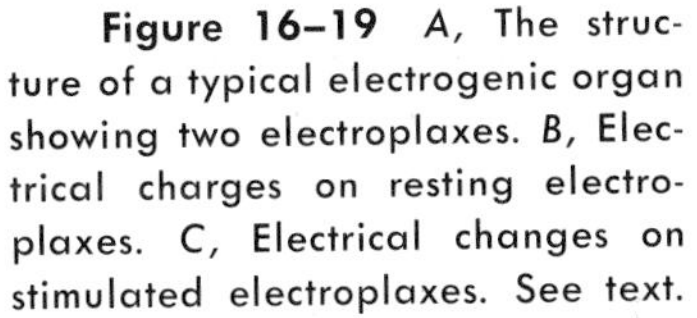
Figure 16–19 *A*, The structure of a typical electrogenic organ showing two electroplaxes. *B*, Electrical charges on resting electroplaxes. *C*, Electrical changes on stimulated electroplaxes. See text.

and positive below. In this configuration the membranes of the column are like multiple plates in a battery that are wired in series.

An ordinary skeletal muscle action potential develops about 90 millivolts of potential. An individual electroplax develops somewhat more—about 150 millivolts. In this series arrangement, however, the voltages are additive. In a stack of 5000 to 10,000 electroplaxes, as is found in the electric eel *Electrophorus*, the total may be 500 volts or more. This fish has about 70 such stacks in its electrogenic organ. The electric ray *Torpedo* has only about 1000 electroplaxes per column, but there are more than 2000 parallel stacks in its electrogenic organ. Thus, while only about 50 volts are generated, an amperage of about 50 and a total power output of more than 2500 watts is developed!

Certainly such a complex adaptation did not appear all at once in evolution; the ancestors of these animals must have developed the electrogenic organs slowly. Therefore, there must have been some selective advantage in developing potentials that were far too low to serve in either predation or defense. It was of great interest, therefore, to discover that certain fishes inhabiting very dark or turbid waters are able to detect objects around them through the use of primitive electrogenic structures. *Gymnarchus*, an African teleost fish, has been much studied in this regard. By means of a small electrogenic organ, this fish generates a pulsating electrical field in the water around itself. By means of lateral line receptors (Sect. 13.13) the fish is able to sense any changes in conductivity of the surrounding medium, and it responds to such objects within its electromagnetic field in purposeful ways. It seems reasonable to suppose that in evolution any increase in the power output of such an organ would have enlarged the scope of the animal's awareness and, thus, would have given it advantages in terms of navigation and food procurement. Finally a level was reached, presumably, at which the prey detection device was strong enough to serve also as an effective weapon.

Since electrogenic organs are composed of modified muscle cells, the total quantity of contractile muscle is reduced. Thus, possession of electrogenic organs is generally correlated with decreased swimming ability.

16.10 LOCOMOTION OF AMPHIBIANS AND REPTILES

A salamander has an elongate, somewhat streamlined body much like that of a fish. Ordinarily it moves by means of its limbs but when frightened it often allows the legs to trail unused and travels on its

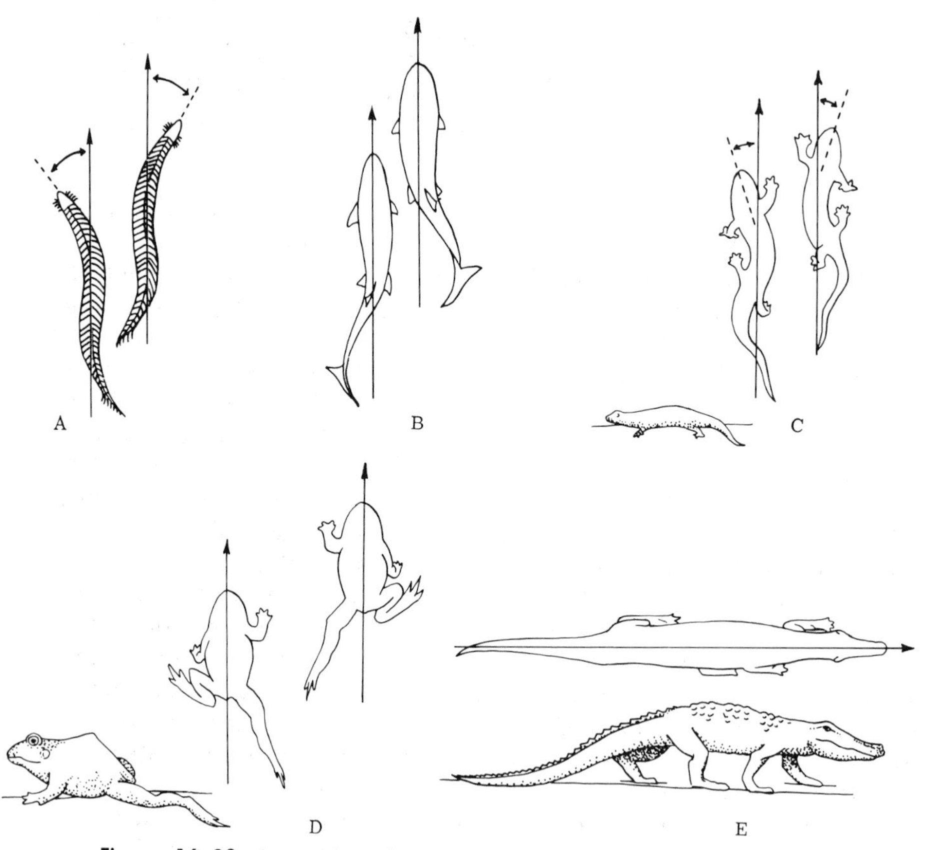

Figure 16–20 Locomotion. *A,* swimming by *Amphioxus; B,* swimming by shark; *C,* walking by salamander; *D,* walking by frog; *E,* walking by reptile. (From Cockrum and McCauley: *Zoology.* Philadelphia, W. B. Saunders Company, 1965.)

belly by means of typical swimming movements. Even when walking on its limbs, successive bending movements of the body occur, and the movement of the limbs is coordinated with this (Fig. 16–20). This pattern of leg movement, arising as it did from swimming activity, remains the pattern in all four-legged vertebrates. Even adult humans, who do not use the "forelimbs" in walking, still swing them at the sides in the same motion pattern. Long spinal reflexes (Sect. 11.2) were established early in vertebrate evolution and continue to be functional.

The development of legs that were more than stiff levers like those of certain fishes required the further modification of the associated muscle groups. Walking on legs places new stresses on the vertebral column, which must now serve as an arch to support the body weight. This, too, required additional modification of myotome arrangement. Such developments eventually resulted in limbs which could move independently and without simultaneous undulations of the body (Fig. 16-17).

The body structure of frogs is modified for jumping. The body cannot be bent to and fro but is rigid so that the power of the hind limbs can be transmitted evenly in jumping. In this case the hind limbs must both act simultaneously, which they usually do when swimming also. Yet, even a frog, when not frightened, is capable of the alternating type of limb movement either on land or in the water (Fig. 16-20).

The locomotion of lizards is much like that of salamanders. The body successively curves from side to side and the limb movements are coordinated with it. In a few cases lizards can run on the hind limbs but only when frightened. Snakes have even more primitive locomotion. It is essentially "swimming on land," as an eel does. A few snakes move by dorsoventral undulations like those of a caterpillar, but much less

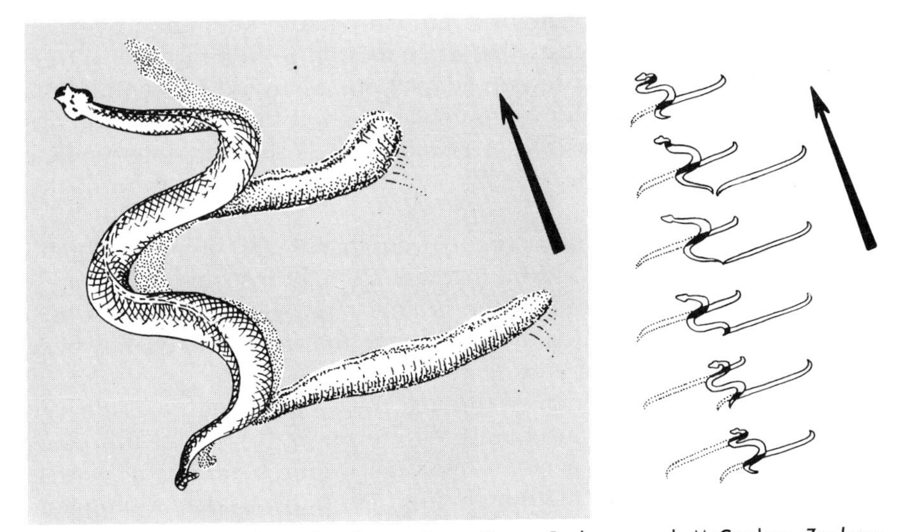

Figure 16–21 Sidewinder locomotion. (From Cockrum and McCauley: *Zoology*. Philadelphia, W. B. Saunders Company, 1965.)

noticeable. The sidewinder rattlesnake *Crotalus cerastes* employs a different variation, one which consists of a series of looping motions (Fig. 16-21) during which only two portions of the snake contact the ground at any one time. This corkscrew-like motion leaves a trail of long parallel marks in the sands of the desert where this snake occurs (southwestern United States and northern Mexico).

Another modification of reptilian locomotion is seen in *Draco*, the flying lizard of India and the Malay region. This animal has large lateral webs supported by ribs and these serve as gliding wings when the lizard leaps from trees.

16.11 LOCOMOTION OF BIRDS

Like some modern lizards, many extinct reptiles were able to move about on the hindlimbs only. When forelimbs are relieved of any important role in walking, they become available for other tasks, and the modification of our own arms and hands reflects this fact. Similarly, in a number of dinosaurs the forelimbs were reduced in size and the forefeet modified as grasping structures to carry food to the mouth. Undoubtedly, bipedal walking preceded the development of wings in the first birds since, as it is believed, birds evolved from reptiles such as these.

Two theories have been offered to explain how the evolution of bird flight may have come about. One theory suggests that a group of bipedal reptiles developed the habit of flapping the forelimbs up and down when running (much as a barnyard hen does). The least bit of lift attained by such flapping would have been advantageous since it would have permitted longer and longer strides. Presumably, then, these strides would have become gliding leaps and finally sustained flight as other adaptive changes took place in the body to permit it.

The second theory of flight is probably better supported by available evidence. This suggests that the ancestral reptile-bird was a tree-dweller that developed forelimb adaptations as an aid in leaping from tree to tree. Progressive development extended these leaps until sustained flight became possible. *Archaeopteryx* is the only intermediate fossil species known (Fig. 16-22), and the reptile-bird nature of this animal is obvious.

Bird locomotion differs markedly from that of most other vertebrates in that the forelimbs are moved together in flight instead of alternately as in bipedal motion. Even on the ground many birds hop rather than walk. Most flightless birds or birds that fly very weakly, however, walk or run in the typical vertebrate fashion.

When a flat plane is moved through the air, the medium is cleaved by it, thus resulting in an equal pressure on both sides of the plane (Fig. 16-23). This parting of the medium requires energy, of course, and the reuniting of the medium behind the trailing edge of the plane creates friction or drag. If the shape of the plane is modified so that its surfaces coincide with the natural pathways of the parted and rejoining

Figure 16–22 Possible ancestry of modern birds. *A*, bipedal reptile; *B*, theoretical bird ancestor; *C*, archaeopteryx; *D*, modern bird. (From Cockrum and McCauley: *Zoology*. Philadelphia, W. B. Saunders Company, 1965.)

air (*i.e.*, a streamlined plane), the changes in pressure on the two sides and the drag are minimized so that less energy is required to move the plane forward.

If the upper and lower surfaces of the plane are not alike, the pressure developed on the two sides will also be different. If, for example, the plane is flat on the underside but streamlined above, the air will be thrown further aside below and thus create more pressure there. This is a typical airfoil; when it is moved forward through the air, the increased pressure below lifts the plane. If the leading edge of the plane is tipped upwards, the lift is still further increased but so is the drag, especially at higher speeds. In any case, at some particular minimum forward speed, the drag becomes maximal and the lift becomes minimal. At this point the plane stalls and falls.

A wing slot in the leading edge of a wing can be so constructed that it directs a layer of air over the wing to reduce drag and, therefore, lower the stalling speed. Birds make use of the airfoil shape of wings and even have leading edge slots. A small structure called the alula (Fig. 16-24) can be moved forward to create such a slot. Unlike aircraft, birds are also able to change the shape and angle of the airfoil (wing) to some extent while in flight and thus vary the lift and drag qualities to meet their immediate needs.

When landing, aircraft commonly slow the forward speed by lower-

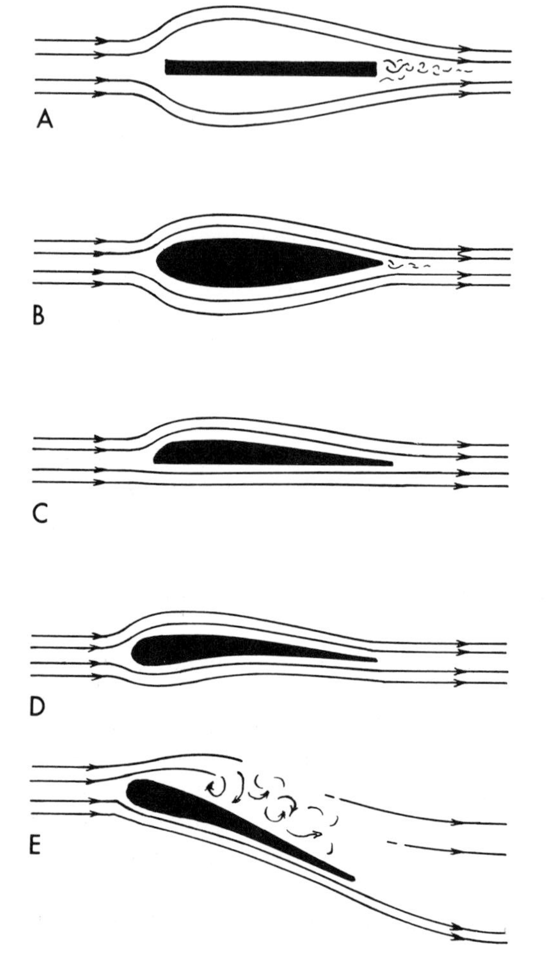

Figure 16–23 Principles of the airfoil. The rounding of the anterior edge reduces turbulence and drag (*A, B,* C). The airfoil in position *D* results in lift, whereas in position *E* it results in drag and stalling. (From Cockrum and McCauley: *Zoology*. Philadelphia, W. B. Saunders Company, 1965.)

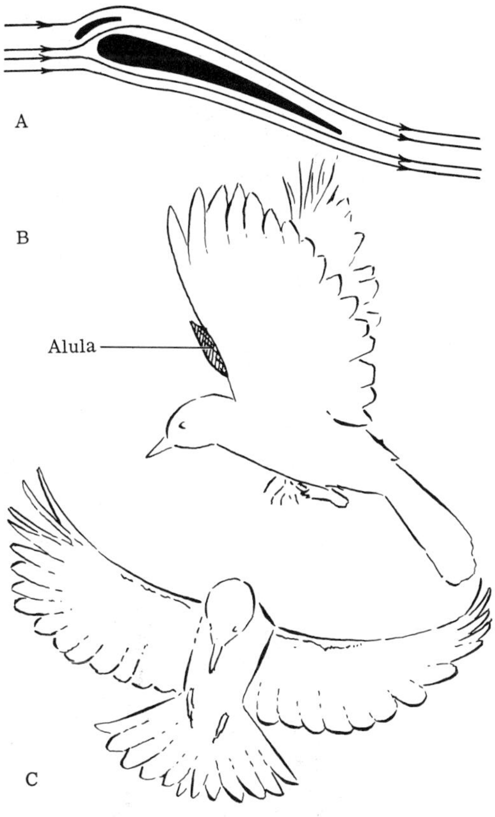

Figure 16–24 Wing slots in an aircraft (*A*) serve the same function as the alula of a bird wing (*B*). The flaps of a landing aircraft serve the same function as the wings and tail of a landing bird (C) (see text). (From Cockrum and McCauley: *Zoology*. Philadelphia, W. B. Saunders Company, 1965.)

Figure 16-25 A series of photographs of the California brown pelican showing successive wing positions as the bird lands. (Photos by L. W. Walker.) (From Cockrum and McCauley: *Zoology*. Philadelphia, W. B. Saunders Company, 1965.)

Figure 16–26 Bird flight. *A*, downstroke; *B*, upstroke; *C*, hovering flight of a hummingbird. (From Cockrum and McCauley: *Zoology*. Philadelphia, W. B. Saunders Company, 1965.)

ing flaps from the trailing edge of the wing. Birds accomplish the same thing by lowering feathers of the trailing edge of the wing as well as feathers of the tail (Fig. 16-25).

It is commonly supposed that the downstroke of a bird's wing is for the purpose of lifting the bird upward; this, however, is not true. Actually the bird obtains lift in the same way that an aircraft does, namely, from the cross-sectional shape of the wings and the forward motion of them through the air. The flapping of the wings is for the purpose of maintaining forward motion. In other words, it is really "swimming" through the air. On the downstroke there is also a strong forward thrust imparted, and in some birds the upstroke also drives the bird forward because of a change in the angle of the wings. In most cases, however, the upstroke is merely a recovery stroke (Fig. 16-26). The wing is folded close to the body and the feathers separate like the strips of a venetian blind so that little air resistance is encountered during the upstroke.

Since in hovering flight, like that of hummingbirds, forward motion is zero, wing flapping must provide all the lift. This is accomplished by bringing the body into an almost vertical position and then flying upward (Fig. 16-26). The bird in hovering flight can be thought of as flying upward at the same rate that it is falling downward. Such flying requires 40 to 80 wing beats per second, depending upon the size of the bird.

Swimming by penguins is literally flying through water and is accomplished in the same way as is flying through air by other birds.

16.12 LOCOMOTION OF MAMMALS

Mammalian locomotion on land typically involves movement of the four limbs in the same sequential pattern as that seen in most lower vertebrates—the pattern established by the swimming motions of fishes (Sects. 11.2 and 16.8). The muscles associated with the limbs are derived, both in evolution and in embryonic development, from segmental myotomes and they are innervated by the same spinal nerves that were associated with those myotomes. Much of the musculature of the dorsal trunk region retains the segmental nature even in the adult.

Various modifications of limbs and their associated muscles occur among mammals, of course, and these are adaptations to various environmental situations. Mammals walk, creep, run, dig, climb trees, and even fly, and each mode of locomotion or activity is associated with modifications in structure.

The wings of bats are formed by webbing between the fingers. A slight curling of the fingers downward produces a curved plane which acts as an airfoil.

The forelimbs of aquatic mammals are reduced and provide, as do the lateral fins of fishes, stability and steering. The hind limbs are also reduced and brought together posteriorly to form a horizontal plane which acts like the vertical tail of a fish. Swimming is accomplished as by fish through undulatory movements of the body, but, because of the horizontal tail, the movements are up and down rather than side-to-side. Those aquatic mammals which also walk on land, such as seals, have limbs which can be used somewhat awkwardly for walking as well as for swimming. Even on land, however, the characteristic movement is a dorsal and ventral undulatory motion rather than the typical alternate movement of the limbs.

CHAPTER 17

ENDOCRINOLOGY

17.1 INTRODUCTION

We have already considered the nervous system and its function in the integration of movement and various body processes. But there is another means by which many body processes are controlled. This control is chemical in nature and involves a series of glands and secretory cells that make up the endocrine system. Endocrine substances (hormones) are chemical "messengers" which are synthesized and released by cells of specific types and which have particular effects, either local or general, upon other cell types. Generally there are specific "target" organs or tissues which respond in predictable ways to any given hormone, but in some cases the effects are more diffuse. Usually hormone production and release is controlled by some sort of feedback mechanism (Sect. 1.6) so that the desired effect is achieved precisely. Hormones are proteins, protein derivatives (amino acids or polypeptides), or steroids and they exert effects upon the metabolic processes of the body. Hormones exert their effects by influencing specific cell permeabilities or other variables of cell metabolism. However, they do not act as catalysts in the way that enzymes do.

All living cells respond in purposeful ways to chemical influences in the environment (Sect. 13.6). Even protozoans respond to the presence of dissolved nutrients, chemical pollutants, and differential gas concentrations in the water. Sometimes the chemistry of the environment is affected by non-biological factors but frequently it is also influenced by the presence of other organisms. One organism may alter gas concentrations and excrete metabolic products to which another organism may react. Similar individual cell responses are also characteristic of the tissue cells of higher organisms. Moreover, the opportunity to react to the metabolic products of other cells is enhanced by their proximity and by the existence of a circulatory system which distributes metabolic products.

Chemoreceptors, as we have seen, are cells specially adapted to react to specific chemical changes in their environment, and in a number of cases these changes are brought about by the metabolic activity of other cells within the same organism. The response of oxygen and carbon dioxide receptors of the carotid sinus (Sect. 3.9) is an example. During the embryonic development of higher animals, cell metabolic products also influence the differentiation of specific cell types in surrounding tissues. Such influence at the embryonic stage is known as induction and is mediated by special chemical inductors, *i.e.*, substances excreted by other nearby cells.

If a chemical substance in the environment has a specific and desirable influence on certain cells of the organism, there is an advantage in the development of special cells to synthesize, store, and release this substance. These cells assure a constant and dependable supply of the material. Moreover, concentrations of such special synthesizing cells can be controlled so that secretion occurs at desirable times and in necessary quantities. There is evidence, for example, that thyroid-like substances were present and played essential roles in nervous activity before the thyroid gland, as such, appeared in the course of evolution. The gland, therefore, may have served merely as a dependable storage place from which a steady release of the substance could be controlled.

Certain vertebrate endocrine glands seem to have precursors in invertebrates, but most do not. Within vertebrate phylogeny, however, endocrine glands of similar type, structure, and location occur in all classes. Unfortunately, far too little research has been carried out on the endocrine functions of submammalian vertebrates. We are still not certain of the roles or even the presence of all mammalian hormones in the lower vertebrates. We are probably warranted in assuming, for lack of evidence to the contrary, that lower vertebrates have endocrine substances and functions similar to those of mammals. That this is not always the case, however, will be shown by examples discussed later in this chapter. Our lack of knowledge of specific instances of this also greatly hampers our ability to put together any connected ideas concerning the evolution of the endocrine system among the vertebrates.

Generally, but not always, target organs respond slowly and over relatively long time periods to endocrine influences. On the other hand, responses to nervous controls are usually rapid and of shorter duration. The separation of the two systems—nervous and endocrine—is somewhat artificial, since the production of certain hormones is under direct nervous control, and in some cases hormones are produced by nervous tissues rather than by endocrine glands as such. The transmitter substances that act on neuron synapses (Sect. 11.4) and neuromuscular junctions could easily be classed as local hormones, yet they are produced by neurons. Still other examples will become apparent in the following sections.

The gastrointestinal principles, which are also hormones, were discussed in Chapter 9 in connection with the controls of the digestive tract. In addition, another family of hormonelike chemical messengers, called the prostaglandins, has recently been discovered and is currently receiving a great deal of study.

17.2 THE HYPOPHYSIS (PITUITARY GLAND)

The mammalian hypophysis (pituitary gland) lies beneath the floor of the brain in a bony cavity of the sphenoid bone, a cavity called the sella turcica or Turkish saddle. It is connected to the brain by a stalk called the infundibulum (Fig. 17-1). The gland itself is composed of an anterior lobe (adenohypophysis), a posterior lobe (neurohypophysis), and, between the two, a smaller region known as the intermediate lobe or pars intermedia. Neurons lying in the hypothalamus of the brain have axons extending into the neurohypophysis and there is also a special vascular portal system which delivers blood from capillaries in the hypothalamus to other capillaries in the adenohypophysis.

Different cell types in the adenohypophysis produce at least six different hormones: (1) somatotrophic hormone (STH) or growth hormone, (2) thyroid stimulating hormone (STH) or thyrotropin, (3) adrenocorticotropic hormone (ACTH) or adrenal cortex stimulating hormone, (4) follicle stimulating hormone (FSH), (5) luteinizing hormone (LH) or interstitial cell stimulating hormone (ICSH), and (6) luteotropic hormone (LTH). The last three in this list are often grouped together as the gonadotropic hormones because they exert their influences on the male and female gonads (the ovaries and testes).

In mammals the neurohypophysis releases two hormones, oxytocin and vasopressin. Vasopressin is also called the antidiuretic hormone (ADH) and was discussed earlier in connection with water balance and renal function (Sect. 8.6).

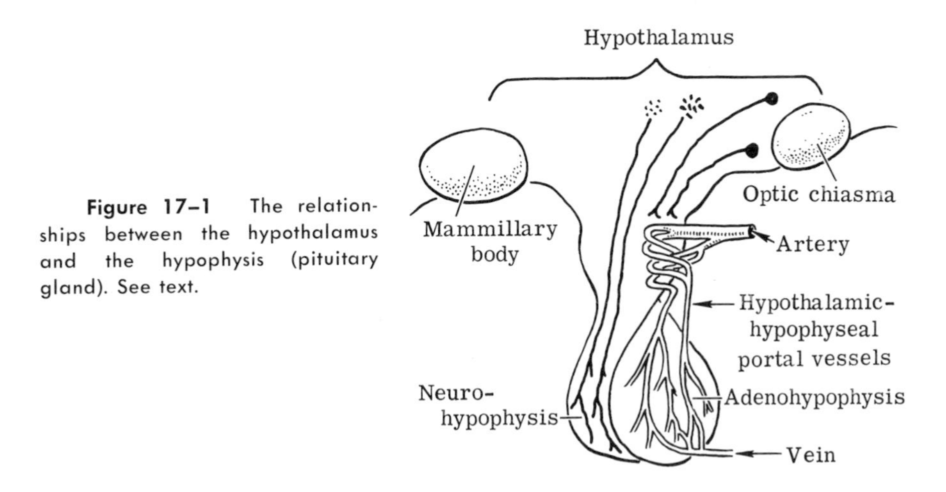

Figure 17–1 The relationships between the hypothalamus and the hypophysis (pituitary gland). See text.

17.3 THE ADENOHYPOPHYSIS

The neuron processes extending from the hypothalamus of the brain into the neurohypophysis may not carry action potentials like other neurons. Instead, as we shall see in the next section, they apparently transport endocrine substances which are stored in and released by the neurohypophysis. Some hypothalamic neurons, instead, carry chemical substances known as releasing factors. These diffuse into the hypothalamic-hypophyseal portal circulation and are carried to the adenohypophysis where they initiate the secretion of specific hormones. In this instance the hormones ultimately released are actually synthesized in the adenohypophysis. Probably all of the releasing factors are short polypeptides, and it may be that some are chemical precursors of the hormones released by the adenohypophysis. With the possible exception of the luteotropic hormone (LTH), there is a separate releasing factor for each anterior lobe secretion.

Somatotropic hormone (STH) acts during the growth of an individual to increase the rate of protein synthesis and other metabolic processes that promote growth. It increases the rate of amino acid transport through cell membranes and the rate of fat utilization as an energy source. The concentration of STH in children is about 300 micrograms per liter of plasma. This drops by about 30 per cent in adults but some secretion continues throughout life. Growth ceases in late adolescence because of the closure of the epiphyseal synchondroses (growth centers) of the long bones of the skeleton, and this closure is also under the influence of hormones. The growth centers cease to exist when sufficient levels of sex hormones are established in the blood.

A deficiency of STH during childhood retards growth and, if severe, may result in ateliotic (or Lorain) dwarfism. This type of dwarf, commonly called a midget, is a normally proportioned but child-sized individual who never attains puberty or the secondary sexual characteristics of the adult. He remains a child both in general appearance and in sexual development.

Excess secretion of STH, on the other hand, produces abnormal growth leading to some degree of giantism. Variations in STH secretion superimposed upon the genetic height of individuals have produced variations in human height ranging from less than three feet to more than nine feet.

In adults deficiencies have little effect because growth and maturation have already taken place and are irreversible. Some loss of weight is the only clear result of this deficiency. Excesses of STH in adults can no longer cause any increase in height because the growth centers of the long bones have ceased to exist. Such excesses can, however, produce increased thickness of the bones, resulting in a condition known as acromegaly. In this syndrome the hands, feet, and hips become abnormally wide and heavy and the supraorbital ridges of the skull and the mandible become noticeably prominent.

The thyroid stimulating hormone (TSH), or thyrotropin as it is sometimes called, specifically stimulates the thyroid gland to synthesize and release its hormone. This is an example of one endocrine gland controlling another, and there are many such examples, especially with regard to the hypophysis. For that reason the pituitary gland is sometimes spoken of as the "master gland" of the endocrine system. The presence of adequate amounts of thyroid hormone in the blood inhibits the production of the thyrotropic releasing factor by the hypothalamus. Therefore, under these conditions less thyroid stimulating hormone is released by the adenohypophysis, and the thyroid gland is less stimulated to produce more of its hormone. When the situation is reversed; *i.e.*, when the blood levels of thyroid hormone are less than adequate, the hypothalamus produces more thyrotropin releasing factor, more thyrotropin is released by the adenohypophysis, and the thyroid is, therefore, stimulated to produce greater quantities of thyroid hormone. This chemical feedback system guarantees that proper amounts of thyroid hormone will always be present in the blood of a normal individual.

In a similar way the adrenocorticotropic hormone (ACTH) stimulates the production of certain hormones by the adrenal cortex. These hormones, in turn, inhibit the production of adrenocorticotropin releasing factor by the hypothalamus. The feedback controlling mechanism is exactly parallel to that of the adenohypophysis-thyroid relationship described in the preceding paragraph. In a similar manner it maintains proper levels of adrenal cortex hormones in the blood of normal individuals at all times.

The follicle stimulating hormone (FSH) stimulates the maturation of sex cells (ova and spermatozoa) in the gonads of the two sexes. In females the luteinizing hormone (LH), along with FSH, stimulates the maturation of ova and their release from the ovary when they are mature. In the male the same substance stimulates certain endocrine cells (interstitial cells of Leydig) which in turn play other roles in male reproduction. For this reason, in the male the counterpart of LH is called the interstitial cell stimulating hormone (ICSH). The luteotropic hormone (LTH) in the female is associated with lactation but apparently has only minor influences, if any, in the male. These hormones, FSH, LH or ICSH, and LTH, are known as the gonadotropic hormones because they affect the gonads (ovaries and testes). They are discussed in greater detail in subsequent sections of this chapter.

Similar adenohypophyseal hormones are apparently present in all vertebrate classes, although the presence of growth hormone (STH) has not been established in birds. In fishes ACTH acts under stress conditions such as crowding to inhibit growth; perhaps it inhibits the secretion of the growth hormone. Aquarium fishes that have apparently attained maximum size will often grow larger if transferred to larger aquaria. Clearly, then, fishes may continue to increase in length even after they attain sexual maturity.

17.4 THE NEUROHYPOPHYSIS

The neurohypophysis serves merely to store and release hormones that are synthesized in the hypothalamus of the brain and transported to the neurohypophysis through the axons of nerve cells. Perhaps these cells also transmit nervous impulses that control the release of the hormones stored in the neurohypophysis; this point, however, has not been demonstrated.

One of the neurohypophyseal hormones, oxytocin, causes rhythmic contractions of the mammalian uterus at the time of childbirth and it may also serve to induce egg laying in some of the lower vertebrates. In mammals the uterus does not respond to oxytocin at any other time because of the inhibiting influence of progesterone, another hormone produced by the ovaries (Sect. 17.12). Oxytocin also plays a role in mammalian lactation (Sect. 17.13).

The other neurohypophyseal hormone, vasopressin, is a blood pressure raising hormone but its more important role is the control of the water balance of the body. For that reason it is also known as the antidiuretic hormone (ADH). It has already been discussed in connection with kidney function (Sect. 8.6).

Oxytocin and vasopressin have similar chemical structures (Fig. 17-2) and each exhibits some of the properties of the other. That is, oxytocin has some antidiuretic effect and vasopressin has some uterus stimulating effect. The vasopressin of the pig and the hippopotamus is different from that of other mammals by one amino acid substitution. The comparable hormone believed to be present in all amphibians, reptiles, and birds as well as in some fishes is another variant called vasotocin. This also differs by a single amino acid. Such simple amino acid replacements result from single mutations and they illustrate one of the mechanisms of biochemical evolution. Some fishes do not have vasotocin but instead have a quite different octapeptide. Elasmobranchs

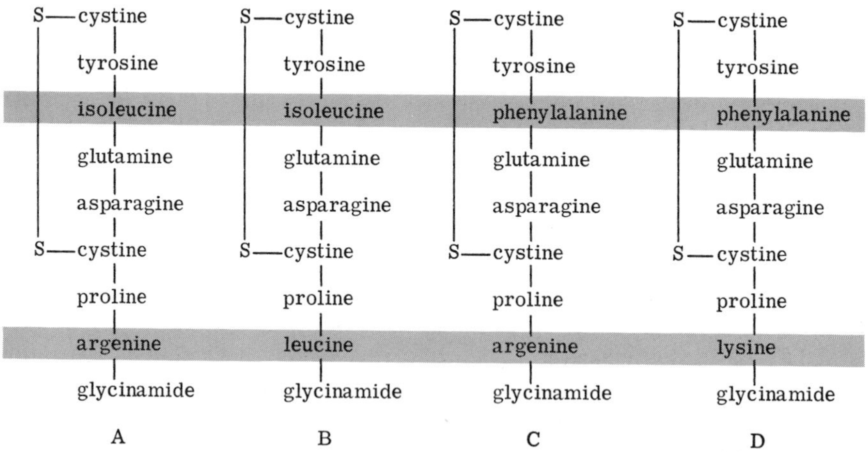

A	B	C	D
S—cystine	S—cystine	S—cystine	S—cystine
tyrosine	tyrosine	tyrosine	tyrosine
isoleucine	isoleucine	phenylalanine	phenylalanine
glutamine	glutamine	glutamine	glutamine
asparagine	asparagine	asparagine	asparagine
S—cystine	S—cystine	S—cystine	S—cystine
proline	proline	proline	proline
argenine	leucine	argenine	lysine
glycinamide	glycinamide	glycinamide	glycinamide

Figure 17–2 Chemical structures of related hormones. *A*, vasotocin; *B*, oxytocin; *C*, argenine vasopressin; *D*, lysine vasopressin.

may have a totally different and unrelated hormone acting in the control of water balance. It will be recalled that elasmobranch water balance is related to urea retention, and in all probability this is under endocrine control (Sect. 8.11).

17.5 THE INTERMEDIATE LOBE

Vertebrate skins contain pigment cells (chromatophores) of three common kinds: melanophores (black), xanthophores (yellow), and erythrophores (red). In some cases other colors may also occur. In many cold-blooded vertebrates the chromatophores have extended processes, and the contained pigment can either be dispersed throughout the cell or concentrated in a small mass at its center (Fig. 15-1). Such pigment migrations result in changes in the general color of the animal and this plays a role in protective camouflage. The pars intermedia, or intermediate lobe of the hypophysis, of such animals produces a melanophore stimulating hormone (MSH) which causes a dispersion of the pigment and a consequent darkening of the animal's appearance. MSH also stimulates the metabolic synthesis of melanin, the pigment in the melanophores. The pigment cells of humans are not constructed in this way, so MSH has little function except, perhaps, to stimulate melanin synthesis. Experimentally introduced MSH and ACTH both cause darkening of human skin, and it has been shown that both hormones have an identical amino acid sequence as a part of the total molecule.

Elasmobranches and teleosts both have MSH but the pigment cells of teleosts are also directly innervated and respond to nerve action potentials. Innervated chromatophores respond quickly, whereas hours may be required for the complete MSH-mediated response to occur. As a result, color changes in teleost fishes are often abrupt and quite spectacular.

The MSH mechanism is best developed in amphibians and has been extensively studied in these animals. Only a few reptiles, notably chameleons, show pronounced color changes. Although MSH is present in birds, there is no intermediate lobe in the avian hypophysis. Instead cells which secrete MSH must be present in other portions of the gland. The function of the hormone in birds is also uncertain since color changes of the feathers are mediated by other controlling mechanisms.

There is evidence that another hormone, melatonin, which is released from the pineal gland, has an effect opposing that of MSH. That is, melatonin may act to cause concentration of pigment within melanophores. At least in the Amphibia the two hormones probably act together to attain various degrees of lightening or darkening of the animal's skin. Epinephrine, the hormone secreted by the adrenal medulla, also causes concentration of melanin in the melanophores.

The intermediate lobe of the mammalian pituitary is very small and it is apparently completely lacking in some members of this group. As we have seen, it plays at best a minor role.

17.6 THE THYROID GLAND

The mammalian thyroid gland is composed of paired lobes which lie on either side of the larynx in the neck. These paired lobes are joined by a narrow anterior bridge of glandular tissue, giving the entire structure a shape something like a bow tie. During embryonic development the gland is derived from a downward growth of oral epithelium, and for a time it retains a connection with the oral cavity through a thyroglossal duct which opens near the base of the tongue. The hormone produced is thyroxin or a related derivative of the amino acid tyrosine (Fig. 17-3), combined with a protein. Diiodotyrosine and triiodotyrosine are very potent derivatives and may also be active forms of the hormone in humans.

Dietary iodine is actively accumulated by the thyroid gland, where it is substituted onto the tyrosine molecule. The resulting thyroxin is stored within follicles of the gland and released to the blood stream as required. The thyrotropic hormone (TSH) of the adenohypophysis stimulates both the synthesis and the release of thyroxin.

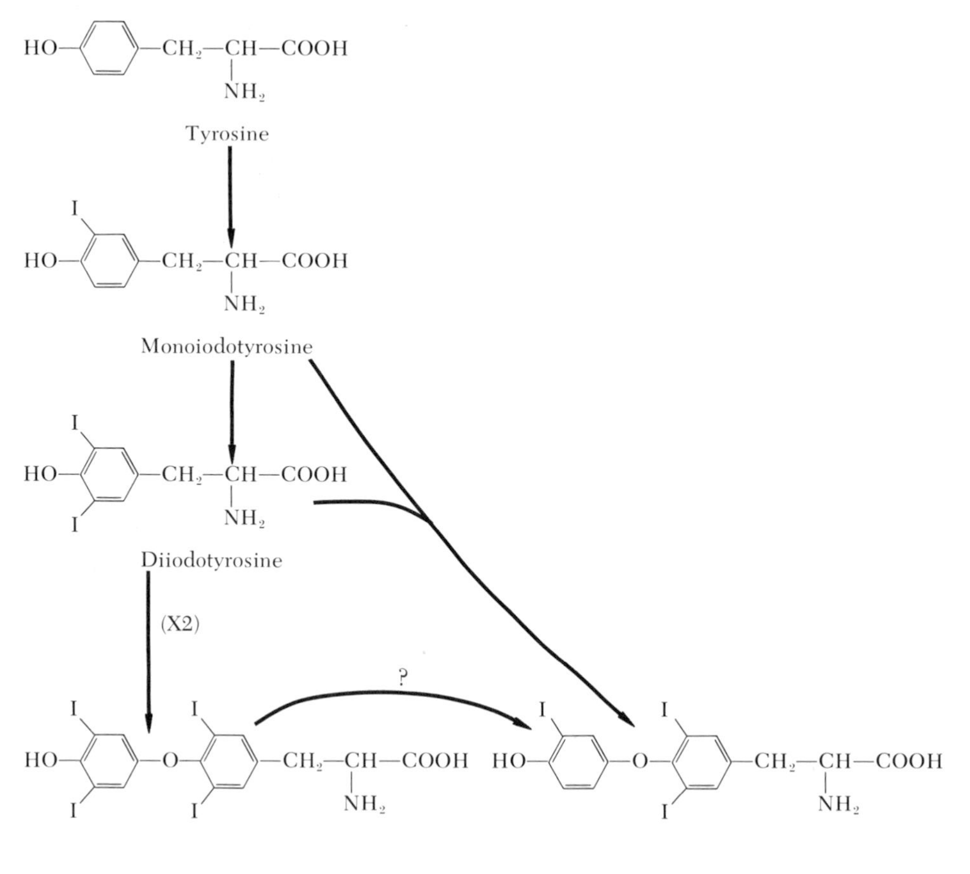

Figure 17–3 The synthesis of thyroid hormones. See text.

In mammals the primary action of the hormone is to increase the general metabolic rate of the animal. The production of energy, the synthesis of protein, the activity of the nervous system, and the activity of other endocrine glands are all generally increased by thyroxin. The mechanism of this action is not yet completely understood. Increased quantities of almost all intracellular enzymes appear when thyroxin is experimentally introduced into an animal for a time. It may be that changes in cell membrane permeabilities are the primary effect of the hormone and that the other observed changes are consequences of these changes.

Abnormally high secretion rates of thyroid hormone commonly result in loss of body weight, but in some cases, the appetite is also increased to such an extent that a weight gain occurs. Such hypersecretion by the thyroid gland can result from a thyroid tumor (excess secreting thyroid tissue) or from overstimulation by the hypophysis. This overstimulation can result, in turn, from a tumor or other abnormality in either the adenohypophysis or the hypothalamus. Conversely, decreased secretion of thyroid hormone generally results in obesity, but in some cases the appetite is sufficiently depressed that this does not occur.

Much more important are the effects of thyroid hyposecretion on mental development and function. Inadequate amounts of thyroxin in children prevent normal mental development. The reduction in metabolic rate also retards general growth, and together these effects result in a condition known as cretinism. Cretins are mentally deficient dwarfs and are easily recognized. Thyroid insufficiency occurring in the adult after normal growth and mental development have already taken place still causes a dulling of mental ability and an accumulation of fluids in the tissues. This condition is known as myxedema.

Goiters result from hypertrophy (enlargement) of the thyroid gland and may accompany either hypo- or hypersecretion of the hormone. When there is an insufficiency of dietary iodine, the hormone cannot, of course, be produced in adequate quantities. In the normal human, then, the shortage of thyroxin leads to the production and release of greater quantities of thyroid stimulating hormone (TSH) from the hypophysis (Sect. 17.3). This stimulates growth of the thyroid gland. At one time such "simple goiters" were common to regions of the United States where the soil was poor in iodine. Because most foods eaten were raised locally, people in these regions frequently showed symptoms of hypothyroidism. It was in response to this situation that iodized salt was first placed on the market. Today, foods eaten in any region come from all parts of the United States and abroad, and the danger of hypothyroidism is not as great.

In cases which involve pathological conditions of the thyroid gland or of the hypothalamus-hypopysis controlling system it is also possible to have hypertrophy of the thyroid gland (goiter) associated with hypersecretion of thyroid hormone. An excess of this hormone results in tremors of the muscles, rapid heart rates, and other signs of increased metabolic and nervous activity. A person suffering from a "toxic goiter," as it is called, sometimes has a peculiar pop-eyed appearance, or exophthalmos.

Since the rate of metabolism determines the rate of heat production in the body, the production of thyrotropin releasing factor by the hypothalamus may be influenced by the environmental temperature. The production of the releasing factor can also be influenced by emotional states. Thus, thyroid release and consequent metabolic rates can be readjusted to meet seasonal demands or those imposed by prolonged periods of emotional stress.

Thyroid hormone is present in all vertebrates, but its function in the control of metabolic rate is apparently confined to the warm-blooded vertebrates. In the others it is of importance in the development and normal function of the nervous system. It is also necessary for the metamorphosis of amphibians. Tadpoles from which the thyroid gland has been removed are unable to metamorphose, unless treated with thyroid hormone, and they simply grow into very large tadpoles. On the other hand, the administration of excess thyroid hormone to young tadpoles induces premature metamorphosis and consequently the production of very small adult frogs.

In the larval lamprey (a cyclostome) a longitudinal groove in the pharyngeal floor elaborates thyroxine. The endostyle, as the groove is called, secretes the hormone into the pharynx where it is swallowed. During digestion the thyroxin is absorbed by the animal and plays its usual role in metabolism. In the adult lamprey, in which the endostyle is closed off from the pharynx, the hormone diffuses directly into the blood stream. The embryonic development of this gland is similar in higher vertebrates. Interestingly, the thyroid hormone is the only one which can be given therapeutically by mouth, and it appears that this was the original route of natural administration in early vertebrates.

17.7 THE PARATHYROID GLANDS

Human parathyroid glands usually consist of four small bodies posterior to and perhaps embedded in the substance of the thyroid gland. Two or more parathyroids are present in all vertebrates more advanced than fishes and they uniformly arise from pharyngeal pouches during the embryonic development of the animal. Fishes have, instead, an ultimobranchial gland of similar embryonic origin. Although it does not look like a parathyroid gland, it may produce a similar hormone. Little is known of the function of the parathyroid glands in submammalian vertebrates, but in the mammals they are known to control the metabolism of calcium and perhaps phosphate.

Bone is composed of inorganic salts to the extent of about 75 per cent, and much of this inorganic material is calcium phosphate. It is laid down by cells called osteoblasts and resorbed by other cells known as osteoclasts. The deposition and resorption are both continuous processes that permit the constant remodeling of bone structure, which accompanies growth and accommodation to changing stresses. Both calcium and phosphorus in bones are in equilibrium with the same ions dissolved in the blood plasma. The parathyroid hormone favors deposition of these salts in the bone and increases the renal retention of calcium ions. The hormone effect at the bone sites and in the kidney probably increases the transport of calcium through cell membranes.

Calcium, like all bivalent ions, is poorly absorbed from the intestinal tract. However, its absorption there is promoted by vitamin D and probably also by parathyroid hormone. The bones provide a large store of calcium, and in the presence of temporary dietary deficiencies they maintain adequate blood levels of ion. Prolonged calcium-free diets or vitamin D deficiencies, however, may result in degenerative bone diseases such as rickets. Low blood calcium levels also cause muscular tetany, which is due to abnormal nervous excitation. High blood calcium levels have the opposite effect, *i.e.*, decreased nervous activity, sluggish reflexes, and so forth. Abnormal variations in the secretion of parathyroid hormone can produce symptoms similar to those described earlier for abnormal blood calcium levels. Normally the level of blood calcium controls the rate of secretion of parathyroid hormone in a feedback manner. It is possible that there is also a parathyrotrophic hormone of the adenohypophysis, but the existence of this hormone has not been demonstrated.

17.8 THE ISLETS OF LANGERHANS

The endocrine portion of the mammalian pancreas consists of small groups of cells lying between the portions of the gland which produce digestive enzymes (acinar portions). These endocrine parts are called the islets of Langerhans (Fig. 17-4) and they contain two cell types, designated as alpha cells and beta cells. The alpha cells produce the hormone glucagon, and the beta cells produce a different hormone known as insulin. Both hormones act to control, in part, the metabolism of carbohydrates (sugars). Carbohydrate metabolism is a complex process that is influenced by many factors, including certain other endocrine glands. Both insulin and glucagon are secreted into the blood of the hepatic portal vein and are carried directly to the liver where they produce their major effects. Some of the hormone passes on through the

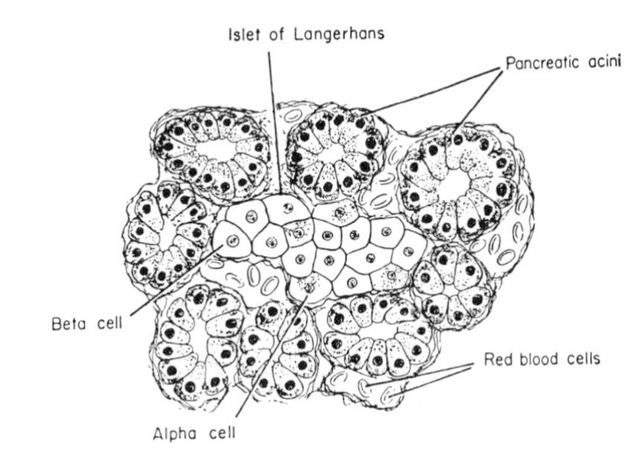

Figure 17–4 Pancreatic tissue showing an islet of Langerhans. The acini produce digestive juices. (From Guyton: *Textbook of Medical Physiology*, 4th Edition. Philadelphia, W. B. Saunders Company, 1971.)

liver and into the general body circulation. This influences the utilization of sugars by all of the cells of the body.

Insulin promotes the transport of glucose through the cell membranes of muscles and, especially, of the liver. Since the liver normally has large stores of glucose in the form of glycogen (Fig. 9-4), it maintains proper levels of blood glucose by either taking it up or releasing it as required. Insulin facilitates this exchange as well as the uptake and utilization of glucose by the muscles and other tissues. The level of blood glucose normally acts in a feedback manner to control the rate of insulin release from the islets of Langerhans.

In the absence of sufficient insulin, both the storage and the utilization of glucose are inhibited and, as a result, dietary glucose accumulates in the blood until its levels may exceed the renal threshold (Sect. 8.5). Then glucose appears in the urine. When insulin levels are low, fat catabolism is increased and fats are converted into glucose. This still further increases the blood glucose levels and also results in the accumulation of ketone bodies (Sect. 10.10). The ketone production in diabetes mellitus, as the condition is called, may eventually lead to actual shifts in blood pH and consequent death.

Since the effect of insulin on glucose uptake by muscles exceeds its effect on liver release of glucose, abnormally large amounts of this hormone can lead to lowered blood glucose levels. The brain and certain other tissues cannot use any energy source other than glucose, and such lowered blood levels can have serious consequences.

Glucagon has effects on blood glucose levels which are, in general, the opposite of the effects of insulin. Glucagon favors the release of glucose from the glycogen stores of the liver and, thus, raises blood glucose levels. A proper balance between insulin and glucagon production is therefore necessary to maintain proper blood glucose concentrations.

Cells similar to beta cells occur in the walls of the cyclostome gut. All other vertebrates have alpha or beta cells associated with the pancreas. In the lower vertebrates there is a general tendency to have one cell type or the other, whereas higher vertebrates commonly have both. In a few animals, such as the non-carnivorous birds and some reptiles, insulin is apparently absent.

17.9 THE ADRENAL GLANDS

In mammals the adrenal glands are pyramidal structures which lie in close association with the kidneys. An adrenal gland is composed of an outer cortex and an inner medulla; these two portions are separate endocrine glands. In lower vertebrates the tissues corresponding to the medullary portions of the mammalian adrenal glands are called chromaffin tissues by reason of their staining affinities for chromium salts. The chromaffin tissues do not occur as a discrete medulla in these lower forms but rather as strands or islets scattered through and around the cortical tissues. Chromaffin cells also occur in all vertebrates in association with the paravertebral sympathetic ganglia (Sect. 12.6). This is not surprising since they arise during embryonic development, as does the mammalian adrenal medulla, from the same cells that give rise to the paravertebral ganglia. Moreover, they all produce similar substances. Norepinephrine is most commonly secreted by chromaffin tissues and it is the transmitter substance at the pre- to postganglionic synapses of sympathetic neurons in the paravertebral ganglia. Epinephrine is the most common constituent in the secretions of the mammalian adrenal medulla. Norepinephrine is the chemical precursor of epinephrine (Fig. 17-5) and both have similar physiological effects.

The cortical tissues of the mammalian adrenal gland are arranged in three layers, the outermost of which is called the zona glomerulosa. This layer produces steroid hormones known as mineralocorticoids because the major effect is upon the metabolism of sodium ions and, indirectly, potassium ions. The second layer, the zona fasciculata, and the third layer, the zona reticularis, produce other steroid hormones known as glucocorticoids. These hormones, as their name implies, play roles in the metabolism of carbohydrates. The adenohypophyseal hormone ACTH (adrenocorticotropic hormone) stimulates the production

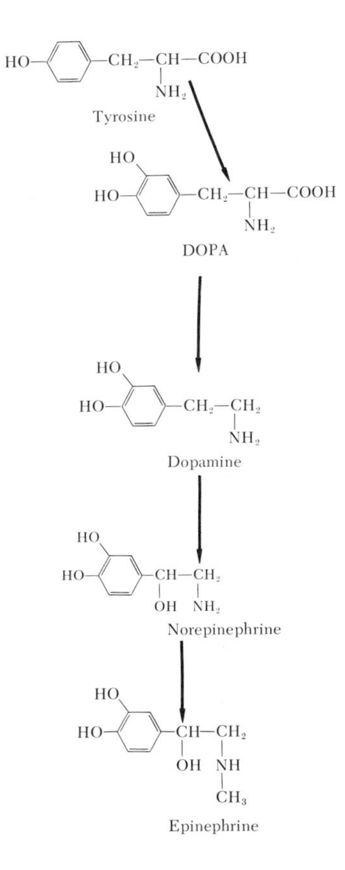

Figure 17–5 The synthesis of epinephrine and norepinephrine. See text.

of glucocorticoids and to a lesser extent of mineralocorticoids. The blood levels of glucocorticoids, in turn, control the production of the ACTH releasing factor from the hypothalamus.

Adrenal cortical tissues are present and have similar appearance in all submammalian vertebrates other than fishes, which have instead interrenal tissues that probably produce similar hormones but have quite a different appearance. As far as is known adrenal cortical hormones have similar functions in all vertebrates.

The major mineralocorticoid hormone is aldosterone (Fig. 17-6). Its most important effect is to promote the resorption of sodium ions from the renal glomerular filtrate (Sect. 8.7). Secondary effects of sodium retention are an increased chloride retention and a decreased potassium retention by the kidneys. The retention of sodium ion also leads to water retention by means of the antidiuretic hormone mechanism (Sect. 8.6). Thus, the control of the levels of blood sodium has far-reaching consequences. Plasma levels of Na^+ and perhaps of K^+ act by way of an unknown feedback pathway to control the rate of aldosterone secretion.

The major glucocorticoids are cortisone and certain closely related steroids (Fig. 17-6). These hormones stimulate the production of glucose from non-carbohydrate sources such as fats and amino acids (Chapter 10). Production of such "new" glucose is called gluconeogenesis. These hormones, thus, promote the reduction of amino acid stores in most tissues and the uptake and catabolism of amino acids in the liver. Glucocorticoids also decrease glucose utilization by tissues in general. All of these effects lead to increased blood glucose levels. Thus, the glucocorticoids are intimately associated with the hormones of the islets of Langerhans (Sect. 17.8) in the controls of blood glucose concentration.

Cortisone also acts as an anti-inflammatory agent; however, the mechanism of this action is not known. During all kinds of stressful situations, cortisone is secreted in increased quantities because of the release of increased amounts of ACTH by the adenohypophysis. In some way cortisone seems to increase the total ability of the animal to withstand a great variety of physical stresses.

In addition, the adrenal cortex normally secretes small amounts of sex hormones, including estrogens (female) and androgens (male). The roles of these hormones from this source are not known because the same hormones are produced in greater quantities by the gonads (ovaries or testes) of mature individuals. Dysfunctions of the adrenal cortex sometimes lead to hypersecretion of sex hormones which may result in such symptoms as masculinization of females and gynecomastia (excessive breast development) in males. It is interesting to note that mice and certain other small mammals have a region of the adrenal cortex called the X-zone which disappears at the time of puberty in males but not in females. In the mature female there is evidence that the X-zone is somehow involved in the control of sex cycles but the way in which it does this is unknown.

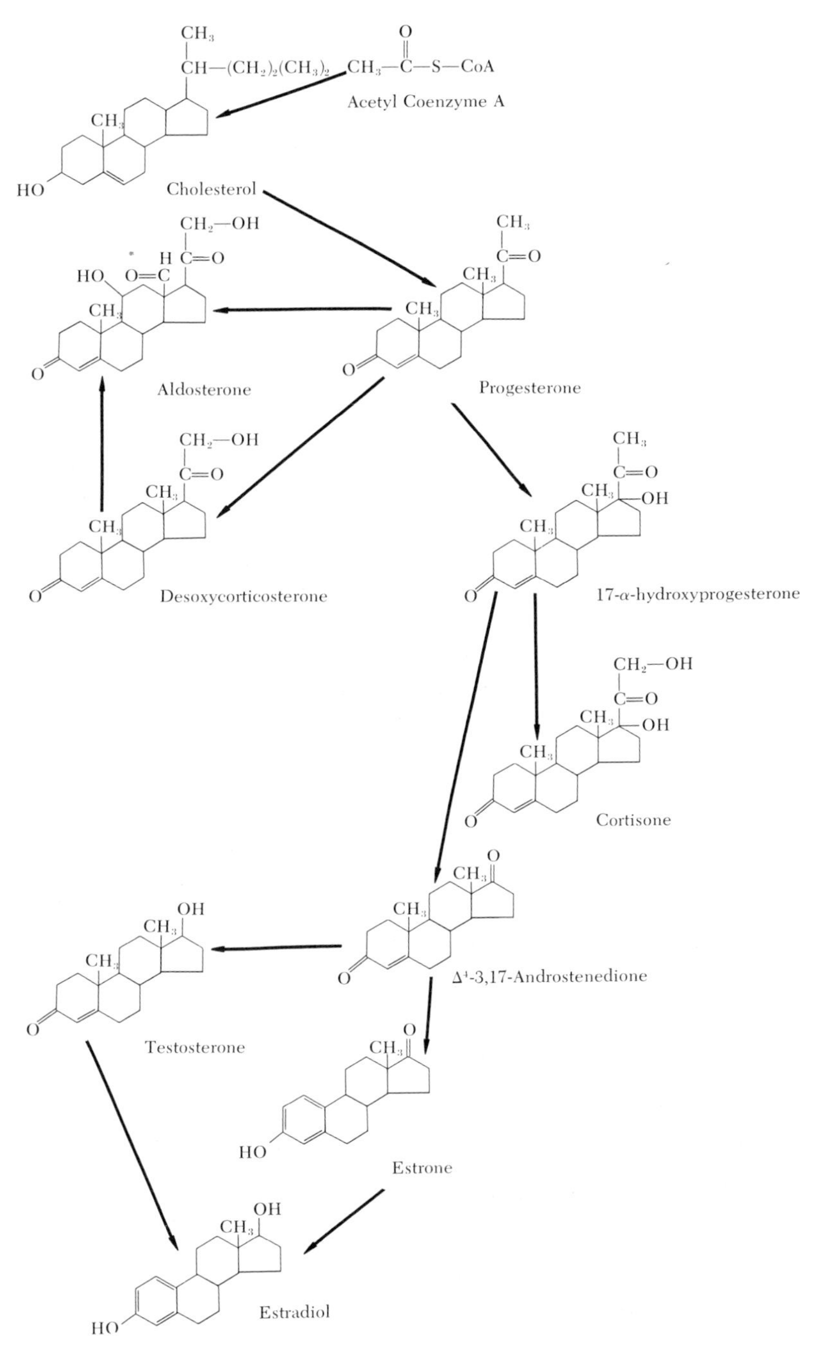

Figure 17–6 Chemical relationships of the steroid hormones. See text.

In some small mammals the adrenal size increases with increasing population density. It is suspected that at such times increased androgen secretion by the adrenal cortex inhibits female fertility and, thus, limits further increase in population size.

REPRODUCTIVE ENDOCRINOLOGY

17.10 THE TESTES

The word "testis" in Latin means "witness." According to Roman law, women and eunuchs (castrated male slaves) could not testify in court. The male gonads, then, as a sign of Roman citizenship were called testes. In mammals they are composed of cords of cells arranged in lobules. Spermatozoa arise from these cords of cells under the influence of FSH secreted from the adenohypophysis. The spermatozoa are transported by way of the epididymis and vas deferens to the ejaculatory duct from which they are expelled, together with accessory glandular secretions, at the time of mating. Between the sperm-producing cords within the testes are the interstitial cells of Leydig (Fig. 17-7), which have an endocrine function. Under the influence of the interstitial cell stimulating hormone (ICSH) from the adenohypophysis these cells produce testosterone, the major sex hormone of the male. Testosterone and other male sex hormones are known collectively as androgens.

Testosterone (Fig. 17-6) causes embryonic development of the male reproductive organs. It also induces the maturation of these organs during childhood and maintains their normal function following puberty. At the time of puberty testosterone also promotes the development of the secondary sex characteristics of males, including physical development, hair distribution, masculine voice, and male behavior.

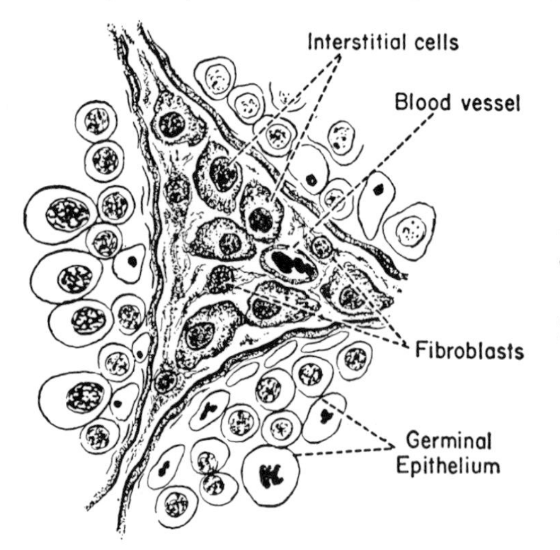

Figure 17-7 Interstitial cells of Leydig located between the cords of spermatogenic tissue in the testis. (From Guyton: *Textbook of Medical Physiology*, 4th Edition. Philadelphia, W. B. Saunders Company, 1971.)

In most mammals the testes are suspended outside the body in a sac called the scrotum. It has been shown that the temperature within the scrotum is maintained somewhat lower than intra-abdominal temperature and that this is necessary for normal spermatozoa production. In humans, special muscles called cremasters in the walls of the scrotum elevate the testes close to the body in cold weather or lower them for more effective cooling in warm weather. In all mammals the testes develop first within the abdominal cavity. In humans they descend into the scrotum, and within a few years after birth the route of descent is obliterated so that they remain permanently there. In seasonally breeding mammals, the testes often return to the intra-abdominal location during non-breeding periods of the year. With the advent of the breeding season, they descend into the scrotum and increase considerably in size. In a few mammals the testes never leave the intra-abdominal site at all (*e.g.*, elephant, whale). The situation regarding temperature and sperm production in these species is not known. The testes of birds are similarly located in the abdominal cavity, and birds have even higher body temperatures than mammals. Avian body temperature fluctuates quite widely, however, being at its lowest in the early morning hours, in those species that have been studied. It has been shown that spermatogenesis occurs almost exclusively during the period of lowest body temperature.

Testosterone is apparently present in all vertebrates and it probably plays similar roles in all.

17.11 THE OVARIES

Mammalian ovaries are oblong organs situated in the pelvic portion of the abdominal cavity. Each is close to the expanded opening of a fallopian tube or oviduct which leads centrally to the opening into the uterus. Another opening from the cervix or neck of the uterus leads to the vagina, and this, in turn, opens to the surface of the body. Each ovary contains a large number of undeveloped sex cells or ova, and as these develop they rise to the surface of the organ and become enclosed in blister-like graafian follicles. At the time of ovulation the follicles rupture and the contained ova, together with some surrounding follicular cells, enter the associated fallopian tube and move through it to the uterus. If mating has occurred and spermatozoa are present, each ovum may be fertilized by the penetration of a sperm cell. The resulting zygote (fertilized ovum) then implants itself into the uterine wall (endometrium) and there undergoes embryonic development. At parturition the developed young are expelled through the vagina. In humans the maturation of a single ovum in one ovary ordinarily inhibits the simultaneous maturation of other ova so that only a single offspring can result. In most mammals, however, multiple births are the rule.

The entire process of ovum maturation and ovulation is a cyclic phenomenon which takes place annually in some vertebrates but more frequently in others. In humans a new ovum matures every 28 days on the average unless pregnancy occurs. The human cycle and that of other higher primates differs in still other ways from the cycles of most mammals although all involve a similar interplay of controlling hormones. The human cycle is called the menstrual cycle, derived from the word "menses," or monthly. The cycle of most mammals is, instead, referred to as an estral (spring) cycle.

17.12 THE MENSTRUAL CYCLE AND ESTRUS

Beginning at puberty and continuing until the menopause at the age of 40 to 50, a new ovum begins to mature in human females every lunar month. Under the influence of the follicle stimulating hormone (FSH) from the adenohypophysis, the ovum grows and becomes enclosed in a graafian follicle on the surface of the ovary. Associated cells of the follicle are endocrine in function and produce a steroid hormone called estrogen. Estrogen has two influences on the adenohypophysis: it inhibits the production of FSH and it promotes the production of the luteinizing hormone (LH). As the follicle grows, the secretion of estrogen increases, and as a result FSH levels in the blood progressively fall and LH levels progressively rise (Fig. 17-8). At some critical ratio of FSH to LH, growth of the follicle ceases and it ruptures (ovulation). As the ovum, together with some of the surrounding follicular cells, moves away through the fallopian tube, the site of the ruptured follicle becomes altered in appearance and function. The remaining follicular cells form a new structure called the corpus luteum. Under the continued influence of LH, the corpus luteum grows and produces another steroid hormone, progesterone (Fig. 17-6). In humans the corpus luteum also produces estrogen. Progesterone and estrogen both inhibit the production of FSH, so no additional ova can mature in the ovaries at this time. If fertilization and uterine implantation occur, the corpus luteum continues this inhibitory function throughout pregnancy and for some time following parturition. If fertilization does not occur, however, the corpus luteum regresses, FSH production is resumed, and a new cycle begins.

The hormones estrogen and progesterone have a number of functions other than those just described. Estrogen appears during embryonic life and stimulates the development of female reproductive organs, just as testosterone stimulates the development of male reproductive organs. At puberty estrogen and progesterone together initiate the cycle of reproductive activity and continue the development of female secondary sex characteristics. These include the development of mammary glands, typical female fat deposition, hair distribution, treble voice, and female behavior.

During the monthly ovarian cycle, estrogen and progesterone cause cyclic changes in the lining mucosa (endometrium) of the uterus. Under the influence of estrogen during the period of ovum maturation the endometrium thickens. Progesterone from the corpus luteum subsequently causes increased activity of endometrial glands and increased

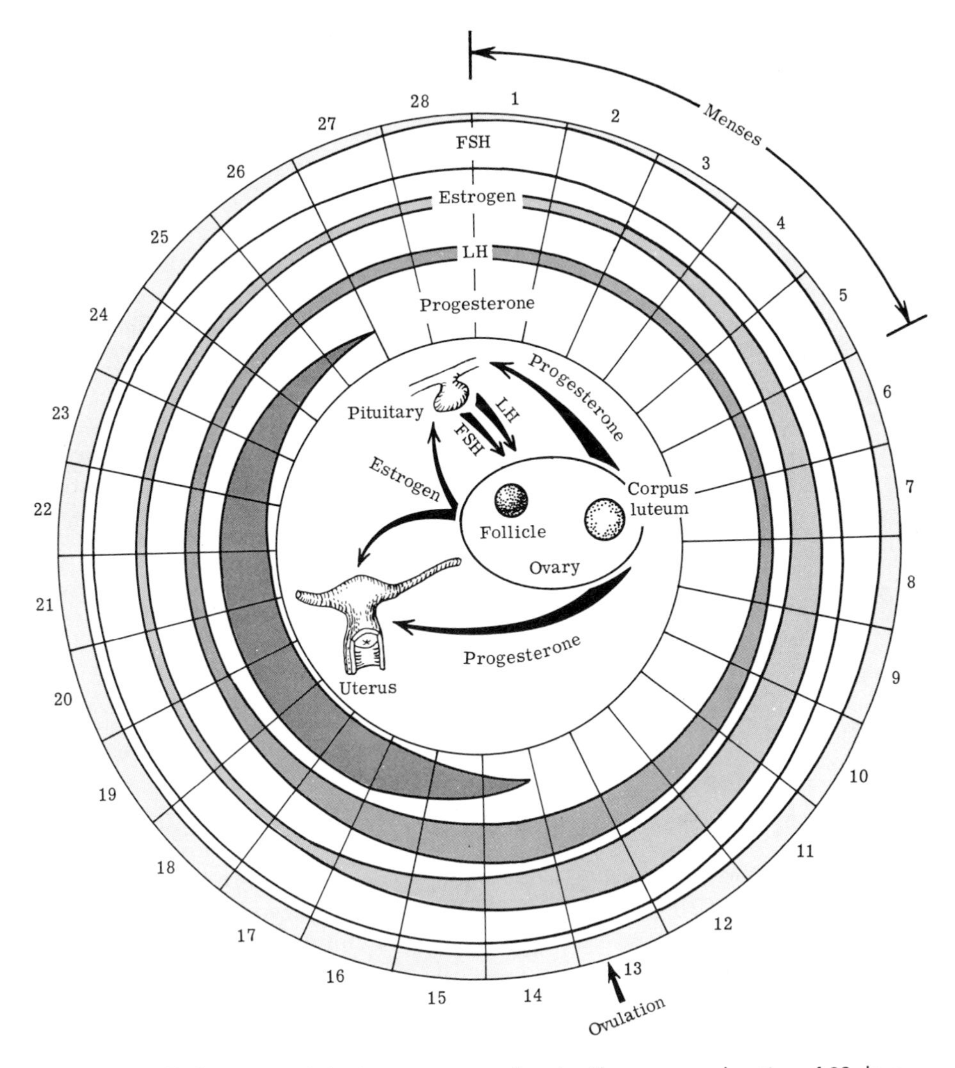

Figure 17–8 Events of the human menstrual cycle. The average duration of 28 days is indicated by numbers around the outside. (From Cockrum and McCauley: *Zoology*. Philadelphia, W. B. Saunders Company, 1965.)

vascularization and sponginess of the tissue. These changes prepare the endometrium to receive and nurture an implanted embryo. If fertilization has not occurred, however, the source of progesterone disappears, the bulk of the endometrium is discarded, and it passes out as the menstrual flow. This menstruation, as it is called, comes about because the blood supply to the endometrium is discontinued and the endometrial tissues die and are sloughed away.

Also, during the monthly cycle estrogen and progesterone cause an increase in the size and development of the mammary glands. If pregnancy occurs, this development continues in preparation for lactation. If pregnancy does not occur, some regression of the breasts takes place until the next cycle begins. Cyclic changes in behavior also occur and are correlated with the waxing and waning production of estrogen and progesterone.

In subprimate mammals no menstrual flow takes place but the interplay of hormones occurs as in primates. At the time of ovulation the female enters a period of estrus, or "heat," at which time she will receive the male. This occurs under the influence of progesterone. Female rabbits do not ovulate spontaneously but remain in a condition of continuous estrus until nervous reflexes induced by mating cause ovulation to occur.

There are several known estrogens but the most potent naturally occurring one is probably estradiole (Fig. 17-6). It, or a closely related hormone, is present in all female vertebrates. Progesterone, however, is apparently present only in mammals and, perhaps, in birds. Estrogens, as well as androgens, are toxic to cold-blooded vertebrates; consequently, experimental introduction of these hormones has not been a fruitful field of research in lower vertebrates. In very low doses they appear to have effects similar to those in mammals, however.

The luteotropic hormone (LTH) was once thought to play a role in the ovarian reproductive cycle, but it now appears that this is true only in rats and mice. In these species LTH augments the effects of LH in the maintenance of the corpus luteum. In other mammals, however, the roles of LTH appear to be confined to the period following reproduction and are concerned, instead, with the care of the young.

17.13 PREGNANCY, PARTURITION, AND LACTATION

When the fertilized ovum (zygote) has implanted itself into the endometrium of the mammalian uterus, extraembryonic tissues developed by the early embryo cooperate with modifications of maternal tissues to produce the placenta. Besides serving as a membrane across which nutrients, gases, and wastes can be exchanged between the maternal and the fetal circulatory systems, certain cells of the placenta also take on an endocrine function. These cells produce and release chorionic gonadotropins, and in humans estrogen and progesterone as well. The chorionic gonadotropic hormones maintain the corpus luteum, inhibit the menstrual cycle from recurring, and stimulate uterine growth. Chorionic gonadotropin in the urine forms the basis for pregnancy testing since other vertebrates react in predictable ways to injections of this hormone or urine samples containing it.

During pregnancy the secretions of a number of other endocrine glands are also increased. Among them are those secreted by the thyroid, parathyroid, and adrenal cortex.

The large quantities of progesterone present during pregnancy inhibit uterine muscular contraction, but shortly before the time of parturition (childbirth) increases in estrogen secretion overcome this inhibitory effect. The uterus then becomes responsive to the stimulatory effect of oxytocin from the neurohypophysis. When it becomes sufficiently sensitive to oxytocin, rhythmic contractions of the uterus begin. These lead to expulsion of the child and soon thereafter of the placenta as well. This abrupt reduction in chorionic gonadotropin, estrogen, and progesterone has major effects upon the mother and sometimes upon the child as well. Most importantly, the mammary glands of the mother are rendered responsive to subsequent influences.

It is interesting to note in passing that the developing child has been exposed to the same hormonal environment as the mother. The mammary glands of both sexes have been stimulated, and it is not unusual for some small amount of lactation to occur from the newborn as well as from the mother. For similar reasons, newborn female children sometimes show some small menstrual flow for a day or so after parturition.

The role of the luteotropic hormone (LTH), sometimes called prolactin, in mammary response is currently in doubt. It was once believed that this hormone initiated milk secretion in a mammary gland which had been prepared anatomically by the combined action of estrogen and progesterone. More recent research has shown, however, that the growth hormone (STH) functions just as effectively as LTH and that the two hormones may be identical in mammals. In any case, following parturition oxytocin causes ejection of milk when the nipple is physically stimulated by the nursing child.

Another hormone, relaxin, has also been shown to play a role in parturition. This hormone is produced by the ovary, placenta, and uterus and it may have some unknown function during pregnancy, since the blood levels of relaxin progressively increase during this period. At the time of parturition, relaxin is responsible for the dilation of the uterine cervix and the vagina. It also causes relaxation of the pelvic ligaments of some mammals so that the bony birth canal is enlarged.

Certain birds, notably the pigeons and doves, feed the young on special secretions (pigeon milk) from the crop (Sect. 9.11). Interestingly, the formation of pigeon milk is caused by the luteotropic hormone. While the function of this hormone in mammals is in doubt, it does act in a number of other ways in birds. It causes brooding behavior in domestic fowls and the development of the brood patch (a featherless area on the bird's breast) in those birds which incubate eggs by sitting on them. It also causes premigration restlessness and increased food intake in migratory species. LTH also stimulates the return to water, for breeding purposes, in certain terrestrial salamanders.

Much work remains to be done to elucidate the presence and probable roles of reproductive hormones in the lower vertebrates.

CHAPTER 18

THERMOREGULATION

18.1 INTRODUCTION

Although certain bacteria live in hot springs at temperatures of up to 70° C, most organisms, and certainly all vertebrate animals, are limited to environments in which the body temperature remains below 40° C. Similarly, animals must all maintain body temperatures that do not fall much below the freezing point of water. Chemical reaction rates are influenced by temperature, and this means that the biochemical processes of life are similarly limited. The rates of most chemical reactions are doubled with each 10° C rise in temperature. Many biochemical compounds, and especially proteins, are heat labile; that is, they are chemically altered by exposure to temperatures above about 40 to 41° C. At the other end of the scale, the formation of ice crystals within tissues generally disrupts cell membranes and in this way destroys life. Thus, even though animals can range over about 40° C, there is a chemical advantage in maintaining temperatures which more closely approach the upper limits of this range, since biochemical processes are favored at these temperatures.

The temperatures that occur in most natural bodies of water are within the range that is compatible with life (Sect. 8.1). Air temperatures, however, fluctuate over much wider ranges. Consequently, the problem of body temperature maintenance is less serious for aquatic organisms than for terrestrial organisms.

An animal can gain heat through metabolic activity (energy production) or by absorption from the environment. Even when the immediate environment (*e.g.*, air) is colder than the animal tissues, it is possible that the animal will absorb the radiant energy from the sun. Conversely, an animal can lose heat by conduction, convection, radiation, or water evaporation. In an aquatic environment conduction is the most important of these, but in the air conduction is of little importance since air is a poor thermal conductor; indeed, it is a good thermal insulator. However, convection, radiation, and water evaporation are of great importance in air.

Animals can be described as being cold-blooded or warm-blooded, depending upon whether they generally feel cool or warm to the touch. While this is certainly not a precise criterion, the terms are often used in reference to major animal groups. Cold-blooded vertebrates, as we have seen in previous chapters, are the fishes, amphibians, and reptiles; the warm-blooded vertebrates are the birds and mammals. More precisely, we can describe animals as being ectothermic or endothermic. Ectothermic animals are primarily dependent upon environmental temperatures to determine their own body temperatures. All animals produce metabolic heat, but the ectotherms are unable to vary heat production or to control heat loss by physiological adjustments. They can only maintain tolerable body temperatures by behavioral means, *i.e.*, they must find warmer or cooler portions of their habitats as required. Endotherms, on the other hand, are able to thermoregulate by both behavioral and physiological means. In general, endotherms maintain body temperatures that are higher than those of ectotherms.

There is still another set of terms used to describe the temperature relationships of vertebrate groups. These terms are poikilothermic, homeothermic, and heterothermic. A poikilotherm is an animal in which the body temperature is always very close to the environmental temperature. The term is, therefore, essentially a synonym for ectotherm. A homeotherm is an animal which maintains a constant or nearly constant body temperature despite wide fluctuations of environmental temperature. Such animals are, of course, endotherms, but not all endotherms are homeothermic. Some endotherms, instead, tolerate fairly wide fluctuations of body temperature and resist further change in body temperature only when it approaches the critical limits compatible with life. Such animals are called heterotherms.

Under experimental conditions many vertebrates can adjust their sensitivities to temperature extremes. That is, animals exposed for a time to temperatures near the critical limits of life become more temperature tolerant, and the critical limits are extended somewhat. If, for example, a species of fish ordinarily dies at a water temperature of 38° C, continuous exposure to 37° C for a few days may render it tolerant of 38° C. Death then may not occur until a temperature of 39° C or more is reached. Such adjustment of tolerance is known as acclimation. Natural conditions, as opposed to experimental conditions, also involve variables other than temperature. Seasonal changes in photoperiod (day-night lengths), diet, pelage, etc. have influences over relatively long time periods on an animal's tolerance. In such cases more profound physiological adjustments must take place, and this process is known as acclimatization.

18.2 FISHES

Fishes have an internal temperature that is slightly higher than that of the surrounding water but the usual difference is slight. The rate of metabolism is slow and the thermal conductivity of both tissues and water is high. Heat loss, therefore, is almost as rapid as heat production. Increased activity of the fish produces more heat, but since it then also requires more ventilation of the gills, the rate of heat loss is also increased. The body temperature of most fishes is generally less than a degree higher than the water temperature. In a few large, active fishes (*e.g.*,marlin) there may be a temperature difference of as much as 5 or 6° C.

Temperature regulation in fishes, then, depends almost entirely on the selection of environmental regions which have appropriate temperatures. When species are trapped in especially warm or cold waters, they are acclimatized in various ways. Some are even able to tolerate sudden temperature changes of considerable extent. For example, certain small cyprinid fishes occur in isolated desert ponds of southwestern Arizona. During the dry season these ponds become very shallow and warm. Then, when the rainy season begins, a storm may increase the water volume by several fold and lower the water temperature by 10° C or more within a matter of minutes. It is interesting to note also that during the dry season the pond may actually have minerals precipitated on the surface; the sudden rain storm, then, produces great dilution of the pond's mineral content. Thus, these fishes are also exposed to sudden and drastic changes in the salinity of their environment. The physiological means for adaptation to such salinity changes have not been thoroughly studied.

18.3 AMPHIBIANS AND REPTILES

Aquatic amphibians, of course, have much the same thermoregulatory relationships as do fishes. They must rely almost entirely upon environment selection to maintain body temperatures within tolerable limits.

Terrestrial amphibians are largely limited to behavioral thermoregulation. In some cases they withstand prolonged low temperatures by hibernation (Sect. 18.6). For many amphibians thermal regulation to high temperature is quite effective by reason of their wet skin and consequent water evaporation, but the resulting loss of body water is a limiting factor. Desert amphibians estivate, a process somewhat like hibernation, by burying themselves in the soil during the hot, dry part of the summer. While semiaquatic amphibians thermoregulate well in warm temperatures, thermoregulation to low temperatures is more difficult. The skin is primarily a respiratory surface and, therefore, loses heat easily.

Because the skins of reptiles are dry, these animals lose heat less easily and also have better control over water loss than amphibians. Improved renal functions also contribute to water conservation (Sect. 8.12). Behavioral adaptations to temperature fluctuations can be more sophisticated than in the Amphibia. Depending on their body temperature at the time, reptiles may select warm or cool surfaces to lie upon. During the cool desert nights, reptiles often lie upon exposed rock or paved highways to absorb the remaining heat from the sun. During the hot day they often bury themselves in sand or crawl under other insulating materials. Reptiles may also position themselves in sun or shade so that a greater or lesser area of skin is exposed to solar radiation.

A number of reptiles are capable of some small amount of physiological thermoregulation, and this indicates the beginnings of homeothermy among the vertebrates. They have a thermoregulatory center in the central nervous system which reflexly induces panting activity or blood pressure changes. Increased blood pressure carries heat to the surface more rapidly where it can be lost by radiation or convection. Some of the larger reptiles are capable of some adjustment of the metabolic rate. The Indian python, for example, even incubates its eggs by actively contracting muscles to produce heat.

BIRDS AND MAMMALS

18.4 ADAPTATIONS TO LOW TEMPERATURE

Acclimation and acclimatization to cold are more pronounced in endothermic vertebrates than in ectotherms. These responses involve adjustments of thyroid and adrenocortical hormone production, which increase the metabolic rate and the general resistance to cold stress (Chapter 17). Moreover, thermal insulation is improved in these animals by the presence of cutaneous fat deposits and pelage (fur or feathers). Fats are used to a greater extent as energy sources since fat produces more heat than does the oxidation of an equal weight of glucose (Sect. 10.10).

Other physiological responses to cold include reflex shivering and pilomotor activity. Shivering is, of course, muscle activity, which produces heat. The contraction of cutaneous pilomotor muscles erects the hair or feathers so as to trap larger insulating layers of air near the skin. The "goose bumps" which appear on human skin result from such pilomotor activity but are of no value because of the lack of sufficient hair. Certain aquatic mammals such as the muskrat retain air layers in the fur even when immersed in water. These animals are able to remain active even when water temperatures approach freezing. Hairless mammals depend, instead, on thick cutaneous fat deposits for heat insulation. Thermal insulation is more important to small birds and mammals than to larger ones because of the different surface to volume ratios involved (Sect. 3.1).

The extremities, because of their relatively large surface areas, may lose particularly large amounts of heat to a cold environment. This factor is often minimized in arctic mammals since they generally have shorter, stockier extremities than do closely related mammals of more temperate regions.

In certain birds and mammals of arctic regions the temperature of the extremities may fall well below body temperature and still receive an adequate blood supply. In such animals the tissues of the extremities are more tolerant of cold than are other body tissues—an example of regional acclimatization. The temperature differential is maintained without undue heat loss from the extremities by means of a countercurrent heat exchange mechanism. This operates on principles similar to those encountered in gill respiration (Sect. 3.3) and swim bladder inflation (Fig. 3-4). Arteries carrying warm blood into the extremities pass close to and parallel with veins carrying cool blood toward the heart. In fact, the veins often form sleevelike networks about the arteries. Much of the heat of the arterial blood is used to warm the venous blood, and as a result little body heat actually enters the extremities where it could be easily lost to the cold air. Whales and seals have such vascular arrangements in the flippers and flukes because these parts, unlike the body, are not protected by insulating fat layers. Penguins have similar vessels in the axillary regions so that heat loss from the wings is reduced.

Countercurrent heat exchangers of this kind are not limited to animals inhabiting cold climates. Certain tropical mammals such as the sloth, the loris, and the more primitive armadillos and anteaters are extremely sensitive to environmental temperatures that are only a little below body temperature. These animals are capable of little thermoregulation by other means, but they have vascular arrangements similar to those discussed in the preceding paragraph. Some countercurrent heat exchange occurs in the extremities of humans as well.

The continued operation of a heat exchanger during the warm summer months would interfere with necessary heat loss for some animals, especially those which retain a heavy pelage. Such animals, therefore, have alternate routes for venous return which avoid close proximity to the arteries of the extremities. They can, therefore, adjust to fairly wide extremes of environmental temperature by selecting the appropriate route for venous return of blood from the extremities.

Deep oceanic waters are very cold at all latitudes, and the cessation of cutaneous circulation by diving vertebrates (Sect. 6.4) also serves to preserve body heat during deep dives. Terrestrial homeotherms also adjust cutaneous circulation in response to both warm and cool air temperatures.

Both birds and mammals, in some cases, avoid low temperatures by undertaking long migrations. Others tolerate decreased body temperature by entering a state of torpor or hibernation (Sect. 18.6). Newly hatched or newborn vertebrates are generally less capable of thermoregulation than are older animals of the same species. They are smaller, have less pelage, and have immature physiological responses. Appropriate nests and parental care are provided as necessary, and breeding places and times are selected so that the young will have the benefit of the best season and climate for their early growth and development. Adult birds and mammals avoid temperature extremes in nests and burrows, and they curl up to reduce exposed skin area or huddle with others of their own kind to preserve body heat.

18.5 ADAPTATIONS TO HIGH TEMPERATURE

Cooling generally requires water evaporation and, therefore, adaptations to control body water loss as well. Some of the physiological and behavioral adaptations of this kind have already been discussed in connection with water balance in Chapter 8. Evaporative cooling is involved in sweating, of course, and in the panting activity of many mammals. Birds engage in gular fluttering, an activity much like panting. Some mammals that do not sweat (*e.g.*, rodents) wet the fur by licking it and thus achieve the same end.

In addition, birds and mammals avoid high temperatures by burrowing or by seeking shade or damp ground to lie on. Birds sometimes fan themselves with their wings or, under certain conditions, dust themselves to improve the thermal conductivity through the feathers. Increased vascularity of the skin promotes heat loss by radiation and convection as well. Under certain conditions insulating air layers within fur or feathers are effective in reducing the gain in heat from the radiant energy of the sun. Humans living in desert regions sometimes wear large, loose robes covering the entire body for the same reason.

18.6 HETEROTHERMIC VERTEBRATES

Some animals do not maintain constant body temperatures even though they are classed as endothermic animals. Instead, they tolerate comparatively large changes in body temperature at certain times and they resist further change in body temperature only when it approaches the critical limits. The ability of camels to tolerate several degrees' change in body temperature has been discussed elsewhere (Sect. 8.12). This ability is beneficial in terms of desert survival, a benefit which completely homeothermic animals do not have. Similarly, many bats are homeothermic when they are awake and active but become heterothermic, almost poikilothermic, when they are asleep. Often they even become torpid or hibernate during certain seasons. In this case the poikilothermy or torpor serve the purpose of reducing the nutrient requirements during sleep or during seasons when the normal food supply (*e.g.*, insects) is deficient. Some species of bats even migrate to higher elevations where temperatures fall low enough to permit true hibernation. Other bats (*e.g.*, the Mexican free tailed bat *Tadarida*) migrate long distances, as do many birds, in order to remain in a temperate climate with a sufficient food supply at all seasons. Certain birds such as the poorwill and some hummingbirds also show torpidity during sleep. Alaskan ground squirrels are homeothermic during the summer but become heterothermic and enter hibernation in the winter. The black bear, the opossum, and several other mammals enter a deep sleep (not a true hibernation) for periods of time and, during these periods, allow the body temperature to fall to fairly low levels.

The physiological changes involved in hibernation are much more profound than those of deep sleep or sleep torpidity. In hibernation there are changes in the circulatory system much like those which occur during a deep dive in those vertebrates capable of such activity (Chapter 6); that is, blood supply is cut off from cutaneous areas and, largely, from skeletal muscles but is maintained to the brain and heart at reduced pressures. Metabolic changes are also associated with hibernation. Moreover, arousal from deep sleep or sleep torpidity is typically accomplished within a few minutes, whereas arousal from true hibernation requires a matter of several hours in most instances.

Estivation is much like hibernation, but it is usually for the purpose of avoiding periods of draught or food shortage rather than cold. The physiology of torpor, hibernation, and estivation is still incompletely understood; much more research is needed in this field of study.

SUPPLEMENTARY READING

Allen, G. M.: *Bats* (paperbound). New York, Dover Publications, Inc., 1939.

Baldwin, E.: *An Introduction to Comparative Biochemistry*, 4th Edition. Cambridge, England, Cambridge University Press, 1964.

Barrington, E. J. W.: *An Introduction to General and Comparative Endocrinology*. London, Oxford University Press, 1963.

Beament, J. W. L. (ed.): *Biological Receptor Mechanism*. Number XVI of Symposia of the Society for Experimental Biology. London, Cambridge University Press, 1962.

Bloom, W. and Fawcett, D.: *Textbook of Histology*, 9th Edition. Philadelphia, W. B. Saunders Company, 1968.

Brodie, B. B. (ed.): *Evolution of Nervous Control*. Publication No. 52 of the American Association for the Advancement of Science. Washington, D.C., 1959.

Brown, M. E. (ed.): *The Physiology of Fishes*. Two volumes. New York, Academic Press, 1957.

Cantarow, A. and Schepartz, B.: *Biochemistry*, 4th Edition. Philadelphia, W. B. Saunders Company, 1967.

Cockrum, E. L. and McCauley, W. J.: *Zoology*. Philadelphia, W. B. Saunders Company, 1965.

Cockrum, E. L., McCauley, W. J., and Younggren, N. A.: *Biology*. Philadelphia, W. B. Saunders Company, 1966.

Cold Spring Harbor Symposia on Quantitative Biology (Volume XXX): *Sensory Receptors*. Cold Spring Harbor, Long Island, N.Y., 1965.

De Robertis, E. D. P., Nowinski, W. W., and Saez, F. A.: *Cell Biology*, 5th Edition. Philadelphia, W. B. Saunders Company, 1970.

Dillet, D. B., *et al.* (eds.): *Adaptation to the Environment*. Section 4 of the Handbook of Physiology (American Physiological Society). Baltimore, Williams and Wilkins Company, 1964.

Dougherty, R. W. (ed.): *Physiology of Digestion in the Ruminant*. Washington, D. C., Butterworth Inc. (printed by Waverly Press, Inc., Baltimore, Md.), 1965.

Florey, E.: *An Introduction to General and Comparative Animal Physiology*. Philadelphia, W. B. Saunders Company, 1966.

Gardner, E.: *Fundamentals of Neurology*, 5th Edition. Philadelphia, W. B. Saunders Company, 1968.

Goodrich, E. S.: *Studies on the Structure and Development of Vertebrates*. Two volumes (paperbound). New York, Dover Publications, Inc., 1958.

Gorbman, A. and Bern, H. A.: *A Textbook of Comparative Endocrinology*. New York, John Wiley & Sons, Inc., 1962.

Gordon, M. S., *et al.*: *Animal Function: Principles and Adaptations*. New York, The MacMillan Company, 1968.

Grollman, S.: *The Human Body*, 2nd Edition. New York, The MacMillan Company, 1969.

Guyton, A. C.: *Textbook of Medical Physiology*. 4th Edition. Philadelphia, W. B. Saunders Company, 1971.

Harper, H. A.: *Review of Physiological Chemistry*, 12th Edition (paperbound). Los Altos, Calif., Lange Medical Publications, 1969.
Hoar, W. S.: *General and Comparative Physiology*. Englewood Cliffs, N.J., Prentice-Hall, Inc., 1966.
Hughes, G. M.: *Comparative Physiology of Vertebrate Respiration*. Cambridge, Mass., Harvard University Press, 1963.
Hughes, G. M. (Ed.): *Homeostasis and Feedback Mechanisms*. Number XVIII of the Symposia of The Society for Experimental Biology. London, Cambridge University Press, 1964.
Kellogg, W. N.: *Porpoises and Sonar*. Chicago, University of Chicago Press, 1961.
Loewy, A. G., and Siekevitz, P.: *Cell Structure and Function*, 2nd Edition (paperbound). New York, Holt, Rinehart, and Winston, 1969.
Marler, P. and Hamilton, W. J.: *Mechanisms of Animal Behavior*. New York, John Wiley & Sons, Inc., 1966.
Marshall, A. J. (ed.).: *Biology and Comparative Physiology of Birds*. Two volumes. New York, Academic Press, 1960.
Mayer, W. V. and Van Gelder, R. G. (eds.): *Physiological Mammalogy*. Two volumes. New York, Academic Press, 1963.
McElroy, W. D.: *Cell Physiology and Biochemistry*, 2nd Edition (paperbound). Englewood Cliffs, N.J., Prentice-Hall, Inc., 1964.
Milthorpe, F. L. (ed.): *Mechanisms in Biological Competition*. Number XV of the Symposia of the Society for Experimental Biology. New York, Academic Press, 1961.
Moore, J. A.: *Physiology of the Amphibia*. New York, Academic Press, 1964.
Nalbandov, A. V.: *Reproductive Physiology*, 2nd Edition. San Francisco, W. H. Freeman and Company, 1964.
Orr, Robert T.: *Vertebrate Biology*, 3rd Edition. Philadelphia, W. B. Saunders Company, 1971.
Prosser, C. L. and Brown, F. A., Jr.: *Comparative Animal Physiology*. 2nd Edition. Philadelphia, W. B. Saunders Company, 1961.
Ranson, S. W. and Clark, S. L.: *The Anatomy of the Nervous System*, 10th Edition. Philadelphia, W. B. Saunders Company, 1959.
Romer, A. S.: *The Vertebrate Story*. Chicago, University of Chicago Press, 1959.
Romer, A. S.: *The Vertebrate Body*, 4th Edition. Philadelphia, W. B. Saunders Company, 1970.
Rose, S.: *The Chemistry of Life* (paperbound). Baltimore, Penguin Books, Inc., 1966.
Rushmer, R. F.: *Cardiovascular Dynamics*, 3rd Edition. Philadelphia, W. B. Saunders Company, 1970.
Scheer, B. T.: *Animal Physiology*. New York, John Wiley & Sons, Inc., 1963.
Schmidt-Nielsen, K.: *Desert Animals*. London, Oxford University Press, 1964.
Sturkie, P. D.: *Avian Physiology*. Ithaca, N.Y., Comstock Publishing Associates, 1954.
Swanson, C. P.: *The Cell* (paperbound). Englewood Cliffs, N.J., Prentice-Hall, Inc., 1960.
Tepperman, J.: *Metabolic and Endocrine Physiology*. Chicago, Year Book Medical Publishers, Inc., 1962.
Turner, C. D.: *General Endocrinology*, 4th Edition. Philadelphia, W. B. Saunders Company, 1966.
Van Tienhoven, A.: *Reproductive Physiology of the Vertebrates*. Philadelphia, W. B. Saunders Company, 1968.
Welty, J. C.: *The Life of Birds*. Philadelphia, W. B. Saunders Company, 1962.
Young, J. Z.: *The Life of Vertebrates*, 2nd Edition. New York, Oxford University Press, 1962.
Young, W. C. (ed.): *Sex and Internal Secretion*. Baltimore, Williams and Wilkins Company, 1961.

INDEX

Note: Page numbers in *italics* refer to illustrations.

Living

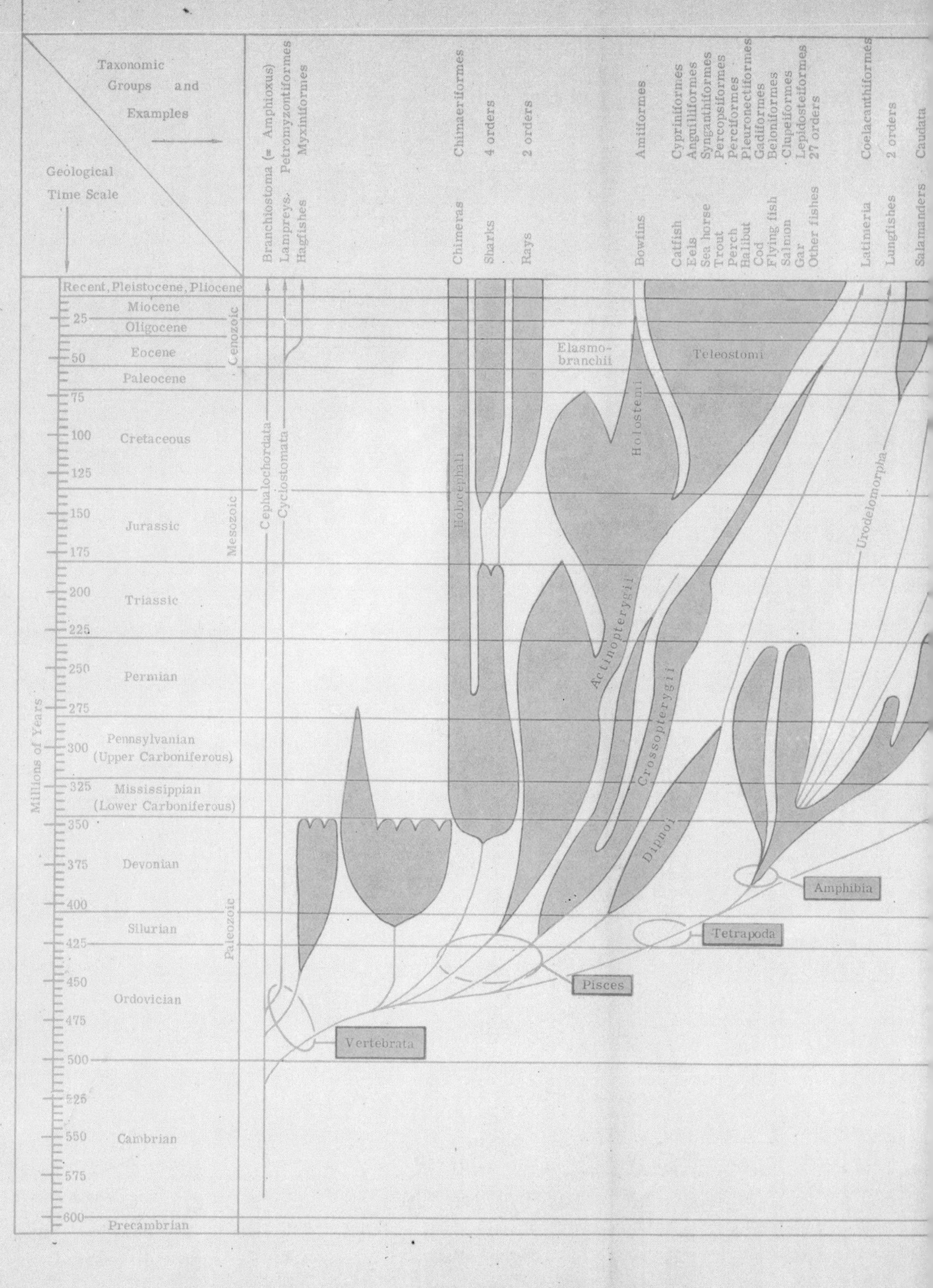

Taxonomic Groups and Examples
Geological Time Scale
Branchiostoma (= Amphioxus)
Lampreys. Petromyzontiformes
Hagfishes Myxiniformes
Chimeras Chimaeriformes
Sharks 4 orders
Rays 2 orders
Bowfins Amiiformes
Catfish Cypriniformes
Eels Anguilliformes
Sea horse Syngnathiformes
Trout Percopsiformes
Perch Perciformes
Halibut Pleuronectiformes
Cod Gadiformes
Flying fish Beloniformes
Salmon Clupeiformes
Gar Lepidosteiformes
Other fishes 27 orders
Latimeria Coelacanthiformes
Lungfishes 2 orders
Salamanders Caudata
Recent, Pleistocene, Pliocene
Miocene
Oligocene
Eocene
Paleocene
Cretaceous
Jurassic
Triassic
Permian
Pennsylvanian (Upper Carboniferous)
Mississippian (Lower Carboniferous)
Devonian
Silurian
Ordovician
Cambrian
Precambrian
Cenozoic
Mesozoic
Paleozoic
Millions of Years
25
50
75
100
125
150
175
200
225
250
275
300
325
350
375
400
425
450
475
500
525
550
575
600
Cephalochordata
Cyclostomata
Holocephali
Elasmobranchii
Holostei
Teleostomi
Actinopterygii
Crossopterygii
Dipnoi
Urodelomorpha
Amphibia
Tetrapoda
Pisces
Vertebrata